Subcellular Biochemistry

Volume 112

The book series SUBCELLULAR BIOCHEMISTRY is a renowned and well recognized forum for disseminating advances of emerging topics in Cell Biology and related subjects. All volumes are edited by established scientists and the individual chapters are written by experts on the relevant topic. The individual chapters of each volume are fully citable and indexed in Medline/Pubmed to ensure maximum visibility of the work.

Giuseppe Legname • Fabio Moda
Editors

Biomarkers and Therapeutical Targets for Prion Diseases

Springer

Editors
Giuseppe Legname
Department of Neuroscience
SISSA
Trieste, Trieste, Italy

Fabio Moda
Department of Medical Biotechnology
and Translational Medicine
Università degli Studi di Milano
Milano, Milan, Italy

This work was supported by University College London.

ISSN 0306-0225 ISSN 2542-8810 (electronic)
Subcellular Biochemistry
ISBN 978-3-031-97054-2 ISBN 978-3-031-97055-9 (eBook)
https://doi.org/10.1007/978-3-031-97055-9

This Springer imprint is published by the registered company Springer Nature Switzerland AG
The registered company address is: Gewerbestrasse 11, 6330 Cham, Switzerland

Contents

Chapter 1
Advancements in Chronic Wasting Disease (CWD) Prion Detection: Moving Beyond the Gold Standards

Paulina Soto, Reece McGinn, and Rodrigo Morales

Abstract Chronic wasting disease (CWD) is perhaps the most problematic prion disease at present, considering its rapid spread in North America, its presence in both captive and wild animals, and its unknown zoonotic potential. Although several strategies have been attempted to contain the spread of CWD, their success appears to be limited. One of the main problems associated with the management of this disease lies in its diagnosis. At present, CWD diagnosis is evaluated using *post-mortem* tissues using techniques of insufficient analytical sensitivity. As a consequence, these techniques do not allow for the identification of infected animals using biological samples that could be collected from live subjects. Along this line, further development of prion amplification methods may fill this much-needed gap. This chapter summarizes the current methods used to diagnose CWD in regulatory and laboratory settings, and provides perspectives on how new technologies may help facilitate the identification of diseased animals.

Keywords Chronic wasting disease (CWD) · CWD detection · PMCA · RT-QuIC

P. Soto
Department of Neurology, The University of Texas Health Science Center at Houston, Houston, TX, USA

Centro Integrativo de Biologia y Quimica Aplicada (CIBQA), Universidad Bernardo O'Higgins, Santiago, Chile

R. McGinn · R. Morales (✉)
Department of Neurology, The University of Texas Health Science Center at Houston, Houston, TX, USA
e-mail: Rodrigo.MoralesLoyola@uth.tmc.edu

G. Legname, F. Moda (eds.), *Biomarkers and Therapeutical Targets for Prion Diseases*, Subcellular Biochemistry 112, https://doi.org/10.1007/978-3-031-97055-9_1

Introduction

Transmissible spongiform encephalopathies (TSEs), or prion diseases, are fatal neurodegenerative diseases that affect humans and some animals. These diseases occur when the normal prion protein (PrP^{C}) misfolds into a pathological and infectious form (PrP^{Sc}), which can self-propagate and induce additional misfolding processes as well as the brain deposition of these deleterious particles (Dearmond and Prusinertt 1995; Orge et al. 2021; Prusiner 1991).

Prion diseases include several disorders, such as Creutzfeldt–Jakob disease (CJD), Gerstmann–Sträussler–Scheinker (GSS) syndrome, fatal familial insomnia (FFI), and kuru in humans (Belay 1999; Collinge 1997, 2001). In animals, scrapie in sheep and goats (Sigurdson et al. 2003), bovine spongiform encephalopathy (BSE) in cattle (Collinge 1997), and chronic wasting disease (CWD) in cervids (Sigurdson 2008) are relevant to highlight considering their impact on animal and human populations.

Prion diseases can manifest in three ways: (I) familial, (II) sporadic, or (III) acquired. In humans, sporadic prion disease is the most common form (about 85–90% of all cases) and is exemplified by sporadic Creutzfeldt–Jakob disease (sCJD). Acquired prion diseases account for less than 1% of cases and include kuru, variant Creutzfeldt–Jakob disease (vCJD), and iatrogenic Creutzfeldt–Jakob disease (iCJD). The iCJD is primarily caused by contaminated dura mater and corneal transplants, as well as treatments with cadaveric human growth hormone and blood transfusions (Belay 1999; Collinge 2001). The remaining percentage of cases are due to mutations in the prion protein that favor the misfolded form of the protein. These familial cases include GSS, FFI, and inherited forms of CJD (Collins et al. 2001; Medori et al. 1992).

Different from humans, most of the natural TSE cases in animals are thought to be transmitted through ingestion, although vertical routes have also been proposed. This is supported by epidemiological and experimental evidence (Gallardo and Delgado 2021). However, it is important to highlight that the cause of transmission remains unclear in some specific cases. Recent examples of this include the emergence of chronic wasting disease (CWD) in the Scandinavian Peninsula (Benestad et al. 2016) and the prion disease affecting camels in Algeria (Babelhadj et al. 2018).

Chronic Wasting Disease (CWD)

CWD affects cervids, including white-tailed deer (*Odocoileus virginianus*), mule deer (*Odocoileus hemionus*), reindeer (*Rangifer tarandus*), red deer (*Cervus elaphus*), elk (*Cervus canadensis*), moose (*Alces alces*), sika deer (*Cervus nippon*), and muntjac deer (*Muntiacus muntjak*) (Escobar et al. 2020). CWD was first observed in 1967 in the state of Colorado, USA (Williams and Young 1980). The escalating threat of CWD in cervid populations is underscored by surveillance data. In the year

2000, CWD was recorded in five states in the United States and one Canadian province. By 2010, the disease had been identified in 17 states and two provinces. As of 2018, CWD was reported in 26 states and three provinces, indicating a significant geographic expansion of the infection in the cervid population (Osterholm et al. 2019). At the moment of writing (February 2025), CWD has been identified in 36 US states, five Canadian provinces (https://www.usgs.gov/centers/nwhc/science/expanding-distribution-chronic-wasting-disease), three Nordic countries, and South Korea.

The exact origin of CWD is unknown; however, several hypotheses have been proposed. One of them suggests that CWD may result from cervids being infected by scrapie-affected sheep that often share the same environment. This hypothesis is supported by experimental evidence indicating that cervids can be infected with scrapie prions derived from sheep. Studies have shown that brain homogenates contaminated with scrapie can affect elk and white-tailed deer (Greenlee et al. 2011). These originally infected animals, in turn, may adapt the infectious particles for efficient transmission within the same and other cervid species. Additional experimental research has further validated this theory, demonstrating that infection of sheep with CWD prions is possible, leading to a disease that resembles conventional scrapie (Cassmann et al. 2021). Another hypothesis posits that some of the CWD cases are of sporadic origin, and these act as the original foci of transmission (Tranulis et al. 2021). This hypothesis finds strong support in the fact that the CWD agents naturally existing in North America and Europe appear to be unrelated and spontaneously generated in each geographic location (Tranulis et al. 2021).

CWD can be transmitted both vertically and horizontally. Experimental evidence collected from muntjac deer (*Muntiacus muntjak*) shows that 80% of the offspring from CWD-positive dams are infected, indicating that prion infection can occur *in utero*, during parturition, or during nursing (Nalls et al. 2013). Additional studies identified PrP^{Sc} in the fetal, gestational, and reproductive tissues of deer infected with CWD (Bravo-Risi et al. 2021). Regardless, horizontal transmission is perhaps the best-accepted mode of transmission for CWD in natural scenarios. This may occur through direct contact between infected animals or indirect contact through environmental elements (Jo Moore et al. 2016). Excreta from animals infected with CWD, including saliva, blood, urine, and feces, can contaminate environmental components such as water, vegetation, and soil, as well as surfaces, and they can persist for years (Carlson et al. 2023; Nichols et al. 2009; Pritzkow et al. 2015; Smith et al. 2011). The contact of naïve animals with these contaminated fomites may, in turn, facilitate the propagation between animals. These assumptions go along with the high prevalence and persistence of prion infectivity in captive settings (Belay et al. 2004).

CWD, like all prion diseases, has a lengthy preclinical phase that can last from several months to years before any visible clinical signs appear. This makes it challenging to differentiate between preclinical CWD-infected and uninfected animals. After an animal is exposed to CWD prions through ingestion, the infectious agents are thought to first replicate in lymphoid tissues associated with the digestive tract. The prions then spread mainly through the circulatory system to other lymphatic centers, including the tonsils and retropharyngeal lymph nodes (Sigurdson et al. 2002). Once prions reach a critical concentration in the body, neuro-invasion is favored and occurs in a retrograde manner, moving from the peripheral nervous system to the central nervous system (CNS). This process occurs specifically through the ascending fibers of the autonomic nervous system, leading to prion deposits and spongiform degeneration in the dorsal motor nucleus of the vagus nerve located in the obex region of the medulla oblongata. Ultimately, prions multiply in the CNS, causing neuronal death, the onset of clinical symptoms, and eventually, the death of the individual (Moreno and Telling 2018).

Multiple researchers and government agencies are actively working to address the CWD problem by implementing diverse strategies. These approaches focus on the proactive detection and removal of affected animals using diagnostic tests, aiming to effectively manage and mitigate the impact of the disease. Additionally, biosecurity protocols are in place to reduce the movement of infected animals to locations not affected by CWD, including the movement of dead animals, prevention of contact of captive animals with wildlife, and/or any risk of animal exposure to possibly CWD-contaminated environments. Due to the significant spread of CWD, there is an urgent need to develop rapid, sensitive, and cost-effective diagnostic tests. These tests should ideally utilize samples collected *ante-mortem*, be minimally invasive, and be easy to access. Currently, there are only a few approved tests for detecting CWD in the United States (the country most affected by CWD), which primarily focus on two types of samples regarded as the "gold standards" (Haley and Richt 2017).

Currently, the official tests for detecting CWD prions in cervids are conducted *post-mortem* (Peters et al. 2000). These tests utilize tissues such as the retropharyngeal lymph nodes and the brain obex region. The retropharyngeal lymph nodes, located in the head of cervids underneath the back of the throat, have been observed to accumulate prions relatively soon after infection in white-tailed deer (one of the most CWD-susceptible species), well before any clinical signs appear. The diagnosis of prion diseases has traditionally relied on detecting the biological marker associated with the disease, PrP^{Sc}.

The Official Methodologies for CWD Detection

The official diagnostic tests for CWD in cervids, also referred to as the "gold standards," include immunohistochemistry (IHC) (Peters et al. 2000) and the enzyme-linked immunosorbent assay (ELISA) (Hibler et al. 2003) (Fig. 1.1). IHC is used to detect PrP^{Sc} deposits in tissues, while the ELISA assay examines fresh tissue homogenate to identify the accumulation of the prion protein. Both assays require antibodies specifically targeting the prion protein. As mentioned before, there are two approved tissue types for official CWD *post-mortem* testing in cervids: the medial retropharyngeal lymph nodes (MRPLNs) and the brainstem (obex) (Haley

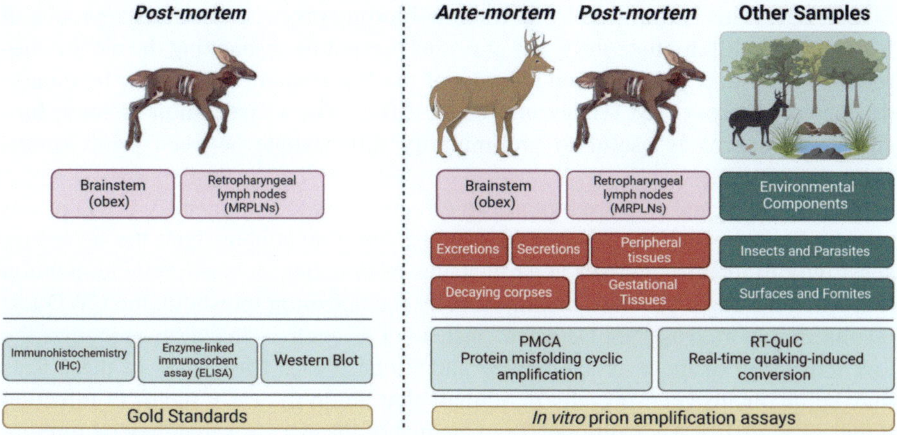

Fig. 1.1 Methods to detect CWD prions. This figure shows some of the current methodologies available to identify CWD prions. Current approved methods use two different tissue types (obex and MRPLNs) coupled with specific techniques (IHC and ELISA, although western blots are also widely accepted). Recently optimized prion amplification technologies allow the detection of CWD prions in a wide variety of samples, including a plethora of tissue and fluid samples. Importantly, these prion amplification assays allow for the *ante-mortem* identification of diseased animals. Moreover, these techniques are amenable to detecting prions in multiple environmental fomites

and Richt 2017). It is essential to analyze both types of tissues to obtain an accurate diagnosis (Benavente et al. 2023).

Immunohistochemistry (IHC) For the immunohistochemistry assay, formalin-fixed tissues are utilized. These tissues are sliced into thin sections and treated with antibodies that bind to prion proteins that have been pretreated with proteases and denatured to help in the exposure of reactive epitopes (Sajnani et al. 2012). The binding of the antibody to infectious prion clumps is indicated by a pink color, which can be observed under a microscope. A key characteristic of infectious prions is their partial resistance to treatment with acids and digestion by proteases (usually proteinase K, or PK). As a result, normal cellular prions can be digested by PK, while the misfolded and infectious protein remains (Guiroy et al. 1991).

ELISA This test uses fresh and homogenized tissues to look for infectious prions. An antibody is used, and the measurement is observed by colorimetry, giving a numerical value, which refers to the intensity of binding between the antibody and possible infectious prions present in the sample. The intensity values that are more significant than a predetermined threshold indicate the presence of prions (Haley and Richt 2017).

Western Blot (WB) Although not part of the "gold standard," this technique is well accepted as a diagnostic tool for CWD prions, as it detects infectious particles when they are already at high concentrations. This methodology requires homogenized tissue that has been digested with an enzyme (PK) and fractionated by size through

SDS-PAGE. This is followed by an immunoblotting process. With this approach, it is possible to differentiate the types of prions present by visualizing the electrophoretic mobility and glycosylation patterns of the PrP protein, which may be mono-, di-, or non-glycosylated (Haley and Richt 2017). The visualization of these biochemical patterns is useful to preliminarily differentiate between prion strains (Morales 2017).

IHC, ELISA, and western blot are all effective methods to identify CWD prions in samples collected *post-mortem*. However, these techniques lack the necessary sensitivity to identify animals at early stages of infection. It is essential to mention that when a sample provides readings below the detection threshold, the CWD test results are reported as "Not Detected" instead of "Negative" to clearly recognize the low sensitivity of these assays (Haley and Richt 2017). The results of these tests may be inconclusive, especially in animals that are in the early stages of infection (Giles et al. 2017). An additional disadvantage of these tests is that they require tissues that can only be collected *post-mortem*. A study conducted in 2019 found that WB and ELISA techniques have detection capabilities ranging from dilutions of 10^{-1} to 10^{-2} when starting from a 10% weight/volume tissue extract. This may result in impracticality when analyzing tissues from animals at early incubation periods: at those stages, animals usually display low and/or variable levels of PrP^{Sc} proteins in these tissues. Very low concentrations of prion deposits reduce the reliability of the tests (McNulty et al. 2019). This limitation makes it challenging to use these detection methods as a control strategy for CWD.

Bioassays to Identify CWD Prions

It is important to note that the first bioassays related to prion diseases were conducted on large animals, but the results were inconclusive, likely due to the insufficient length of the observation periods. Specifically, between 1936 and 1939, Cuille and Chelle (Brown and Bradley 1998) reported the transmission of scrapie to sheep more than 1 year after inoculation. Other early experimental transmissions of TSE include the inoculation of the kuru and CJD agents into chimpanzees, with observations lasting approximately 2 years (Gibbs Jr 1968; Gajdusek et al. 1966). While large animal experiments yielded invaluable insights, the prototypes developed from these studies were often unwieldy and prohibitively expensive.

The evolution of bioassays increasingly favors the use of transgenic mice that overexpress the physiological prion protein, PrP^C, from diverse animal species. Genetic technologies facilitate the elimination of the endogenous mouse PrP ($PrP^{0/0}$) (Telling et al. 1995) in certain instances and introduce PrP from multiple, larger animals. This enormously facilitates bioassays for different strains of natural prions, enabling the study of interspecies prion transmissions and the subsequent adaptation of the infectious agent in the new hosts (Watts and Prusiner 2014).

When considering bioassays as a detection methodology, we must recognize the ethical issues involving the use of animals and the extensive incubation periods necessary to reach conclusive results. These experiments may also be quite expensive (Giles et al. 2017). Regardless, bioassays are sensitive as they allow for the detection of CWD prions at high dilutions (10^{-6}) (McNulty et al. 2019). Another advantage of bioassays is that they allow the study of the pathophysiology of CWD, as well as being responsive to the specific strain identity of the agent present in the problem sample (Beck et al. 2012).

In Vitro Prion Replication Assays to Detect CWD Prions

In the past two decades, innovative in vitro prion replication assays such as the protein misfolding cyclic amplification (PMCA) (Castilla et al. 2006) and the real-time quaking-induced conversion (RT-QuIC) (Green 2019) have emerged as powerful tools in the quest to identify prions. These assays partially simulate the mechanism of prion replication in vitro, allowing the amplification of trace amounts of PrP^{Sc} while consuming larger amounts of normally folded prion proteins. This is achieved through a cyclic amplification procedure involving repeated cycles of incubation and fragmentation. Both PMCA and RT-QuIC are highly sensitive techniques for detecting CWD prions in various biological samples, such as blood, urine, saliva, feces, tears, gestational tissues, nasal mucosa, peripheral tissues, and decaying corpses, and additionally include environmental samples (Fig. 1.1) (Bartz et al. 2024; Benavente et al. 2023; Bravo-Risi et al. 2021, 2023; Carlson et al. 2023; Davenport et al. 2018; Denkers et al. 2020; Haley et al. 2009, 2011; Henderson et al. 2015; Inzalaco et al. 2024; Jo Moore et al. 2016; Johnson et al. 2006; Kraft et al. 2023; Kramm et al. 2020; Nichols et al. 2009; Pulford et al. 2012; Smith et al. 2011; Soto et al. 2023, 2024; Yuan et al. 2022).

Protein Misfolding Cyclic Amplification (PMCA) The PMCA technique involves alternating incubation and sonication cycles of a mixture including a sample suspected to contain PrP^{Sc} and the normal cellular prion protein PrP^{C} provided in the brain extract of prion-free animals (Soto et al. 2002). This technique aims to convert PrP^{C} to PrP^{Sc} in vitro. The sonication steps included in the PMCA protocol aim to break PrP^{Sc} into smaller pieces, which enhances the growth of active PrP^{Sc} nuclei. During incubation, these smaller particles recruit and convert more PrP^{C} to PrP^{Sc}. Due to the amplification of PrP^{Sc} using this technique, it is possible to identify infectious particles in samples with a small number of prions through immunoblot techniques (Morales et al. 2012). According to McNulty and collaborators (McNulty et al. 2019), detection using this technique can be extremely sensitive, being 10,000 times more sensitive than IHC, ELISA, and WB. However, other groups have described increased sensitivities using this assay (Bartz et al. 2024).

The PMCA technique has been extensively used to study CWD in terms of disease mechanisms and diagnosis. It uses a substrate derived from the cellular prion protein obtained from the brains of transgenic mice that overexpress the prion protein found in cervids, such as white-tailed deer, elk, and moose. This approach provides an abundant source of PrP^C, which can be converted in vitro. In the case of the PMCA protocol to detect CWD prions, the "substrate" (source of the PrP^C pool) is initially supplemented with EDTA, digitonin, and a protease inhibitor cocktail. This mixture is then combined with the "sample" (suspected to contain PrP^{Sc}). The combined mixture is subjected to cycles of incubation and sonication for 72 hours at 37 °C. After this process is completed (a "round"), a portion of the resulting product is transferred into a new PrP^C brain homogenate preparation. Typically, three PMCA rounds are performed to assess the presence of CWD prions in a given sample, although additional rounds may be added for samples inhibiting the reaction, such as soils, plants, and others (Soto et al. 2024). Finally, the product obtained from the final PMCA round is subjected to digestion with PK. The final concentration of PK added can vary, depending on the specific strain of prions being studied. After the PK treatment, this final product is evaluated using WB (Castilla et al. 2006; Morales et al. 2012).

As of 2025, the CWD-adapted PMCA protocol has been used on biological samples (Benavente et al. 2023), including body fluids (Haley et al. 2011), excretions (Bravo-Risi et al. 2023), decaying carcasses (Soto et al. 2023), and various tissues (Bartz et al. 2024). Additionally, due to its high sensitivity, PMCA can also detect prions in environmental samples, including different types of soil (Smith et al. 2011), plants (Carlson et al. 2023), water (Nichols et al. 2009), grass (Pritzkow et al. 2015; Soto et al. 2024), insects and parasites (Pritzkow et al. 2021; Soto et al. 2024), and surfaces (Pritzkow et al. 2018).

Real-Time Quaking-Induced Conversion (RT-QuIC) RT-QuIC is an effective prion replication method (Atarashi et al. 2011) that has been extensively used to detect CWD prions in multiple sample types. RT-QuIC was introduced in 2010 as a novel technique for detecting small amounts of PrP^{Sc}. This method takes advantage of the ability of the recombinant prion protein to misfold, leading to the formation of recombinant PrP fibrils. The aggregation process is monitored in real-time through the binding of a fluorescent dye called thioflavin T. This dye intercalates within the growing amyloid fibrils, exhibiting a different spectrum compared to that if the dye were free in solution. According to previous reports (McNulty et al. 2019) where different detection techniques are compared for a pool of infectious material with CWD, the RT-QuIC technique has a detection power of one log higher compared to PMCA. The latter is still contested, considering other available articles that describe a higher efficiency of PMCA compared with RT-QuIC (Benavente et al. 2023). RT-QuIC has been used for the detection of CWD prions in multiple sample types, such as muscles (Li et al. 2021), secretions, excretions (Davenport et al. 2018; Henderson et al. 2015), and many others (Burgener et al. 2022).

Summary of Pros and Cons of the In Vitro Prion Amplification Assays RT-QuIC and PMCA have been shown to provide false negative results (Jones et al. 2023; Lacroux et al. 2014; Vijaywargiya et al. 2024). With the emergence of CWD, detection has been an issue for labs, and determining a standard approach for testing has not happened yet. The latter can lead to variable results depending on the lab, equipment, and specific protocols used (Rowden et al. 2023). Another negative aspect of these techniques involves the concern of working with biohazard samples (Bistaffa et al. 2017).

A positive aspect of RT-QuIC involves its rapid turnover. Most RT-QuIC tests take around 90 hours or more to complete (Green 2019). This is considerable compared with bioassays that can take 250+ days to complete (Notari et al. 2012). RT-QuIC substantially reduces this time to just a few hours, depending on the protocol used in different laboratories (Tewari et al. 2021; Yilmaz et al. 2024). RT-QuIC also has an extremely high sensitivity, with around 95% accuracy (Fiorini et al. 2020) and up to 98% specificity (Rossi et al. 2020; Bistaffa et al. 2019). For testing, RT-QuIC can use a wide variety of samples, as observed for humans and multiple other animal species. The use of RT-QuIC is not limited to *post-mortem* specimens (Yuan et al. 2022). The RT-QuIC products are considered not to be infectious, a fact that is advantageous in terms of sample manipulation but a disadvantage to studying the properties of the input infectious agent (Bartz et al. 2024). Considering the latter, RT-QuIC is not able to discriminate between prion strains, importantly limiting its diagnostic use (Bartz et al. 2024).

PMCA faithfully replicates infectious prions, creating biosafety concerns as additional infectivity is generated. Besides this negative aspect, the PMCA products can be further studied to specifically assess the infectivity and strain properties of the infectious particles present in a given sample. In addition, PMCA can partially mimic interspecies transmissions (Bartz et al. 2024). Technical aspects are also a challenge. Due to the high sensitivity of PMCA, contamination of the samples can occur, and specific training, as well as the inclusion of multiple positive and negative controls, must be considered. Along with these technical aspects, the PMCA technique requires specific equipment that may challenge its implementation.

Multiple types of samples can be interrogated by PMCA for their prion content (Benavente et al. 2023). These include environmental samples like soils and water (Nichols et al. 2009). An optimized PMCA method can detect a single prion particle of PrP^{Sc} from a sample (Wang et al. 2022), making this technique a highly sensitive platform when compared to other forms of detection. PMCA can also differentiate between prion strains (Castilla et al. 2008). Strain specificity is extremely valuable because strains can cause different disease phenotypes that can have a wide range of incubation periods and clinical symptoms (Crowell et al. 2015).

Perspectives

An innovative, noninvasive test to identify live animals infected with CWD prions before symptoms appear could significantly change the way we combat this devastating disease. By detecting infected deer early, we can greatly reduce the risk of environmental prion contamination and limit both horizontal and vertical transmission of the disease.

Fortunately, recent advances in in vitro prion replication assays, specifically in the PMCA and RT-QuIC techniques, show great potential for detecting prions in the biological fluids of cervids affected by CWD. By utilizing these innovative techniques, which allow for noninvasive sample collection from living animals and can be easily accessed, we can significantly enhance our efforts to control CWD in wildlife and ecosystems.

Acknowledgments The writing of this article was supported by a grant from NIH (1R01AI132695) to RM.

References

Atarashi R, Sano K, Satoh K, Nishida N (2011) Real-time quaking-induced conversion: a highly sensitive assay for prion detection. Prion 5(3):150–153. https://doi.org/10.4161/pri.5.3.16893

Babelhadj B, Di Bari MA, Pirisinu L, Chiappini B, Gaouar SBS, Riccardi G, Marcon S, Agrimi U, Nonno R, Vaccari G (2018) Prion disease in dromedary camels, Algeria. Emerg Infect Dis 24(6):1029–1036. https://doi.org/10.3201/eid2406.172007

Bartz JC, Benavente R, Caughey B, Christensen S, Herbst A, Hoover EA, Mathiason CK, McKenzie D, Morales R, Schwabenlander MD, Walsh DP (2024) Chronic Wasting disease: state of the science. Pathogens 13(2). Multidisciplinary Digital Publishing Institute (MDPI). https://doi.org/10.3390/pathogens13020138

Beck KE, Vickery CM, Lockey R, Holder T, Thorne L, Terry LA, Denyer M, Webb P, Simmons MM, Spiropoulos J (2012) The interpretation of disease phenotypes to identify TSE strains following murine bioassay: characterisation of classical scrapie. http://www.veterinaryresearch.org/content/43/1/77

Belay ED (1999) Transmissible Spongiform Encephalopathies in humans. https://doi.org/10.1146/annurev.micro.53.1.283

Belay ED, Maddox RA, Williams ES, Miller MW, Gambetti P, Schonberger LB (2004) Chronic Wasting disease and potential transmission to humans. Emerg Infect Dis 10:977. https://doi.org/10.3201/eid1006.031082

Benavente R, Reed JH, Lockwood M, Morales R (2023) PMCA screening of retropharyngeal lymph nodes in white-tailed deer and comparisons with ELISA and IHC. Sci Rep 13(1):20171. https://doi.org/10.1038/s41598-023-47105-9

Benestad SL, Mitchell G, Simmons M, Ytrehus B, Vikøren T (2016) First case of chronic wasting disease in Europe in a Norwegian free-ranging reindeer. Vet Res 47(1):88. https://doi.org/10.1186/s13567-016-0375-4

Bistaffa E, Rossi M, De Luca CMG, Moda F (2017) Biosafety of Prions. Prog Mol Biol Transl Sci 150:455–485). Elsevier B.V. https://doi.org/10.1016/bs.pmbts.2017.06.017

Bistaffa E, Vuong TT, Cazzaniga FA et al (2019) Use of different RT-QuIC substrates for detecting CWD prions in the brain of Norwegian cervids. Sci Rep 9:18595. https://doi.org/10.1038/s41598-019-55078-x

Bravo-Risi F, Soto P, Eckland T, Dittmar R, Ramírez S, Catumbela CSG, Soto C, Lockwood M, Nichols T, Morales R (2021) Detection of CWD prions in naturally infected white-tailed deer fetuses and gestational tissues by PMCA. Sci Rep 11(1):18385. https://doi.org/10.1038/s41598-021-97737-y

Bravo-Risi F, Soto P, Benavente R, Nichols TA, Morales R (2023) Dynamics of CWD prion detection in feces and blood from naturally infected white-tailed deer. Sci Rep 13(1):20170. https://doi.org/10.1038/s41598-023-46929-9

Brown P, Bradley R (1998) Clinical review 1755 and all that: a historical primer of transmissible spongiform encephalopathy. www.bmj.com

Burgener KR, Lichtenberg SS, Lomax A, Storm DJ, Walsh DP, Pedersen JA (2022) Diagnostic testing of chronic wasting disease in white-tailed deer (Odocoileus virginianus) by RT-QuIC using multiple tissues. PLoS One 17:e0274531. https://doi.org/10.1371/journal.pone.0274531

Carlson CM, Thomas S, Keating MW, Soto P, Gibbs NM, Chang H, Wiepz JK, Austin AG, Schneider JR, Morales R, Johnson CJ, Pedersen JA (2023) Plants as vectors for environmental prion transmission. IScience 108428:108428. https://doi.org/10.1016/j.isci.2023.108428

Cassmann ED, Frese RD, Greenlee JJ (2021) Second passage of chronic wasting disease of mule deer to sheep by intracranial inoculation compared to classical scrapie. J Vet Diagn Invest 33(4):711–720. https://doi.org/10.1177/10406387211017615

Castilla J, Saá P, Morales R, Abid K, Maundrell K, Soto C (2006) Protein Misfolding cyclic amplification for diagnosis and Prion propagation studies. Methods Enzymol 412:3–21. https://doi.org/10.1016/S0076-6879(06)12001-7

Castilla J, Morales R, Saá P, Barria M, Gambetti P, Soto C (2008) Cell-free propagation of prion strains. EMBO Journal 27:2557–2566. https://doi.org/10.1038/emboj.2008.181

Collinge J (1997) Human prion diseases and bovine spongiform encephalopathy (BSE). Hum Mol Genet 6(10):1699. Oxford University Press

Collinge J (2001) Prion diseases of humans and animals: their causes and molecular basis. Annu Rev Neurosci 24:519–550. https://doi.org/10.1146/annurev.neuro.24.1.519

Collins S, McLean CA, Masters CL (2001) Gerstmann-Sträussler-Scheinker syndrome, fatal familial insomnia, and kuru: a review of these less common human transmissible spongiform encephalopathies. J Clin Neurosci 8(5):387–397). Churchill Livingstone. https://doi.org/10.1054/jocn.2001.0919

Crowell J, Hughson A, Caughey B, Bessen RA (2015) Host determinants of prion strain diversity independent of Prion protein genotype. J Virol 89(20):10427–10441. https://doi.org/10.1128/jvi.01586-15

Davenport KA, Hoover CE, Denkers ND, Mathiason CK, Hoover EA (2018) Modified protein misfolding cyclic amplification overcomes real-time quaking-induced conversion assay inhibitors in Deer Saliva to detect chronic wasting disease prions. J Clin Microbiol 56:e00947

Dearmond SJ, Prusinertt SB (1995) Etiology and pathogenesis of Prion diseases. Am J Pathol 146(4):785

Denkers ND, Hoover CE, Davenport KA, Henderson DM, McNulty EE, Nalls AV, Mathiason CK, Hoover EA (2020) Very low oral exposure to prions of brain or saliva origin can transmit chronic wasting disease. PLoS One 15:e0237410. https://doi.org/10.1371/journal.pone.0237410

Escobar LE, Pritzkow S, Winter SN, Grear DA, Kirchgessner MS, Dominguez-Villegas E, Machado G, Townsend Peterson A, Soto C (2020) The ecology of chronic wasting disease in wildlife. Biol Rev 95(2):393–408. https://doi.org/10.1111/brv.12568

Fiorini M, Iselle G, Perra D, Bongianni M, Capaldi S, Sacchetto L, Ferrari S, Mombello A, Vascellari S, Testi S, Monaco S, Zanusso G (2020) High diagnostic accuracy of RT-QuIC assay in a prospective study of patients with suspected sCJD. Int J Mol Sci 21(3). https://doi.org/10.3390/ijms21030880

Gajdusek DC, Gibbs CJ, Alpers M (1966) Experimental transmission of a Kuru-like syndrome to chimpanzees.Nature 209(5025):794–796. https://doi.org/10.1038/209794a0

Gallardo MJ, Delgado FO (2021) Animal prion diseases: a review of intraspecies transmission. Open Veterin J 11(4):707–723). Faculty of Veterinary Medicine, University of Tripoli. https://doi.org/10.5455/OVJ.2021.v11.i4.23

Gibbs Jr CJ, Gajdusek DC, Asher DM, Alpers MP, Beck E, Daniel PM, Matthews WB (1968) Creutzfeldt-Jakob disease (Spongiform Encephalopathy): transmission to the Chimpanzee. Science 161(3839):388–389. https://doi.org/10.1126/science.161.3839.388

Giles K, Woerman AL, Berry DB, Prusiner SB (2017) Bioassays and inactivation of prions. Cold Spring Harb Perspect Biol 9(8). https://doi.org/10.1101/cshperspect.a023499

Green AJE (2019) RT-QuIC: a new test for sporadic CJD. Pract Neurol 19(1):49–55. https://doi.org/10.1136/practneurol-2018-001935

Greenlee JJ, Smith JD, Kunkle RA (2011) White-tailed deer are susceptible to the agent of sheep scrapie by intracerebral inoculation. Vet Res 42(1):107. https://doi.org/10.1186/1297-9716-42-107

Guiroy DC, Williams ES, Yanagihara R, Gajdusek DC (1991) Neuropatholostca topographic distribution of scrapie amyloid-immunoreactive plaques in chronic wasting disease in captive mule deer (Odocoileus hemionus hemionus). Acta Neuropathol 81:475

Haley NJ, Richt JA (2017) Evolution of diagnostic tests for chronic wasting disease, a naturally occurring prion disease of cervids. Pathogens 6(3). MDPI AG. https://doi.org/10.3390/pathogens6030035

Haley NJ, Mathiason CK, Zabel MD, Telling GC, Hoover EA (2009) Detection of sub-clinical CWD infection in conventional test-negative deer long after oral exposure to urine and feces from CWD+ deer. PLoS One 4(11):e7990. https://doi.org/10.1371/journal.pone.0007990

Haley NJ, Mathiason CK, Carver S, Zabel M, Telling GC, Hoover EA (2011) Detection of chronic wasting disease prions in salivary, urinary, and intestinal tissues of deer: potential mechanisms of Prion shedding and transmission. J Virol 85(13):6309–6318. https://doi.org/10.1128/jvi.00425-11

Henderson DM, Denkers ND, Hoover CE, Garbino N, Mathiason CK, Hoover EA (2015) Longitudinal detection of Prion shedding in saliva and urine by chronic wasting disease-infected deer by real-time quaking-induced conversion. J Virol 89(18):9338–9347. https://doi.org/10.1128/jvi.01118-15

Hibler CP, Wilson KL, Spraker TR, Miller MW, Zink RR, Debuse LL, Andersen E, Schweitzer D, Kennedy JA, Baeten LA, Smeltzer JF, Salman MD, Powers BE (2003) Field validation and assessment of an enzyme-linked immunosorbent assay for detecting chronic wasting disease in mule deer (Odocoileus hemionus), white-tailed deer (Odocoileus virginianus), and Rocky Mountain elk (Cervus elaphus nelsoni). J Vet Diagn Invest 15:311

Inzalaco HN, Brandell EE, Wilson SP, Hunsaker M, Stahler DR, Woelfel K, Walsh DP, Nordeen T, Storm DJ, Lichtenberg SS, Turner WC (2024) Detection of prions from spiked and free-ranging carnivore feces. Sci Rep 14(1):3804. https://doi.org/10.1038/s41598-023-44167-7

Jo Moore S, Kunkle R, West Greenlee MH, Nicholson E, Richt J, Hamir A, Ray Waters W, Greenlee J (2016) Horizontal transmission of chronic wasting disease in Reindeer. Emerg Infect Dis 22(12):2142–2145. https://doi.org/10.3201/eid2212.160635

Johnson CJ, Phillips KE, Schramm PT, McKenzie D, Aiken JM, Pedersen JA (2006) Prions adhere to soil minerals and remain infectious. PLoS Pathog 2(4):296–302. https://doi.org/10.1371/journal.ppat.0020032

Jones SM, Lazar EB, Porter AL, Prusinski CC, Brier MR, Bucelli RC, Day GS (2023) Real-time quaking-induced conversion assays for prions: applying a sensitive but imperfect test in clinical practice. Eur J Neurol 30(7):1854–1860. https://doi.org/10.1111/ene.15795

Kraft CN, Denkers ND, Mathiason CK, Hoover EA (2023) Longitudinal detection of prion shedding in nasal secretions of CWD-infected white-tailed deer. J Gen Virol 104(1). https://doi.org/10.1099/jgv.0.001825

Kramm C, Soto P, Nichols TA, Morales R (2020) Chronic wasting disease (CWD) prion detection in blood from pre-symptomatic white-tailed deer harboring PRNP polymorphic variants. Sci Rep 10(1):19763. https://doi.org/10.1038/s41598-020-75681-7

Lacroux C, Comoy E, Moudjou M, Perret-Liaudet A, Lugan S, Litaise C, Simmons H, Jas-Duval C, Lantier I, Béringue V, Groschup M, Fichet G, Costes P, Streichenberger N, Lantier F, Deslys JP, Vilette D, Andréoletti O (2014) Preclinical detection of variant CJD and BSE Prions in blood. PLoS Pathog 10(6):e1004202. https://doi.org/10.1371/journal.ppat.1004202

Li M, Schwabenlander MD, Rowden GR, Schefers JM, Jennelle CS, Carstensen M, Seelig D, Larsen PA (2021) RT-QuIC detection of CWD prion seeding activity in white-tailed deer muscle tissues. Sci Rep 11(1):16759. https://doi.org/10.1038/s41598-021-96127-8

McNulty E, Nalls AV, Mellentine S, Hughes E, Pulscher L, Hoover EA, Mathiason CK (2019) Comparison of conventional, amplification and bio-assay detection methods for a chronic wasting disease inoculum pool. PLoS One 14(5):e0216621. https://doi.org/10.1371/journal.pone.0216621

Medori R, Tritschler H-J, LeBlanc A, Villare F, Manetto V, Chen HY, Xue R, Leal S, Montagna P, Cortelli P, Tinuper P, Avoni P, Mochi M, Baruzzi A, Hauw JJ, Ott J, Lugaresi E, Autilio-Gambetti L, Gambetti P (1992) Fatal familial insomnia, a Prion disease with a mutation at Codon 178 of the Prion protein gene. N Engl J Med 326(7):444–449. https://doi.org/10.1056/nejm199202133260704

Morales R (2017) Prion strains in mammals: different conformations leading to disease. PLoS Pathog 13(7). Public Library of Science. https://doi.org/10.1371/journal.ppat.1006323

Morales R, Duran-Aniotz C, Diaz-Espinoza R, Camacho MV, Soto C (2012) Protein misfolding cyclic amplification of infectious prions. Nat Protoc 7(7):1397–1409. https://doi.org/10.1038/nprot.2012.067

Moreno JA, Telling GC (2018) Molecular mechanisms of chronic wasting disease prion propagation. Cold Spring Harb Perspect Med 8(6). https://doi.org/10.1101/cshperspect.a024448

Nalls AV, McNulty E, Powers J, Seelig DM, Hoover C, Haley NJ, Hayes-Klug J, Anderson K, Stewart P, Goldmann W, Hoover EA, Mathiason CK (2013) Mother to offspring transmission of Chronic Wasting disease in Reeves' Muntjac Deer. PLoS One 8(8). https://doi.org/10.1371/journal.pone.0071844

Nichols TA, Pulford B, Wyckoff AC, Meyerett C, Michel B, Gertig K, Hoover EA, Jewell JE, Telling GC, Zabel MD (2009) Detection of protease-resistant cervid prion protein in water from a CWD-endemic area. Prion 3(3):171–183. https://doi.org/10.4161/pri.3.3.9819

Notari S, Qing L, Pocchiari M, Dagdanova A, Hatcher K, Dogterom A, Groisman JF, Lumholtz IB, Puopolo M, Lasmezas C, Chen SG, Kong Q, Gambetti P (2012) Assessing prion infectivity of human urine in sporadic Creutzfeldt-Jakob disease. Emerg Infect Dis 18(1):21–28. https://doi.org/10.3201/eid1801.110589

Orge L, Lima C, Machado C, Tavares P, Mendonça P, Carvalho P, Silva J, de Pinto ML, Bastos E, Pereira JC, Gonçalves-Anjo N, Gama A, Esteves A, Alves A, Matos AC, Seixas F, Silva F, Pires I, Figueira L et al (2021) Neuropathology of animal prion diseases. Biomol Ther 11(3):1–29). MDPI AG. https://doi.org/10.3390/biom11030466

Osterholm MT, Anderson CJ, Zabel MD, Scheftel JM, Moore KA, Appleby BS (2019) Chronic wasting disease in cervids: implications for prion transmission to humans and other animal species. MBio 10(4). https://doi.org/10.1128/mBio.01091-19

Peters J, Miller JM, Jenny AL, Peterson TL, Carmichael KP (2000) Immunohistochemical diagnosis of chronic wasting disease in preclinically affected elk from a captive herd. Brief Commun J Vet Diagn Invest 12:579

Pritzkow S, Morales R, Moda F, Khan U, Telling GC, Hoover E, Soto C (2015) Grass plants bind, retain, uptake, and transport infectious prions. Cell Rep 11(8):1168–1175. https://doi.org/10.1016/j.celrep.2015.04.036

Pritzkow S, Morales R, Lyon A, Concha-Marambio L, Urayama A, Soto C (2018) Efficient prion disease transmission through common environmental materials. J Biol Chem 293(9):3363–3373. https://doi.org/10.1074/jbc.M117.810747

Pritzkow S, Morales R, Camacho M, Soto C (2021) Uptake, retention, and excretion of infectious prions by experimentally exposed earthworms. Emerg Infect Dis 27(12):3151–3154. https://doi.org/10.3201/eid2712.204236

Prusiner SB (1991) Molecular biology of Prion diseases. https://doi.org/10.1126/science.1675487

Pulford B, Spraker TR, Wyckoff AC, Meyerett C, Bender H, Ferguson A, Wyatt B, Lockwood K, Powers J, Telling GC, Wild MA, Zabel MD (2012) Detection of PrPCWD in feces from naturally exposed Rocky Mountain elk (Cervus elaphus nelsoni) using protein misfolding cyclic amplification. J Wildl Dis 48(2):425–434. https://doi.org/10.7589/0090-3558-48.2.425

Rossi M, Candelise N, Baiardi S, Capellari S, Giannini G, Orrù CD, Antelmi E, Mammana A, Hughson AG, Calandra-Buonaura G, Ladogana A, Plazzi G, Cortelli P, Caughey B, Parchi P (2020) Ultrasensitive RT-QuIC assay with high sensitivity and specificity for Lewy body-associated synucleinopathies. Acta Neuropathol 140(1):49–62. https://doi.org/10.1007/s00401-020-02160-8

Rowden GR, Picasso-Risso C, Li M, Schwabenlander MD, Wolf TM, Larsen PA (2023) Standardization of data analysis for RT-QuIC-based detection of Chronic Wasting disease. Pathogens 12(2). https://doi.org/10.3390/pathogens12020309

Sajnani G, Silva CJ, Ramos A, Pastrana MA, Onisko BC, Erickson ML, Antaki EM, Dynin I, Vázquez-Fernández E, Sigurdson CJ, Carter JM, Requena JR (2012) PK-sensitive PrPSc is infectious and shares basic structural features with PK-resistant PrPSc. PLoS Pathog 8(3):e1002547. https://doi.org/10.1371/journal.ppat.1002547

Sigurdson CJ (2008) A prion disease of cervids: Chronic wasting disease. Vet Res 39(4):41. https://doi.org/10.1051/vetres:2008018

Sigurdson CJ, Barillas-Mury C, Miller MW, Oesch B, Van Keulen LJM, Langeveld JPM, Hoover EA (2002) Printed in Great Britain PrP CWD lymphoid cell targets in early and advanced chronic wasting disease of mule deer. In. J Gen Virol 83:2617

Sigurdson CJ, Miller MW, Sigurdson C (2003) Other animal prion diseases. Br Med Bull 66:199–212. https://doi.org/10.1093/bmb/dg66.199

Smith CB, Booth CJ, Pedersen JA (2011) Fate of Prions in soil: a review. J Environ Qual 40(2):449–461. https://doi.org/10.2134/jeq2010.0412

Soto C, Saborio GP, Anderes L (2002) Cyclic amplification of protein misfolding: application to prion-related disorders and beyond. http://tins.trends.com

Soto P, Bravo-Risi F, Benavente R, Lichtenberg S, Lockwood M, Reed JH, Morales R (2023) Identification of chronic wasting disease prions in decaying tongue tissues from exhumed white-tailed deer. MSphere 8(5):e0027223. https://doi.org/10.1128/msphere.00272-23

Soto P, Bravo-Risi F, Kramm C, Gamez N, Benavente R, Bonilla DL, Reed JH, Lockwood M, Spraker TR, Nichols T, Morales R (2024) Nasal bots carry relevant titers of CWD prions in naturally infected white-tailed deer. EMBO Rep 25:334. https://doi.org/10.1038/s44319-023-00003-7

Telling GC, Scott M, Mastrianni J, Gabizon R, Marilyn Torchia LI, Cohen FE, De Armond SJ, Prusiner SB (1995) Prion propagation in mice expressing human and Chimeric PrP transgenes implicates the interaction of cellular PrP with another protein. Cell 83:79

Tewari D, Steward D, Fasnacht M, Livengood J (2021) Detection by real-time quaking-induced conversion (RT-QuIC), ELISA, and IHC of chronic wasting disease prion in lymph nodes from Pennsylvania white-tailed deer with specific PRNP genotypes. J Vet Diagn Invest 33(5):943–948. https://doi.org/10.1177/10406387211021411

Tranulis MA, Gavier-Widén D, Våge J, Nöremark M, Korpenfelt SL, Hautaniemi M, Pirisinu L, Nonno R, Benestad SL (2021) Chronic wasting disease in Europe: new strains on the horizon. Acta Vet Scand 63(1)). BioMed Central Ltd:48. https://doi.org/10.1186/s13028-021-00606-x

Vijaywargiya D, Abdelhafiz S, Chabrashvili T (2024) Sporadic CJD with false negative RT-QUIC testing (P3-9.011). Neurology 102(17_supplement_1). https://doi.org/10.1212/WNL.0000000000206340

Wang F, Pritzkow S, Soto C (2022) PMCA for ultrasensitive detection of prions and to study disease biology. Cell Tissue Res 392:307. https://doi.org/10.1007/s00441-022-03727-5. Springer Science and Business Media Deutschland GmbH

Watts JC, Prusiner SB (2014) Mouse models for studying the formation and propagation of prions. J Biol Chem 289(29):19841–19849). American Society for Biochemistry and Molecular Biology Inc. https://doi.org/10.1074/jbc.R114.550707

Williams ES, Young S (1980) Chronic Wasting disease of Captive Mule Deer: a spongiform encephalopathy. J Wildl Dis 16(1). http://meridian.allenpress.com/jwd/article-pdf/16/1/89/2228141/0090-3558-16_1_89.pdf

Yilmaz G, Morrill T, Pilot W, Ward C, Mitchell G, Soutyrine A, Dan H, Lin M, Guan J (2024) Optimization of RT-QuIC assay duration for screening chronic wasting disease in white-tailed deer. Veterin Sci 11(2). https://doi.org/10.3390/vetsci11020060

Yuan Q, Rowden G, Wolf TM, Schwabenlander MD, Larsen PA, Bartelt-Hunt SL, Bartz JC (2022) Sensitive detection of chronic wasting disease prions recovered from environmentally relevant surfaces. Environ Int 166:107347. https://doi.org/10.1016/j.envint.2022.107347

Chapter 2
Diagnosis of Prion Diseases

Tayyaba Saleem, Anna-Lisa Fischer, Sezgi Canaslan, Susana Da Silva Correia, Peter Hermann, Matthias Schmitz, Angela Da Silva Correia, and Inga Zerr

Abstract Prion diseases are rapidly progressive and fatal neurodegenerative disorders caused by misfolded prion proteins. Accurate and early diagnosis is essential to distinguish these conditions from treatable dementias and to prevent iatrogenic transmission. While definitive confirmation still depends on *postmortem* neuropathological techniques such as immunohistochemistry and western blot, recent advances have significantly improved *antemortem* diagnostic capabilities. The *antemortem* diagnosis combines clinical evaluation, neuroimaging, electroencephalography, and cerebrospinal fluid biomarkers. The development of real-time quaking-induced conversion (RT-QuIC) has enhanced the detection of misfolded prion proteins with high specificity, complementing existing diagnostic methods. Although advancements in biomarkers and diagnostic methodologies have improved the early detection of prion diseases, challenges remain. Continued research is crucial for enhancing early identification, tracking disease progression, optimizing patient management, and further elucidating disease pathogenesis.

Keywords Prion diseases · Prion diagnosis · CJD · TSEs · Biomarkers

Authors "Tayyaba Saleem, Anna-Lisa Fischer, Sezgi Canaslan, Susana Da Silva Correia" share first authorship. Authors "Angela Da Silva Correia, Inga Zerr" share last authorship

T. Saleem · A.-L. Fischer · S. Canaslan · S. D. S. Correia · P. Hermann · M. Schmitz
A. D. S. Correia (✉) · I. Zerr
Department of Neurology, University Medical Center, Georg-August University, Goettingen, Germany
e-mail: angela.silva-correia@med.uni-goettingen.de

G. Legname, F. Moda (eds.), *Biomarkers and Therapeutical Targets for Prion Diseases*, Subcellular Biochemistry 112, https://doi.org/10.1007/978-3-031-97055-9_2

Introduction

Early and precise diagnosis of Creutzfeldt–Jakob disease (CJD) is crucial for distinguishing this incurable condition from treatable, rapidly progressive dementias and preventing iatrogenic transmission (Zanusso et al. 2016; Paterson et al. 2012). As per the diagnostic criteria established by the Creutzfeldt–Jakob Disease International Surveillance Network (Diagnostic criteria for surveillance of CJD 2017; Hermann and Zerr 2022), conventional neuropathological methods, such as immunohistochemistry and western blot, which confirm misfolded prion protein (PrP^{Sc}) in the brain tissue, are considered valid for definitively diagnosing prion diseases. Typically, these diagnostic procedures are conducted as part of *postmortem* examinations or through biopsy, posing challenges in clinical practice (Brown and Farrell 2015; Chitravas et al. 2011). Patients may receive a diagnosis of probable or possible prion disease through *antemortem* assessment if they meet the clinical diagnostic criteria for prion disease (Diagnostic criteria for surveillance of CJD 2017) and other neurodegenerative conditions have been excluded. Supporting investigations such as magnetic resonance imaging (MRI), electroencephalography (EEG), and conventional cerebrospinal fluid (CSF) biomarker analyses aid in the diagnostic process, and real-time quaking-induced conversion (RT-QuIC) assays have improved diagnostic accuracy; however, these analyses do not provide a definite diagnosis (Collins et al. 2006; Zanusso et al. 2003). The most effective method for obtaining a *premortem* diagnosis of prion disease currently involves a combination of CSF laboratory findings and MRI results. An early and confident diagnosis of prion disease is not only beneficial for infection control measures but also plays a crucial role in planning patient care decisions (Connor et al. 2019).

Sporadic CJD (sCJD)

sCJD is the most prevalent form of prion disease in humans, comprising around 85% to 95% of all prion cases (Ladogana et al. 2005; Masters et al. 1979). Its origin remains unknown, as it is not associated with any mutations in the *PRNP* gene. However, studies indicate that individuals with a family history of CJD, a medical history of psychosis, a record of multiple surgical procedures, or prolonged residence on a farm (more than 10 years) are significantly more prone to developing sCJD (De Villemeur 2013). sCJD exhibits a broad range of phenotypic variations, linked to polymorphisms of codon 129 in the *PRNP* gene and the size of the protease-resistant core of the abnormal prion protein (PrPres), typically around 21 KDa (type 1) or 19 KDa (type 2). Codon 129 of the *PRNP* gene encodes either methionine (M) or valine (V) amino acids on each allele. Consequently, individuals with CJD can manifest MM, MV, or VV genotypes in the protease-resistant core. These variations, encompassing both genotypic and core size differences (type 1 and 2), contribute to distinct phenotypes of prion disease (Cali et al. 2006). As a

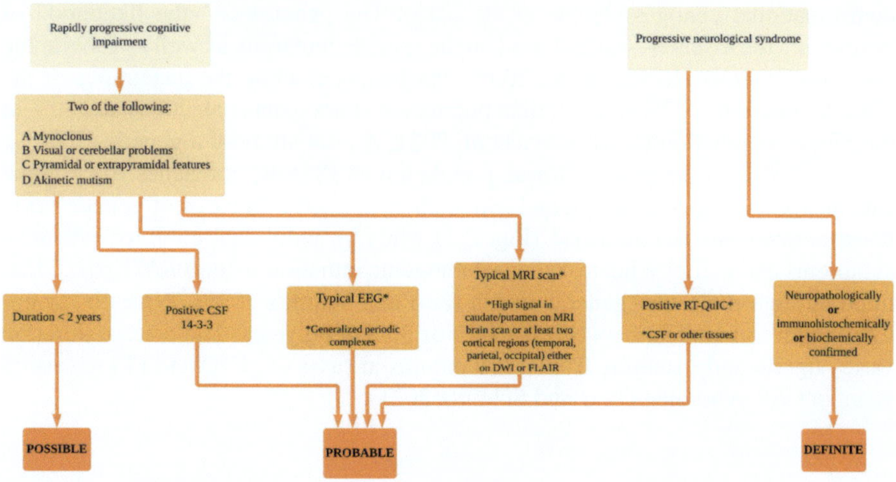

Fig. 2.1 sCJD Diagnostic Criteria. Definite, probable, and possible. Based on diagnostic criteria established by the Creutzfeldt–Jakob Disease International Surveillance Network. (Adapted from Diagnostic criteria for surveillance of CJD (2017))

result, sCJD is classified into six distinct subtypes: sCJD MM1, sCJD MM2, sCJD MV1, sCJD MV2, sCJD VV1, and sCJD VV2 (Cali et al. 2009; Puoti et al. 2012).

The clinical diagnosis of sCJD prior to death is based on a combination of characteristic neuropsychiatric symptoms, electroencephalography (EEG), magnetic resonance imaging (MRI), CSF 14-3-3 protein analysis, and real-time quaking-induced conversion (RT-QuIC) testing (Fig. 2.1). Additional biomarkers, such as elevated total Tau (t-Tau) levels in CSF, may improve the accuracy of *premortem* diagnosis. However, discrepancies among studies have raised controversies surrounding the clinical usefulness of certain established surrogate biomarkers (Abu-Rumeileh et al. 2019).

Genetic Prion Diseases

Genetic forms of prion diseases are relatively uncommon, comprising approximately 10% to 15% of all cases of prion diseases. This group includes the familiar CJD (fCJD), also known as genetic CJD (gCJD), Gerstmann–Sträussler–Scheinker syndrome (GSS), and fatal familial insomnia (FFI) (Zerr and Schmitz 2021; Ghetti et al. 2018; Budka et al. 1995; Uttley et al. 2020; Kovács et al. 2005). These disorders are linked to autosomal dominant pathogenic mutations in the *PRNP*, which may involve point mutations resulting in amino acid substitutions, premature stop codons, or insertions/deletions of octapeptide repeats (OPRs) in the N-terminal region of the *PRNP* (Schmitz et al. 2017). These mutations induce structural alterations in the normal PrP^{C}, facilitating its transformation into PrP^{Sc} through conformational changes (Noble et al. 2015). The penetrance—the likelihood of developing the disease—varies based on the specific mutation, as well as geographic and ethnic factors (Kovács et al. 2005). For instance, while the penetrance of the E200K mutation is 67% in the Italian population (Ladogana et al. 2005) and 60% in the Slovak population (Kosorinova et al. 2021; A pilot study of a genetic CJD risk factor (E200K) in the general Slovak population on JSTOR), it reaches 100% in the Libyan Jewish population in Israel (Spudich et al. 1995). In cases of genetic prion disease, diagnosis is confirmed (Fig. 2.2) when an individual exhibits indicative symptoms and carries a heterozygous pathogenic mutation in the *PRNP* gene, identified through molecular genetic testing (Zerr and Schmitz 2021). Typically, symptoms begin to manifest between the ages of 50 and 60 years. The disease duration varies significantly, ranging from a few months in cases of gCJD and FFI to several years in GSS syndrome (Zerr and Schmitz 2021).

Iatrogenic CJD (iCJD)

Iatrogenic transmission of CJD arises from specific surgical and medical procedures. The primary sources of iCJD include the utilization of contaminated growth hormone (hGH) and dura mater grafts obtained from human cadavers with undiagnosed CJD (Kobayashi et al. 2018; Will 2003). Other sources of iCJD include corneal transplants, treatment involving cadaveric pituitary-derived gonadotropin, and the utilization of CJD-contaminated electroencephalogram (EEG) depth electrodes and neurosurgical instruments (Kobayashi et al. 2018). Although iCJD diagnosis typically relies on identifying a known iatrogenic source (Fig. 2.3), it is important to acknowledge the possibility of transmission through unrecognized mechanisms (Llorens et al. 2020a). The clinical presentation of iCJD can vary depending on the specific route of exposure, and it is often characterized by an unusual young age at onset. Exposure through neurosurgical instruments or dura mater grafting typically mirrors sCJD, with comparable illness duration. While iCJD linked to contaminated hGH exposure may show predominant cerebellar syndrome and slightly longer disease duration (Liberski et al. 2023).

Variant CJD (vCJD)

In March 1996, the UK witnessed the first cases of vCJD, a newly recognized neurological disorder. The disease was linked to the consumption of beef from cattle affected by Bovine Spongiform Encephalopathy (BSE), commonly referred to as "mad cow" disease (Creutzfeldt-Jakob disease - NHS). vCJD can also be transmitted through blood transfusion, although this occurrence has been documented only five times in the UK (Creutzfeldt-Jakob disease - NHS). The duration of the incubation period remains unclear. vCJD exhibits distinct clinical features (Fig. 2.4).

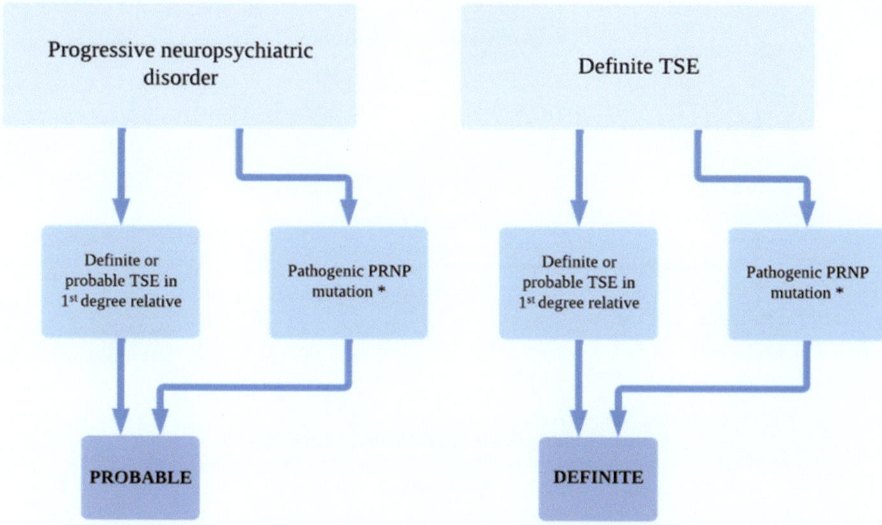

***MUTATIONS**

PRNP mutations associated with **GSS** neuropathological phenotype: P102L, P105L, A117V, G131V, F198S, D202N, Q212P, Q217R, M232T, 192 bpi

PRNP mutations associated with **CJD** neuropathological phenotype: D178N-129V, V180I, V180I+M232R, T183A, T188A, E196K, E200K, V203I, R208H, V210I, E211Q, M232R, 96 bpi, 120 bpi, 144 bpi, 168 bpi, 48 bpdel

PRNP mutations associated with **FFI** neuropathological phenotype: D178N-129M

PRNP mutation associated with vascular prp amyloid: Y145s

PRNP mutations associated with proven but unclassified prion disease: H187R, 216 bpi

Mutations associated with neuro-psychiatric disorder but not proven prion disease: I138M, G142S, Q160S, T188K, M232R, 24 bpi, 48 bpi, 48 bpi + nucleotide substitution in other octapeptides

ADDITIONAL LIST OF MUTATIONS

PRNP mutations without clinical and neuropathological data: T188R, P238S

PRNP polymorphisms with established influence on phenotype: M129V

PRNP polymorphisms with suggested influence on phenotype: N171S, E219K, 24 bp deletion

PRNP polymorphisms without established influence on phenotype: P68P, A117A, G124G, V161V, N173N, H177H, T188T, D202D, Q212Q, R228R, S230S

Fig. 2.2 Genetic TSE diagnostic criteria. Definite, probable, and mutation list. Based on diagnostic criteria established by the Creutzfeldt–Jakob Disease International Surveillance Network. (Adapted from Diagnostic criteria for surveillance of CJD (2017))

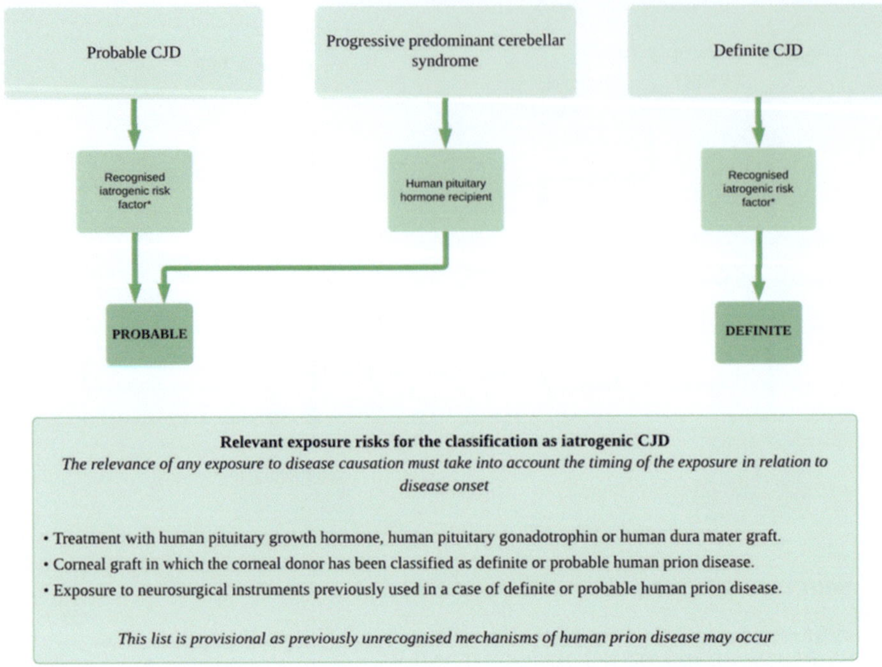

Fig. 2.3 Accidentally transmitted TSE diagnostic criteria. Definite, probable, and relevant exposure risks. Based on diagnostic criteria established by the Creutzfeldt–Jakob Disease International Surveillance Network. (Adapted from Diagnostic criteria for surveillance of CJD (2017))

Psychiatric symptoms, including behavioral changes, depression, and anxiety, typically appear first. Additionally, many patients experience limb or joint pain, as well as painful paresthesia or dysesthesia. Neurological deficits, myoclonus, and akinetic mutism develop progressively. Unlike sCJD, vCJD is significantly longer (~14 months), and its onset occurs in younger individuals (median age of onset is 29 years) (Zerr and Poser 2002).

Overview of Established Biomarkers

In 1998, a "biomarker" was defined by the National Institutes of Health Biomarker's Definition Working Group (Atkinson et al. 2001) as "a characteristic that is objectively measured and evaluated as an indicator of normal biological processes, pathogenic processes, or pharmacologic responses to a therapeutic intervention" (Atkinson et al. 2001). Biomarkers are objective and quantifiable representations of biological processes. Most of the biomarkers presently utilized in clinical settings are based on genomics, proteomics, and metabolomics (Ginsburg and Willard 2009). Biomarkers are essential for disease diagnosis, prognosis, and therapeutic monitoring.

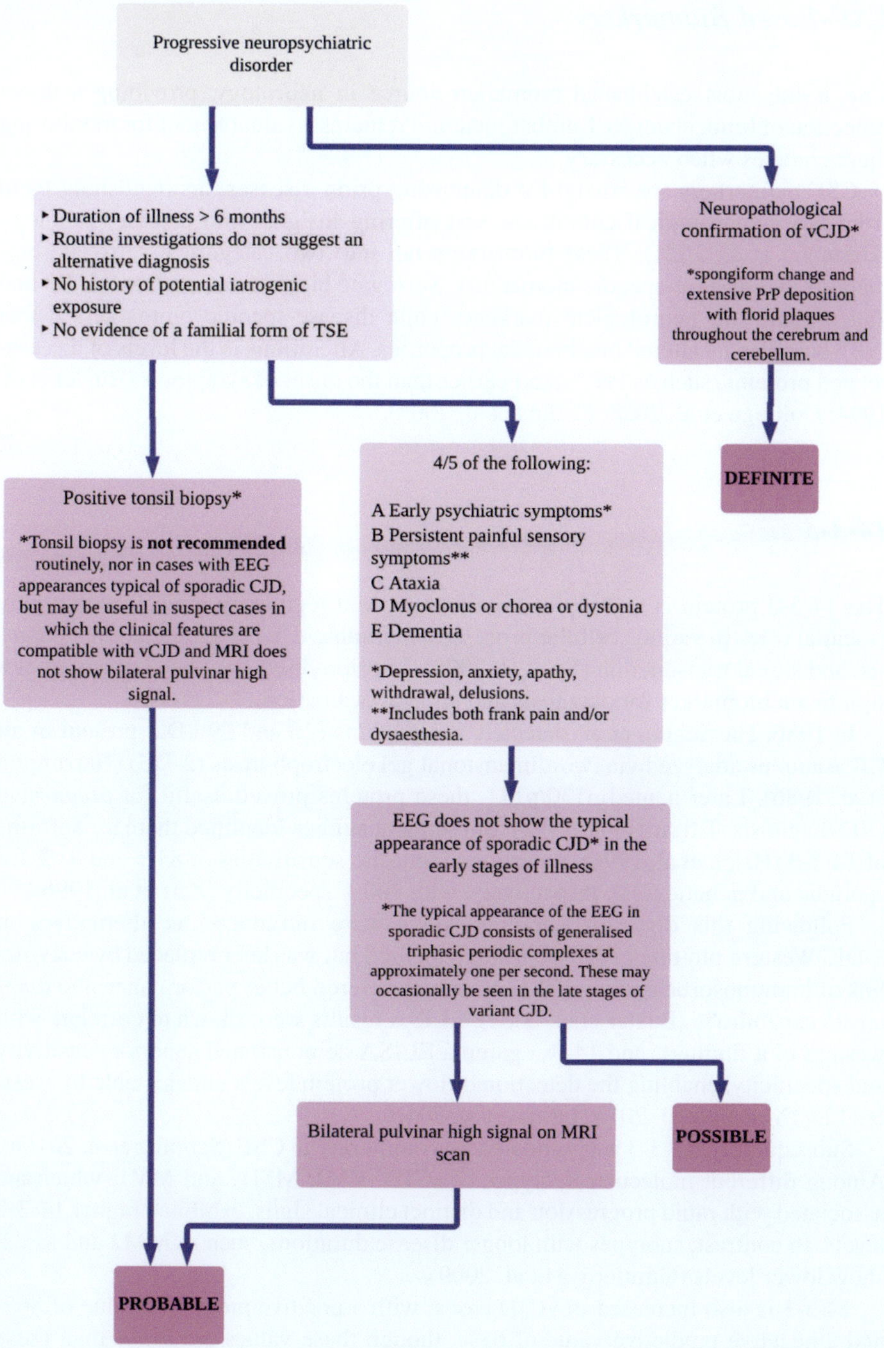

Fig. 2.4 Variant CJD diagnostic criteria. Definite, probable, and possible. Based on diagnostic criteria established by the Creutzfeldt–Jakob Disease International Surveillance Network. (Adapted from Diagnostic criteria for surveillance of CJD (2017))

CSF-Based Biomarkers

CSF is the most established biomarker source in neurology, providing a direct reflection of brain changes. Lumbar puncture remains a valuable tool for monitoring these changes when necessary.

CSF biomarkers are crucial for diagnosing prion diseases, distinguishing them from other neurological conditions, and offering insights into disease pathology (Hermann et al. 2021). These biomarkers fall into two categories: surrogate biomarkers and disease-specific biomarkers. Surrogate biomarkers, such as 14-3-3 and Tau protein, are neurological markers, while disease-specific biomarkers detect PrP^{Sc} based on its altered biochemical properties. Alterations in the levels of disease-related proteins, such as PrP^{Sc}, start earlier than the onset of symptoms (Büeler et al. 1994; Collinge et al. 2008; Collins et al. 2004).

14-3-3

The 14-3-3 protein is a group of highly conserved regulatory molecules that play essential roles in various cellular processes, including cell cycle regulation, apoptosis, and signal transduction (Fu et al. 2000). In prion diseases, 14-3-3 emerged as a significant biomarker for diagnosis and clinical evaluation.

In 1986, Harrington et al. detected two proteins (26 and 29 kDa) present in all CJD samples analyzed via two-dimensional gel electrophoresis (2-DE) (Harrington et al. 1986). Later named p130/p131, these proteins proved useful for *premortem* CJD diagnosis (Blisard et al. 1990). Subsequent studies identified them as isoforms of 14-3-3 (Hsich et al. 1996), reporting diagnostic sensitivities of 84% and 81% for sporadic and genetic CJD, respectively, with 100% specificity (Zerr et al. 1996).

Following this discovery, immunoassays were introduced as alternatives to 2-DE. Western blotting became a routine method but was later replaced by enzyme-linked immunosorbent assay (ELISA), which offered better performance and diagnostic capabilities (Baxter et al. 2002). ELISA results were shown to correlate with western blot findings, and 14-3-3 gamma ELISA demonstrated superior sensitivity and specificity, enabling the detection of lower protein levels undetectable by western blot (Matsui et al. 2011; Leitão et al. 2016).

Subsequently, 14-3-3 was validated as a biomarker in CSF (Schmitz et al. 2016a). Among different molecular subtypes of sCJD, VV2, MM1, and MV1, which are associated with rapid progression and distinct clinical signs, exhibited higher 14-3-3 levels. In contrast, subtypes with longer disease durations, such as MM2 and MV2, show lower levels (Gmitterová et al. 2009).

14-3-3 is also increased in vCJD cases, with a positive predictive value of 86% and a negative predictive value of 63%, though these values are lower than those observed in sCJD (Green et al. 2001).

Tau

Tau is a microtubule-associated protein predominantly expressed in neuronal cells, with lower expression levels in glial cells (Weingarten et al. 1975). In CSF, total Tau (t-Tau) concentrations are thought to reflect the pace of axonal degeneration across various neurological disorders (Blennow et al. 2010). In the diagnosis of sCJD, which is the most common type, changes in the levels of CSF 14-3-3 and t-Tau are key indicators, both of which are considered neurodegenerative markers (Schmitz et al. 2016a; Otto et al. 2002; Sanchez-Juan et al. 2006; Zerr et al. 1998).

In sCJD, t-Tau protein levels are elevated (Otto et al. 1997), with an optimal cutoff range of 1300–1400 pg/mL for diagnosis (Sanchez-Juan et al. 2006; Otto et al. 1997; Blennow et al. 2005). In contrast, vCJD shows elevated t-Tau levels compared to controls, with a cutoff above 500 pg/mL, similar to levels observed in AD but different from those seen in sCJD (Green et al. 2001). t-Tau levels in CJD are also influenced by the prion protein type, specifically the codon 129 genotype on the *PRNP* gene (Karch et al. 2015).

There is a significant correlation between t-Tau levels and disease duration, severity, and cognitive decline in CJD, suggesting that CSF Tau concentrations are an indicator of neuronal damage in the brain (Meiner et al. 2011; Cohen et al. 2016). Conversely, phosphorylated Tau (p-Tau) levels are elevated in AD but remain stable across other progressive neurological conditions (Riemenschneider et al. 2003; Baldeiras et al. 2009). The predictive capabilities of t-Tau levels and the t-Tau/p-Tau ratio enhance their utility, particularly when combined with other diagnostic approaches, for identifying CJD (Bahl et al. 2009; Skinningsrud et al. 2008). Tau has a positive predictive value of 93% and a negative predictive value of 81%. The combination of CSF 14-3-3 and Tau offers even greater diagnostic accuracy, with a positive predictive value of 91% and a negative predictive value of 84% (Green et al. 2001).

Another option to differentiate between CJD and AD is the application of non-phosphorylated tau (non-p-tau) as a biomarker due to its distinctive patterns in cerebrospinal fluid (CSF). In CJD, non-p-tau levels are typically found to be significantly lower compared to AD, where phosphorylated tau (p-tau) levels are markedly elevated (Llorens et al. 2020b; Ermann et al. 2018). This difference in non-p-tau, combined with other diagnostic markers, offers a valuable tool in distinguishing these two neurodegenerative conditions, as CJD often presents with rapidly progressive dementia and specific prion-related biomarkers, whereas AD is characterized by amyloid plaques and a progressive buildup of p-tau. The use of non-p-tau, along with traditional diagnostic methods, can improve the accuracy of differential diagnosis, allowing for more targeted therapeutic strategies and better prognosis for patients (Llorens et al. 2020b; Ermann et al. 2018).

Neuron-Specific Enolase (NSE) and S100 Calcium Binding Protein B (S100b)

NSE and S100b were among the most studied proteins after 14-3-3 and t-Tau. An increase in neuron-specific enolase is linked with rapid nerve cell loss, observed in CJD cases where NSE levels were elevated during disease progression (Kropp et al. 1999). A prospective study exploring the NSE levels found that with a threshold of 35 ng/ml, the sensitivity was 80%, and the specificity was 92% for the diagnosis of CJD (Weber et al. 1997). In a simultaneous comparison of 14-3-3, NSE, and S-100 in a suspected CJD cohort, the sensitivity of 14-3-3, NSE, and S-100 was 89.8%, 79.7%, and 94.2%, while their specificity was 100%, 91.5%, and 85.4%, respectively. Among the tested markers, 14-3-3 alone exhibited the highest specificity and sensitivity (Beaudry et al. 1999). Although S100b has been extensively studied, it was not as accurate or successful compared to 14-3-3.

Alpha-Synuclein

Alpha-synuclein, a neuronal protein involved in synaptic function, has emerged as a promising biomarker for Creutzfeldt–Jakob disease (CJD) due to its markedly elevated levels in the CSF of affected patients compared to those with other neurodegenerative disorders. The adoption of Meso Scale Discovery (MSD) technology in its measurement has enhanced the precision of quantification, offering a robust platform for detecting subtle variations in protein concentration. In the study by Llorens et al., an electrochemiluminescence (ECL)-based alpha-synuclein enzyme-linked immunosorbent assay (ELISA) implemented on the MSD system was used to accurately measure CSF alpha-synuclein levels (Kruse et al. 2018; Llorens et al. 2017a, 2018a). Their findings demonstrated high diagnostic accuracy—with sensitivity and specificity values that effectively distinguished CJD cases from non-CJD controls—thereby underscoring the assay's potential as a reliable diagnostic tool for CJD diagnosis.

Protein Misfolding Cyclic Amplification (PMCA)

The detection of misfolded prion protein in CSF is considered more specific for prion diseases. Aggregation assays originated with the development of PMCA, which mimics the conversion of PrP^{C} into PrP^{Sc} in an in vitro environment. This process involves incubation and sonication to generate more PrP^{Sc}. PMCA enables the detection of minute levels of PrP^{Sc} by exponentially amplifying its presence to detectable levels using a standard western blot. This capability has led to its adoption in prion disease diagnosis (Gonzalez-Montalban et al. 2011).

The application of PMCA has enabled the detection of prions in a range of biological fluids and tissues. Specifically, prions have been detected in brain (Saborio et al. 2001), CSF (Barria et al. 2018), blood (Lacroux et al. 2014; Concha-Marambio et al. 2016), urine (Moda et al. 2014), and several other tissues and fluids of patients with vCJD. Additionally, prions have been detected in olfactory mucosa (OM) (Cazzaniga et al. 2022) and urine (Pritzkow et al. 2023) of patients with sCJD and in the olfactory mucosa of FFI patients (Redaelli et al. 2017).

RT-QuIC

Building on the principles of PMCA, an advanced technology for the amplification of misfolded PrP was later developed and is known as RT-QuIC. In RT-QuIC, recombinant prion protein (rec PrP) substrate is converted into misfolded proteins and monitored using amyloid-sensitive thioflavin-T (Th-T) dye. This assay, which is plate-based, utilizes the amyloid-binding dye as a key component. The emission spectrum of this dye shifts upon binding to amyloid structures, enabling the real-time monitoring of the aggregation kinetics of the assay (Atarashi et al. 2011). The RT-QuIC assay is a significant breakthrough in the *antemortem* diagnosis of prion diseases, with high reproducibility between different diagnostic laboratories (Atarashi et al. 2011; McGuire et al. 2012; Cramm et al. 2016). The RT-QuIC assay (Fig. 2.5) mimics the seeded conversion process of PrP^{C} in vitro, amplifying minuscule amounts of PrP^{Sc} derived from brain, CSF, olfactory mucosa (OM) (Zanusso et al. 2003), tear fluid (Schmitz et al. 2023), or other biological samples to a detectable level (Atarashi et al. 2008). CSF is a well-established biological fluid for the detection of PrP^{Sc} in the RT-QuIC with a high specificity of approximately 100% (Hermann et al. 2021). The sensitivity of this method depends on the kind of prion disease; for sporadic forms of prion diseases, the sensitivity varies between 80 and 96% for sporadic CJD (sCJD) and around 60% for sporadic fatal insomnia (sFI) (Abu-Rumeileh et al. 2018a). For genetic forms of prion diseases, the sensitivity ranges between 62% and 88% for gCJD E200K patients (Xiao et al. 2019), 16.7% and 83.3% for FFI patients (Cramm et al. 2016; Xiao et al. 2020; Schmitz et al. 2022a), and 33% and 81.8% for GSS (Schmitz et al. 2022a; Sano et al. 2013). The variation in the sensitivity of the RT-QuIC assay can depend on the different prion diseases that are caused by different PrP^{Sc} strains, which exhibit different biochemical properties, including the structure and stability of the PrP^{Sc}. The disease stage may also influence the sensitivity of the assay; the amount of PrP^{Sc} present in CSF and other biological samples may vary depending on the disease progression, which could be influenced by the abundance of PrP^{Sc} in the sample, with higher sensitivity typically observed in later stages of disease when PrP^{Sc} levels are higher.

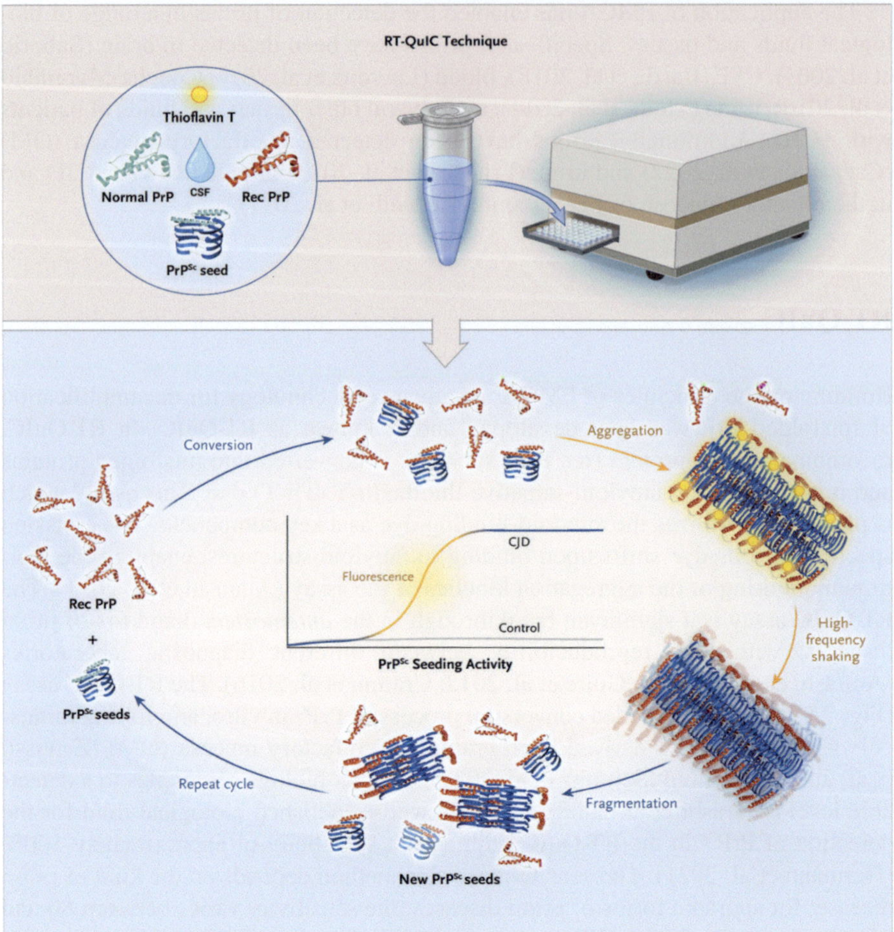

Fig. 2.5 Principle of RT-QuIC technique. Misfolding of the recombinant prion protein (rec PrP) after contact with PrP^{Sc} seeds (conversion) is followed by aggregation into larger structures, which then are fragmented by high-frequency shaking. New seeds provide the basis for further conversion of rec PrP in repeated cycles of incubation and shaking. The results are visualized by measuring the thioflavin T fluorescence that is emitted when the dye binds to aggregates of misfolded protein. (Adapted from Zerr (2022). Reprinted with permission Zerr (2022))

Neuroimaging and Electrophysiological Diagnostic Methods

MRI

MRI is a valuable tool for diagnosing prion diseases by detecting characteristic brain abnormalities. The combination of different MRI pulse sequences has emerged as an additional parameter for establishing a "probable" diagnosis of sCJD, particularly when ruling out other potential neurodegenerative diseases such as ischemia,

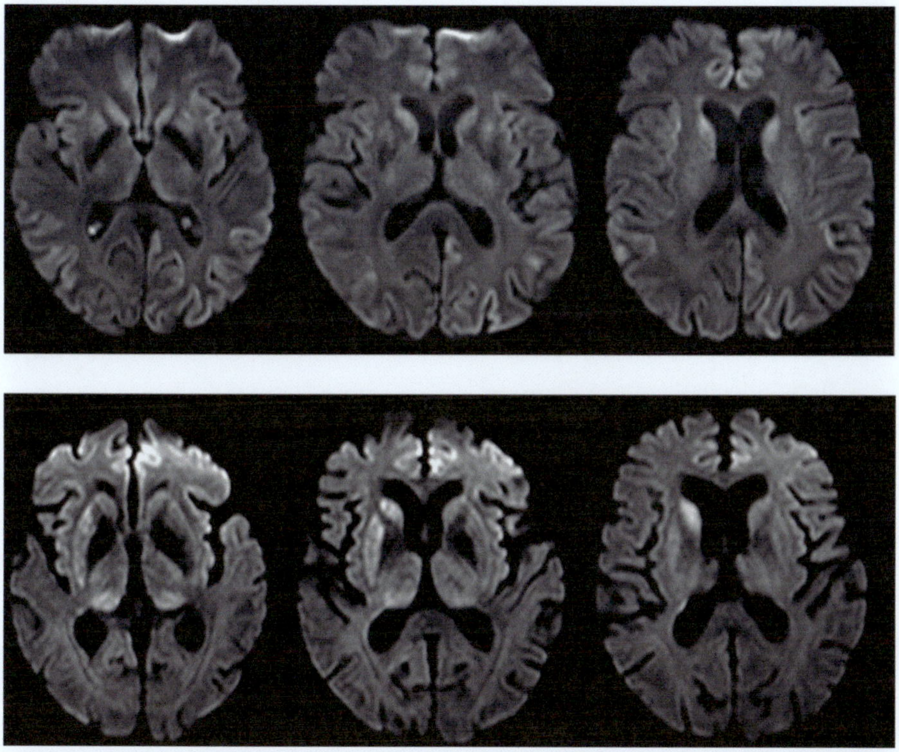

Fig. 2.6 Transverse cMRI series from two CJD patients. The upper series, representing an MM type, shows bilateral hyperintensities in the temporoparietal and temporooccipital cortex as well as in the caudate nucleus and putamen. The lower series, representing a VV type, shows hyperintensities in the caudate nucleus, putamen, and thalamus (pulvinar). (Source: National Reference Center for TSE)

encephalitis, and neoplasia in the patient's diagnosis. Diffusion-weighted imaging (DWI) and fluid-attenuated inversion recovery (FLAIR) sequences often reveal hyperintense signals in key regions such as the cerebral cortex, basal ganglia, and thalamus. These signals can appear as characteristic patterns like "cortical ribboning" or the "pulvinar sign," which are especially significant in vCJD. Additionally, in clinical variants, signal increases may be localized differently—for example, the Heidenhain variant shows an occipital emphasis that correlates with early visual disturbances (Fig. 2.6).

Consequently, in 2009, DWI and FLAIR MRI sequences were proposed to be incorporated into the criteria (Hermann et al. 2021; Zerr et al. 2009). As per these criteria, a positive MRI brain scan indicative of sCJD diagnosis is characterized by high signal intensity in at least two cortical regions (temporal, parietal, occipital) or at the caudate nucleus and putamen, observed on either DWI or FLAIR sequence (Meissner et al. 2009). The combination of those MRI sequences was shown to be highly specific (96%) and have a higher sensitivity (80%) for sCJD patients

(Hermann et al. 2018; Rosenbloom et al. 2015). In genetic prion diseases, the sensitivity ranges between 50% and 88% for gCJD E200K patients, 50% and 88% for GSS, and 0% for FFI patients (Connor et al. 2019).

These characteristic imaging findings not only help differentiate prion diseases like CJD from other neurodegenerative conditions but also contribute to a timelier diagnosis when interpreted alongside clinical assessments and cerebrospinal fluid biomarkers.

Positron Emission Tomography

Positron Emission Tomography (PET) using [18F] fluoro-2-deoxy-D-glucose as a tracer (FDG-PET) is a powerful imaging tool that can detect alterations in glucose metabolism in the brain. In particular, FDG-PET is effective in identifying decreased glucose metabolism in the cortical regions of patients with sCJD. The value of FDG-PET in the differential diagnosis is limited due to the lack of distinct, consistent metabolic patterns across all prion disease subtypes (Renard et al. 2017). Nonetheless, FDG-PET remains a promising tool, particularly in the early stages of sCJD. One notable finding is reduced glucose metabolism in the thalamus observed in patients with a rare MM2T subtype of sporadic fatal insomnia (Abu-Rumeileh et al. 2018a).

Electroencephalography (EEG)

Periodic sharp-wave complexes (PSWCs) observed on electroencephalography (EEG) recordings (Fig. 2.7), with a frequency of 1 Hz, represent a hallmark finding throughout the course of illness in patients with sCJD, exhibiting a sensitivity of 64% and a specificity of 91% (Steinhoff et al. 2004). PSWC serves not only as a pivotal indicator of sCJD but also boasts excellent interobserver reliability across different EEG evaluations, ensuring consistent and trustworthy EEG assessments.

The diagnosis of PSWCs on EEG recordings entails the presence of strictly periodic cerebral potentials, predominantly lasting between 100 and 600 milliseconds, with inter-complex intervals falling between 500 and 2000 milliseconds. These criteria are established based on EEG studies of periodic complexes in pathologically confirmed cases of CJD. The detection of PSWCs on EEG varies depending on the sCJD subtype. For instance, PSWCs may be detected in a very early stage of the disease (1–4 months after onset) in classic sCJD patients with MM1 subtype or only in the later stages of the disease in slow progression sCJD patients such as MM2C and MV2K subtype (Steinhoff et al. 2004; Neufeld and Korczyn 1992; Burger et al. 1972; Lee and Blair 1973; Wieser et al. 2006). For genetic forms of prion diseases, the sensitivity is approximately 50% for gCJD E200K patients (Gao

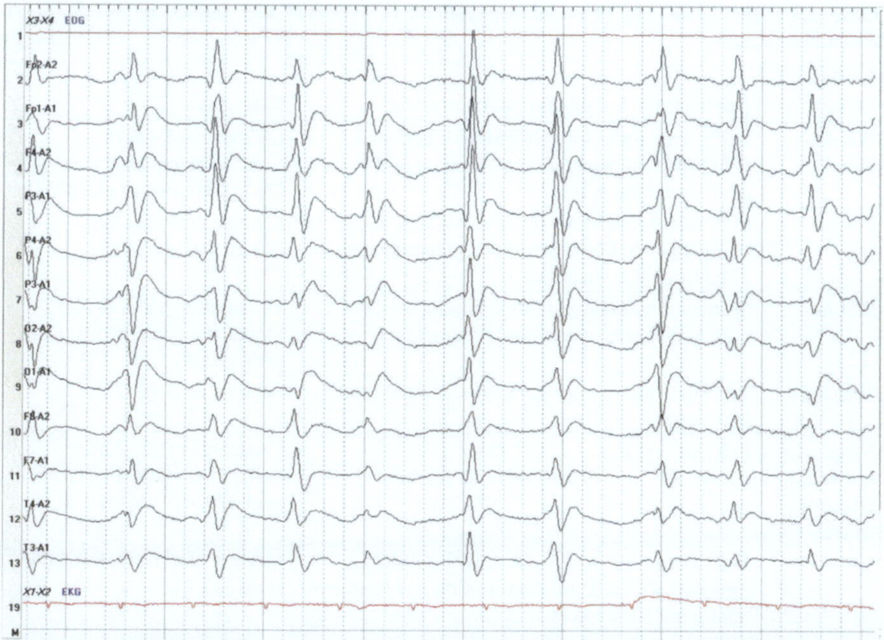

Fig. 2.7 Electroencephalogram in CJD. EEG of a patient with sporadic Creutzfeldt–Jakob disease (sCJD). The findings show periodic bi- and triphasic complexes, which are observed in about two-thirds of sCJD patients

et al. 2019) and is rarely observed in FFI patients with a specificity of about 3% (Chu et al. 2022).

Biomarker Under Development

Recent technological advancements have enabled the detection of disease-related proteins in various tissues and fluids, including olfactory mucosa, skin, blood, urine, and tear fluid (Table 2.1) (Moda et al. 2014; Pritzkow et al. 2023; Schmitz et al. 2023; Maass et al. 2020; Bongianni et al. 2017, 2022; Okuzumi et al. 2023; Schmitz et al. 2022b; Orrú et al. 2014, 2017).

The exploration of molecules associated with synaptic degeneration stands as one of the most extensively studied approaches in the quest for preclinical biomarkers of TSEs (Table 2.2). Neurofilament light chain protein (NfL) is recognized as an indicator of neuronal damage and is routinely subjected to analysis (Schmitz et al. 2016a; Zerr et al. 1998; Otto et al. 2002; Thompson and Mead 2019). Additional markers of neuronal death, such as alpha-synuclein (Llorens et al. 2018b), S100b protein (Chohan et al. 2010), glial fibrillary acidic protein (GFAP), neuron-specific

Table 2.1 Established and potential applications of biofluids and tissues in prion disease biomarker research

Biofluid/ tissue	Applications (established and potential)	References
Whole blood	PrP^{Sc} detection (vCJD by PMCA); potential RT-QuIC implementation and source of additional biomarkers	Lacroux et al. (2014), Concha-Marambio et al. (2016), Thomas et al. (2023)
Plasma	PrP^{Sc} detection (vCJD by PMCA); source of neurodegeneration biomarkers; potential source of additional biomarkers.	Schmitz et al. (2022b), Bougard et al. (2016)
Serum	Source of neurodegeneration biomarkers; potential source of additional biomarkers.	Shimamura et al. (2024), Steinacker et al. (2016), Chen et al. (2016)
Tear fluid	PrP^{Sc} detection; potential source of additional biomarkers	Schmitz et al. (2023)
Nasal brushings / Olfactory tissue	PrP^{Sc} detection; potential source of additional biomarkers	Cazzaniga et al. (2022), Redaelli et al. (2017), Orrú et al. (2014)
Urine	PrP^{Sc} detection (vCJD and sCJD by PMCA); potential RT-QuIC implementation and source of additional biomarkers	Moda et al. (2014), Pritzkow et al. (2023)
Skin biopsies	PrP^{Sc} detection; potential source of additional biomarkers	Orrú et al. (2017)
Saliva	PrP^{Sc} detection in salivary gland (vCJD by PMCA); potential RT-QuIC implementation and source of additional biomarkers	Douet et al. (2017)

Table 2.2 Additional potential CSF biomarkers for the differential diagnosis of sCJD

Biomarker	Diagnostic accuracy	References
Malate dehydrogenase 1 (MDH1)	Sensitivity: 83% / Specificity: 85%[a]	Schmitz et al. (2016b)
α-, β-, and γ-Synuclein	Increased in sCJD[b]	Oeckl et al. (2016)
Non-phosphorylated Tau (non-P-Tau)	Differentiates sCJD from AD (AUC 0.99) and from other neurological diseases (AUC 0.99)	Llorens et al. (2020b), Ermann et al. (2018)
Neurogranin	Differentiates sCJD from other neurological diseases (AUC 0.96) and AD (AUC 0.85)	Blennow et al. (2019)
Ubiquitin	Differentiates prion diseases from healthy controls (AUC 0.95), AD (AUC 0.85), and FTD (AUC 0.95)	Abu-Rumeileh et al. (2020)
Bone morphogenetic protein and activin membrane-bound inhibitor (BAMBI)	Increased in sCJD[a]	López-Pérez et al. (2020)
Cell-free mitochondrial DNA (mtDNA)	Increased in sCJD[a]	Li et al. (2019)

Adapted from: Hermann Hermann et al. (2021)
[a]Compared to other neurological diseases
[b]Compared to other neurological diseases and non-neurodegenerative controls

enolase (NSE) (Thompson and Mead 2019), and neurogranin (Blennow et al. 2019), exist, yet their diagnostic utility in prion diseases is not fully established.

NfL

NfL serves as a constituent of the neuronal cytoskeleton and is released in response to neuronal damage across various conditions, rendering it a highly reliable biomarker for neurodegeneration (Zerr et al. 2018). Increased levels of NfL are found in all subtypes of sCJD, even in those typically associated with low levels of t-tau and negative protein 14-3-3, such as sCJD MV2K, MM2C, and gCJD E200K (Blennow et al. 2019; Kanata et al. 2019). NfL has exceptional sensitivity in detecting sCJD (Schmitz et al. 2022b), higher than 95%, but demonstrates a low specificity of 43.1% (Abu-Rumeileh et al. 2019; Abu-Rumeileh et al. 2018b). Increased plasma levels of NfL have been observed in CJD, significantly higher than in other neurodegenerative dementias such as AD, Lewy body dementia (LBD), or frontotemporal dementia (FTD), as well as compared to control groups (Zerr et al. 1998). A study demonstrated an AUC of 0.93 for distinguishing CJD from non-CJD dementias (Zerr et al. 2021), while another reported a sensitivity of 100% and specificity of 85.5% in diagnosing various prionopathies, including sporadic, genetic, and iatrogenic CJD, as well as GSS (Steinacker et al. 2016; Thompson and Mead 2019).

YLK-40

Activation of microglial and astrocyte cells represents a pivotal characteristic in numerous neurodegenerative conditions (Porter et al. 1973). In prion diseases, however, there is an absence of a typical immune or inflammatory response, and it remains unclear whether the activation of these mechanisms contributes to the neurodegenerative process or emerges as a consequence of the disease's pathogenesis (Kercher et al. 2007). Nevertheless, neuroinflammation phenomena are widely documented in these disorders, prompting exploration into the potential of glia activation and neuroinflammation biomarkers (Aguzzi et al. 2013). Among the investigated markers of inflammation and astroglial activation, the glycoprotein YKL-40 emerges as particularly promising. Elevated levels of YKL-40 in CSF have been demonstrated in sCJD compared to other neurodegenerative diseases (Llorens et al. 2017b; Villar-Piqué et al. 2019). Furthermore, recent studies have quantified this glycoprotein in plasma from various patients with neurodegenerative disorders, emphasizing notably higher levels in individuals with CJD (Llorens et al. 2017b; Villar-Piqué et al. 2019).

Advances in Biomarker Discovery

Advances in technology have greatly contributed to the development of new detection methods in science. When examining the historical methodologies used in diagnostics, the progression of new detection methods is characterized by increased accuracy, ease of use, heightened sensitivity, and enhanced user-friendliness. The detection of 14-3-3 protein biomarkers, which began with 2-DE, has evolved, and currently, ELISA stands out as the most prominent and user-friendly methodology. ELISA, as an immunoassay, leverages the catalytic characteristics of enzymes for the detection and quantification of reactions. The heightened sensitivity of ELISA facilitates the utilization of various body fluids for diagnostic purposes.

Single-molecule array (SIMOA) is an ultrasensitive technology that is better than ELISA in its detection capability, allowing measurement up to 10^{-19} M (Rissin et al. 2010). The advantages of SIMOA are reduced reaction volume and the ability to differentiate femtomolar differences in protein concentrations with fluorescence-based detection. With the advancement in detection methodologies, blood and other noninvasive body fluids can replace CSF as a biomarker resource in the near future. There are studies that compared the detectability and efficiency of the same marker in different fluids like CSF and blood. Neurofilament light (NfL), which is an axonal damage marker, is highly increased in both plasma and CSF of sCJD patients in comparison to control groups that are non-neurodegenerative and neurodegenerative controls (Zerr et al. 2018; Abu-Rumeileh et al. 2018b; Kanata et al. 2019; Zerr et al. 2021; Schmitz et al. 2022b). ROC curve analysis showed a diagnostic accuracy of CSF by AUC of 0.92 with 80% specificity and 87% sensitivity, indicating that it is in the comparable range with other markers like 14-3-3 or RT-QuIC exhibiting specificity and sensitivity between 80 and 99% (Schmitz et al. 2016a; Cramm et al. 2016; Schmitz et al. 2022b). Independently replicated studies from other laboratories using SIMOA for blood NfL with high sensitivity and specificity in sCJD support that NfL is a suitable blood-based biomarker (Zerr et al. 2021; Steinacker et al. 2016; Thompson et al. 2018, 2021). It makes not only monitoring disease progression easier but also allows for screening individuals for early diagnosis.

References

A pilot study of a genetic CJD risk factor (E200K) in the general Slovak population on JSTOR. https://www.jstor.org/stable/43775010

Abu-Rumeileh S et al (2018a) Sporadic Fatal Insomnia in Europe: phenotypic features and diagnostic challenges. Ann Neurol 84:347–360

Abu-Rumeileh S et al (2018b) The CSF neurofilament light signature in rapidly progressive neurodegenerative dementias. Alzheimers Res Ther 10:3

Abu-Rumeileh S et al (2019) Diagnostic value of surrogate CSF biomarkers for Creutzfeldt-Jakob disease in the era of RT-QuIC. J Neurol 266:3136–3143

Abu-Rumeileh S et al (2020) CSF Ubiquitin levels are higher in Alzheimer's disease than in frontotemporal dementia and reflect the molecular subtype in Prion disease. Biomol Ther 10:497

Aguzzi A, Nuvolone M, Zhu C (2013) The immunobiology of prion diseases. Nat Rev Immunol 13:888–902

Atarashi R et al (2008) Simplified ultrasensitive prion detection by recombinant PrP conversion with shaking. Nat Methods 5:211–212

Atarashi R, Sano K, Satoh K, Nishida N (2011) Real-time quaking-induced conversion: a highly sensitive assay for prion detection. Prion 5:150

Atkinson AJ et al (2001) Biomarkers and surrogate endpoints: preferred definitions and conceptual framework. Clin Pharmacol Ther 69:89–95

Bahl JMC et al (2009) The diagnostic efficiency of biomarkers in sporadic Creutzfeldt-Jakob disease compared to Alzheimer's disease. Neurobiol Aging 30:1834–1841

Baldeiras IE et al (2009) Diagnostic value of CSF protein profile in a Portuguese population of sCJD patients. J Neurol 256:1540–1550

Barria MA, Lee A, Green AJE, Knight R, Head MW (2018) Rapid amplification of prions from variant Creutzfeldt-Jakob disease cerebrospinal fluid. J Pathol Clin Res 4:86–92

Baxter HC et al (2002) Specific 14-3-3 isoform detection and immunolocalization in prion diseases. Biochem Soc Trans 30:387–391

Beaudry P et al (1999) 14-3-3 protein, neuron-specific enolase, and S-100 protein in cerebrospinal fluid of patients with Creutzfeldt-Jakob disease. Dement Geriatr Cogn Disord 10:40–46

Blennow K, Johansson A, Zetterberg H (2005) Diagnostic value of 14-3-3beta immunoblot and T-tau/P-tau ratio in clinically suspected Creutzfeldt-Jakob disease. Int J Mol Med 16:1147–1149

Blennow K, Hampel H, Weiner M, Zetterberg H (2010) Cerebrospinal fluid and plasma biomarkers in Alzheimer disease. Nat Rev Neurol 6:131–144

Blennow K et al (2019) CSF neurogranin as a neuronal damage marker in CJD: a comparative study with AD. J Neurol Neurosurg Psychiatry 90:846–853

Blisard KS et al (1990) Pre-mortem diagnosis of Creutzfeldt-Jakob disease by detection of abnormal cerebrospinal fluid proteins. J Neurol Sci 99:75–81

Bongianni M et al (2017) Diagnosis of human Prion disease using real-time quaking-induced conversion testing of Olfactory Mucosa and cerebrospinal fluid samples. JAMA Neurol 74:155–162

Bongianni M et al (2022) Olfactory swab sampling optimization for α-synuclein aggregate detection in patients with Parkinson's disease. Transl Neurodegener 11:37

Bougard D et al (2016) Detection of prions in the plasma of presymptomatic and symptomatic patients with variant Creutzfeldt-Jakob disease. Sci Transl Med 8:370ra182

Brown P, Farrell M (2015) A practical approach to avoiding iatrogenic Creutzfeldt-Jakob disease (CJD) from invasive instruments. Infect Control Hosp Epidemiol 36:844–848

Budka H et al (1995) Neuropathological diagnostic criteria for Creutzfeldt-Jakob disease (CJD) and other human spongiform encephalopathies (prion diseases). Brain Pathol 5:459–466

Büeler H et al (1994) High prion and PrPSc levels but delayed onset of disease in scrapie-inoculated mice heterozygous for a disrupted PrP gene. Mol Med 1:19–30

Burger LJ, Rowan AJ, Goldensohn ES (1972) Creutzfeldt-Jakob disease: an electroencephalographic study. Arch Neurol 26:428–433

Cali I et al (2006) Classification of sporadic Creutzfeldt-Jakob disease revisited. Brain 129:2266–2277

Cali I et al (2009) Co-existence of scrapie prion protein types 1 and 2 in sporadic Creutzfeldt–Jakob disease: its effect on the phenotype and prion-type characteristics. Brain 132:2643

Cazzaniga FA et al (2022) PMCA-based detection of prions in the Olfactory Mucosa of patients with Sporadic Creutzfeldt-Jakob disease. Front Aging Neurosci 14:848991

Chen S et al (2016) Serum uric acid and lipid profiles in sporadic Creutzfeldt-Jakob disease. Clin Biochem 49:292–294

Chitravas N et al (2011) Treatable neurological disorders misdiagnosed as Creutzfeldt-Jakob disease. Ann Neurol 70:437–444

Chohan G et al (2010) The role of cerebrospinal fluid 14-3-3 and other proteins in the diagnosis of sporadic Creutzfeldt-Jakob disease in the UK: a 10-year review. J Neurol Neurosurg Psychiatry 81:1243–1248

Chu M et al (2022) Proposal of new diagnostic criteria for fatal familial insomnia. J Neurol 269:4909–4919
Cohen OS et al (2016) CSF tau correlates with CJD disease severity and cognitive decline. Acta Neurol Scand 133:119–123
Collinge J et al (2008) A clinical study of kuru patients with long incubation periods at the end of the epidemic in Papua New Guinea. Philos Trans R Soc B Biol Sci 363:3725
Collins SJ, Lawson VA, Masters CL (2004) Transmissible spongiform encephalopathies. Lancet 363:51–61
Collins SJ et al (2006) Determinants of diagnostic investigation sensitivities across the clinical spectrum of sporadic Creutzfeldt–Jakob disease. Brain 129:2278–2287
Concha-Marambio L et al (2016) Detection of prions in blood from patients with variant Creutzfeldt-Jakob disease. Sci Transl Med 8:370ra183
Connor A, Wang H, Appleby BS, Rhoads DD (2019) Clinical laboratory tests used to aid in diagnosis of human prion disease. J Clin Microbiol 57:10–1128
Cramm M et al (2016) Stability and reproducibility underscore utility of RT-QuIC for diagnosis of Creutzfeldt-Jakob disease. Mol Neurobiol 53:1896
Creutzfeldt-Jakob disease - NHS. https://www.nhs.uk/conditions/creutzfeldt-jakob-disease-cjd/
De Villemeur TB (2013) Creutzfeldt-Jakob disease. Handb Clin Neurol 112:1191–1193
Diagnostic criteria for surveillance of CJD from 1 January 2017 | Creutzfeldt-Jakob Disease International Surveillance Network. https://www.eurocjd.ed.ac.uk/node/833
Douet JY et al (2017) Distribution and quantitative estimates of variant Creutzfeldt-Jakob disease Prions in tissues of clinical and asymptomatic patients. Emerg Infect Dis 23(6):946–956
Ermann N et al (2018) CSF nonphosphorylated Tau as a biomarker for the discrimination of AD from CJD. Ann Clin Transl Neurol 5:883–887
Fu H, Subramanian RR, Masters SC (2000) 14-3-3 proteins: structure, function, and regulation. Annu Rev Pharmacol Toxicol 40:617–647
Gao L-P et al (2019) The genetic Creutzfeldt-Jakob disease with E200K mutation: analysis of clinical, genetic and laboratory features of 30 Chinese patients. Sci Rep 9:1836. https://doi.org/10.1038/s41598-019-38520-y
Ghetti B, Piccardo P, Zanusso G (2018) Dominantly inherited prion protein cerebral amyloidoses - a modern view of Gerstmann-Sträussler-Scheinker. Handb Clin Neurol 153:243–269
Ginsburg GS, Willard HF (2009) Genomic and personalized medicine: foundations and applications. Transl Res 154:277–287
Gmitterová K et al (2009) 14-3-3 CSF levels in sporadic Creutzfeldt-Jakob disease differ across molecular subtypes. Neurobiol Aging 30:1842–1850
Gonzalez-Montalban N et al (2011) Highly efficient protein misfolding cyclic amplification. PLoS Pathog 7:e1001277
Green AJE et al (2001) Use of 14-3-3 and other brain-specific proteins in CSF in the diagnosis of variant Creutzfeldt-Jakob disease. J Neurol Neurosurg Psychiatry 70:744–748
Harrington MG, Merril CR, Asher DM, Gajdusek DC (1986) Abnormal proteins in the cerebrospinal fluid of patients with Creutzfeldt–Jakob disease. N Engl J Med 315:279–283
Hermann P, Zerr I (2022) Rapidly progressive dementias — aetiologies, diagnosis and management. Nat Rev Neurol 18:363
Hermann P et al (2018) Validation and utilization of amended diagnostic criteria in Creutzfeldt-Jakob disease surveillance. Neurology 91:e331–e338
Hermann P et al (2021) Biomarkers and diagnostic guidelines for sporadic Creutzfeldt-Jakob disease. Lancet Neurol 20:235–246. https://doi.org/10.1016/S1474-4422(20)30477-4
Hsich G, Kenney K, Gibbs CJ, Lee KH, Harrington MG (1996) The 14-3-3 brain protein in cerebrospinal fluid as a marker for transmissible Spongiform Encephalopathies. N Engl J Med 335:924–930
Kanata E et al (2019) Cerebrospinal fluid neurofilament light in suspected sporadic Creutzfeldt-Jakob disease. J Clin Neurosci 60:124–127
Karch A et al (2015) Cerebrospinal fluid tau levels are a marker for molecular subtype in sporadic Creutzfeldt-Jakob disease. Neurobiol Aging 36:1964–1968

Kercher L, Favara C, Striebel JF, LaCasse R, Chesebro B (2007) Prion protein expression differences in Microglia and Astroglia influence Scrapie-induced neurodegeneration in the retina and brain of transgenic mice. J Virol 81:10340
Kobayashi A, Kitamoto T, Mizusawa H (2018) Iatrogenic Creutzfeldt-Jakob disease. Handb Clin Neurol 153:207–218
Kosorinova D, Belay G, Zakova D, Stelzer M, Mitrova E (2021) Genetic risk factors of Creutzfeldt-Jakob disease in the population of newborns in Slovakia. Pathogens 10:435
Kovács GG et al (2005) Genetic prion disease: the EUROCJD experience. Hum Genet 118:166–174
Kropp S et al (1999) Increase of neuron-specific enolase in patients with Creutzfeldt-Jakob disease. Neurosci Lett 261:124–126
Kruse N et al (2018) Interlaboratory validation of cerebrospinal fluid α-synuclein quantification in the diagnosis of sporadic Creutzfeldt-Jakob disease. Alzheimers Dement (Amst) 10:461–470
Lacroux C et al (2014) Preclinical detection of variant CJD and BSE prions in blood. PLoS Pathog 10:e1004202
Ladogana A et al (2005) High incidence of genetic human transmissible spongiform encephalopathies in Italy. Neurology 64:1592–1597
Lee RG, Blair RDG (1973) Evolution of EEG and visual evoked response changes in Jakob-Creutzfeldt disease. Electroencephalogr Clin Neurophysiol 35:133–142
Leitão MJ et al (2016) Sporadic Creutzfeldt-Jakob disease diagnostic accuracy is improved by a new CSF ELISA 14-3-3?? Assay. Neuroscience 322:398–407
Li J et al (2019) Detection of cell-free mitochondrial DNA in cerebrospinal fluid of creutzfeldt-jakob patients. Front Neurol 10:461888
Liberski PP, Ritchie DL, Sikorska B (2023) Prion diseases. In: Neurobiology of brain disorders: biological basis of neurological and psychiatric disorders, 2nd edn, pp 361–378. https://doi.org/10.1016/B978-0-323-85654-6.00004-6
Llorens F et al (2017a) Evaluation of α-synuclein as a novel cerebrospinal fluid biomarker in different forms of prion diseases. Alzheimers Dement 13:710–719
Llorens F et al (2017b) YKL-40 in the brain and cerebrospinal fluid of neurodegenerative dementias. Mol Neurodegener 12:83
Llorens F et al (2018a) Validation of α-Synuclein as a CSF biomarker for Sporadic Creutzfeldt-Jakob disease. Mol Neurobiol 55:2249–2257
Llorens F et al (2018b) Cerebrospinal Fluid Prion disease biomarkers in pre-clinical and clinical naturally occurring Scrapie. Mol Neurobiol 55:8586–8591
Llorens F et al (2020a) Diagnostic accuracy of Prion disease biomarkers in Iatrogenic Creutzfeldt-Jakob disease. Biomol Ther 10:290
Llorens F et al (2020b) Cerebrospinal fluid non-phosphorylated tau in the differential diagnosis of Creutzfeldt-Jakob disease: a comparative prospective study with 14-3-3. J Neurol 267:543–550
López-Pérez Ó et al (2020) BAMBI and CHGA in Prion diseases: neuropathological assessment and potential role as disease biomarkers. Biomol Ther 10:706
Maass F et al (2020) Increased alpha-synuclein tear fluid levels in patients with Parkinson's disease. Sci Rep 10:8507. https://doi.org/10.1038/s41598-020-65503-1
Masters CL et al (1979) Creutzfeldt-Jakob disease: patterns of worldwide occurrence and the significance of familial and sporadic clustering. Ann Neurol 5:177–188
Matsui Y et al (2011) High sensitivity of an ELISA kit for detection of the gamma-isoform of 14-3-3 proteins: usefulness in laboratory diagnosis of human prion disease. BMC Neurol 11:120
McGuire LI et al (2012) Real time quaking-induced conversion analysis of cerebrospinal fluid in sporadic Creutzfeldt-Jakob disease. Ann Neurol 72:278
Meiner Z et al (2011) Tau and 14-3-3 of genetic and sporadic Creutzfeldt-Jakob disease patients in Israel. J Neurol 258:255–262
Meissner B et al (2009) MRI lesion profiles in sporadic Creutzfeldt-Jakob disease. Neurology 72:1994–2001
Moda F et al (2014) Prions in the urine of patients with variant Creutzfeldt-Jakob disease. N Engl J Med 371:530–539

Neufeld MY, Korczyn AD (1992) Topographic distribution of the periodic discharges in Creutzfeldt-Jakob Disease (CJD). Brain Topogr 4:201

Noble GP, Walsh DJ, Miller MB, Jackson WS, Supattapone S (2015) Requirements for mutant and wild-type prion protein misfolding in vitro. Biochemistry 54:1180

Oeckl P et al (2016) Alpha-, Beta-, and Gamma-synuclein quantification in cerebrospinal fluid by multiple reaction monitoring reveals increased concentrations in Alzheimer's and Creutzfeldt-Jakob disease but no alteration in Synucleinopathies. Mol Cell Proteomics 15:3126–3138

Okuzumi A et al (2023) Propagative α-synuclein seeds as serum biomarkers for synucleinopathies. Nat Med 29:1448–1455

Orrú CD et al (2014) A Test for Creutzfeldt–Jakob Disease Using Nasal Brushings. N Engl J Med 371:519–529

Orrú CD et al (2017) Prion seeding activity and infectivity in skin samples from patients with sporadic Creutzfeldt-Jakob disease. Sci Transl Med 9:eaam7785

Otto M et al (1997) Elevated levels of tau-protein in cerebrospinal fluid of patients with Creutzfeldt-Jakob disease. Neurosci Lett 225:210–212

Otto M et al (2002) Tau protein and 14-3-3 protein in the differential diagnosis of Creutzfeldt-Jakob disease. Neurology 58:192–197

Paterson RW et al (2012) Differential diagnosis of Jakob-Creutzfeldt disease. Arch Neurol 69:1578–1582

Porter DD, Porter HG, Cox NA (1973) Failure to demonstrate a humoral immune response to Scrapie infection in mice. J Immunol 111:1407

Pritzkow S et al (2023) Detection of prions in the urine of patients affected by sporadic Creutzfeldt-Jakob disease. Ann Clin Transl Neurol 10:2316–2323

Puoti G et al (2012) Sporadic human prion diseases: molecular insights and diagnosis. Lancet Neurol 11:618–628

Redaelli V et al (2017) Detection of prion seeding activity in the olfactory mucosa of patients with Fatal Familial Insomnia. Sci Rep 7:1–8

Renard D, Castelnovo G, Collombier L, Thouvenot E, Boudousq V (2017) FDG-PET in Creutzfeldt-Jakob disease: analysis of clinical-PET correlation. Prion 11:440

Riemenschneider M et al (2003) Phospho-tau/total tau ratio in cerebrospinal fluid discriminates Creutzfeldt-Jakob disease from other dementias. Mol Psychiatry 8:343–347

Rissin DM et al (2010) Single-molecule enzyme-linked immunosorbent assay detects serum proteins at subfemtomolar concentrations. Nat Biotechnol 28:595. https://doi.org/10.1038/nbt.1641

Rosenbloom MH et al (2015) Metabolic disorders with clinical and radiologic features of sporadic Creutzfeldt-Jakob disease. Neurol Clin Pract 5:108

Saborio GP, Permanne B, Soto C (2001) Sensitive detection of pathological prion protein by cyclic amplification of protein misfolding. Nature 411:810–813

Sanchez-Juan P et al (2006) CSF tests in the differential diagnosis of Creutzfeldt-Jakob disease. Neurology 67:637–643

Sano K et al (2013) Early detection of abnormal Prion protein in genetic human prion diseases now possible using real-time QUIC assay. PLoS One 8:e54915

Schmitz M et al (2016a) Validation of 14-3-3 protein as a marker in Sporadic Creutzfeldt-Jakob disease diagnostic. Mol Neurobiol 53:2189–2199

Schmitz M et al (2016b) Regulation of human cerebrospinal fluid malate dehydrogenase 1 in sporadic Creutzfeldt-Jakob disease patients. Aging 8:2927–2935

Schmitz M et al (2017) Hereditary human prion diseases: an update. Mol Neurobiol 54:4138–4149

Schmitz M et al (2022a) Diagnostic accuracy of cerebrospinal fluid biomarkers in genetic prion diseases. Brain 145:700–712

Schmitz M et al (2022b) Validation of plasma and CSF Neurofilament light chain as an early marker for Sporadic Creutzfeldt-Jakob disease. Mol Neurobiol 59:1–9

Schmitz M et al (2023) Detection of Prion protein seeding activity in tear fluids. N Engl J Med 388:1816–1817

Shimamura M et al (2024) Neurofilament light chain levels in serum and cerebrospinal fluid do not correlate with survival times in patients with prion disease. Biomol Ther 15:8
Skinningsrud A, Stenset V, Gundersen AS, Fladby T (2008) Cerebrospinal fluid markers in Creutzfeldt-Jakob disease. Cerebrospinal Fluid Res 5:14
Spudich S et al (1995) Complete penetrance of Creutzfeldt-Jakob disease in Libyan Jews carrying the E200K mutation in the prion protein gene. Mol Med 1:607
Steinacker P et al (2016) Neurofilaments in blood and CSF for diagnosis and prediction of onset in Creutzfeldt-Jakob disease. Sci Rep 6:1–6
Steinhoff BJ et al (2004) Diagnostic value of periodic complexes in Creutzfeldt-Jakob disease. Ann Neurol 56:702–708
Thomas CM et al (2023) Development of a sensitive real-time quaking-induced conversion (RT-QuIC) assay for application in prion-infected blood. PLoS One 18:e0293845
Thompson AGB, Mead SH (2019) Review: fluid biomarkers in the human prion diseases. Mol Cell Neurosci 97:81–92
Thompson AGB et al (2018) Neurofilament light chain and tau concentrations are markedly increased in the serum of patients with sporadic Creutzfeldt-Jakob disease, and tau correlates with rate of disease progression. J Neurol Neurosurg Psychiatry 89:955–961
Thompson AGB et al (2021) Evaluation of plasma tau and neurofilament light chain biomarkers in a 12-year clinical cohort of human prion diseases. Mol Psychiatry 26:5955–5966
Uttley L, Carroll C, Wong R, Hilton DA, Stevenson M (2020) Creutzfeldt-Jakob disease: a systematic review of global incidence, prevalence, infectivity, and incubation. Lancet Infect Dis 20:e2–e10
Villar-Piqué A et al (2019) Plasma YKL-40 in the spectrum of neurodegenerative dementia. J Neuroinflammation 16:1–5
Weber T, Otto M, Bodemer M, Zerr I (1997) Diagnosis of Creutzfeldt-Jakob disease and related human spongiform encephalopathies. Biomed Pharmacother 51:381–387
Weingarten MD, Lockwood AH, Hwo SY, Kirschner MW (1975) A protein factor essential for microtubule assembly. Proc Natl Acad Sci USA 72:1858–1862
Wieser HG, Schindler K, Zumsteg D (2006) Invited review EEG in Creutzfeldt-Jakob disease. https://doi.org/10.1016/j.clinph.2005.12.007
Will RG (2003) Acquired prion disease: iatrogenic CJD, variant CJD, kuru. Br Med Bull 66:255–265
Xiao K et al (2019) T188K-familial Creutzfeldt-Jacob disease, predominant among Chinese, has a reactive pattern in CSF RT-QuIC different from D178N-Fatal Familial Insomnia and E200K-Familial CJD. Neurosci Bull 35:519
Xiao K, Shi Q, Zhou W, Dong X-P (2020) Different post-mortem brain regions from three Chinese FFI patients induce different reactive profiles both in the first and second generation RT-QuIC assays. Prion 14:163. https://doi.org/10.1080/19336896.2020.1782809
Zanusso G et al (2003) Detection of pathologic prion protein in the olfactory epithelium in sporadic Creutzfeldt-Jakob disease. N Engl J Med 348:711–719
Zanusso G, Monaco S, Pocchiari M, Caughey B (2016) Advanced tests for early and accurate diagnosis of Creutzfeldt–Jakob disease. Nat Rev Neurol 12:325–333
Zerr I (2022) Laboratory diagnosis of Creutzfeldt–Jakob disease. N Engl J Med 386:1345–1350
Zerr I, Poser S (2002) Clinical diagnosis and differential diagnosis of CJD and vCJD. APMIS 110:88–98
Zerr I, Schmitz M (2021) Genetic Prion Disease. The order of appearance and/or predominance of these features and other associated neurologic and psychiatric findings vary 5–7 TT-years; in rare instances, >10 years
Zerr I et al (1996) Diagnosis of Creutzfeldt-Jakob disease by two-dimensional gel electrophoresis of cerebrospinal fluid. Lancet 348:846–849
Zerr I et al (1998) Detection of 14-3-3 protein in the cerebrospinal fluid supports the diagnosis of Creutzfeldt-Jakob disease. Ann Neurol 43:32–40

Zerr I et al (2009) Updated clinical diagnostic criteria for sporadic Creutzfeldt-Jakob disease. Brain 132:2659–2668

Zerr I et al (2018) Cerebrospinal fluid neurofilament light levels in neurodegenerative dementia: evaluation of diagnostic accuracy in the differential diagnosis of prion diseases. Alzheimers Dement 14:751–763

Zerr I et al (2021) Diagnostic and prognostic value of plasma neurofilament light and total-tau in sporadic Creutzfeldt-Jakob disease. Alzheimers Res Ther 13:86

Chapter 3
New Therapeutic Modalities in Prion Diseases

Alberto Pettinari, Elisa Uliassi, and Maria Laura Bolognesi

Abstract Prion diseases are fatal neurodegenerative disorders for which no effective therapies exist. Despite decades of drug discovery efforts, progress in developing disease-modifying treatments has been slow. However, recent advances have introduced novel therapeutic modalities targeting key aspects of prion pathology, including prion protein biogenesis, aggregation, and degradation. Advancements in diagnostic tools and highly sensitive prion detection methods are also playing a crucial role in enabling early and accurate diagnosis, which is essential for the timely application of emerging therapeutics. This chapter explores novel therapeutic modalities for prion diseases, focusing on small-molecule theranostics and compounds promoting prion protein degradation, RNA-based therapeutics, and gene therapy approaches. We critically assess the advantages and limitations of these therapeutic strategies, considering their development, efficacy, and translational potential. By leveraging these innovative modalities, the therapeutic toolbox for prion diseases is expanding, offering hope for the development of effective treatments.

Keywords Prion diseases · Scrapie prion protein (PrP^{Sc}) · Drug discovery · New therapeutic modalities · Small-molecule theranostics · Prion protein (PrP) · Cellular prion protein (PrP^{C}) · PrP degradation · RNA-based therapeutics · Antisense oligonucleotides · Short interfering RNA · RNA aptamers · Gene therapy

A. Pettinari · E. Uliassi · M. L. Bolognesi (✉)
Department of Pharmacy and Biotechnology, Alma Mater Studiorum - University of Bologna, Bologna, Italy
e-mail: marialaura.bolognesi@unibo.it

G. Legname, F. Moda (eds.), *Biomarkers and Therapeutical Targets for Prion Diseases*, Subcellular Biochemistry 112, https://doi.org/10.1007/978-3-031-97055-9_3

Introduction

Human prion diseases are rare and incurable neurodegenerative diseases. Despite the rarity, there are undoubtedly unmet medical needs where current therapeutic offerings are not even supportive in symptom management. As of January 2025, the presence of just one active clinical study in clinicaltrials.gov clearly underscores that prion diseases are an unmet therapeutic challenge. This is mainly due to limited knowledge of disease etiology (Legname 2017), limited research (compared to other neurodegenerative diseases) because of the relative rarity, and the potential to be transmitted within and across species as well as their consequences for human and animal health (leading to strict safety protocols to avoid laboratory/personnel infections).

In addition, prion diseases present significant obstacles to treatment due to their unique variability and pathophysiology (Zerr et al. 2024). Different mutations in the PRNP gene lead to genetic prion diseases, including genetic Creutzfeldt–Jakob disease (gCJD), Gerstmann–Sträussler–Scheinker disease (GSS), and fatal familial insomnia (FFI), accounting for about 15% of cases. Sporadic prion diseases occur spontaneously in about 85% of patients, with no known environmental or genetic trigger. Human-to-human transmission of these diseases has occurred due to iatrogenic exposure, and zoonotic forms of prion diseases are linked to bovine disease. These diseases vary in onset age, progression, and clinical symptoms. Variability in disease progression has been observed even within families with the same mutation (Baiardi et al. 2021), which further complicates the development of effective therapeutic treatments (Uliassi et al. 2023).

In spite of these differences, all forms of prion diseases involve the conversion of normal cellular prion protein (PrP^C) into its misfolded, pathogenic, and transmissible form (PrP^{Sc}). This process leads to the accumulation of toxic aggregates, resulting in widespread neuronal damage and brain dysfunction. Prion strains are encoded in distinct PrP conformations, which exhibit different aggregation patterns. Furthermore, PrP^C is required for prion propagation and mediates neurotoxicity. Although PrP^C has been widely investigated, its physiological roles have not been entirely elucidated yet (Legname 2017). Similarly, prion protein misfolding has been extensively studied, but the precise triggers and mechanisms are still not fully understood. However, some good news appeared in gloomy times. Indeed, the discovery of such a paradigm of templated prion protein misfolding has had a striking influence on the "prion-like" misfolding and propagation of other protein aggregates (Eid et al. 2024), which is now recognized as a common mechanism in other neurodegenerative disorders, such as Alzheimer's disease (AD) and Parkinson's (PD) disease.

Hence, genetics and the central role of PrP at the molecular level have provided a strong therapeutic hypothesis for prion diseases.

Drug discovery efforts have been mainly focused on stabilizing the normal structure of PrP^C or disrupting the interactions that promote its conversion to PrP^{Sc} and aggregation. These principally include the prevention of PrP^C unfolding, the inhibition of PrP^{Sc} polymer and oligomer formation, and the blockade of PrP^{C} recruitment by PrP^{Sc4}.

Notably, Kuwata has advanced to in vivo studies with macaques a PrP^{C} molecular chaperone that stabilizes PrP^{C}, avoiding PrP^{Sc} conversion (Yamaguchi et al. 2019). In spite of promising results, no human clinical studies have been started yet. Another well-explored strategy is the inhibition of PrP^{Sc} polymer and oligomer formation, which has demonstrated disappointing results (Giles et al. 2015). Likewise, compounds inhibiting PrP^{Sc} aggregation showed critical issues related to the selectivity toward prion aggregates, among others.

Additionally, repurposing approaches have been largely explored in prion diseases. These allowed us to identify flupirtine (Otto et al. 2004), quinacrine (Geschwind et al. 2013), and doxycycline (Haïk et al. 2014), which have been tested in clinical trials, showing no clinical benefits to the patients.

Similarly, several monoclonal antibodies (mABs), which target PrP to either block its misfolding or clear aggregates from the brain, have been developed (White et al. 2003). Encouraging results from a prion protein mAB, PRN100, were presented in the first human trial on CJD patients (Mead et al. 2022).

However, no symptomatic and disease-modifying treatment still exists for prion diseases.

Given the failure of PrP-centric therapeutic approaches and the established multifactorial nature of prion diseases, we and others have envisioned the development of compounds endowed with other therapeutic properties beyond PrP modulation, the so-called "multi-target-directed ligands" (MTDLs), as effective therapeutic tools to counteract prion-mediated neurotoxicity (Klingenstein et al. 2006). Despite promising, we had not moved forward with preclinical evidence from animal studies.

Also, diagnosis of prion diseases is difficult, as it relies on advanced disease-stage biomarkers. In fact, PrP^{Sc} represents a good biomarker to be used in the diagnosis of prion diseases via different techniques with high sensitivity and specificity (Cazzaniga et al. 2021). However, early diagnosis is urgently needed, particularly due to the rapid clinical course of these devastating disorders.

After decades of drug discovery efforts during which no therapies emerged, recent progress has been made in the development of disease-modifying therapies for prion diseases (Vallabh et al. 2020). Recent studies have explored the potential of novel therapeutic modalities (Blanco and Gardinier 2020) targeting prion biogenesis, PrP^{Sc} aggregation, and PrP degradation to provide an effective treatment for prion diseases. These therapeutic modalities span small molecules with unique features (i.e., theranostics and compounds promoting PrP degradation) and RNA-based therapeutics (i.e., ASO, siRNA, and aptamers), as well as gene therapy (i.e., CRISPR-Cas). In the following, we will describe and critically discuss each of them, highlighting advantages and limitations.

Novel Therapeutic Modalities for Prion Diseases

Theranostics

The Diagnosis of Prion Diseases

Diagnosing prion diseases presents several challenges due to their rarity, rapid progression, and similarities to other neurodegenerative disorders (Wu et al. 2021). Currently, the definitive diagnosis of human prion diseases relies on identifying histological features and detecting PrP^{Sc} in brain tissue obtained through biopsy or autopsy, which is highly invasive or only possible postmortem (Wu et al. 2021). In contrast, clinical diagnosis is primarily based on a combination of the patient's symptoms, magnetic resonance imaging (MRI) and electroencephalography (EEG) changes (Zerr et al. 2009), and the presence of surrogate markers like the 14-3-3 protein in cerebrospinal fluid (CSF) (Hsich et al. 1996). While this approach demonstrates high accuracy in the symptomatic stage, its sensitivity in the early stages is limited, as 14-3-3 protein detection is difficult due to its low levels, and clinical symptoms may overlap with other neurological disorders. Additionally, the absence of specific antibodies for PrP^C and PrP^{Sc} makes it difficult for traditional techniques like western blotting (WB) to differentiate between them. The WB method detects PrP^{Sc} after PrP^C is removed with proteinase K, but it lacks sufficient sensitivity for routine biological samples. However, recent advancements in in vitro cell-free conversion techniques, such as protein misfolding cyclic amplification (PMCA) and real-time quaking-induced conversion (RT-QuIC), have significantly improved PrP^{Sc} detection, enabling earlier and less invasive diagnosis of human prion diseases. PMCA is designed to amplify misfolded PrP^{Sc} by mimicking the natural prion replication process (Saborio et al. 2001). It works by repeatedly exposing PrP^C to misfolded PrP^{Sc} in the presence of sonication, which fragments PrP^{Sc} aggregates, increasing their surface area and accelerating conversion. These cycles of incubation and sonication enhance PrP^{Sc} accumulation, allowing for highly sensitive detection, even in low quantities. PMCA has been instrumental in detecting prions in biological fluids such as CSF, blood, and urine, significantly improving early diagnosis. RT-QuIC is another highly sensitive assay that detects prion diseases by amplifying PrP^{Sc}-induced misfolding (Atarashi et al. 2011). Unlike PMCA, RT-QuIC uses shaking rather than sonication to promote the conversion of recombinant PrP^C into a misfolded form. A fluorescent dye, such as thioflavin T, binds to the newly formed amyloid fibrils, producing a fluorescence signal that increases as misfolded PrP^{Sc} accumulates. RT-QuIC provides a rapid, specific, and quantitative means of detecting prions in CSF and other tissues, making it a valuable tool for early and minimally invasive prion disease diagnosis.

However, in the early stages, all these diagnostic methods are not definitive. Furthermore, different prion strains cause variations in disease presentation and progression, complicating diagnosis (Parchi et al. 2011). Thus, the need for less invasive yet reliable early diagnostic tools remains a critical challenge.

Small-Molecule Theranostics

In the past decade, advances in diagnostic and therapeutic technologies, along with innovative medicinal chemistry and synthetic methodologies, have opened up more opportunities for developing innovative and creative small-molecule drugs for prion diseases. In addition to therapeutic treatment, small molecules can leverage diagnosis and monitoring of prion diseases from a precision medicine perspective (Aulić et al. 2013). This integrated approach, referred to as the theranostic approach (Aulić et al. 2013), combining therapeutic and diagnostic properties, offers a promising paradigm for tackling prion diseases, which, by progressing silently, make early diagnosis and intervention extremely challenging. Theranostics aims to detect prion diseases early and simultaneously deliver targeted treatments, tailoring interventions based on individual prion strains, hence enabling precision medicine approaches (Aulić et al. 2013). A theranostic also allows real-time monitoring of its delivery to the desired target, as well as the visualization of molecular changes associated with its therapeutic effect.

The design of small-molecule theranostics requires careful consideration of molecular properties, target specificity, and functionality (Bolognesi et al. 2016). Molecular properties should be properly optimized to combine diagnostic and therapeutic properties into a single molecule without compromising either function and to ensure that the molecule can cross the blood–brain barrier (BBB). To enhance target specificity, structure–activity relationship (SAR) studies should be performed to optimize binding affinity and selectivity for the biomarker and the target while minimizing off-target effects. Equally important is the proper selection of the diagnostic technology. These include fluorescence imaging (e.g., fluorophores), positron emission tomography (PET), or single-photon emission computed tomography (SPECT) using radiolabeled isotopes (Bolognesi et al. 2016). PET ligands are imaging tools with established clinical applications in neurodegenerative diseases (Nordberg 2004). Though they exhibit short half-lives, they require the on-site synthesis of PET tracers and access to radiochemistry equipment and a cyclotron. Thus, they would not be the preferred choice for the development of prion-directed small-molecule theranostics. By contrast, fluorescence imaging has emerged as the preferred diagnostic technology to be used in theranostic development for its nonradioactive nature, high versatility and sensitivity, and cost-effectiveness, being easier to synthesize and administer with respect to PET/SPECT imaging tools (Bolognesi et al. 2016; Staderini et al. 2015). Particularly, near-infrared (NIR) fluorescent molecules (e.g., phthalocyanines and cyanine dyes) with emissions ranging from 650 to 950 nm have many advantages over conventional fluorophores. The NIR irradiation is sufficiently penetrating, has low background noise, and is noninvasive and harmless. Additionally, it is a versatile, sensitive, low-cost, and easy-to-obtain diagnostic tool, especially for preclinical research in neurodegenerative diseases (Staderini et al. 2015; Sarabia-Vallejo et al. 2023). Despite the different diagnostic technologies, theranostics have been frequently obtained by conjugating two individual therapeutic and diagnostic fragments (Kelkar and Reineke 2011). However, this approach typically provides larger and more complex molecules with

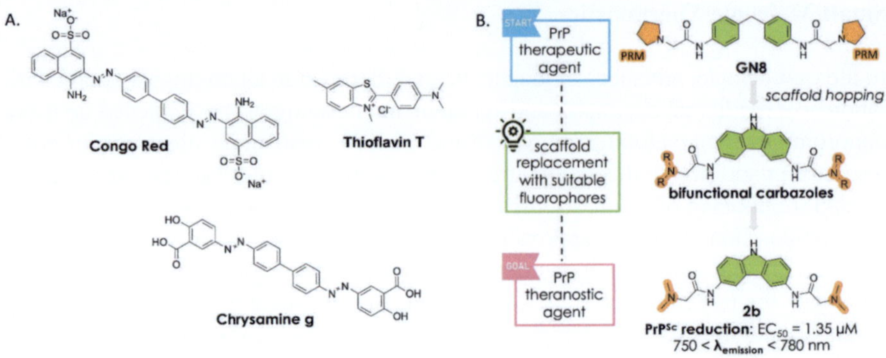

Fig. 3.1 (**a**) Chemical structures of amyloid-like protein fluorescent probes. (**b**) Design strategy for new anti-prion theranostics

poor drug-likeness. This aspect is of particular importance for theranostics acting in the central nervous system (CNS).

PrP^{Sc} formation and aggregation have been selected as a suitable biomarker and therapeutic target for the development of prion-directed theranostics. By taking advantage of PrP's dual role, the first fluorescent imaging agents for PrP aggregates (e.g., Congo red and thioflavin T, Fig. 3.1a) can not only serve as imaging probes to detect prion aggregates but also as anti-aggregation molecules. In fact, there are several types of small molecules with potential theranostic activity, despite not being deliberately designed as theranostics. The structure of thioflavin T or Congo red (Fig. 3.1a), both of which bind to Aβ due to their planar and π-conjugated systems able to intercalate into the β-sheet structure, has been the source of many fluorescent probes able to bind and label PrP^{Sc}. Congo red and its analog chrysamine G (Fig. 3.1a) have also shown therapeutic potential through their ability to inhibit prion aggregation.

Examples of theranostic agents for prion diseases remain limited (Staderini et al. 2013, 2023), as the field is still emerging. An interesting example of fluorescent theranostics designed to simultaneously deliver diagnostic and therapeutic properties for prion diseases has been developed by us (Staderini et al. 2023). The founding idea was based on the use of fluorescent molecules that can label PrP and visualize its distribution while exerting anti-aggregation effects. To do that, a scaffold hopping approach has been applied to the central diphenylmethane core of GN8. GN8 binds with high affinity to PrP^{C}, resulting in a promising therapeutic treatment. Nevertheless, its diphenylmethane scaffold is not characterized by fluorescent properties for subsequent theranostic development. Hence, based on GN8's bivalent nature endowed with two identical protein recognition motifs (PRMs, highlighted in orange in Fig. 3.1b) essential for PrP engagement, the core motif of GN8 has been replaced with a tricyclic carbazole (Staderini et al. 2023) (highlighted in green in Fig. 3.1b). The planar and aromatic structure of the carbazole ring, along with its extended π-conjugated system, is a key characteristic of its fluorescence. Additionally, these are important features for binding to and interacting with amyloid-like proteins, as fluorescence changes have been associated with binding to abnormally folded proteins. Furthermore, this design strategy, by incorporating the fluorescent nucleus into the structure of GN8, allows for maintaining a lower molecular weight and hence a potentially more favorable BBB permeation compared to conventional conjugation approaches. Thus, a focused library of bifunctional carbazole derivatives has been designed and synthesized through a four-step process. The synthesized compounds underwent cytotoxicity assessment, followed by testing on Rocky Mountain Laboratory (RML)-infected ScGT1 neuronal cells to monitor levels of protease-resistant PrP^{Sc}. The presence of small dialkylamino groups was demonstrated to be a key feature for both fluorescence and therapeutic properties. Among them, the bis-(dimethylaminoacetamido)carbazole derivative **2b** (Fig. 3.1b) has been selected for further studies due to its promising profile. Compound **2b** effectively reduced PrP^{Sc} levels in two cell lines infected with the mouse-adapted RML strain (ScGT1 and ScN2a). Unlike GN8, **2b** did not affect PrP^{C} levels, suggesting a potential advantage in reducing toxicity. In amyloid seeding assay experiments, **2b** delayed the aggregation of recombinant mouse PrP. Additionally, it interfered with the amplification of the scrapie RML strain in the PMCA assay more effectively than GN8. However, **2b** did not inhibit the amplification of human vCJD prions. Notably, **2b** exhibited native fluorescence, including emission in the NIR, and successfully stained PrP^{Sc} aggregates in living cells, confirming its potential as a fluorescent imaging agent. Furthermore, a positive BBB permeability was predicted for **2b** by the in vitro parallel artificial membrane permeation assay (PAMPA-BBB) (Staderini et al. 2023).

Compound **2b** emerges as a promising lead for developing theranostic agents capable of both treating and imaging prion diseases, offering hope for these currently incurable conditions. Its ability to reduce prion aggregation without affecting normal prion protein levels, combined with its fluorescent properties, highlights its potential for further medicinal chemistry optimization toward strain-independent anti-prion therapies.

Prion Protein Degradation: A New Therapeutic Opportunity for Prion Diseases

Prion Protein Degradation

Degradation of the prion protein occurs primarily through two key cellular pathways: the lysosomal system and the ubiquitin-proteasome system (UPS) (Fig. 3.2). Understanding these two interplaying degradation systems is necessary for unraveling the pathological mechanisms underlying prion diseases and for the development of potentially effective treatments (Appleby et al. 2019). These systems are, in fact, compromised by prion-driven dysfunctions (Hutti et al. 2020). For example, the lysosomal hydrolases effectively degrade native PrP^{C}, but PrP^{Sc} resists proteolysis and accumulates in lysosomes, impairing their function and leading to cytotoxicity

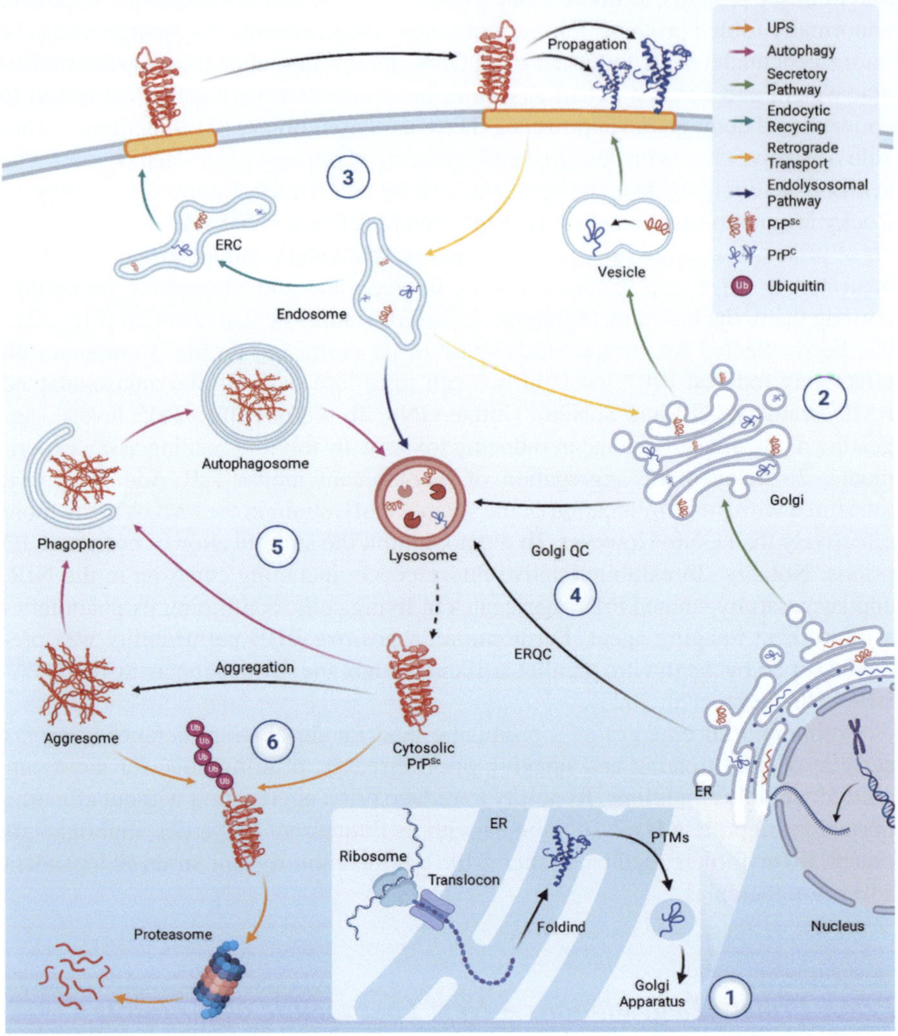

Fig. 3.2 Schematic representation of cellular prion trafficking: (1) PrP synthesis and PTMs, (2) vesicular trafficking system and Golgi apparatus, (3) endo-lysosomal pathway, (4) ER and Golgi quality controls, (5) autophagy, (6) ubiquitin-proteasome system

(Borchelt et al. 1992). Similarly, the UPS effectively degrades soluble PrP^C, but it is unable to remove the large, insoluble aggregates of PrP^{Sc} (McKinnon et al. 2016), exacerbating cellular proteostasis dysfunction.

Additionally, the interplay between these pathways is central to the cellular response to prion infection. When the UPS is overwhelmed, misfolded proteins are often rerouted to the lysosomal pathway for degradation (Scotter et al. 2014). Good news, but in the late stages of the disease, when aggregates impair autophagic functionality, this feedback may lead to cellular toxicity (Raffeiner et al. 2023). Prion diseases are further complicated by the ability of PrP^{Sc} to spread between cells. Aggregates released into the extracellular space can seed the conversion of PrP^{C} to PrP^{Sc} in neighboring cells, perpetuating the disease. Lysosomal exocytosis and other secretory mechanisms are thought to contribute to this intercellular propagation, making the degradation pathways not only essential for cellular homeostasis but also for limiting disease spread (Fehlinger et al. 2017). Importantly, it is also possible to exploit these degradation systems to promote PrP degradation, thus lowering its cellular levels (Büeler et al. 1993; Goold et al. 2015). Before describing small molecules able to induce PrP degradation, we will briefly describe prion protein secretory, recycling, and degradation pathways.

As illustrated in Fig. 3.2, PrP^{C} follows the classic secretory pathway: (1) it is synthesized by ribosomes on the surface of the endoplasmic reticulum (ER) from the PRNP mRNA. The nascent polypeptide enters the ER lumen through the translocon, where it folds and undergoes several post-translational modifications (PTMs), including glycosylation and the addition of a glycophosphatidylinositol (GPI) anchor (Benetti and Legname 2009). (2) The folded protein is transported to the plasma membrane via the vesicular trafficking system of the Golgi apparatus. Then, PrP^{C} undergoes internalization and recycling, which involves endosome-mediated degradation and autophagy (Appelqvist et al. 2013; Lawrence and Zoncu 2019). (3) The endolysosomal pathway consists of the formation of an endosome where PrP^{C} can either be recycled back to the plasma membrane via endocytic recycling (ERC) (green arrows) or to the Golgi via retrograde transport (yellow arrows). Alternatively, late endosomes can mature into lysosomes (dark blue arrow), where cathepsins and proteolytic enzymes digest PrP^{C} (Trivedi et al. 2020). In addition, (4) ER and Golgi quality controls (ERQC and Golgi QC, respectively) direct PrP^{C} toward lysosomes (Fig. 3.2). (5) Autophagy is another degradation system contributing to PrP homeostasis and removal of PrP^{Sc} aggregates (Griffey and Yamamoto 2022). Autophagy is characterized by the formation of an autophagosome, which initiates with the nucleation of the phagophore, an open double-membrane structure whose edges close to form a sealed autophagosome containing the misfolded proteins (Vargas et al. 2023). The autophagosomes then merge with lysosomes (purple arrows), resulting in the degradation of PrP^{Sc}. (6) The UPS is the primary pathway for the degradation of cytosolic prion proteins in mammalian cells, safeguarding against misfolded or damaged ones. UPS employs a sequential ubiquitination process involving three enzyme classes (Ciechanover and Schwartz 1998): E1 (ubiquitin-activating enzymes), E2 (ubiquitin-conjugating enzymes), and E3 (ubiquitin ligases). Ubiquitin tags, composed of at least four units, are recognized by the 26S proteasome, which then proceeds to degradation (orange arrows).

Disruptions in PrP degradation allow PrP^{Sc} to accumulate, leading to neuronal dysfunction and death. Hence, therapeutic approaches targeting PrP degradation (e.g., enhancing lysosomal or autophagic pathways) are being explored to counteract prion diseases. In the following, an overview of small molecules promoting PrP degradation via lysosomal or autophagic mechanisms will be discussed. It should be noted that the deliberate design of compounds targeting degradation pathways—the so-called degraders (e.g., proteolysis targeting chimeras, PROTACs)—to intentionally lower PrP levels is an unexplored field; rather, small molecules have been retrospectively discovered to promote PrP degradation.

Small Molecules Promoting Autophagy and Lysosomal Degradation of Prion Protein

Harnessing degradation systems to reduce PrP cellular levels is considered one of the most promising strategies (Goold et al. 2015; Shim et al. 2022).

Autophagy-inducing agents have shown significant promise for prion diseases. Rapamycin (Cortes et al. 2012) and trehalose (Aguib et al. 2009) (Fig. 3.3) were found to activate autophagy, clearing PrP^{Sc} in cell models and extending lifespan in prion-infected mice. Lithium (Relaño-Ginés et al. 2018) and natural compounds (Shim et al. 2022), such as FK506 (Nakagaki et al. 2020), also enhanced autophagy, reduced PrP^{Sc} levels, and increased survival. Alternative strategies include promoting lysosomal degradation through autophagy-independent pathways. Branched polyamines facilitate PrP^{Sc} breakdown by enhancing lysosomal activity (Supattapone et al. 1999).

UPS-targeting therapies have been less successful but remain a focus of research. IU1, a deubiquitinase inhibitor, and its derivatives increase proteasome activity and enhance the degradation of aggregation-prone proteins (Boselli et al. 2017; Kiprowska et al. 2017).

Interestingly, a derivative of celecoxib, named AR-12 (Fig. 3.3), down-regulates the cellular chaperone machinery, thereby inducing autophagy and facilitating the clearance of intracellular viruses and/or unfolded proteins (Booth et al. 2016).

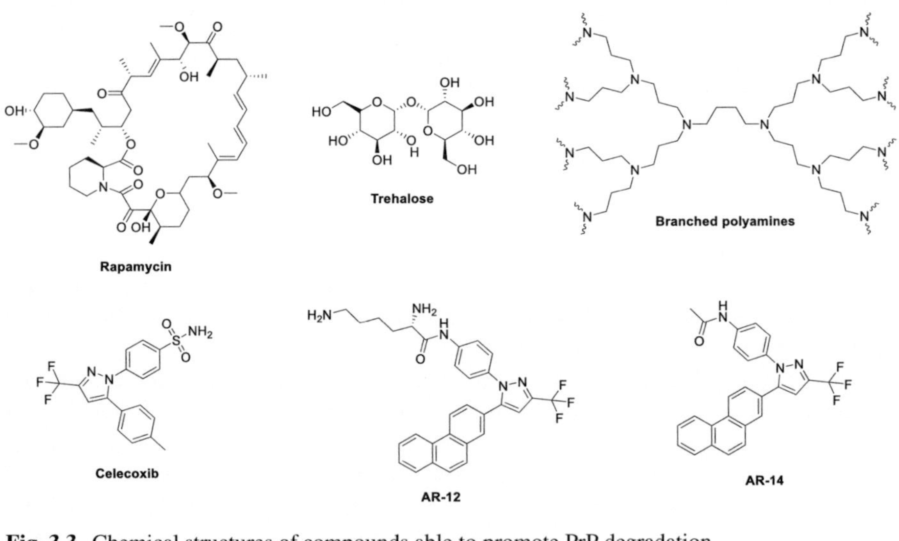

Fig. 3.3 Chemical structures of compounds able to promote PrP degradation

Abdurahman et al. demonstrated that both AR-12 and its derivative, AR-14 (Fig. 3.3), exhibit potent anti-prion properties in neuronal and nonneuronal cell models (Abdulrahman et al. 2017). These compounds significantly reduced PrP^{Sc} levels within 72 hours of treatment in cell lines like ScN2a and scrapie-infected mouse embryonic fibroblasts (ScMEFs) in a time- and concentration-dependent manner, leading to PrP^{Sc} degradation and loss of prion conversion activity (Abdulrahman et al. 2017). Mechanistically, AR-12 and AR-14 induce autophagy, as shown by the upregulation of autophagy markers, including LC3-II. The anti-prion effects of AR-12 were demonstrated as autophagy-dependent in ATG5-deficient cells, in which autophagy function is impaired. Accordingly, the anti-prion effect of AR-12 was diminished in autophagy-compromised Atg5-KO cells. Importantly, long-term treatment with AR-12 enhanced the clearance of PrP^{Sc} from prion-infected ScN2a and ScMEF cells, as confirmed by immunoblot and RT-QuIC analyses, with no reappearance of PrP^{Sc} after drug discontinuation (Abdulrahman et al. 2017). This confirms that autophagy plays a crucial role in PrP^{Sc} clearance and can be exploited for the development of autophagy-directed anti-prion small molecules.

Recently, small-molecule-based therapeutic intervention for prion diseases has taken a decisive turn. A fresh take on the treatment for diseases involving protein misfolding has been proposed by Biasini's group (Spagnolli et al. 2021). This approach capitalizes on the unique and often transient but stable conformational states that proteins adopt during their folding process, i.e., the structurally distinct folding intermediates (Fig. 3.4a). It is referred to as Pharmacological Protein Inactivation by Folding Intermediate Targeting (PPI-FIT). PPI-FIT provides a pathway to selectively modulate the stability of protein folding intermediates and eventually promote degradation using small molecules. At the heart of PPI-FIT lies computational modeling, which identifies druggable sites within intermediate folding states of PrP. This strategy enabled researchers to pinpoint specific conformations vulnerable to therapeutic intervention starting from molecular dynamics studies of the human PrP. These PrP intermediate structures have been the starting point for a virtual screening campaign with a library of $\sim 3.2 \times 10^5$ commercially available compounds. Four hit candidate molecules were identified, with compound SM875 emerging as the most effective (Fig. 3.4b). By stabilizing PrP folding intermediates, SM875 channels these transient forms toward degradation via the autophagy-lysosomal pathway. SM875 selectively acts on nascent PrP without interfering with mature PrP or unrelated GPI-anchored proteins, ensuring high specificity. Experimental validation showed that SM875 significantly reduced PrP^{Sc} levels in prion-infected cell lines, including ScN2a and ScMEF. Mechanistically, SM875 induces PrP aggregation into partially unfolded states, which are more amenable to lysosomal degradation. Autophagy-dependent degradation was demonstrated by co-administration of SM875 with bafilomycin A1, an autophagy inhibitor, which prevented the degradation. This approach not only prevents toxic PrP^{Sc} accumulation but also utilizes the cell's natural quality control systems to restore PrP proteostasis.

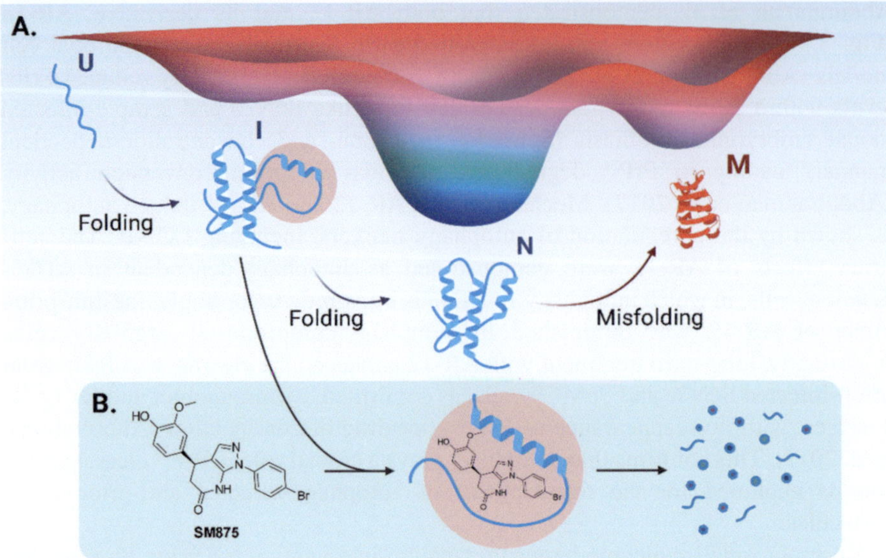

Fig. 3.4 (**a**) Graphical representation of the free energy landscape of prion protein from unfolded (U) to native conformation (N) through the targeted intermediate (I) and misfolding process leading to misfolded conformation (M); (**b**) PPI-FIT process for SM875 identification. SM875 stabilizes the conformational intermediate, leading to its subsequent degradation by the autophagy-lysosomal system

As new protein degradation mechanisms are uncovered, endogenous degradation systems may be exploited to develop new therapeutics for prion diseases and other neurodegenerative diseases associated with protein misfolding phenomena (Uliassi et al. 2025).

RNA-Based Therapeutics

RNA-based therapeutics utilize RNA molecules to regulate gene expression, modulate protein production, or correct disease-causing genetic errors. These therapies have gained significant attention for their potential to treat genetic disorders, cancers, and neurodegenerative diseases. These are also opening new frontiers for therapeutic solutions specifically tailored to prion diseases (Minikel et al. 2020; Bender et al. 2019; Proske et al. 2002). By silencing toxic proteins, modulating gene expression, or stabilizing proper protein conformations, RNA-based approaches hold the promise of reshaping the therapeutic landscape for prion disorders.

As illustrated in Fig. 3.5, we focus on three RNA-based therapeutics: antisense oligonucleotides (ASOs), RNA interference (RNAi)/siRNAs, and RNA aptamers (Hu et al. 2020; Lauffer et al. 2024; Ni et al. 2021). While ASOs hybridize with mRNA to induce degradation or block translation, and siRNAs induce silencing of

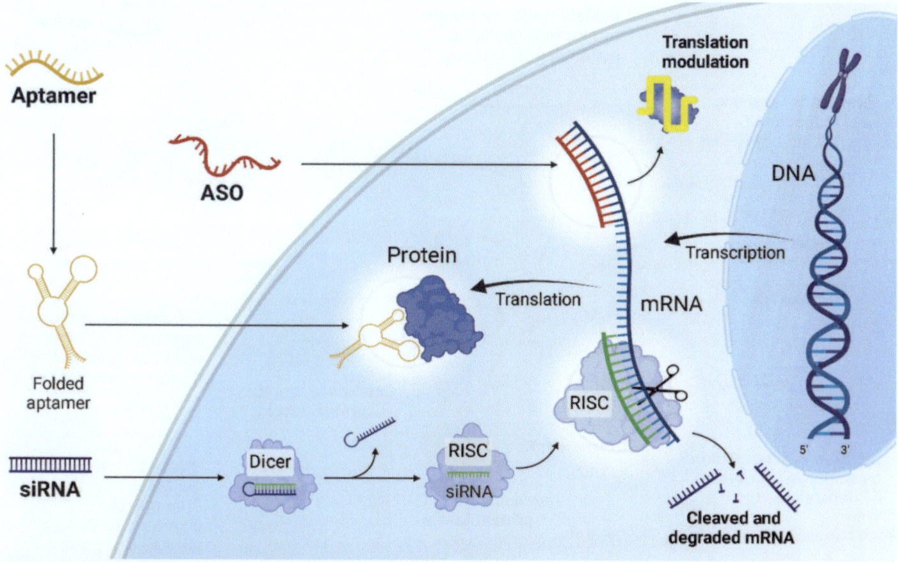

Fig. 3.5 Mechanism of action of RNA-based therapeutics: Aptamers bind target proteins to inhibit their function, antisense oligonucleotides (ASOs) hybridize with mRNA to induce degradation or block translation, and small interfering RNAs (siRNAs) utilize the RNA-induced silencing complex (RISC) to degrade specific mRNA, reducing protein expression

specific mRNA, RNA aptamers bind target proteins to inhibit their functions, acting similarly to antibodies. In the following, each of these approaches will be discussed, highlighting their design, pharmacology, and application to prion-specific challenges.

Antisense Oligonucleotides (ASOs)

Antisense oligonucleotides (ASOs) are synthetic, single-stranded nucleotides (typically consisting of 13–25 oligonucleotides) that bind RNAs via Watson–Crick base pairing to modulate protein expression (Leavitt and Tabrizi 2020). They act by inducing RNase H1 activity to degrade RNA–DNA hybrids (Vickers et al. 2003), inhibiting translation via steric hindrance (Tallet-Lopez 2003), or altering pre-mRNA stability by modifying 5′ cap formation or 3′ polyadenylation (Baker et al. 1997) (Fig. 3.6). ASOs can enhance protein expression by targeting upstream open reading frames (uORFs) or suppressing nonproductive splicing (Liang et al. 2016).

ASOs have already reached the clinic, showing success in spinal muscular atrophy with nusinersen, which modifies SMN2 splicing to restore functional protein (Hoy 2017). Similarly, tofersen induces degradation of SOD1 mRNA in amyotrophic lateral sclerosis (Saini and Chawla 2024). Nusinersen and tofersen employ 2′-OMe and MOE to improve stability. Tominersen is undergoing clinical trials for

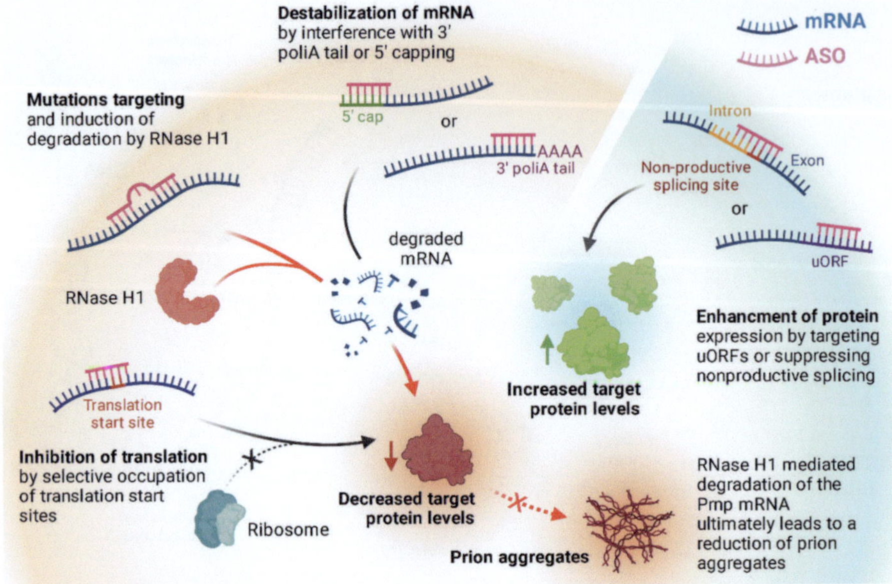

Fig. 3.6 Antisense Oligonucleotides (ASOs) mechanisms of action. Red arrows indicate the mechanism of action for anti-prion ASOs (Minikel et al. 2020; Nazor Friberg et al. 2012; Raymond et al. 2019). The formation of a DNA/RNA hybrid with the target mRNA induces its degradation by RNase H1

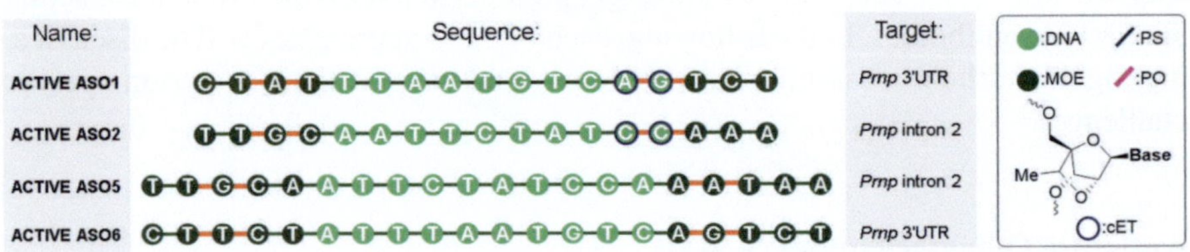

Fig. 3.7 Reported ASOs for prion disease treatment in animal models (Minikel et al. 2020; Raymond et al. 2019)

Huntington's disease for its ability to lower the mutant huntingtin (McColgan et al. 2023).

Starting from the notion that lowering PrP is a safe and effective strategy for the treatment of prion diseases (Mallucci et al. 2003), ASOs represent a promising therapeutic strategy. Several ASOs have been developed for targeting PRNP mRNA, with promising results in prion animal models. ASO1-6 (Fig. 3.7) allows RNase H1-mediated cleavage of the PRNP mRNA-ASO RNA–DNA hybrid, reducing the production of PrP and limiting its pathogenic potential. Early studies demonstrated survival extensions of up to 40% in prion-infected mice when ASO treatment began early (Nazor Friberg et al. 2012), highlighting their potential for altering disease progression.

Prion-targeting ASOs incorporate several medicinal chemistry modifications to enhance their pharmacological profile, as shown in Fig. 3.7. Phosphorothioate backbones increase nuclease resistance and improve bioavailability by promoting serum protein binding, while 2′ modifications, such as 2′-O-methoxyethyl (MOE) and 2′,4′-constrained ethyl (cEt) (Seth et al. 2010), further stabilize the oligonucleotide and enhance binding affinity to mRNA. Gapmer ASOs, with central DNA gaps flanked by modified RNA-like nucleotides, induce efficient RNase H1-mediated mRNA degradation. Additionally, 5-methylcytosine modifications have been shown to improve binding specificity (Wan and Seth 2016).

Intrathecal administration has emerged as the preferred delivery method (Raymond et al. 2019) for ASOs targeting prion diseases. This approach bypasses the BBB, ensuring effective CNS distribution and durable therapeutic effects. Periodic bolus dosing, particularly via intracerebroventricular injection, has been shown to achieve sustained reductions in PrP mRNA and protein levels (Raymond et al. 2019; Mortberg et al. 2023).

ION717 is a leading investigational ASO for prion diseases. The ongoing Phase 1/2a clinical trial (NCT06153966) is evaluating the safety, tolerability, and efficacy of intrathecal administration of ION717 in patients with symptomatic prion diseases (Ionis Pharmaceuticals, Inc 2024). The trial employs CSF PrP as a biomarker for therapeutic response, providing critical insights into the drug's pharmacodynamics. This study offers new hope for people dying from gCJD, GSS, and FFI after the previous clinical setbacks (Shim et al. 2022).

Short Interfering RNA (siRNA)

Small interfering RNAs (siRNAs) are double-stranded RNA molecules, typically 20–25 nucleotides in length, that play a central role in the RNA interference (RNAi) pathway, a highly conserved cellular mechanism for regulating gene expression. Discovered in the late 1990s, siRNAs are now recognized as powerful tools for sequence-specific gene silencing, offering therapeutic potential across a wide range of diseases, including viral infections, cancers, and neurodegenerative disorders (Fire et al. 1998; Elbashir et al. 2001). The mechanism of siRNA action (Fig. 3.8) involves their incorporation into the RNA-induced silencing complex (RISC), where the antisense strand guides the complex to complementary mRNA targets. Once bound, the Argonaute protein within RISC cleaves the mRNA, leading to its degradation and subsequent suppression of protein translation (Alterman et al. 2019; Hutvagner and Simard 2008).

Clinical success has been achieved with siRNA therapies like patisiran, which targets transthyretin mRNA to treat hereditary amyloidosis (Adams 2018). Notably, patisiran features 2′-OMe and MOE modifications (Adams 2018; Gangopadhyay and Gore 2022), to enhance its pharmacokinetic profile. Preclinical studies have further shown promise for RNAi in neurodegenerative conditions, including targeting tau and alpha-synuclein aggregation pathways in AD (Zhou et al. 2020) and PD (Titze-de-Almeida et al. 2020).

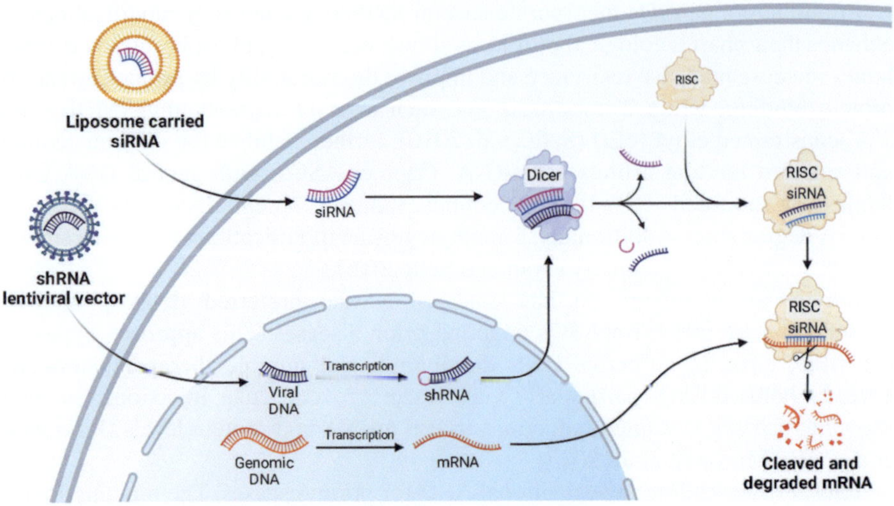

Fig. 3.8 Mechanism of action for siRNAs (Meister and Tuschl 2004) and shRNAs. Both siRNAs and shRNAs guide the RNA-induced silencing complex (RISC) to complementary mRNA targets, leading to mRNA cleavage and degradation. While siRNAs are directly delivered as short duplexes, shRNAs are expressed within cells and processed by Dicer into functional siRNA-like duplexes. These mechanisms result in sequence-specific gene silencing, effectively reducing the expression of targeted proteins

Advancements in siRNA-based therapies have shown great promise in the treatment of prion diseases, focusing on chemical modifications and innovative delivery systems that address the challenges of CNS-targeted interventions (Bender et al. 2019; Lehmann et al. 2014; Pulford et al. 2010; White and Mallucci 2009). These efforts aim to overcome key barriers such as efficient delivery across the BBB, durability of therapeutic effects, and minimization of off-target interactions.

A landmark preclinical study from the Broad Institute introduced divalent siRNA technology as a promising approach to prion disease treatment (Gentile et al. 2024). This innovative modality uses chemically modified siRNAs connected by a linker, forming a larger molecule designed for enhanced potency and durability (Alterman et al. 2019). Delivered via bolus intrathecal injection—a method previously validated for ASO therapies (Minikel et al. 2020)—divalent siRNAs effectively targeted both murine and human PRNP RNA in transgenic models. Key results include up to 83% knockdown of PRNP RNA in specific brain regions of humanized PRNP BAC transgenic mice and sustained effects lasting up to six months after a single dose. Chronic dosing further improved outcomes, with PrP levels reduced to as low as 14% of baseline, underscoring the potential of divalent siRNAs for the long-term management of prion diseases.

The chemical modifications employed in these siRNA therapies are pivotal for their success. As shown in Fig. 3.9, modifications such as 5′-vinylphosphonate (5′-(E)-VP) enhance RISC loading, while ribose alterations, including 2′-OMe and 2′-F, improve nuclease resistance and binding affinity. Phosphorothioate (PS)

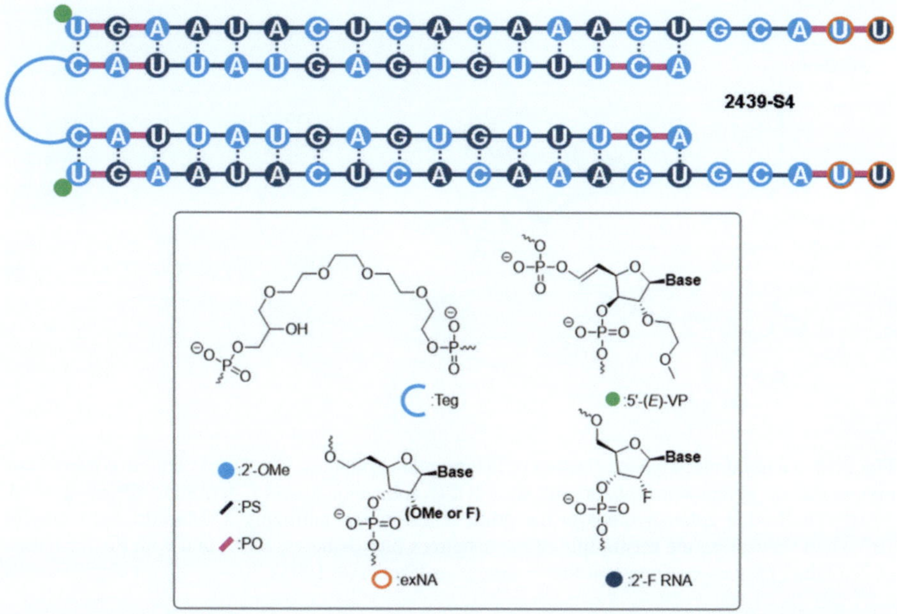

Fig. 3.9 Identity of 2439-s4 (Khvorova and Kennedy, 2021) with chemical modifications. Abbreviations: 5′-(E)-VP 5′-vinyl phosphonate, 2′-OMe 2′-O-methyl, 2′-F 2′-fluoro, exNA extended nucleic acid, PS phosphorothioate, PO phosphodiester, TEG tetraethylene glycol

linkages at strand termini increase stability and cellular uptake, and phosphodiester (PO) linkages maintain structural compatibility. A notable advancement, extended nucleic acids (exNAs), provides additional durability and potency by stabilizing specific regions of the siRNA molecule, enabling deeper and more sustained PrP knockdown with fewer modifications.

RNA Aptamers

Aptamers are short, single-stranded DNA or RNA molecules that fold into intricate 3D structures, enabling high-affinity and specific interactions with target molecules (Fig. 3.10) similar to antibodies. Through the SELEX (Systematic Evolution of Ligands by Exponential Enrichment) process, aptamers are selected from a library of 10^{15} RNAs for their ability to bind specific targets with high precision (Nimjee et al. 2017). The selected sequences are enriched and subjected to iterative rounds, typically up to twelve cycles, resulting in aptamers with K_d in the nanomolar or even picomolar range. These highly selective binders have extensive applications in diagnostics and therapeutic development. For example, RNA aptamers such as E22P-AbD43 target $A\beta_{42}$ protofibrils, halting nucleation and mitigating neurotoxicity through a G-quadruplex structure.

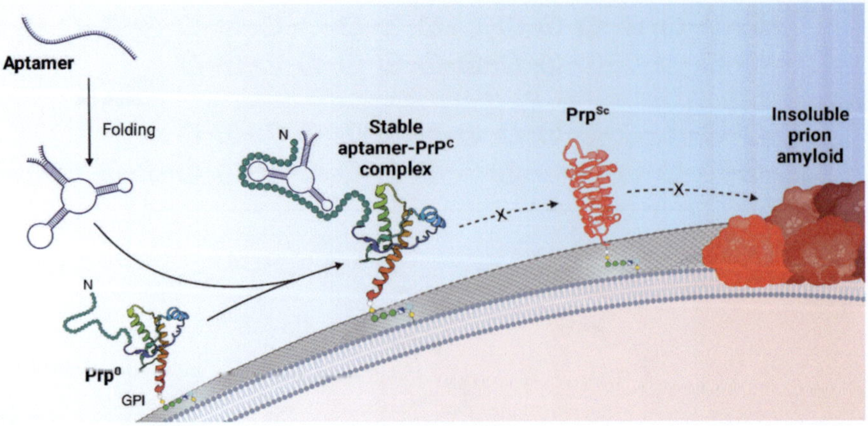

Fig. 3.10 General mechanism of action of PrP-targeted aptamers. The PrP^C is bound to the plasma membrane by glycosylphosphatidylinositol (GPI) connected to its C-terminus (Prusiner et al. 1998). The folded aptamer binds to the prion N-terminus, stabilizing it (Macedo and Cordeiro 2017) and preventing the misfolding of the α-helices into ß-sheets, associated with the formation of insoluble, proteasome-resistant PrP^{Sc} aggregates (Pan et al. 1993)

Diverse chemical modifications have been introduced to improve aptamer stability, bioavailability, and binding efficacy. Sugar modifications, including 2′-F (Ruckman et al. 1998) and 2′-OMe (Freund et al. 2023) substitutions, and backbone modifications such as PS linkages, also prolong serum half-life and stabilize aptamer structures, favoring CNS access (Wang et al. 2024). Conjugations, including terminal capping with inverted thymidine or polyethylene glycol (PEG), further protect against exonuclease activity and reduce renal clearance, enhancing pharmacokinetics.

Several aptamers for prion diseases have been reported and patented. RNA aptamers have shown the potential to bind PrP with high specificity, stabilizing its native conformation and preventing conversion into PrP^{Sc} (Macedo and Cordeiro 2017). Structural features like G-quadruplexes significantly improve rigidity and specificity, exemplified by the R12 (Mashima et al. 2020) aptamer's robust binding to PrP's N-terminal domain. As assessed by NMR studies, aptamers forming G-quadruplex structures provide additional stabilization mechanisms, influencing aggregation dynamics and misfolding (Olsthoorn 2014).

R12 exhibited a dissociation constant (K_d) of 8.5 nM for PrP^C and significantly reduced PrP^{Sc} levels in infected mouse neuronal cells, with an IC_{50} value of 10 μM (Mashima et al. 2013). R24, designed by tandemly linking two R12 sequences, achieved an IC_{50} of approximately 100 nM (Mashima et al. 2020), demonstrating two orders of magnitude higher activity than monovalent R12. In in vitro models, R24 at 5 μM concentration nearly abolished PrP^{Sc} formation, reducing it to approximately 1% of control levels. Related constructs, such as R12-A-R12 (Mashima et al. 2020), also showed potent anti-prion activity with an IC_{50} of approximately 500 nM.

Efficient delivery of RNA aptamers to the brain remains a major hurdle. Therapeutic applications have been evaluated using delivery systems for enhanced targeting. For example, aptamers conjugated to transferrin receptor-binding moieties demonstrated the ability to cross the BBB (Vasconcelos et al. 2024).

Gene Therapy

Gene Therapy Approaches

Gene therapy is a therapeutic approach that involves modifying or manipulating genetic material to treat or prevent a disease (Ingusci et al. 2019). It typically works by introducing, replacing, repairing, or regulating genes within a patient's cells to correct genetic disorders. Gene therapy can use various techniques, such as delivering functional copies of faulty genes, silencing harmful genes, or editing specific DNA sequences using tools like clustered regularly interspaced short palindromic repeats (CRISPRs). According to a broader definition, gene therapy encompasses any drug containing or consisting of recombinant nucleic acids (DNA or RNA), with the aim of correcting, repairing, replacing, adding, or removing a gene sequence (Ingusci et al. 2019). Instead of treating symptoms, gene therapy strives to directly address the genetic drivers of the disease. As such, it is, in principle, particularly suitable for genetic prion diseases.

As depicted in Fig. 3.11, gene therapy consists of two main approaches: gene addition (A) and targeted genome editing (B) approaches. Gene addition relies on the ex vivo or in vivo transfer of genetic information into a patient's cells. During gene addition, viral vectors, including lentiviruses, adenoviruses, and adeno-associated viruses (AAVs), and nonviral vectors deliver a whole gene of interest with promoter or enhancer elements and polyadenylation signals. However, gene addition requires precise regulation of the introduced gene's activity to ensure proper function. Gene expression is indeed influenced not only by promoters and enhancers but also by genome architecture, neighboring coding and noncoding sequences, and the gene's spatial positioning within the nucleus, further complicating the development of gene addition strategies. Unlike gene addition methods, genome editing—among which CRISPR-Cas9 is the most widely used and versatile tool—enables precise, site-specific modifications of DNA sequences within living cells by using programmable nucleases. It involves recognizing and binding to specific genomic sequences via effector DNA-binding domains (DBDs). This is followed by the introduction of double-strand breaks (DSBs) in the target DNA by restriction endonucleases such as Cas. The resulting breaks are then repaired to facilitate precise gene correction or insertion or to lead to gene disruption or small insertions and deletions.

While gene therapy offers high specificity and potential disease correction compared to classical small molecules, it faces challenges like high costs, immune responses, potential off-target gene modifications, and delivery complexities. Unlike

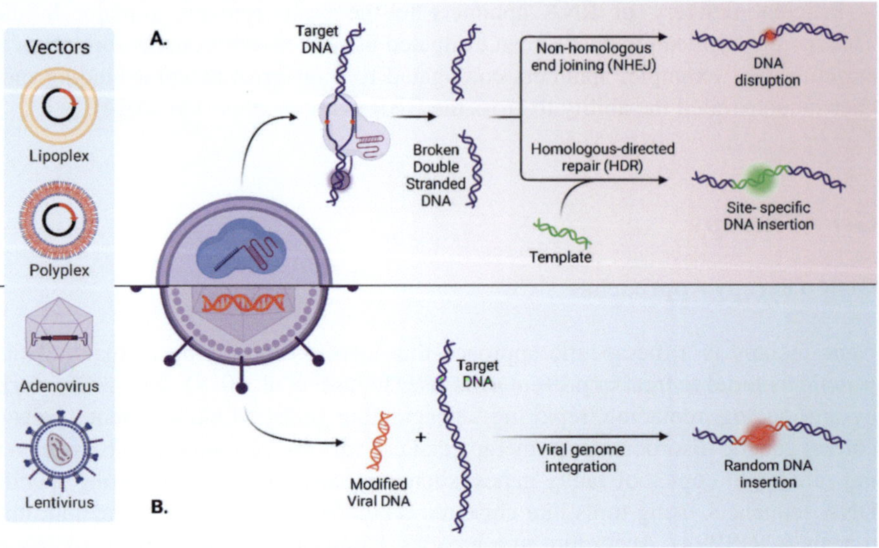

Fig. 3.11 Gene therapy consists of two main approaches: (**a**) targeted modification and (**b**) non-specific addition

gene addition, genome editing tools, such as CRISPR-Cas9, do not inherently require gene vectors, but vectors are often used to efficiently deliver these tools into target cells. The choice of a delivery system for gene therapy is therefore crucial. A suitable vector should meet key criteria, which include (i) ease of manipulation to engineer the vector for recombination and propagation in appropriate host cells; (ii) minimal invasiveness to avoid unintended genomic alterations; (iii) target specificity to be expressed exclusively in the intended target cells; (iv) low immunogenicity; and (v) long-term stability (Ingusci et al. 2019).

CHARM: A Targeted Genome Editing Tool for Prion Diseases

Given all the challenges, both gene therapy approaches have been developed solely in response to rare monogenic diseases (e.g., lipoprotein lipase deficiency and transfusion-dependent beta-thalassemia) classified as unmet medical needs (Schambach et al. 2024). Similarly, genetic prion diseases, being a class of rapidly lethal gain-of-function disorders that depend upon the expression of a single gene whose knockout appears well-tolerated, are an ideal target for gene therapy based on CRISPR/Cas9 technology. As a way to deplete PrP and slow or stop prion diseases, knocking out PRNP with CRISPR/Cas9 has potential advantages over RNAi or ASOs. Unlike RNAi or ASOs, which require continuous administration to suppress PRNP expression, CRISPR/Cas9 enables a permanent gene knockout, eliminating the need for repeated treatments. Additionally, it completely eliminates, not merely reduces, PrP production in cells. Current CRISPR-based DNA-editing

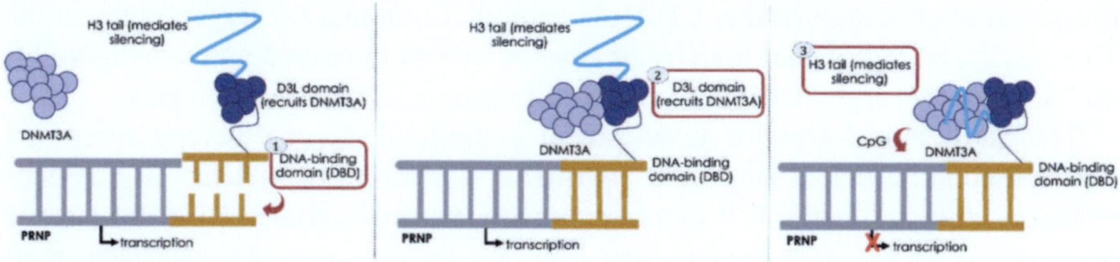

Fig. 3.12 Schematic drawing of CHARM functioning. The CHARM construct is constituted by specific DBD fused to D3L domain and the tail of histone H3 tail. D3L recruits de novo methyltransferase DNMT3A, which is then activated by the H3 tail. The resulting heterodimer methylates the PRNP target gene's promoter, silencing its transcription

technologies are complex, large molecules that are challenging to deliver, especially into the brain, and have been associated with unintended editing outcomes. Therefore, an epigenetic editing approach to permanently silence PrP expression in the brain without modifying the DNA sequence or causing the production of an altered mRNA or protein has been developed by Weissman et al. (Neumann et al. 2024). This strategy relies on DNA methylation to achieve long-term transcriptional repression because the human PRNP promoter contains a cytosine-guanine dinucleotide (CpG) site that serves as a substrate for DNA methylation. However, existing epigenetic editors can be cytotoxic in certain conditions and are often too large to be packaged into AAV vectors, the preferred delivery system for the CNS. To overcome these limitations, the authors designed CHARM (Coupled Histone tail for Autoinhibition Release of Methyltransferase), a compact, enzyme-free epigenetic editor (Fig. 3.12). The CHARM construct consists of a fusion between the histone H3 tail and the noncatalytic Dnmt3L domain. This design enables CHARM to recruit and activate endogenous DNA methyltransferases (DNMT3A) present within the cell, leading to targeted DNA methylation and gene silencing. By utilizing components already existing in the cell, CHARM avoids the need for introducing exogenous enzymes (i.e., the programmable nucleases), thereby reducing potential cytotoxicity and minimizing the overall size of the construct.

To direct CHARM to specific genomic loci, it is coupled with a DNA-binding domain, such as a zinc finger protein, engineered to recognize and bind to the promoter region of the target PRNP gene. Delivery of the CHARM construct was achieved using an AAV vector, which is well-suited for CNS applications due to its ability to transduce neurons efficiently. CHARM resulted in up to 80% brainwide reduction in neuronal PrP—far surpassing the minimum threshold for therapeutic efficacy and demonstrating successful AAV-mediated delivery and efficient target gene suppression. PRNP silencing was persistent over time, indicating the stability of CHARM-mediated epigenetic modifications. DNA methylation at the target promoter was confirmed, supporting the long-term transcriptional repression of PRNP without altering the DNA sequence. Additionally, the authors developed a self-silencing CHARM construct that deactivates itself, limiting prolonged expression to reduce potential antigenicity and off-target effects. Whole-genome analysis and functional assays showed that CHARM exhibited minimal off-target methylation. The absence of significant toxicity or adverse effects in treated animals highlights the safety of this approach compared to other gene-silencing technologies.

This study marks a major advancement in the field of gene therapy, offering a highly specific, efficient, and safer alternative to existing approaches. By demonstrating stable, brainwide PrP suppression with minimal toxicity, CHARM has the potential to be a transformative therapy for prion diseases and other neurodegenerative disorders.

Conclusions and Future Perspectives

In the last decade, we have witnessed remarkable advancements in therapeutic modalities for tackling prion diseases. In this chapter, we have explored these emerging strategies, highlighting selected examples that demonstrate their potential impact while addressing the challenges they face. The application of these approaches to prion and other neurodegenerative diseases has the potential to accelerate development timelines and mitigate the risk of early clinical failures. While RNA-based therapeutics have already progressed to investigational drugs in clinical trials, small-molecule theranostics, prion-targeting degradation compounds, and gene-editing tools have thus far yielded only hit compounds, lead molecules, or drug candidates that have been validated at the in vivo proof-of-concept stage.

Extensive basic research and a growing understanding of the mechanisms underlying prion diseases have facilitated the development of multiple preclinical candidates, some of which may ultimately translate into disease-modifying treatments. Future research will be critical to advancing these strategies toward clinical application.

References

Abdulrahman BA et al (2017) The celecoxib derivatives AR-12 and AR-14 induce autophagy and clear prion-infected cells from prions. Sci Rep 7:17565

Adams D et al (2018) Patisiran, an RNAi therapeutic, for Hereditary Transthyretin Amyloidosis. N Engl J Med 379:11–21

Aguib Y et al (2009) Autophagy induction by trehalose counter-acts cellular prion-infection. Autophagy 5:361–369

Alterman JF et al (2019) A divalent siRNA chemical scaffold for potent and sustained modulation of gene expression throughout the central nervous system. Nat Biotechnol 37:884–894

Appelqvist H, Wäster P, Kågedal K, Öllinger K (2013) The lysosome: from waste bag to potential therapeutic target. J Mol Cell Biol 5:214–226

Appleby BS, Connor A, Wang H (2019) Therapeutic strategies for prion disease: a practical perspective. Curr Opin Pharmacol 44:15–19

Atarashi R et al (2011) Ultrasensitive human prion detection in cerebrospinal fluid by real-time quaking-induced conversion. Nat Med 17:175–178

Aulić S, Bolognesi ML, Legname G (2013) Small-molecule Theranostic Probes: a promising future in neurodegenerative diseases. Int J Cell Biol 2013:1–19
Baiardi S et al (2021) Phenotypic diversity of genetic Creutzfeldt–Jakob disease: a histo-molecular-based classification. Acta Neuropathol 142:707–728
Baker BF et al (1997) 2′-O-(2-Methoxy)ethyl-modified Anti-intercellular Adhesion Molecule 1 (ICAM-1) oligonucleotides selectively increase the ICAM-1 mRNA level and inhibit formation of the ICAM-1 translation initiation complex in human umbilical vein endothelial cells. J Biol Chem 272:11994–12000
Bender H et al (2019) PrPC knockdown by liposome-siRNA-peptide complexes (LSPCs) prolongs survival and normal behavior of prion-infected mice immunotolerant to treatment. PLoS One 14:e0219995
Benetti F, Legname G (2009) De novo mammalian prion synthesis. Prion 3:213–219
Blanco M-J, Gardinier KM (2020) New chemical modalities and strategic thinking in early drug discovery. ACS Med Chem Lett 11:228–231
Bolognesi ML, Gandini A, Prati F, Uliassi E (2016) From companion diagnostics to Theranostics: a new avenue for Alzheimer's disease?: Miniperspective. J Med Chem 59:7759–7770
Booth L et al (2016) AR-12 inhibits multiple chaperones concomitant with stimulating autophagosome formation collectively preventing virus replication. J Cell Physiol 231:2286–2302
Borchelt DR, Taraboulos A, Prusiner SB (1992) Evidence for synthesis of scrapie prion proteins in the endocytic pathway. J Biol Chem 267:16188–16199
Boselli M et al (2017) An inhibitor of the proteasomal deubiquitinating enzyme USP14 induces tau elimination in cultured neurons. J Biol Chem 292:19209–19225
Büeler H et al (1993) Mice devoid of PrP are resistant to scrapie. Cell 73:1339–1347
Cazzaniga FA et al (2021) Sporadic Creutzfeldt-Jakob disease: Real-Time Quaking Induced Conversion (RT-QuIC) assay represents a major diagnostic advance. Eur J Histochem 65:3298
Ciechanover A, Schwartz AL (1998) The ubiquitin-proteasome pathway: the complexity and myriad functions of proteins death. Proc Natl Acad Sci USA 95:2727–2730
Cortes CJ, Qin K, Cook J, Solanki A, Mastrianni JA (2012) Rapamycin delays disease onset and prevents PrP plaque deposition in a Mouse Model of Gerstmann–Sträussler–Scheinker disease. J Neurosci 32:12396–12405
Eid S et al (2024) The importance of prion research. Biochem Cell Biol 102:448–471
Elbashir SM et al (2001) Duplexes of 21-nucleotide RNAs mediate RNA interference in cultured mammalian cells. Nature 411:494–498
Fehlinger A et al (2017) Prion strains depend on different endocytic routes for productive infection. Sci Rep 7:6923
Fire A et al (1998) Potent and specific genetic interference by double-stranded RNA in Caenorhabditis elegans. Nature 391:806–811
Freund N et al (2023) A two-residue nascent-strand steric gate controls synthesis of 2′-O-methyl- and 2′-O-(2-methoxyethyl)-RNA. Nat Chem 15:91–100
Gangopadhyay S, Gore KR (2022) Advances in siRNA therapeutics and synergistic effect on siRNA activity using emerging dual ribose modifications. RNA Biol 19:452–467
Gentile JE et al (2024) Divalent siRNA for prion disease. Preprint at https://doi.org/10.1101/2024.12.05.627039
Geschwind MD et al (2013) Quinacrine treatment trial for sporadic Creutzfeldt-Jakob disease. Neurology 81:2015–2023
Giles K et al (2015) Different 2-Aminothiazole therapeutics produce distinct patterns of Scrapie Prion neuropathology in mouse brains. J Pharmacol Exp Ther 355:2–12
Goold R, McKinnon C, Tabrizi SJ (2015) Prion degradation pathways: potential for therapeutic intervention. Mol Cell Neurosci 66:12–20
Griffey CJ, Yamamoto A (2022) Macroautophagy in CNS health and disease. Nat Rev Neurosci 23:411–427
Haïk S et al (2014) Doxycycline in Creutzfeldt-Jakob disease: a phase 2, randomised, double-blind, placebo-controlled trial. Lancet Neurol 13:150–158

Hoy S (2017) M. Nusinersen: first global approval. Drugs 77:473–479

Hsich G, Kenney K, Gibbs CJ, Lee KH, Harrington MG (1996) The 14-3-3 brain protein in cerebrospinal fluid as a marker for transmissible Spongiform Encephalopathies. N Engl J Med 335:924–930

Hu B et al (2020) Therapeutic siRNA: state of the art. Sig Transduct Target Ther 5:101

Hutti CR, Welle KA, Hryhorenko JR, Ghaemmaghami S (2020) Global analysis of protein degradation in prion infected cells. Sci Rep 10:10800

Hutvagner G, Simard MJ (2008) Argonaute proteins: key players in RNA silencing. Nat Rev Mol Cell Biol 9:22–32

Ingusci S, Verlengia G, Soukupova M, Zucchini S, Simonato M (2019) Gene therapy tools for brain diseases. Front Pharmacol 10:724

Ionis Pharmaceuticals, Inc (2024) PrProfile: a study to assess the safety, tolerability, pharmacokinetics and pharmacodynamics of ION717

Kelkar SS, Reineke TM (2011) Theranostics: combining imaging and therapy. Bioconjug Chem 22:1879–1903

Khvorova A, Kennedy Z (2021) Oligonucleotides for prnp modulation, US20210317460.

Kiprowska MJ et al (2017) Neurotoxic mechanisms by which the USP14 inhibitor IU1 depletes ubiquitinated proteins and Tau in rat cerebral cortical neurons: relevance to Alzheimer's disease. Biochim Biophys Acta (BBA) - Mol Basis Dis 1863:1157–1170

Klingenstein R et al (2006) Tricyclic antidepressants, quinacrine and a novel, synthetic chimera thereof clear prions by destabilizing detergent-resistant membrane compartments. J Neurochem 98:748–759

Lauffer MC, Van Roon-Mom W, Aartsma-Rus A, N = 1 Collaborative (2024) Possibilities and limitations of antisense oligonucleotide therapies for the treatment of monogenic disorders. Commun Med 4:6

Lawrence RE, Zoncu R (2019) The lysosome as a cellular centre for signalling, metabolism and quality control. Nat Cell Biol 21:133–142

Leavitt BR, Tabrizi SJ (2020) Antisense oligonucleotides for neurodegeneration. Science 367:1428–1429

Legname G (2017) Elucidating the function of the prion protein. PLoS Pathog 13:e1006458

Lehmann S et al (2014) Systemic delivery of siRNA down regulates brain Prion protein and Ameliorates neuropathology in Prion disorder. PLoS One 9:e88797

Liang X et al (2016) Translation efficiency of mRNAs is increased by antisense oligonucleotides targeting upstream open reading frames. Nat Biotechnol 34:875–880

Macedo B, Cordeiro Y (2017) Unraveling Prion protein interactions with Aptamers and other PrP-binding nucleic acids. IJMS 18:1023

Mallucci G et al (2003) Depleting neuronal PrP in prion infection prevents disease and reverses spongiosis. Science 302:871–874

Mashima T et al (2013) Anti-prion activity of an RNA aptamer and its structural basis. Nucleic Acids Res 41:1355–1362

Mashima T et al (2020) Development and structural determination of an anti-PrPC aptamer that blocks pathological conformational conversion of prion protein. Sci Rep 10:4934

McColgan P et al (2023) Tominersen in adults with manifest Huntington's disease. N Engl J Med 389:2203–2205

McKinnon C et al (2016) Prion-mediated neurodegeneration is associated with early impairment of the ubiquitin–proteasome system. Acta Neuropathol 131:411–425

Mead S et al (2022) Prion protein monoclonal antibody (PRN100) therapy for Creutzfeldt–Jakob disease: evaluation of a first-in-human treatment programme. Lancet Neurol 21:342–354

Meister G, Tuschl T (2004) Mechanisms of gene silencing by double-stranded RNA. Nature 431:343–349

Minikel EV et al (2020) Prion protein lowering is a disease-modifying therapy across prion disease stages, strains and endpoints. Nucleic Acids Res 48:10615–10631

Mortberg MA et al (2023) A single-cell map of antisense oligonucleotide activity in the brain. Nucleic Acids Res 51:7109–7124

Nakagaki T et al (2020) Administration of FK506 from late stage of disease prolongs survival of human Prion-inoculated mice. Neurotherapeutics 17:1850–1860

Nazor Friberg K et al (2012) Intracerebral infusion of antisense Oligonucleotides into Prion-infected mice. Mol Ther Nucl Acids 1:e9

Neumann EN et al (2024) Brainwide silencing of prion protein by AAV-mediated delivery of an engineered compact epigenetic editor. Science 384:ado7082

Ni S et al (2021) Recent progress in Aptamer discoveries and modifications for therapeutic applications. ACS Appl Mater Interfaces 13:9500–9519

Nimjee SM, White RR, Becker RC, Sullenger BA (2017) Aptamers as therapeutics. Annu Rev Pharmacol Toxicol 57:61–79

Nordberg A (2004) PET imaging of amyloid in Alzheimer's disease. Lancet Neurol 3:519–527

Olsthoorn RCL (2014) G-quadruplexes within prion mRNA: the missing link in prion disease? Nucleic Acids Res 42:9327–9333

Otto M et al (2004) Efficacy of flupirtine on cognitive function in patients with CJD: a double-blind study. Neurology 62:714–718

Pan KM et al (1993) Conversion of alpha-helices into beta-sheets features in the formation of the scrapie prion proteins. Proc Natl Acad Sci USA 90:10962–10966

Parchi P, Strammiello R, Giese A, Kretzschmar H (2011) Phenotypic variability of sporadic human prion disease and its molecular basis: past, present, and future. Acta Neuropathol 121:91–112

Proske D et al (2002) Prion-protein-specific Aptamer reduces PrPSc formation. Chembiochem 3:717

Prusiner SB, Scott MR, DeArmond SJ, Cohen FE (1998) Prion protein biology. Cell 93:337–348

Pulford B et al (2010) Liposome-siRNA-peptide complexes cross the blood-brain barrier and significantly decrease PrPC on neuronal cells and PrPRES in infected cell cultures. PLoS One 5:e11085

Raffeiner M, Zhu S, González-Fuente M, Üstün S (2023) Interplay between autophagy and proteasome during protein turnover. Trends Plant Sci 28:698–714

Raymond GJ et al (2019) Antisense oligonucleotides extend survival of prion-infected mice. JCI Insight 4:e131175

Relaño-Ginés A et al (2018) Lithium as a disease-modifying agent for prion diseases. Transl Psychiatry 8:163

Ruckman J et al (1998) 2′-Fluoropyrimidine RNA-based Aptamers to the 165-amino acid form of vascular endothelial growth factor (VEGF165). J Biol Chem 273:20556–20567

Saborio GP, Permanne B, Soto C (2001) Sensitive detection of pathological prion protein by cyclic amplification of protein misfolding. Nature 411:810–813

Saini A, Chawla PA (2024) Breaking barriers with tofersen: enhancing therapeutic opportunities in amyotrophic lateral sclerosis. Eur J Neurol 31:e16140

Sarabia-Vallejo Á, López-Alvarado P, Menéndez JC (2023) Small-molecule theranostics in Alzheimer's disease. Eur J Med Chem 255:115382

Schambach A et al (2024) A new age of precision gene therapy. Lancet 403:568–582

Scotter EL et al (2014) Differential roles of the ubiquitin proteasome system (UPS) and autophagy in the clearance of soluble and aggregated TDP-43 species. J Cell Sci. https://doi.org/10.1242/jcs.140087

Seth PP et al (2010) Synthesis and biophysical evaluation of 2′,4′-constrained 2′ *O* -Methoxyethyl and 2′,4′-constrained 2′ *O* -ethyl nucleic acid analogues. J Org Chem 75:1569–1581

Shim KH, Sharma N, An SSA (2022) Prion therapeutics: lessons from the past. Prion 16:265–294

Spagnolli G et al (2021) Pharmacological inactivation of the prion protein by targeting a folding intermediate. Commun Biol 4:62

Staderini M et al (2013) A fluorescent Styrylquinoline with combined therapeutic and diagnostic activities against Alzheimer's and Prion diseases. ACS Med Chem Lett 4:225–229

Staderini M, Martín MA, Bolognesi ML, Menéndez JC (2015) Imaging of β-amyloid plaques by near infrared fluorescent tracers: a new frontier for chemical neuroscience. Chem Soc Rev 44:1807–1819

Staderini M et al (2023) Bifunctional carbazole derivatives for simultaneous therapy and fluorescence imaging in prion disease murine cell models. Eur J Med Chem 245:114923

Supattapone S, Nguyen H-OB, Cohen FE, Prusiner SB, Scott MR (1999) Elimination of prions by branched polyamines and implications for therapeutics. Proc Natl Acad Sci USA 96:14529–14534

Tallet-Lopez B (2003) Antisense oligonucleotides targeted to the domain IIId of the hepatitis C virus IRES compete with 40S ribosomal subunit binding and prevent in vitro translation. Nucleic Acids Res 31:734–742

Titze-de-Almeida SS et al (2020) The promise and challenges of developing miRNA-based therapeutics for Parkinson's disease. Cells 9:841

Trivedi PC, Bartlett JJ, Pulinilkunnil T (2020) Lysosomal biology and function: modern view of cellular Debris Bin. Cells 9:1131

Uliassi E, Nikolic L, Bolognesi ML, Legname G (2023) Therapeutic strategies for identifying small molecules against prion diseases. Cell Tissue Res 392:337–347

Uliassi E, Bolognesi ML, Milelli A (2025) Targeting Tau protein with proximity inducing modulators: a new frontier to Combat Tauopathies. ACS Pharmacol Transl Sci acsptsci.4c00733:654. https://doi.org/10.1021/acsptsci.4c00733

Vallabh SM, Minikel EV, Schreiber SL, Lander ES (2020) Towards a treatment for genetic prion disease: trials and biomarkers. Lancet Neurol 19:361–368

Vargas JNS, Hamasaki M, Kawabata T, Youle RJ, Yoshimori T (2023) The mechanisms and roles of selective autophagy in mammals. Nat Rev Mol Cell Biol 24:167–185

Vasconcelos D, Pina A, Habib N, Sousa S (2024) In silico analysis of aptamer-RNA conjugate interactions with human transferrin receptor. Biophys Chem 314:107308

Vickers TA et al (2003) Efficient reduction of target RNAs by small interfering RNA and RNase H-dependent antisense agents. J Biol Chem 278:7108–7118

Wan WB, Seth PP (2016) The medicinal chemistry of therapeutic Oligonucleotides. J Med Chem 59:9645–9667

Wang B et al (2024) Aptamer technologies in neuroscience, neuro-diagnostics and neuro-medicine development. Molecules 29:1124

White MD, Mallucci GR (2009) Therapy for prion diseases: insights from the use of RNA interference. Prion 3:121–128

White AR et al (2003) Monoclonal antibodies inhibit prion replication and delay the development of prion disease. Nature 80:422

Wu J, Chen D, Shi Q, Dong X (2021) Protein amplification technology: new advances in human prion disease diagnosis. Biosafety Health 3:325–332

Yamaguchi K et al (2019) A designer molecular chaperone against transmissible spongiform encephalopathy slows disease progression in mice and macaques. Nat Biomed Eng 3:206–219

Zerr I et al (2009) Updated clinical diagnostic criteria for sporadic Creutzfeldt-Jakob disease. Brain 132:2659–2668

Zerr I et al (2024) Creutzfeldt–Jakob disease and other prion diseases. Nat Rev Dis Primers 10:14

Zhou Y et al (2020) Blood-brain barrier–penetrating siRNA nanomedicine for Alzheimer's disease therapy. Sci Adv 6:eabc7031

Chapter 4
Prion Disease Diagnostic Biomarker Utility in Pre-symptomatic Disease

Laura J. Ellett, Matteo Senesi, Steven J. Collins, and Victoria Lewis

Abstract A typical feature of human prion diseases (PrDs) is the rapid decline to terminal illness that patients experience after symptom onset, with the most common phenotype, sporadic Creutzfeldt–Jakob disease (sCJD), frequently progressing from full independence to requiring palliative care over the course of weeks. A similar disease course is often observed in the much less common genetic CJD, especially when associated with the more common pathogenic mutations E200K and D178N. Therefore, the temporal therapeutic window is greatly reduced in PrDs compared with other dementias. There are currently no recognised reliable indicators of imminent or prodromal disease preceding the onset of overt, rapid, and currently unalterable decline. The advent of disease-modifying therapies will further underscore the need to expedite the time taken to achieve an accurate diagnosis in order to improve patient outcomes, highlighting the importance of detecting PrDs as early as possible in their clinical evolution. This review discusses what we currently know about pre-symptomatic and prodromal PrD derived from incidental case

L. J. Ellett
The Florey Institute of Neuroscience and Mental Health, The University of Melbourne, Parkville, VIC, Australia

M. Senesi · V. Lewis
Australian National Creutzfeldt-Jakob Disease Registry (ANCJDR), The Florey Institute of Neuroscience and Mental Health, The University of Melbourne, Parkville, VIC, Australia

The Department of Medicine, Royal Melbourne Hospital, The University of Melbourne, Parkville, VIC, Australia

S. J. Collins (✉)
The Florey Institute of Neuroscience and Mental Health, The University of Melbourne, Parkville, VIC, Australia

Australian National Creutzfeldt-Jakob Disease Registry (ANCJDR), The Florey Institute of Neuroscience and Mental Health, The University of Melbourne, Parkville, VIC, Australia

The Department of Medicine, Royal Melbourne Hospital, The University of Melbourne, Parkville, VIC, Australia
e-mail: s.collins@unimelb.edu.au

G. Legname, F. Moda (eds.), *Biomarkers and Therapeutical Targets for Prion Diseases*, Subcellular Biochemistry 112, https://doi.org/10.1007/978-3-031-97055-9_4

reports, limited preclinical cohort studies, and large-scale retrospective tissue screening programmes, contextualising the utility of current diagnostic tools and biomarkers for the detection of PrDs at these nascent disease stages.

Keywords Preclinical · Prodromal · Biomarkers · RT-QuIC · Magnetic resonance imaging · Cerebrospinal fluid · Asymptomatic · Longitudinal

Introduction

Prion diseases (PrDs) are clinically multifarious and invariably fatal, with no disease-modifying therapeutics currently available to patients. All PrDs arise due to the misfolding of the normal, host-encoded form of the prion protein (PrP^{C}) into disease-associated conformers (PrP^{Sc}), with the accumulation of PrP^{Sc} leading to degeneration of the brain (Prusiner 1998). The most common phenotype of PrD is sporadic Creutzfeldt–Jakob disease (sCJD: ~85% of all PrD), arising without apparent explanation. Inherited forms of PrD associated with coding mutations in the prion protein gene (*PRNP)* are much less common, accounting for only ~10–15% of all PrD. The most common inherited form is genetic CJD (gCJD), which is often indistinguishable from sCJD clinically and pathologically and is most frequently associated with the pathogenic E200K *PRNP* mutation, though it is also linked to many other point, as well as insert mutations (Appleby et al. 2022): More protracted illnesses are usually observed in the rarer genetic PrD phenotypes such as Gerstmann–Sträussler–Scheinker (GSS) syndrome, which, like gCJD, has been associated with many different *PRNP* mutations, or fatal familial insomnia (FFI), which is always associated with a pathogenic D178N *PRNP* mutation. Transmitted forms of PrD, including medically acquired iatrogenic CJD (iCJD) and the zoonosis variant CJD (vCJD) due to bovine spongiform encephalopathy (BSE), are extremely rare, accounting for less than 1% of all PrD cases (Gao et al. 2024).

PrD pathogenesis appears driven by the misfolding of PrP^{C} and accumulation of PrP^{Sc}; therefore, markedly decreasing the expression of the substrate PrP^{C} can prevent disease occurrence or arrest disease progression and even reverse initial pathological changes in animal models (Mallucci et al. 2003; An et al. 2025; Büeler et al. 1993). Antisense oligonucleotide (ASO) therapeutics are one approach to attenuating PrP^{C} expression levels, harnessing novel disease-modulating strategies through their ability to impact protein expression at the level of the messenger RNA. Building on previous preclinical ASO animal studies (Nazor Friberg et al. 2012; Raymond et al. 2019; Minikel et al. 2020; Büeler et al. 1993), this promising treatment strategy has commenced human clinical trials, namely the IONIS "PrProfile" Phase 1/2a clinical trial for early symptomatic disease that is currently nearing completion (Ionis Pharmaceuticals 2024).

Acknowledging previous therapeutic studies in animal prion disease models and drawing on the broader experience in neurodegenerative disorders, it is apparent that a key component of therapeutic success for prion diseases is likely to be early intervention to minimise irreversible neuronal damage. Three major features of prion disease make early intervention a challenge: their onset without a clear-cut or distinctive prodrome, rapid progression, and rare occurrence. In comparison to other neurodegenerative diseases such as Alzheimer's disease (AD), PrDs are typically extremely rapid in their progression. The average duration of illness for sCJD is 6.3 months, with a median of 3.9 months (Stehmann et al. 2023). As PrDs are clinically heterogeneous and uncommon, 83% of PrD patients are initially misdiagnosed, most often thought to represent other dementias (Appleby et al. 2014), and on average, PrD is misdiagnosed four times before a correct diagnosis of PrD is made (Paterson et al. 2012). Once disease-modifying therapies are available, the limited timeframe between symptom onset and terminal disease means that delays in accurate diagnosis will have significant negative impacts on patient outcomes.

It is known from animal biochemical and neuropathological studies that PrD-related changes are present and detectable before symptom onset (Wang et al. 2019; Brazier et al. 2006; Masters et al. 1975, 1984) and that subtle behavioural and cognitive changes can be detected before more typical clinical features are extant if appropriate, sensitive metrics are employed (Senesi et al. 2023a). A key challenge is the ability of existing diagnostic methods and biomarkers to detect PrDs at the earliest stage possible, especially at prodromal stages, to allow the best opportunity for meaningful therapeutic interventions to improve patient outcomes. Herein, we describe the currently available diagnostic tools that are utilised by clinicians in the investigation of symptomatic patients wherein PrD may be considered a differential diagnosis, followed by a review of the literature concerning pre-symptomatic or prodromal PrD-related biomarker changes in people, found either incidentally during investigations for other illnesses or during studies of asymptomatic individuals at increased risk of prion disease, such as due to their *PRNP* mutation status.

Currently Available Clinical Diagnostics and Biomarkers for Prion Disease

Clinical Profile

Clinically, PrD can be a challenge to diagnose early, as initial symptoms are often non-specific and are common to several other infectious, vascular, autoimmune, neurological, or neurodegenerative conditions, some of which may be reversible. Illustrative of this, evaluation to identify early features of prion disease revealed symptoms such as headache and fatigue (Krasnianski et al. 2014), while mood disorder may also be a prodromal feature (Wurm et al. 2025). For typical CJD cases, the distinctive clinical profile of rapidly progressive dementia, in combination with other features such as myoclonus and cerebellar ataxia, although strongly suggestive may take some time to evince during which significant damage to the brain has occurred. Further complicating the early evaluation of overt CJD is the existence of different molecular subtypes of sCJD, chiefly influenced by the combination of PrP^{Sc} glycoform type (Type 1 or Type 2) and genotype at the polymorphic *PRNP* codon 129 (methionine (M) or valine (V) homozygote or heterozygote). As a consequence, the clinical spectrum is more diverse, and patients may present with psychiatric symptoms, movement disorder, or visual complaints rather than cognitive issues, and the tempo of disease progression may be much slower than usual (Parchi et al. 2011). Ultimately, PrD classification can only be "definite" with neuropathological confirmation of the features of the disease or "probable" when neuropathological confirmation is not possible but where the clinical profile and investigational studies such as electroencephalogram (EEG), magnetic resonance imaging (MRI), and cerebrospinal fluid (CSF) biomarkers are supportive (CJD Diagnostic Criteria 2024).

A range of autosomal dominantly inherited *PRNP* mutations are recognised, including missense, nonsense, and octapeptide repeat insertions (OPRIs) that result in genetic forms of prion disease (Minikel et al. 2016). Although there have been dozens of reported *PRNP* sequence variations, recent evidence suggests many of these may in fact be "benign" polymorphisms or carry a much lower level of disease penetrance, with lifetime risk as low as 0.1–10% (Minikel et al. 2016). It is assumed that the presence of a disease-causing *PRNP* mutation greatly pre-disposes PrP^{C} to spontaneously misfold to PrP^{Sc} at some point during an individual's lifetime, leading to the development of overt PrD, although the precise influences and mechanism/s behind this have not yet been determined. Different *PRNP* mutations can also be influenced by the prion molecular subtype, especially the polymorphic codon 129 genotype, thereby eliciting clinically and neuropathologically distinct and often quite variable phenotypes (Baiardi et al. 2021; Ladogana and Kovacs 2018), such as the effect of coding for a valine or methionine at codon 129 in cis with the D178N mutation. If D178N is in cis with a methionine at codon 129, the mutation typically leads to an FFI clinical presentation, with an onset typically between the fourth and sixth decade, progressive sleep and autonomic nervous system impairment, and other symptoms associated with severe atrophy of the medial thalamic nuclei and negative or not suggestive investigation results (CSF 14-3-3, MRI, and EEG). If D178N is in cis with valine at codon 129, then the associated disease phenotype is usually CJD (Cracco et al. 2018). The most common E200K mutation displays high penetrance and usually presents as classical CJD with rapidly progressive dementia of under a year and positive biomarker investigation results such as positive CSF 14-3-3 or RT-QuIC seeding and a suggestive MRI (Appleby et al. 2022). In contrast, mutations associated with GSS, for example, P102L, typically have more prominent cerebellar ataxia for an extended period before significant cognitive decline, with negative biomarker investigation results, significantly longer illness durations up to several years, and disease onsets anywhere from the third to the sixth decade (Chen et al. 2024; Ghetti et al. 2018). The wide variation in gPrD presentations observed between kindreds carrying the same *PRNP* mutation can even be observed within kindreds, including symptomatology, age at onset, and disease duration (McLean et al. 1997; Yamada et al. 1999; Mead et al. 2006; Webb et al. 2008).

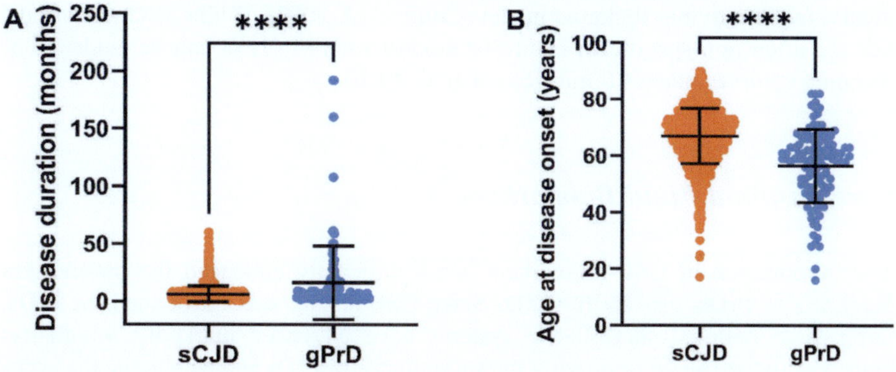

Fig. 4.1 Differences in disease features between sporadic (sCJD; includes data from two cases of sporadic fatal insomnia) and genetic prion disease (gPrD). Average prion disease (**a**) duration and (**b**) age at disease onset are significantly different (****) across the different disease aetiologic subtypes. (Novel data obtained from the Australian National CJD Registry (ANCJDR) from its inception in 1993 to Dec 2024)

As illustrated by Fig. 4.1a, whilst the average illness duration of sCJD is approximately 6 months, gPrDs, especially non-CJD phenotypes, generally have much longer durations spanning up to a few years, and some patients with rare forms of genetic PrD can remain symptomatic for over a decade prior to death. Similarly, whilst typically the age at onset for sCJD is in the seventh decade, some individuals manifest PrD much earlier, especially some forms of genetic PrD, as demonstrated in Fig. 4.1b.

Imaging

Brain Magnetic Resonance Imaging

Brain MRI is the leading non-invasive tool for the identification of PrDs and exclusion of mimics with characteristic changes officially included in the diagnostic criteria in 2009 (Zerr et al. 2009). The MRI features most characteristic of sCJD are high fluid-attenuated inversion recovery (FLAIR) or diffusion-weighted imaging (DWI) signal intensities occurring in the corpus striatum (caudate nucleus and/or putamen) or in at least two cortical (excluding the frontal) regions; significant cortical swelling and contrast enhancement are not considered features of sCJD. MRI changes may advance with the clinical progression of the disease (Kulkarni 2015). MRI features have also been found to vary between sCJD subtypes when imaged early in disease progression (Bizzi et al. 2020). sCJD types MM/V1 and MM/V2C are less likely to present with DWI abnormalities in the striatum, while VV2 and MV2K are unlikely to demonstrate changes in the cerebral cortex (Bizzi et al. 2020). Other prion diseases demonstrate their own unique MRI features, such as the "pulvinar sign" in vCJD, which presents as bilateral FLAIR or DWI hyperintensities involving the pulvinar thalamic nuclei (Collie et al. 2003). While MRI abnormalities are often non-specific in nature or modest for FFI, there can be evidence of thalamic volume changes (Grau-Rivera et al. 2016).

Cerebrospinal Fluid Biomarkers

The introduction of CSF biomarkers has significantly enhanced the pre-mortem diagnosis of PrDs, especially sCJD. Since CSF testing was introduced for PrDs, referrals to national surveillance systems have increased markedly, which has played a crucial role in improving the surveillance of PrDs and enhancing the accuracy of pre-mortem diagnosis. This section will explore several key biomarkers in detail, including their mechanisms, diagnostic values, and recent scientific advancements.

14-3-3 Proteins

14-3-3 proteins are a family of regulatory molecules involved in numerous cellular processes, including signal transduction, cell cycle control, and apoptosis. These proteins are present at high levels in the central nervous system (CNS) and are considered non-specific markers for neuronal damage. They are released into the CSF following neuronal injury or death, which can occur in various neurological and pathological conditions, such as prion diseases but also stroke, prolonged seizures, and viral encephalitis (Muayqil et al. 2012; Stoeck et al. 2012). Despite their non-specificity, 14-3-3 proteins are valuable in a clinical setting when appropriately utilised in the context of a patient's illness. Employed correctly in a high pre-test likelihood setting, the presence of 14-3-3 proteins in CSF has high sensitivity (~85–95%) and specificity (~80–90%) for sCJD, making them highly useful for diagnostic purposes and therefore included in the World Health Organisation (WHO) manual for PrD surveillance (WHO 1998) (Lattanzio et al. 2017; Senesi et al. 2023b; Chohan et al. 2010). For gCJD due to the most common *PRNP* mutations, such as E200K and V210I, the sensitivity is similar to sCJD, with sensitivity around 70–100%, depending on the assay platform utilised (Schmitz et al. 2022). In other genetic PrD phenotypes such as FFI and GSS, the sensitivity of 14-3-3 proteins is variable, often considerably lower than in sCJD, with FFI showing sensitivities around 10–40% and GSS around 45% (Chen et al. 2019; Schmitz et al. 2022). The reason(s) behind such variability in 14-3-3 CSF levels in gPrD is incompletely resolved but at least partly relates to the rapidity of disease progression, CSF sampling at different stages of disease progression, and possibly the location of the mutation (before or after amino acid 83) (Chen et al. 2019).

Total-Tau Protein

Tau is a microtubule-associated protein crucial for the assembly and maintenance of microtubules in axons, playing a significant role in maintaining neuronal structure and function. In the context of prion diseases, Tau is released in high concentrations, particularly during the late stages of the disease. Similar to 14-3-3 proteins, elevated levels of total Tau (tTau) in the CSF are indicative of neuronal damage but are also a non-specific marker for PrDs, as elevated tTau can also be observed in other rapidly progressing dementias, inflammatory diseases, or following seizures (Skillback et al. 2014).

In sCJD, tTau shows a sensitivity and specificity of approximately 80–90% (Senesi et al. 2023b; Lattanzio et al. 2017). For gCJD, sensitivity and specificity are reported to be slightly lower on average, greatly depending on the specific *PRNP* mutation, with sensitivity ranging from 0 for mutations such as G114V, D178N(V), and K194E to >80% for more common mutations such as E200K and a 4-OPRI insertion (Schmitz et al. 2022; Ladogana et al. 2009; Lattanzio et al. 2017). In FFI, the sensitivity of tTau is low (7–18%), while in GSS it is approximately 40–50%. Higher levels of CSF tTau protein in FFI patients seem to correlate with the clinical presentation of myoclonus, indicating greater neuronal damage (Chen et al. 2019). Lattanzio et al. (2017) highlighted the utility of the ratio of phosphorylated Tau at serine 181 (pTau) to tTau in improving the differentiation between PrDs and other neurodegenerative diseases such as AD.

Real-Time Quaking-Induced Conversion Assay

The real-time quaking-induced conversion (RT-QuIC) assay has revolutionised the clinical diagnosis of PrDs by exploiting the ability of minuscule quantities of PrP^{Sc} to faithfully "seed" the misfolding of PrP^{C} into nascent PrP^{Sc} readily detected by fluorophores in vitro (Schmitz et al. 2016). Originally developed for CSF samples, RT-QuIC has expanded its utility to include tissues such as olfactory mucosa (Orru et al. 2014) and skin biopsy (Orrú et al. 2017), as well as tears (Schmitz et al. 2023).

RT-QuIC uses recombinant prion protein (rPrP) as a substrate which, when mixed with a biospecimen containing PrP^{Sc} and subjected to cyclic shaking, amplifies the small amounts of misfolded prion protein present in the sample. Since its inception with human rPrP as a substrate to amplify the reaction (Atarashi et al. 2011), other substrates have been employed in RT-QuIC, such as full-length hamster (Peden et al. 2012), truncated hamster, and bank vole (Watts et al. 2014; Orru et al. 2015; Orrú et al. 2015). Different substrates, especially truncated hamsters combined with altered assay conditions such as higher assay temperatures, often referred to as "second generation" or "improved" RT-QUIC assays, have been found to provide the advantage of increased sensitivity whilst also (Orrú et al. 2015) shortening total assay duration and therefore results turnaround times.

In sCJD, the RT-QuIC assay generally exhibits high sensitivity (~80–90%) and almost 100% specificity, making it an extremely robust diagnostic tool. Analogous to other biomarkers, the application of RT-QuIC to gPrD has demonstrated varying sensitivities and specificities. For example, symptomatic *PRNP* mutation carriers with mutations associated with a gCJD phenotype typically display similar sensitivities and specificities to sCJD, depending on the mutation present (Schmitz et al. 2022; Lattanzio et al. 2017; Rhoads et al. 2020). For other gPrDs such as FFI and GSS, RT-QuIC testing of CSF from symptomatic individuals, even with assay modifications such as employing the "universal substrate" bank vole recombinant PrP, generally shows lower sensitivities (0–30% for FFI and 40–60% for GSS), although this can vary across the specific disease-associated mutations (Schmitz et al. 2022; Lattanzio et al. 2017; Mok et al. 2021; Franceschini et al. 2017; Dong and Satoh 2021; Green 2019; Cramm et al. 2016; Sano et al. 2013). Moreover, due to the low incidence of FFI and GSS, these studies often report only low case numbers, which can also affect the sensitivity and specificity calculated across laboratories.

As mentioned, recent studies have reported an expanded diagnostic ability of RT-QuIC beyond CSF to include other tissues and fluids. Olfactory mucosa (Orru et al. 2014), skin biopsy samples (Orrú et al. 2017), and tears (Schmitz et al. 2023) have been successfully used as alternative biological sources for RT-QuIC testing. These samples have demonstrated varying degrees of sensitivity and specificity but still maintain high diagnostic accuracy, comparable to CSF-based assays, with the advantage of being less invasive than lumbar puncture and offering the potential for greater convenience in longitudinal and preclinical studies. The versatility of RT-QuIC across different bio-sample sources underscores its potential as a robust tool for early diagnosis, monitoring of disease progression, and assessment of therapeutic efficacy in PrD.

Biopotential Recordings

Electroencephalogram

Typical EEG findings were included by the World Health Organisation (WHO) in 1998 as a diagnostic criterion for "probable" sCJD. In these criteria, the presence of periodic sharp-wave complexes (PSWCs) on EEG was recognised as a strongly supportive or hallmark feature for the diagnosis of probable sCJD when combined with clinical symptoms such as rapidly progressive dementia and at least two other typical clinical signs (e.g. myoclonus, visual disturbances, ataxia, etc.).

PSWCs are characterised by strictly periodic cerebral potentials occurring at a frequency of 0.5–2 Hz. PSWCs are typically absent during sleep and sometimes diminished by psychotropic medications. Their occurrence varies according to the molecular subtype of sCJD and age (Kovács et al. 2005; Wieser et al. 2006; Collins et al. 2006), with an overall sensitivity of ≤65% and specificity of 90%. Additionally, the main background frequency with increases in delta/theta frequencies in quantitative EEG appears to correlate with clinical progression, making EEG a potential tool for monitoring PrD progression in the late symptomatic phases (Collins et al. 2001).

In early PrD stages, non-specific changes such as diffuse slowing and frontal intermittent rhythmic delta activity (FIRDA) are more common (Ladogana et al. 2009). Similar EEG patterns have been reported in other neurodegenerative conditions, such as AD and Lewy body dementia (LBD), though less frequently. Furthermore, the likelihood of detecting PSWCs is much lower in other PrDs such as GSS, as well as in certain molecular subtypes of sCJD (Puoti et al. 2012).

Updates to the diagnostic criteria for probable sCJD have continued to include EEG findings, though advancements in other diagnostic tools and biomarkers have somewhat reduced the reliance on an EEG (Hermann et al. 2018). Nevertheless, the EEG remains a valuable tool in the evaluation of potential sCJD, particularly in settings where advanced imaging or biochemical testing is contraindicated or unavailable.

Polysomnography

In FFI, or sporadic fatal insomnia, polysomnography (PSG) frequently offers objective identification of early and progressive reductions in total sleep time, the absence of sleep spindles and K-complexes, disrupted sleep structure, and sleep fragmentation. Although these changes, in combination with brief REM sleep episodes that may involve dream enactment behaviours, can be envisaged as strongly supportive of fatal insomnia (Montagna et al. 2003), they are much less common in other forms of PrD, severely limiting their overall diagnostic utility for PrDs.

Pre-symptomatic Prion Disease

Animal models of prion disease employing a range of hosts, including ovine (Thomas et al. 2024; Pérez-Lázaro et al. 2024), rodent (Wang et al. 2019; Brazier et al. 2006), and cervid (Denkers et al. 2024; Kraft et al. 2023) systems, have provided compelling evidence that the agent of transmission ("prions") and pathological and biological changes are present prior to overt clinical onset. This supports the proposition that detection of disease at the pre-symptomatic phase should be possible; however, corroborating this in the human context is challenging for various reasons. Clearly demarcating pre-symptomatic from prodromal or early symptomatic PrD is not necessarily straightforward, as animal models have demonstrated that subtle behavioural and cognitive changes can be discerned using appropriate techniques in advance of traditional features (Senesi et al. 2023a; Cunningham et al. 2003, 2005). The completely unpredictable and spontaneous onset of a rare disorder such as sCJD (accounting for the vast majority of PrD cases) renders the systematic forward planning of preclinical or even prodromal screening impossible to conduct (Minikel et al. 2019). Nonetheless, as described in the following sections, the concepts of detectable sub-clinical, pre-symptomatic, and subliminal prodromal phases are supported by a growing number of case reports and systematic studies across the spectrum of PrD. These studies give valuable insight into the processes that may be at play prior to the onset of clear-cut symptoms and valuable direction for future preclinical biomarker development.

Retrospective Large-Scale Archived Tissue Analysis

Large-scale, anonymised, retrospective surveys of tonsil and appendix specimens have revealed evidence of disease-associated prion protein deposition in a small number of asymptomatic people in the United Kingdom (Gill et al. 2013; Hilton et al. 2004). These studies were conducted following the epidemic of bovine spongiform encephalopathy (BSE) after it was confirmed that BSE had crossed the species barrier into humans as the zoonosis vCJD (Will et al. 1996; Collinge et al. 1996). vCJD has been responsible for 233 deaths to date, with 178 of these occurring in the UK. Characteristic features distinguishing sCJD from vCJD are younger age at disease onset and deposition of PrP^{Sc} in lymphoreticular tissues. Epidemiological modelling studies were undertaken during the height of the vCJD epidemic to predict the burden of disease that would develop due to transmission of BSE through consumption of contaminated meat (Ghani et al. 2000). Furthermore, concerns regarding asymptomatic people harbouring vCJD and the potential for secondary transmissions, particularly through donated blood products, for example, led to large-scale retrospective immunohistological surveys of lymphoreticular tissues for the detection of disease-associated prion protein deposition. These studies generated estimates that 1 in 2000 people in the UK were likely pre-symptomatically infected with vCJD prions and potentially at risk of developing and secondarily transmitting vCJD (Gill et al. 2013). Fortunately, however, so far over the decade since these findings were published, only one person has succumbed to vCJD in the UK (Mok et al. 2017), suggesting that the presence of detectable prions in lymphoreticular tissue samples was an overestimate or may not completely reflect the likelihood of developing clinical disease. It has been hypothesised that lymphoreticular tissues may have increased permissiveness to prions when compared to central nervous system tissue, leading to the possibility of indefinite sub-clinical disease rather than representing pre-symptomatic disease (Beringue et al. 2012). Of interest, the most recent case of vCJD in the UK reported by Mok et al. (2017) was the first case of vCJD in a person heterozygous at codon 129, raising the possibility of longer incubation periods in these individuals. The full pathobiological implications of the deposition of PrP^{Sc} in lymphoreticular tissues in asymptomatic persons apparently harbouring vCJD remain unresolved, but the development of sensitive and specific tests suitable for mass screening, such as a blood-based assay, could potentially advance our understanding of the pre-symptomatic and prodromal phases of PrD. Indeed, the in vitro PrP^{Sc} amplification technique known as protein misfolding cyclic amplification (PMCA), with the addition of some pre-analytical processing steps, has successfully detected PrP^{Sc} in the blood of symptomatic and even pre-symptomatic (14 and 31 months prior to clinical disease) vCJD patients (Concha-Marambio et al. 2016; Bougard et al. 2016).

Incidental Post-mortem Prion Disease Detection in Apparently Asymptomatic People

Although rare, there have now been several reported instances of the detection of diagnostic prion disease brain neuropathological changes in people who were believed to be asymptomatic for prion disease at the time of death. An adolescent who had received cadaveric pituitary growth hormone treatments was found to have neuropathological features of PrD at brain autopsy following death due to viral respiratory infection (New et al. 1988). There was no clinical suspicion of CJD prior to her succumbing to pneumonia. Ghoshal et al. (2015) described finding prion disease neuropathology in a "cognitively normal" study participant. Nakagaki et al. (2022) reported a positive RT-QuIC result from a cadaver used for anatomical practice, having commenced routine screening of anatomical specimens using RT-QuIC the year prior. A robust medical history was unable to be retrieved beyond the cause of death being determined as aspiration pneumonia. Such cases clearly support the likelihood of established PrD in the human body, especially the brain, prior to convincing clinical features but do not clarify whether these findings represent long-term sub-clinical disease or a pre-symptomatic phase with the onset of overt disease imminent.

Incidental Pre-mortem Detection of Pre-symptomatic Prion Disease

PrD case studies have been reported wherein a diagnostic test has incidentally suggested asymptomatic PrD in a person being investigated for other reasons. To date, 11 such case studies of patients ultimately succumbing to prion disease have been published in the literature (summarised in Table 4.1), with the majority (all but two) being cases of presumed sporadic CJD. All cases had pre-symptomatic brain MRI DWI findings broadly compatible with PrD, and approximately half had undergone other tests prior to symptom onset, such as EEG or CSF analysis.

Table 4.1 Summary of investigation results relevant to prion disease diagnosis conducted before and after the onset of symptoms across 11 incidentally discovered prion disease cases at the pre-symptomatic stage of the disease

		Diagnostic test result				
References	Test results in relation to prion symptom onset	MRI DWI hyperintensity	tTau	14-3-3	RT-QuIC	EEG PSWC
Satoh et al. 2011	Before:	+	–	–	NP	NP
	After:	+	+	–[a]	NP	+
Terasawa et al. (2012)	Before:	+	NP	NP	NP	–
	After:	+	NP	NP	NP	–
Verde et al. (2016)	Before:	+	NP	NP	NP	NP
	After:	+	+	+	+	–[b]
Suzuki et al. (2016)	Before:	+	NP	NP	NP	NP
	After:	+	+	+	NP	–[b]
Zanusso et al. (2016)	Before:	+	NP	NP	NP	NP
	After:	+	+	+	+[c]	+
Iwasaki et al. (2017)	Before:	+	NP	NP	NP	NP
	After:	+	+	+	NP	+
Novi et al. (2018)	Before:	+	–	–	–	–[b]
	After:	+	–	–	+	–[b]
Maeda et al. (2019)	Before:	+	–	–	–	–
	After:	+	+	+	+	+
Koizumi et al. (2021)	Before:	+	NP	NP	NP	NP
	After:	+	+	+	–	–
Hamada et al. (2022)	Before:	+	–	–	–	–
	After:	+	+	+	–	+
Yasuda et al. (2022)	Before:	+	–	–	–	–
	After:	+	+	+	–	+

PSWC periodic sharp-wave complexes, *NP* not performed
+ positive/present
– Negative/absent
[a]progressed to positive 4 weeks post-symptom onset
[b]Abnormal recording but not diagnostic for CJD
[c]positive RT-QuIC results in both CSF and olfactory mucosa

Incidental Pre-symptomatic EEG Findings

To date, five studies have reported EEG investigations prior to the onset of typical clinical PrD (Terasawa et al. 2012; Maeda et al. 2019; Novi et al. 2018; Hamada et al. 2022; Yasuda et al. 2022). All studies were negative for diagnostic PSWC at the pre-symptomatic stage of the disease. Abnormal EEG results consisting of diffuse 6–7 Hz theta activity were recorded in only one study a year prior to the onset of behavioural disturbances (Novi et al. 2018); however, this co-occurred with neurological deficits in face recognition and clock drawing and therefore could indicate an early clinical feature. EEG assessed a year later at the time the patient was admitted for behavioural changes was unchanged.

Non-specific increases in EEG slower frequencies have been noted in several studies during the early symptomatic stages of PrD (Suzuki et al. 2016; Verde et al.

2016; Satoh et al. 2011). One study described initial excess slow waves during the early symptomatic stage, which then progressed to typical PSWCs (Satoh et al. 2011). More usually, however, PSWCs detected at the symptomatic stage of the disease had normal EEG findings prior to symptom onset, making EEG unlikely to be a useful preclinical investigation tool (Maeda et al. 2019; Yasuda et al. 2022; Hamada et al. 2022).

Incidental Pre-symptomatic CSF Biomarker Findings

In six of the 11 case reports (Table 4.1), CSF biomarker testing for PrD was only performed after the onset of overt clinical disease. In the five case studies that did include pre-symptomatic CSF biomarker testing, all of which were eventually classified as sCJD, none of the investigations were supportive of prion disease. Surprisingly, given the ability of the test to amplify very low levels of PrP^{Sc}, this included negative RT-QuIC results in four patients even though DWI hyperintensity on MRI was present (Hamada et al. 2022; Satoh et al. 2011; Maeda et al. 2019; Novi et al. 2018; Yasuda et al. 2022). Novi et al. (2018) reported RT-QuIC positivity but negative 14-3-3 and tTau results at the symptomatic stage. As 14-3-3 and tTau proteins are non-specific indicators of neuro-axonal damage in the brain, negative results obtained during early symptomatic disease may indicate that significant neuronal death or injury has not yet occurred or the evolution of the disease is slower than usual. This may indicate a potentially important diagnostic window for treatment introduction prior to severe irreversible neuronal damage or loss.

Somewhat surprisingly, CSF RT-QuIC results only demonstrated positive results after the onset of symptoms in two of the four reports with sequential assay studies (Table 4.1). They both described cases of sCJD, and RT-QuIC positivity was only demonstrated in CSF collected after overt disease (Maeda et al. 2019; Novi et al. 2018). Reported findings have also shown a disconnection between 14-3-3 and tTau positivity and RT-QuIC positivity. Yasuda et al. (2022), Hamada et al. (2022), and Koizumi et al. (2021) showed diagnostic positivity of 14-3-3 and tTau without a positive RT-QuIC. The explanation for these discrepancies is unclear but may partly reflect differences in RT-QuIC sensitivity between sporadic molecular subtypes (Hermann et al. 2024) or that PrD conversion-competent seeds do not enter the CSF until significant neuronal damage has occurred, in contrast to 14-3-3 and tTau. Alternative diagnostic tissue and bio-fluid sample sources may prove more useful in the early diagnosis of PrD, including during the pre-symptomatic stage. Nasal brushings can directly sample olfactory neurons that may have seeding capacity prior to that detected in CSF (Orru et al. 2014). Skin has been shown in animal studies to have seeding capacity during asymptomatic disease (Wang et al. 2019). As RT-QuIC is currently the only PrD-specific test used routinely in a clinical setting, it would be ideal to optimise the tissue or bio-fluid source to enable this assay to reliably identify symptomatic PrD as early as possible or even at the pre-symptomatic stage to maximise treatment windows before significant neuronal damage and loss have occurred.

Table 4.2 Summary of incidentally detected pre-symptomatic prion disease case reports providing preclinical diagnostic test information

Reference	Age at initial MRI (years)/sex	Reason for MRI	Time from initial suggestive MRI (DWI hyperintensity) to overt symptom onset (months)	Clinical course following symptom onset	Aetiology
Satoh et al. (2011)	68/M	Patient request	3	Slow; >1 year	Sporadic
Terasawa et al. (2012)	68/F	Headache with nausea	4	Slow; >7 months	Genetic; V180I
Verde et al. (2016)	65/F	Headache	12	Slow; >1 year	Sporadic
Suzuki et al. (2016)	69/M	Coronary artery stenosis monitoring	29	Slow; >1 year	Sporadic
Zanusso et al. (2016)	74/F	Carotid body tumour	14	Rapid; 3 months	Sporadic
Iwasaki et al. (2017)	77/M	Dizziness	8	Slow; 5 months	Sporadic
Novi et al. (2018)	64/M	Lightheadedness	16	Slow; >1 year	Sporadic
Maeda et al. (2019)	67/F	Health check	12	Slow; 6 months	Sporadic
Koizumi et al. (2021)	64/F	Headache	6	Slow; >1 year	Genetic; V180I
Hamada et al. (2022)	63/F	Orofacial dystonia	2	Not reported	Sporadic
Yasuda et al. (2022)	47/M	Facial numbness	27	Slow; 5 months	Sporadic

Incidental Pre-Symptomatic MRI Findings

Table 4.2 summarises MRI results in incidentally discovered pre-symptomatic or early symptomatic PrD patients. The majority of cases had initial MRI imaging in the 7th decade of life, which correlates with the average age of onset of sCJD (Fig. 4.1b). The imaging was performed an average of 12 months (SD ±9 months) prior to the onset of clinical features of PrD.

The clinical indications for the initial MRI referral were variable but did not highlight any distinct symptomatology strongly suggestive of early overt PrD, although occasional overlap with prodromal symptoms of PrD, such as headache, cannot be entirely excluded (Table 4.2).

All incidental PrD case reports found brain DWI hyperintensity prior to symptom onset (Table 4.1), which, when present in typical anatomical locations such as the corpus striatum and cerebral cortex, are known distinctive features of sCJD and gCJD (Zerr et al. 2009). This was often initially construed as related to an ischaemic event (Verde et al. 2016; Suzuki et al. 2016; Iwasaki et al. 2017; Koizumi et al.

2021), and it can be challenging to differentiate a sub-clinical ischaemic event from changes due to pre-symptomatic PrD. Suzuki et al. (2016), Iwasaki et al. (2017), and Verde et al. (2016) gave anti-platelet therapy based on initial MRI findings, with patients remaining asymptomatic for 17, 12, and 8 months, respectively, until presenting to the hospital again. Koizumi et al. (2021) described treatment with aspirin associated with improvement in DWI hyperintensity, suggesting the finding may have been due to an ischaemic event and highlighting the possibility that PrD may coincide with other treatable morbidities. The eventual onset of typical clinical features and follow-up MRIs showing the evolution of characteristic DWI hyperintensity in spite of any coincidental treatment clarified the development of PrD.

Longitudinal Investigation of *PRNP* Mutation Carriers at Increased Risk of Developing Prion Disease

As previously discussed, the unpredictable and spontaneous appearance of sCJD, representing the majority of PrD, makes the forward planning of pre-symptomatic screening almost impossible to conduct. Although predicting disease onset in carriers of highly penetrant *PRNP* mutations is also challenging, people known to be "at-risk" of genetic PrD offer a more tractable situation (Minikel et al. 2019). Asymptomatic *PRNP* mutation carriers, especially the highly penetrant mutations such as E200K, D178N, and P102L (Minikel et al. 2016), are still the best opportunity for investigating any potential pre-symptomatic or prodromal changes related to PrD, as they are the only group where progression to symptomatic PrD during life is highly likely.

Studies assessing diagnostic investigations in *PRNP* mutation carriers can be separated and loosely categorised as either retrospective or prospective. Retrospective investigations are primarily those involving incidental findings (as described above) when individuals have been specifically investigated for an unrelated illness and/or non-specific symptoms not considered to be indicative of PrD but which may actually be early/prodromal features of PrD when *PRNP* mutation status is uncovered during the course of the investigations. Retrospective studies may also involve the thorough re-analysis of prior investigation results collected pre-symptomatically from known *PRNP* mutation carriers for unrelated reasons. Prospective studies, including longitudinal studies, involve cohorts of known mutation carriers (and often *PRNP* mutation-negative family members as controls), whereby diagnostic tests are carried out, and in the case of longitudinal studies, systematically repeated over several years in order to detect the earliest development of diagnostic positivity and potentially predict symptom onset. Given the long period of time often required to obtain these datasets and the rarity of this family of diseases, there are few such systematic pre-symptomatic longitudinal studies for gPrD reported in the literature.

Longitudinal Imaging Studies

Magnetic Resonance Imaging

As already mentioned, characteristic MRI changes in symptomatic PrD, especially sCJD, are well described with very good sensitivity and specificity and have been detected incidentally in pre-symptomatic individuals who later developed PrD. As such, the majority of prospective gPrD studies have included MRI investigations. To date, however, the longitudinal studies that have been conducted have not shown significant or convincing MRI DWI changes in asymptomatic gPrD carriers. In G114V carriers, grey matter volume was reduced compared to healthy controls at baseline imaging, but grey matter volume was shown to increase when comparing asymptomatic carriers at baseline to imaging performed 2 years later (Lu et al. 2020). Asymptomatic OPRI carriers showed mild cerebellar atrophy on MRI, and E200K carriers showed some brain volume changes on MRI, but these findings were subtle and unlikely to be usable as pre-symptomatic diagnostic markers (Cohen et al. 2015; Townley et al. 2020).

18Fluorine-Fluorodeoxyglucose (18F-FDG) Positron Emission Tomography (PET)

18fluorine-fluorodeoxyglucose positron emission tomography (18F-FDG-PET) is not a commonly employed diagnostic tool for PrD but may hold potential niche utility: Hypometabolism of glucose in cortical and subcortical brain regions has been described (comprehensively reviewed in Mattoli et al. (2024). PET is an imaging modality that utilises radioactive tracers to monitor bodily processes such as metabolism, blood flow, and the presence of chemicals of interest. 18F-FDG is a radioactive glucose analogue used in conjunction with PET imaging to measure alterations in cellular metabolism in body tissues. 18F-FDG-PET has demonstrated cerebral hypometabolism in symptomatic sCJD when other imaging and biopotential recording techniques were unremarkable in the early stages of the disease (Boero et al. 2024). Hypometabolism has also been shown in brain regions during early-stage genetic PrD, as well as the pre-symptomatic stage in a person harbouring the D178N mutation at risk for FFI, and therefore may be a relevant tool for understanding pre-symptomatic gPrD (Lu et al. 2020; Cortelli et al. 2006), as illustrated by hypometabolism in the thalami evident at 13 months prior to the onset of symptoms typical of FFI (Cortelli et al. 2006).

In carriers of the G114V mutation associated with gCJD, there was no definite hypometabolism during asymptomatic disease (Lu et al. 2020; Chu et al. 2022; Kong et al. 2025). The lack of progression to symptomatic disease in G114V patients in this longitudinal study to date means it has not yet been able to capture the progression from the asymptomatic to the symptomatic stage. Data collected by this same team on symptomatic sCJD and E200K patients was able to show significant reductions in brain FDG uptake, but it is not yet known if the G114V patients will display this characteristic at the symptomatic stage of the disease. Different PrDs are already known to present different findings across varied diagnostic modalities, and not all PrDs have been shown to display hypometabolism on 18F-FDG-PET. 18F-FDG-PET hypometabolism was "not well visualised" in those carrying a seven OPRI mutation even during symptomatic disease, confounded by cerebral hypometabolism increasing with age (Townley et al. 2020). Consequently, given the likely ongoing small sample sizes, substantial pathological differences between gPrDs and age as a confounding factor, 18F-FDG-PET will require considerably more evaluation before determining its utility across the different gPrD mutations before it can be used as a pre-symptomatic PrD marker.

Longitudinal Analysis of CSF and Other Fluid Biomarkers

RT-QuIC Assay

A caveat in relation to longitudinal, pre-symptomatic PrD studies in genetic carriers is that, as alluded to earlier, many genetic forms of PrD do not test positive in the classic diagnostic RT-QuIC CSF assay even when symptomatic. CSF from symptomatic patients carrying the E200K *PRNP* mutation can, however, seed classic RT-QuIC assays using recombinant hamster prion protein as the substrate (Mok et al. 2021; Xiao et al. 2019) as well as in a modified assay using a hamster-sheep recombinant prion protein chimaera (Orru et al. 2015; Mok et al. 2021; Cramm et al. 2015). In the National Prion Monitoring Cohort (NPMC) study, three pre-symptomatic E200K carriers were reported to show positive RT-QuIC assay results: two at least 3 years prior to symptom onset, while one displayed positive results only approximately 2 months prior to clinical onset (Mok et al. 2023; Vallabh et al. 2024). Similarly, in a separate, single-centre, longitudinal study, RT-QuIC seeding activity was detected in CSF from three pre-symptomatic E200K carriers at 1 year, 2.5 years, and 3.1 years prior to onset (Vallabh et al. 2024), confirming a potentially identifiable pre-symptomatic PrP^{Sc} deposition/neurodegeneration window. Unfortunately, similar to observations of lower sensitivity when RT-QuIC testing CSF from symptomatic rarer phenotypes of inherited prion disease cases (Schmitz et al. 2022), in the pre-symptomatic cohort of carriers of other gPrD *PRNP* mutations, including P102L, 6-OPRI, A117V, and D178N, only one asymptomatic P102L carrier tested positive in a bespoke CSF RT-QuIC (Mok et al. 2023; Vallabh et al. 2024). It is clear that CSF, or other bio-fluid testing with RT-QuIC, especially when applied to pre-symptomatic *PRNP* mutation carriers, will likely require significant assay re-development with specific modifications such as changes in incubation temperature, reaction buffer components like salts or anionic detergents, and especially the recombinant protein substrate utilised to suit the particular *PRNP* mutation, and even then, may remain unsuccessful (Mok et al. 2023).

NfL and GFAP

As the screening of neurodegenerative diseases moves forward, biomarkers sourced from less invasive bio-fluids, such as blood, are envisaged as much more desirable. Intracellular fibrillar proteins such as neurofilament light chain (NfL) and glial fibrillary acidic protein (GFAP) are being investigated for their potential to identify neurodegenerative processes. NfL is part of a group of proteins involved specifically in the construction of axons, and levels in both CSF and blood are reliable indicators of neurodegeneration. CSF NfL levels vary significantly amongst CJD molecular subtypes with increases significantly greater than tTau in cases with subcortical pathology, although the diagnostic value decreases markedly when considered within a heterogeneous group of rapidly progressive dementia patients when compared to tTau or 14-3-3 (Abu-Rumeileh and Parchi 2021). GFAP is strongly associated with astrocytes and can be released into the CSF and blood if astrocytes are damaged, which may happen during neurodegenerative processes. NfL can be considered a marker of neurodegeneration, while GFAP can be considered a marker of neuroinflammation.

Plasma from asymptomatic mutation carriers in the NPMC study was analysed using single molecule array (Simoa) technology, showing a potential utility of GFAP and NfL as biomarkers in the 2 years before the onset of symptoms in persons carrying "slow" gPrD mutations such as P102L (Thompson et al. 2021). However, a subsequent study on a very small cohort of E200K and P102L carriers found that NfL and GFAP gave inconsistent results during the asymptomatic phase of the disease (Vallabh et al. 2024). More work is required to better understand the utility of these biomarkers in pre-symptomatic genetic PrDs.

Conclusion

As PrD research enters an era of industry-led, disease-modifying, therapeutic trials, it has become obvious that, for optimal efficacy, treatments will need to be applied as early as possible in the disease course. This means it is vital to establish early and preferably pre-symptomatic diagnostic markers of PrD to identify disease processes at the earliest possible stage and to also facilitate the monitoring of drug effectiveness during clinical trials. Given the various aetiologies and phenotypes of PrDs, finding sensitive biomarkers shared or common to all forms of PrD will remain challenging. Fortunately, most diagnostic tools work well for the most common PrD phenotype, sCJD, but not necessarily for less common gPrDs, which may present an issue in translating gPrD findings as keys to unlocking prodromal or pre-symptomatic biomarkers of sCJD. The sCJD molecular subtype and the specific *PRNP* mutation in gPrD will all undoubtedly influence the development and progression of pre-symptomatic disease biomarkers. There is a strong possibility that pre-symptomatic clinical markers of PrDs will differ between the different aetiologies of PrD, and studies of asymptomatic mutation carriers may find biomarkers that are only relevant to specific mutation types.

Based on the current literature analysis, brain imaging techniques, especially MRI, are the most likely to detect early and even pre-symptomatic PrD compared to other current diagnostic tests; PET imaging is likely to offer a more niche utility. Cerebral hypometabolism in 18F-FDG-PET is not specific to PrD but is commonly seen in neurodegenerative dementias and does hold potential utility for specific entities such as FFI.

Surprisingly, there was limited evidence to support CSF RT-QuIC analysis as a useful tool for pre-symptomatic sCJD but some modest evidence for utility in gCJD caused by the E200K mutation. Future pre-symptomatic clinical studies may need to incorporate alternative biological sample sites for tests such as RT-QuIC, as CSF may not be the ideal specimen for pre-symptomatic detection, especially in sCJD. Brain autopsy studies have demonstrated that incidental pre-symptomatic PrD neuropathology can occur, but it remains unresolved whether such changes imply the ineluctable development of PrD or perhaps long-term sub-clinical PrD.

References

Abu-Rumeileh S, Parchi P (2021) Cerebrospinal fluid and blood neurofilament light chain protein in Prion disease and other rapidly progressive dementias: current state of the art. Front Neurosci 15:648743

An M, Davis JR, Levy JM, Serack FE, Harvey JW, Brauer PP, Pirtle CP, Berríos KN, Newby GA, Yeh W-H, Kamath N, Mortberg MA, Lian Y, Howard MJ, DeSouza-Lenz K, Guzman K, Thai A, Graffam S, Laversenne V, Coffey AA, Frei J, Pierce SE, Safar JG, Deverman BE, Minikel EV, Vallabh SM, Liu DR (2025) In vivo base editing extends lifespan of a humanized mouse model of prion disease. Nat Med 31:1319

Appleby BS, Rincon-Beardsley TD, Appleby KK, Crain BJ, Wallin MT (2014) Initial diagnoses of patients ultimately diagnosed with prion disease. J Alzheimers Dis 42:833–839

Appleby BS, Shetty S, Elkasaby M (2022) Genetic aspects of human prion diseases. Front Neurol 13:1003056

Atarashi R, Satoh K, Sano K, Fuse T, Yamaguchi N, Ishibashi D, Matsubara T, Nakagaki T, Yamanaka H, Shirabe S, Yamada M, Mizusawa H, Kitamoto T, Klug G, McGlade A, Collins SJ, Nishida N (2011) Ultrasensitive human prion detection in cerebrospinal fluid by real-time quaking-induced conversion. Nat Med 17:175–178

Baiardi S, Rossi M, Mammana A, Appleby BS, Barria MA, Cali I, Gambetti P, Gelpi E, Giese A, Ghetti B, Herms J, Ladogana A, Mikol J, Pal S, Ritchie DL, Ruf V, Windl O, Capellari S, Parchi P (2021) Phenotypic diversity of genetic Creutzfeldt-Jakob disease: a histo-molecular-based classification. Acta Neuropathol 142:707–728

Beringue V, Herzog L, Jaumain E, Reine F, Sibille P, Le Dur A, Vilotte JL, Laude H (2012) Facilitated cross-species transmission of prions in extraneural tissue. Science 335:472–475

Bizzi A, Pascuzzo R, Blevins J, Grisoli M, Lodi R, Moscatelli MEM, Castelli G, Cohen ML, Schonberger LB, Foutz A, Safar JG, Appleby BS, Gambetti P (2020) Evaluation of a new criterion for detecting Prion disease with diffusion magnetic resonance imaging. JAMA Neurol 77:1141–1149

Boero G, Lauriero F, Fusillo D, Calvello R, Cianciulli A, Panaro MA, Moda P (2024) 18F-FDG PET/CT in early phase of sporadic Creutzfeldt-Jacob disease: a case report. Clin Case Reports 12:e8974

Bougard D, Brandel JP, Bélondrade M, Béringue V, Segarra C, Fleury H, Laplanche JL, Mayran C, Nicot S, Green A, Welaratne A, Narbey D, Fournier-Wirth C, Knight R, Will R, Tiberghien P, Haïk S, Coste J (2016) Detection of prions in the plasma of presymptomatic and symptomatic patients with variant Creutzfeldt-Jakob disease. Sci Transl Med 8:370ra182

Brazier MW, Lewis V, Ciccotosto GD, Klug GM, Lawson VA, Cappai R, Ironside JW, Masters CL, Hill AF, White AR, Collins S (2006) Correlative studies support lipid peroxidation is linked to PrPres propagation as an early primary pathogenic event in prion disease. Brain Res Bull 68:346–354

Büeler H, Aguzzi A, Sailer A, Greiner RA, Autenried P, Aguet M, Weissmann C (1993) Mice devoid of PrP are resistant to scrapie. Cell 73:1339–1347

Chen C, Hu C, Shi Q, Zhou W, Mao K, Wang Y, Liu L, Chen J, Xia Y, Dong XP (2019) Profiles of 14-3-3 and Total Tau in CSF samples of Chinese patients of different genetic Prion diseases. Front Neurosci 13:934–934

Chen ZY, Kong Y, Zhang J, Zou WQ, Wu LY (2024) Genetic and pathological features encipher the phenotypic heterogeneity of Gerstmann-Straussler-Scheinker disease. Neurobiol Dis 195:6497–6497

Chohan G, Pennington C, Mackenzie JM, Andrews M, Everington D, Will RG, Knight RSG, Green AJE (2010) The role of cerebrospinal fluid 14-3-3 and other proteins in the diagnosis of sporadic Creutzfeldt-Jakob disease in the UK: a 10-year review. J Neurol Neurosurg Psychiatry 81:1243–1248

Chu M, Chen ZY, Nie BB, Liu L, Xie KX, Cui Y, Chen KW, Rosa-Neto P, Wu LY (2022) A longitudinal F-18-FDG PET/MRI study in asymptomatic stage of genetic Creutzfeldt-Jakob disease linked to G114V mutation. J Neurol 269:6094–6103

CJD Diagnostic Criteria (2024) U.S. Centers for Disease Control and Prevention. https://www.cdc.gov/creutzfeldt-jakob/hcp/clinical-overview/diagnosis.html

Cohen OS, Chapman J, Korczyn AD, Nitsan Z, Appel S, Hoffmann C, Rosenmann H, Kahana E, Lee H (2015) Familial Creutzfeldt-Jakob disease with the E200K mutation: longitudinal neuroimaging from asymptomatic to symptomatic CJD. J Neurol 262:604–613

Collie DA, Summers DM, Sellar RJ, Ironside JW, Cooper S, Zeidler M, Knight R, Will RG (2003) Diagnosing variant Creutzfeldt-Jakob disease with the pulvinar sign: MR imaging findings in 86 neuropathologically confirmed cases. AJNR Am J Neuroradiol 24:1560–1569

Collinge J, Sidle KCL, Meads J, Ironside J, Hill AF (1996) Molecular analysis of prion strain variation and the aetiology of 'new variant' CJD. Nature 383:685–690

Collins S, McLean CA, Masters CL (2001) Gerstmann-Sträussler-Scheinker syndrome, fatal familial insomnia, and kuru: a review of these less common human transmissible spongiform encephalopathies. J Clin Neurosci 8:387–397

Collins SJ, Sanchez-Juan P, Masters CL, Klug GM, van Duijn C, Poleggi A, Pocchiari M, Almonti S, Cuadrado-Corrales N, de Pedro-Cuesta J, Budka H, Gelpi E, Glatzel M, Tolnay M, Hewer E, Zerr I, Heinemann U, Kretszchmar HA, Jansen GH, Olsen E, Mitrova E, Alperovitch A, Brandel JP, Mackenzie J, Murray K, Will RG (2006) Determinants of diagnostic investigation sensitivities across the clinical spectrum of sporadic Creutzfeldt-Jakob disease. Brain 129:2278–2287

Concha-Marambio L, Pritzkow S, Moda F, Tagliavini F, Ironside JW, Schulz PE, Soto C (2016) Detection of prions in blood from patients with variant Creutzfeldt-Jakob disease. Sci Transl Med 8:370ra183

Cortelli P, Perani D, Montagna P, Gallassi R, Tinuper P, Federica P, Avoni P, Ferrillo F, Anchisi D, Moresco RM, Fazio F, Parchi P, Baruzzi A, Lugaresi E, Gambetti P (2006) Pre-symptomatic diagnosis in fatal familial insomnia: serial neurophysiological and 18FDG-PET studies. Brain 129:668–675

Cracco L, Appleby BS, Gambetti P (2018) Chapter 15 - fatal familial insomnia and sporadic fatal insomnia. In: Pocchiari M, Manson J (eds) Handbook of clinical neurology. Elsevier, Amsterdam, Netherlands

Cramm M, Schmitz M, Karch A, Zafar S, Varges D, Mitrova E, Schroeder B, Raeber A, Kuhn F, Zerr I (2015) Characteristic CSF prion seeding efficiency in humans with prion diseases. Mol Neurobiol 51:396–405

Cramm M, Schmitz M, Karch A, Mitrova E, Kuhn F, Schroeder B, Raeber A, Varges D, Kim YS, Satoh K, Collins S, Zerr I (2016) Stability and reproducibility underscore utility of RT-QuIC for diagnosis of Creutzfeldt-Jakob disease. Mol Neurobiol 53:1896–1904

Cunningham C, Deacon R, Wells H, Boche D, Waters S, Diniz CP, Scott H, Rawlins JNP, Perry VH (2003) Synaptic changes characterize early behavioural signs in the ME7 model of murine prion disease. Eur J Neurosci 17:2147–2155

Cunningham C, Deacon RMJ, Chan K, Boche D, Rawlins JNP, Perry VH (2005) Neuropathologically distinct prion strains give rise to similar temporal profiles of behavioral deficits. Neurobiol Dis 18:258–269

Denkers ND, McNulty EE, Kraft CN, Nalls AV, Westrich JA, Hoover EA, Mathiason CK (2024) Temporal characterization of Prion shedding in Secreta of White-tailed deer in longitudinal study of Chronic Wasting disease, United States. Emerg Infect Dis 30:2118–2127

Dong TT, Satoh K (2021) The latest research on RT-QuIC assays-a literature review. Pathogens 10:305

Franceschini A, Baiardi S, Hughson AG, McKenzie N, Moda F, Rossi M, Capellari S, Green A, Giaccone G, Caughey B, Parchi P (2017) High diagnostic value of second generation CSF RT-QuIC across the wide spectrum of CJD prions. Sci Rep 7:10655–10655

Gao LP, Tian TT, Xiao K, Chen C, Zhou W, Liang DL, Cao RD, Shi Q, Dong XP (2024) Updated global epidemiology atlas of human prion diseases. Front Public Health 12:1411489

Ghani AC, Ferguson NM, Donnelly CA, Anderson RM (2000) Predicted vCJD mortality in Great Britain. Nature 406:583–584

Ghetti B, Piccardo P, Zanusso G (2018) Chapter 14 - Dominantly inherited prion protein cerebral amyloidoses—a modern view of Gerstmann–Sträussler–Scheinker. In: Pocchiari M, Manson J (eds) Handbook of clinical neurology. Elsevier, Amsterdam, Netherlands

Ghoshal N, Perry A, McKeel D, Schmidt RE, Carter D, Norton J, Zou WQ, Xiao XZ, Puoti G, Notari S, Gambetti P, Morris JC, Cairns NJ (2015) Variably protease-sensitive prionopathy in an apparent cognitively normal 93-year-old. Alzheimer Dis Assoc Disord 29:173–176

Gill ON, Spencer Y, Richard-Loendt A, Kelly C, Dabaghian R, Boyes L, Linehan J, Simmons M, Webb P, Bellerby P, Andrews N, Hilton DA, Ironside JW, Beck J, Poulter M, Mead S, Brandner S (2013) Prevalent abnormal prion protein in human appendixes after bovine spongiform encephalopathy epizootic: large scale survey. BMJ 347:f5675

Grau-Rivera O, Calvo A, Bargallo N, Monte GC, Nos C, Llado A, Molinuevo JL, Gelpi E, Sanchez-Valle R (2016) Quantitative magnetic resonance abnormalities in Creutzfeldt-Jakob disease and fatal insomnia. J Alzheimers Dis 55:431–443

Green AJE (2019) RT-QuIC: a new test for sporadic CJD. Pract Neurol 19:49–55

Hamada Y, Deguchi K, Tachi K, Kita M, Nonaka W, Takata T, Kobara H, Touge T, Satoh K, Masaki T (2022) Significance of cortical ribboning as a biomarker in the prodromal phase of sporadic Creutzfeldt-Jakob disease. Intern Med 61:2667–2670

Hermann P, Laux M, Glatzel M, Matschke J, Knipper T, Goebel S, Treig J, Schulz-Schaeffer W, Cramm M, Schmitz M, Zerr I (2018) Validation and utilization of amended diagnostic criteria in Creutzfeldt-Jakob disease surveillance. Neurology 91:E331–EE38

Hermann P, Goebel S, Zerr I (2024) Clinical characteristics and diagnostics of human spongiform encephalopathies: an update. Nervenarzt 95:376–384

Hilton DA, Ghani AC, Conyers L, Edwards P, McCardle L, Ritchie D, Penney M, Hegazy D, Ironside JW (2004) Prevalence of lymphoreticular prion protein accumulation in UK tissue samples. J Pathol 203:733–739

Ionis Pharmaceuticals, Inc (2024) PrProfile: a study to assess the safety, tolerability, pharmacokinetics and pharmacodynamics of ION717, ClinicalTrials.gov, Accessed 10/4/2025. https://clinicaltrials.gov/study/NCT06153966?cond=prion&rank=2

Iwasaki Y, Mori K, Ito M, Mimuro M, Kitamoto T, Yoshida M (2017) An autopsied case of MM1 + MM2-cortical with thalamic-type sporadic Creutzfeldt-Jakob disease presenting with hyperintensities on diffusion-weighted MRI before clinical onset. Neuropathology 37:78–85

Koizumi R, Ueda N, Mugita A, Kimura K, Kishida H, Tanaka F (2021) Case report: extremely early detection of preclinical magnetic resonance imaging abnormality in Creutzfeldt-Jakob disease with the V180I mutation. Front Neurol 12:751750

Kong Y, Chen Z, Zhang J, Wang Y, Chu M, Nan H, Cui Y, Jiang D, Liyong W (2025) Reconfigured metabolism brain network in asymptomatic Creutzfeldt-Jakob disease. Neurobiol Dis 206:106805

Kovács GG, Puopolo M, Ladogana A, Pocchiari M, Budka H, van Duijn C, Collins SJ, Boyd A, Giulivi A, Coulthart M, Delasnerie-Laupretre N, Brandel JP, Zerr I, Kretzschmar HA, de Pedro-Cuesta J, Calero-Lara M, Glatzel M, Aguzzi A, Bishop M, Knight R, Belay G, Will R, Mitrova E (2005) Genetic prion disease: the EUROCJD experience. Hum Genet 118:166–174

Kraft CN, Denkers ND, Mathiason CK, Hoover EA (2023) Longitudinal detection of prion shedding in nasal secretions of CWD-infected white-tailed deer. J Gen Virol 104:1825–1825

Krasnianski A, Kaune J, Jung K, Kretzschmar HA, Zerr I (2014) First symptom and initial diagnosis in sporadic CJD patients in Germany. J Neurol 261:1811–1817

Kulkarni KM (2015) Progressive MRI findings in Creutzfeldt-Jakob disease. J Neuroophthalmol 35:96–98

Ladogana A, Kovacs GG (2018) Genetic Creutzfeldt-Jakob disease. Handb Clin Neurol 153:219–242

Ladogana A, Sanchez-Juan P, Mitrova E, Green A, Cuadrado-Corrales N, Sanchez-Valle R, Koscova S, Aguzzi A, Sklaviadis T, Kulczycki J, Gawinecka J, Saiz A, Calero M, van Duijn CM, Pocchiari M, Knight R, Zerr I (2009) Cerebrospinal fluid biomarkers in human genetic transmissible spongiform encephalopathies. J Neurol 256:1620–1628

Lattanzio F, Abu-Rumeileh S, Franceschini A, Kai H, Amore G, Poggiolini I, Rossi M, Baiardi S, McGuire L, Ladogana A, Pocchiari M, Green A, Capellari S, Parchi P (2017) Prion-specific and surrogate CSF biomarkers in Creutzfeldt-Jakob disease: diagnostic accuracy in relation to molecular subtypes and analysis of neuropathological correlates of p-tau and Aβ42 levels. Acta Neuropathol 133:559–578

Lu H, Jing D, Chen Y, Cui C, Gao R, Wang L, Liang Z, Chen K, Wu L (2020) Metabolic changes detected by 18F-FDG PET in the preclinical stage of familial Creutzfeldt-Jakob disease. J Alzheimers Dis 77:1–9

Maeda K, Sugihara Y, Shiraishi T, Hirai A, Satoh K (2019) Cortical Hyperintensity on diffusion-weighted images as the presymptomatic marker of Sporadic Creutzfeldt-Jakob disease. Intern Med 58:727–729

Mallucci G, Dickinson A, Linehan J, Klohn PC, Brandner S, Collinge J (2003) Depleting neuronal PrP in prion infection prevents disease and reverses spongiosis. Science 302:871–874

Masters CL, Kakulas BA, Alpers MP, Gajdusek DC, Gibbs CJ Jr (1975) Preclinical lesions and their progression in the experimental spongiform encephalopathies (kuru and Creutzfeldt-Jakob disease) in primates. In: Tariska YS, Gosztonyi G (eds) Proceedings of the seventh international congress of neuropathology. Excerpta Medica, Amsterdam, Netherlands; Budapest, Hungary, pp 23–26

Masters CL, Rohwer RG, Franko MC, Brown P, Gajdusek DC (1984) The sequential development of spongiform change and gliosis of scrapie in the golden Syrian hamster. J Neuropathol Exp Neurol 43:242–252

Mattoli MV, Giancipoli RG, Cocciolillo F, Calcagni ML, Taralli S (2024) The role of PET imaging in patients with Prion disease: a literature review. Mol Imaging Biol 26:195

McLean CA, Storey E, Gardner RJ, Tannenberg AE, Cervenakova L, Brown P (1997) The D178N (cis-129M) "fatal familial insomnia" mutation associated with diverse clinicopathologic phenotypes in an Australian kindred. Neurology 49:552–558

Mead S, Poulter M, Beck J, Webb TE, Campbell TA, Linehan JM, Desbruslais M, Joiner S, Wadsworth JD, King A, Lantos P, Collinge J (2006) Inherited prion disease with six octapeptide repeat insertional mutation—molecular analysis of phenotypic heterogeneity. Brain 129:2297–2317

Minikel EV, Vallabh SM, Lek M, Estrada K, Samocha KE, Sathirapongsasuti JF, McLean CY, Tung JY, Yu LP, Gambetti P, Blevins J, Zhang S, Cohen Y, Chen W, Yamada M, Hamaguchi T, Sanjo N, Mizusawa H, Nakamura Y, Kitamoto T, Collins SJ, Boyd A, Will RG, Knight R, Ponto C, Zerr I, Kraus TF, Eigenbrod S, Giese A, Calero M, de Pedro-Cuesta J, Haik S, Laplanche JL, Bouaziz-Amar E, Brandel JP, Capellari S, Parchi P, Poleggi A, Ladogana A, O'Donnell-Luria AH, Karczewski KJ, Marshall JL, Boehnke M, Laakso M, Mohlke KL, Kahler A, Chambert K, McCarroll S, Sullivan PF, Hultman CM, Purcell SM, Sklar P, van der Lee SJ, Rozemuller A, Jansen C, Hofman A, Kraaij R, van Rooij JG, Ikram MA, Uitterlinden AG, van Duijn CM, Consortium Exome Aggregation, Daly MJ, MacArthur DG (2016) Quantifying prion disease penetrance using large population control cohorts. Sci Transl Med 8:322ra9

Minikel EV, Vallabh SM, Orseth MC, Brandel JP, Haik S, Laplanche JL, Zerr I, Parchi P, Capellari S, Safar J, Kenny J, Fong JC, Takada LT, Ponto C, Hermann P, Knipper T, Stehmann C, Kitamoto T, Ae R, Hamaguchi T, Sanjo N, Tsukamoto T, Mizusawa H, Collins SJ, Chiesa R, Roiter I, de Pedro-Cuesta J, Calero M, Geschwind MD, Yamada M, Nakamura Y, Mead S (2019) Age at onset in genetic prion disease and the design of preventive clinical trials. Neurology 93:E125–EE34

Minikel EV, Zhao HT, Le J, O'Moore J, Pitstick R, Graffam S, Carlson GA, Kavanaugh MP, Kriz J, Kim JB, Ma JY, Wille H, Aiken J, McKenzie D, Doh-Ura K, Beck M, O'Keefe R, Stathopoulos J, Caron T, Schreiber SL, Carroll JB, Kordasiewicz HB, Cabin DE, Vallabh SM (2020) Prion protein lowering is a disease-modifying therapy across prion disease stages, strains and endpoints. Nucleic Acids Res 48:10615–10631

Mok T, Jaunmuktane Z, Joiner S, Campbell T, Morgan C, Wakerley B, Golestani F, Rudge P, Mead S, Jager HR, Wadsworth JDF, Brandner S, Collinge J (2017) Variant Creutzfeldt-Jakob disease in a patient with heterozygosity at PRNP Codon 129. N Engl J Med 376:292–294

Mok TH, Nihat A, Luk C, Sequeira D, Batchelor M, Mead S, Collinge J, Jackson GS (2021) Bank vole prion protein extends the use of RT-QuIC assays to detect prions in a range of inherited prion diseases. Sci Rep 11:5231–5231

Mok TH, Nihat A, Majbour N, Sequeira D, Holm-Mercer L, Coysh T, Darwent L, Batchelor M, Groveman BR, Orr CD, Hughson AG, Heslegrave A, Laban R, Veleva E, Paterson RW, Keshavan A, Schott JM, Swift IJ, Heller C, Rohrer JD, Gerhard A, Butler C, Rowe JB, Masellis M, Chapman M, Lunn MP, Bieschke J, Jackson GS, Zetterberg H, Caughey B, Rudge P, Collinge J, Mead S (2023) Seed amplification and neurodegeneration marker trajectories in individuals at risk of prion disease. Brain 146:2570–2583

Montagna P, Gambetti P, Cortelli P, Lugaresi E (2003) Familial and sporadic fatal insomnia. Lancet Neurol 2:167–176

Muayqil T, Gronseth G, Camicioli R (2012) Evidence-based guideline: diagnostic accuracy of CSF 14-3-3 protein in sporadic Creutzfeldt-Jakob disease: report of the guideline development subcommittee of the American Academy of Neurology. Neurology 79:1499–1506

Nakagaki T, Kaneko M, Satoh K, Murai K, Saiki K, Matsumoto G, Ogami-Takamura K, Ikematsu K, Akagi A, Iwasaki Y, Tsurumoto T, Nishida N (2022) Detection of Prions in a Cadaver for anatomical practice. N Engl J Med 386:2245–2246

Nazor Friberg K, Hung G, Wancewicz E, Giles K, Black C, Freier S, Bennett F, Dearmond SJ, Freyman Y, Lessard P, Ghaemmaghami S, Prusiner SB (2012) Intracerebral infusion of antisense Oligonucleotides into Prion-infected mice. Mol Ther Nucleic Acids 1:e9

New MI, Brown P, Temeck JW, Owens C, Hedly-Whyte ET, Richardson EP (1988) Preclinical Creutzfeldt-Jakob disease discovered at autopsy in a human growth hormone recipient. Neurology 38:1133–1134

Novi G, Canosa A, Nobili F, Bongianni M, Zanusso G, Balestrino M, Roccatagliata L (2018) Longitudinal brain magnetic resonance imaging and real-time quaking induced conversion analysis in presymptomatic Creutzfeldt-Jakob disease. Eur J Neurol 25:E127–EE28

Orru CD, Bongianni M, Tonoli G, Ferrari S, Hughson AG, Groveman BR, Fiorini M, Pocchiari M, Monaco S, Caughey B, Zanusso G (2014) A test for Creutzfeldt-Jakob disease using Nasal brushings. N Engl J Med 371:519–529

Orrú CD, Groveman BR, Hughson AG, Zanusso G, Coulthart MB, Caughey B (2015) Rapid and sensitive RT-QuIC detection of human Creutzfeldt-Jakob disease using cerebrospinal fluid. MBio 6:e02451

Orru CD, Groveman BR, Raymond LD, Hughson AG, Nonno R, Zou WQ, Ghetti B, Gambetti P, Caughey B (2015) Bank Vole Prion protein as an apparently universal substrate for RT-QuIC-based detection and discrimination of Prion strains. PLoS Pathog 11.4983–4983

Orrú CD, Yuan J, Appleby BS, Li B, Li Y, Winner D, Wang Z, Zhan YA, Rodgers M, Rarick J, Wyza RE, Joshi T, Wang GX, Cohen ML, Zhang S, Groveman BR, Petersen RB, Ironside JW, Quinones-Mateu ME, Safar JG, Kong Q, Caughey B, Zou WQ (2017) Prion seeding activity and infectivity in skin samples from patients with sporadic Creutzfeldt-Jakob disease. Sci Transl Med 9:eaam7785

Parchi P, Strammiello R, Giese A, Kretzschmar H (2011) Phenotypic variability of sporadic human prion disease and its molecular basis: past, present, and future. Acta Neuropathol 121:91–112

Paterson RW, Torres-Chae CC, Kuo AL, Ando T, Nguyen EA, Wong K, Dearmond SJ, Haman A, Garcia P, Johnson DY, Miller BL, Geschwind MD (2012) Differential diagnosis of Jakob-Creutzfeldt disease. Arch Neurol 69:1–5

Peden AH, McGuire LI, Appleford NE, Mallinson G, Wilham JM, Orru CD, Caughey B, Ironside JW, Knight RS, Will RG, Green AJ, Head MW (2012) Sensitive and specific detection of sporadic Creutzfeldt-Jakob disease brain prion protein using real-time quaking-induced conversion. J Gen Virol 93:438–449

Pérez-Lázaro S, Barrio T, Bravo SB, Sevilla E, Otero A, Chantada-Vázquez MDP, Martín-Burriel I, Requena JR, Badiola JJ, Bolea R (2024) New preclinical biomarkers for prion diseases in the cerebrospinal fluid proteome revealed by mass spectrometry. Vet Q 44:1–15

Prusiner SB (1998) The prion diseases. Brain Pathol 8:499–513

Puoti G, Bizzi A, Forloni G, Safar JG, Tagliavini F, Gambetti P (2012) Sporadic human prion diseases: molecular insights and diagnosis. Lancet Neurol 11:618–628

Raymond GJ, Zhao HT, Race B, Raymond LD, Williams K, Swayze EE, Graffam S, Le J, Caron T, Stathopoulos J, O'Keefe R, Lubke LL, Reidenbach AG, Kraus A, Schreiber SL, Mazur C, Cabin DE, Carroll JB, Minikel EV, Kordasiewicz H, Caughey B, Vallabh SM (2019) Antisense oligonucleotides extend survival of prion-infected mice. JCI Insight 4:31175–31175

Rhoads DD, Wrona A, Foutz A, Blevins J, Glisic K, Person M, Maddox RA, Belay ED, Schonberger LB, Tatsuoka C (2020) Diagnosis of prion diseases by RT-QuIC results in improved surveillance. Neurology 95:e1017–e1e26

Sano K, Satoh K, Atarashi R, Takashima H, Iwasaki Y, Yoshida M, Sanjo N, Murai H, Mizusawa H, Schmitz M, Zerr I, Kim Y-S, Nishida N (2013) Early detection of abnormal Prion protein in genetic human prion diseases now possible using real-time QUIC assay. PLoS One 8:e54915

Satoh K, Nakaoke R, Nishiura Y, Tsujino A, Motomura M, Yoshimura T, Sasaki K, Shigematsu K, Shirabe S, Eguchi K (2011) Early detection of sporadic CJD by diffusion-weighted MRI before the onset of symptoms. J Neurol Neurosurg Psychiatry 82:942–943

Schmitz M, Cramm M, Llorens F, Muller-Cramm D, Collins S, Atarashi R, Satoh K, Orru CD, Groveman BR, Zafar S, Schulz-Schaeffer WJ, Caughey B, Zerr I (2016) The real-time quaking-induced conversion assay for detection of human prion disease and study of other protein misfolding diseases. Nat Protoc 11:2233–2242

Schmitz M, Villar-Pique A, Hermann P, Escaramis G, Calero M, Chen C, Kruse N, Cramm M, Golanska E, Sikorska B, Liberski PP, Pocchiari M, Lange P, Stehmann C, Sarros S, Marti E, Baldeiras I, Santana I, Zakova D, Mitrova E, Dong XP, Collins S, Poleggi A, Ladogana A, Mollenhauer B, Kovacs GG, Geschwind MD, Sanchez-Valle R, Zerr I, Llorens F (2022) Diagnostic accuracy of cerebrospinal fluid biomarkers in genetic prion diseases. Brain 145:700–712

Schmitz M, Correia SS, Hermann P, Maass F, Goebel S, Bunck T, Correia A, Lingor P, Fischer A, Zerr I (2023) Detection of Prion protein seeding activity in tear fluids. N Engl J Med 388:1816–1817

Senesi M, Lewis V, Adlard PA, Finkelstein DI, Kim JH, Collins SJ (2023a) Tailored behavioural tests reveal early and progressive cognitive deficits in M1000 prion disease. Neurobiol Dis 180:106075

Senesi M, Lewis V, Varghese S, Stehmann C, McGlade A, Doecke JD, Ellett L, Sarros S, Fowler CJ, Masters CL, Li Q-X, Collins SJ (2023b) Diagnostic performance of CSF biomarkers in a well-characterized Australian cohort of sporadic Creutzfeldt-Jakob disease. Front Neurol 14:1072952

Skillback T, Rosen C, Asztely F, Mattsson N, Blennow K, Zetterberg H (2014) Diagnostic performance of cerebrospinal fluid total Tau and phosphorylated Tau in Creutzfeldt-Jakob disease results from the Swedish Mortality Registry. JAMA Neurol 71:476–483

Stehmann C, Senesi M, Sarros S, McGlade A, Lewis V, Ellett L, Barber D, Simpson M, Klug G, McLean CA, Masters CL, Collins SJ (2023) Creutzfeldt-Jakob disease surveillance in Australia: update to 31 December 2022. Commun Dis Intell 2018:47

Stoeck K, Sanchez-Juan P, Gawinecka J, Green A, Ladogana A, Pocchiari M, Sanchez-Valle R, Mitrova E, Sklaviadis T, Kulczycki J, Slivarichova D, Saiz A, Calero M, Knight R, Aguzzi A, Laplanche JL, Peoc'h K, Schelzke G, Karch A, van Duijn CM, Zerr I (2012) Cerebrospinal fluid biomarker supported diagnosis of Creutzfeldt-Jakob disease and rapid dementias: a longitudinal multicentre study over 10 years. Brain 135:3051–3061

Suzuki K, Kawasaki A, Nagashima T, Hirata K (2016) Diffusion-weighted MRI abnormalities antedate the onset of sporadic Creutzfeldt-Jakob disease. Neurology 87:843–845

Terasawa Y, Fujita K, Izumi Y, Kaji R (2012) Early detection of familial Creutzfeldt-jakob disease on diffusion-weighted imaging before symptom onset. J Neurol Sci 319:130–132

Thomas CM, Salamat MKF, Almela F, Cooper JK, Ladhani K, Arnold ME, Bougard D, Andréoletti O, Houston EF (2024) Longitudinal detection of prion infection in preclinical sheep blood samples compared using 3 assays. Blood 144:1962–1973

Thompson AGB, Anastasiadis P, Druyeh R, Whitworth I, Nayak A, Nihat A, Mok TH, Rudge P, Wadsworth JDF, Rohrer J, Schott JM, Heslegrave A, Zetterberg H, Collinge J, Jackson GS, Mead S (2021) Evaluation of plasma tau and neurofilament light chain biomarkers in a 12-year clinical cohort of human prion diseases. Mol Psychiatry 26:5955–5966

Townley RA, Polsinelli AJ, Fields JA, Machulda MM, Jones DT, Graff-Radford J, Kantarci KM, Lowe VJ, Rademakers RV, Baker MC, Kumar N, Boeve BF (2020) Longitudinal clinical, neuropsychological, and neuroimaging characterization of a kindred with a 12-octapeptide repeat insertion inPRNP: the next generation. Neurocase 26:211–219

Vallabh SM, Mortberg MA, Allen SW, Kupferschmid AC, Kivisakk P, Hammerschlag BL, Bolling A, Trombetta BA, Devitte-McKee K, Ford AM, Sather LE, Duffy G, Rivera A, Gerber J, McManus AJ, Minikel EV, Arnold SE (2024) Fluid biomarkers in individuals at risk for genetic prion disease up to disease conversion. Neurology 103:9506–9506

Verde F, Ticozzi N, Messina S, Calcagno N, Girotti F, Maderna L, Moda F, Scola E, Falini A, Tagliavini F, Silani V (2016) MRI abnormalities found 1 year prior to symptom onset in a case of Creutzfeldt-Jakob disease. J Neurol 263:597–599

Wang Z, Manca M, Foutz A, Camacho MV, Raymond GJ, Race B, Orru CD, Yuan J, Shen P, Li B, Lang Y, Dang J, Adornato A, Williams K, Maurer NR, Gambetti P, Xu B, Surewicz W, Petersen RB, Dong X, Appleby BS, Caughey B, Cui L, Kong Q, Zou WQ (2019) Early preclinical detection of prions in the skin of prion-infected animals. Nat Commun 10:247

Watts JC, Giles K, Patel S, Oehler A, DeArmond SJ, Prusiner SB (2014) Evidence that Bank Vole PrP is a universal acceptor for Prions. PLoS Pathog 10:e1003990

Webb TEF, Poulter M, Beck J, Uphill J, Adamson G, Campbell T, Linehan J, Powell C, Brandner S, Pal S, Siddique D, Wadsworth JD, Joiner S, Alner K, Petersen C, Hampson S, Rhymes C, Treacy C, Storey E, Geschwind MD, Nemeth AH, Wroe S, Collinge J, Mead S (2008) Phenotypic heterogeneity and genetic modification of P102L inherited prion disease in an international series. Brain 131:2632–2646

WHO (1998) Global surveillance, diagnosis and therapy of human transmissible spongiform encephalopathies : report of a WHO consultation, Geneva, Switzerland, 9–11 February 1998. World Health Organization, Geneva

Wieser HG, Schindler K, Zumsteg D (2006) EEG in Creutzfeldt-Jakob disease. Clin Neurophysiol 117:935–951

Will RG, Ironside JW, Zeidler M, Cousens SN, Estibeiro K, Alperovitch A, Poser S, Pocchiari M, Hofman A, Smith PG (1996) A new variant of Creutzfeldt-Jakob disease in the UK. Lancet 347:921–925

Wurm R, Klotz S, Erber A, Gruber F, Leitner S, Reichardt B, Stögmann E, Schernhammer E, Gelpi E, Cetin H (2025) Mood alterations in the prodromal phase of Sporadic Creutzfeldt-Jakob disease. JAMA Neurol 82:185–192

Xiao K, Shi Q, Zhou W, Zhang B-Y, Wang Y, Chen C, Ma Y, Gao C, Dong X-P (2019) T188K-familial Creutzfeldt–Jacob disease, predominant among Chinese, has a reactive pattern in CSF RT-QuIC different from D178N-Fatal Familial Insomnia and E200K-Familial CJD. Neurosci Bull 35:519–521

Yamada M, Tomimitsu H, Yokota T, Tomi H, Sunohara N, Mukoyama M, Itoh Y, Suematsu N, Otomo E, Okeda R, Matsushita M, Mizusawa H (1999) Involvement of the spinal posterior horn in Gerstmann-Sträussler- Scheinker disease (PrP P102L). Neurology 52:260–265

Yasuda M, Sugiyama A, Hokkoku H, Suichi T, Ito K, Satoh K, Kitamoto T, Kuwabara S (2022) Propagation of diffusion-weighted MRI abnormalities in the preclinical stage of Sporadic Creutzfeldt-Jakob disease. Neurology 99:699–702

Zanusso G, Camporese G, Ferrari S, Santelli L, Bongianni M, Fiorini M, Monaco S, Manara R, Cagnin A (2016) Long-term preclinical magnetic resonance imaging alterations in Sporadic Creutzfeldt-Jakob disease. Ann Neurol 80:629–632

Zerr I, Kallenberg K, Summers DM, Romero C, Taratuto A, Heinemann U, Breithaupt M, Varges D, Meissner B, Ladogana A, Schuur M, Haik S, Collins SJ, Jansen GH, Stokin GB, Pimentel J, Hewer E, Collie D, Smith P, Roberts H, Brandel JP, van Duijn C, Pocchiari M, Begue C, Cras P, Will RG, Sanchez-Juan P (2009) Updated clinical diagnostic criteria for sporadic Creutzfeldt Jakob disease. Brain 132:2659

Chapter 5
Therapeutic Trajectories in Human Prion Diseases

Maria Letizia Barreca and Emiliano Biasini

Abstract Prion diseases are rare yet devastating neurodegenerative disorders that result from the misfolding of the cellular prion protein, PrP^C, into its infectious and pathogenic isoform, PrP^{Sc}. These diseases are marked by progressive neuronal damage, leading to irreversible cognitive and motor impairments and, ultimately, death. Despite extensive research into their underlying mechanisms, effective treatments for prion diseases remain elusive. Such a lack of effective therapies mainly arises from several challenges, including delayed diagnosis and the complex and poorly understood biology of prion neurotoxicity.

This chapter provides an in-depth exploration of current and emerging therapeutic strategies to treat prion diseases. One promising approach involves using small molecules to inhibit prion replication by destabilizing PrP^{Sc} or modulating PrP^C homeostasis, possibly avoiding previously observed strain-dependent drug resistance. In parallel, immunotherapeutic approaches, including passive and active immunization, have shown potential in targeting prions. However, challenges related to brain penetration and potential neurotoxicity remain significant hurdles to their successful clinical application. Developing advanced genetic tools, such as RNA interference (RNAi) and CRISPR-based technologies, has opened up new avenues for therapeutic intervention. These approaches selectively target and reduce PrP^C expression, thereby preventing the formation and accumulation of PrP^{Sc}. The chapter also highlights the progress in clinical trials, such as the PrProfile trial for ION717, which tests a novel treatment based on an antisense oligonucleotide (ASO). As we look toward the future, the chapter underscores the need for a multifaceted approach to treating prion diseases. Furthermore, early detection methods,

M. L. Barreca (✉)
Department of Pharmaceutical Sciences, University of Perugia, Perugia, Italy
e-mail: maria.barreca@unipg.it

E. Biasini (✉)
Department of Cellular, Computational and Integrative Biology (CIBIO), University of Trento, Trento, Italy
e-mail: emiliano.biasini@unitn.it

G. Legname, F. Moda (eds.), *Biomarkers and Therapeutical Targets for Prion Diseases*, Subcellular Biochemistry 112, https://doi.org/10.1007/978-3-031-97055-9_5

innovative drug delivery systems, and collaborative interdisciplinary research efforts will be essential for translating scientific discoveries into practical clinical breakthroughs.

Keywords Prion diseases · Protein misfolding · Neurodegeneration · Therapeutic strategies · Drug discovery

Introduction to Prion Disease Therapeutics

Prion diseases are invariably fatal neurodegenerative disorders characterized by the misfolding of the normal cellular prion protein (PrP^C) into PrP^{Sc}, a disease-associated form. PrP^{Sc} is an infectious protein (prion), capable of replicating by templating the misfolding of PrP^C molecules. This group of disorders, which includes human conditions like Creutzfeldt–Jakob disease (CJD), fatal familial insomnia (FFI), and Gerstmann–Sträussler–Scheinker syndrome (GSS), as well as animal diseases such as bovine spongiform encephalopathy (BSE) and scrapie, presents considerable challenges to both public health and the livestock sector globally (Prusiner 1998). Despite extensive research, treating prion diseases has proven difficult due to complex diagnostic challenges, the complicated nature of disease progression, and the limited efficacy of existing therapeutic strategies (Barreca et al. 2018; Jurcau et al. 2024).

This chapter explores the varied landscape of prion disease treatments, from historical methods to the most advanced approaches currently under development. We begin by examining the primary challenges in treating prion diseases, such as significant diagnostic delays that hinder early intervention, the complex biological mechanisms underlying these disorders, and the limitations of existing treatment options. A deeper understanding of these obstacles is essential for developing more effective treatments and early disease detection methods, which are critical for managing and potentially halting disease progression. We then discuss therapeutic strategies using small molecules to tackle prion pathogenesis. We highlight approaches to inhibit prion replication, destabilize and remove PrP^{Sc}, and target PrP^C through novel paradigms, including recent advancements in cutting-edge computer-aided methods. We also examine immunotherapy, another promising research field that holds hope for prion disease treatment through both passive and active immunization strategies. Finally, we review advanced genetic approaches like RNA interference (RNAi), antisense oligonucleotides (ASOs), and other genetic tools that represent the forefront of research and offer significant promise to target the genetic and molecular bases of prion diseases directly. We conclude by considering future directions in prion disease therapeutics, discussing novel targets and strategies, and integrating multi-modal treatment approaches and their potential to provide innovative and effective solutions. The chapter emphasizes the urgent need and opportunity to advance treatment options, setting the stage for future research in this challenging field.

Current Challenges in Prion Disease Treatment

The fast progression of prion diseases poses a major challenge for therapeutic interventions, as these diseases typically have a long asymptomatic incubation period followed by rapid clinical decline once symptoms manifest. Such a quick progression leaves only a narrow window for therapeutic intervention, as considerable neurological damage is often already present at the time of diagnosis. Initial symptoms like cognitive decline, memory disturbances, and behavioural changes are non-specific (Llorens et al. 2017; Geschwind 2015; Brown and Mastrianni 2010). They can be misdiagnosed as other neurodegenerative disorders, further delaying accurate diagnosis and reducing the adequate timeframe for effective treatment (Zerr et al. 2024; Kishida et al. 2023). By the time prion diseases are typically diagnosed, extensive neurological damage caused by PrP^{Sc} accumulation has usually occurred, leading to irreversible neurodegeneration, synaptic damage, and neuronal loss. These outcomes significantly limit the effectiveness of therapies, which typically aim to slow disease progression rather than reversing existing damage. The precise molecular mechanisms driving the conversion of normal prion protein into its disease-causing form and its spread within the central nervous system (CNS) are not fully understood, complicating the development of targeted therapies that could interrupt or reverse the disease process (Spagnolli et al. 2020). Additionally, the blood–brain barrier (BBB) presents a significant obstacle in treating any brain disorder, including prion diseases, as many promising therapeutic agents cannot cross it in sufficient concentrations to achieve efficacy. Given these challenges, research continues to focus on developing early detection methods, understanding the fundamental mechanisms of prion propagation, and creating therapies that can effectively cross the BBB to intervene before extensive brain damage occurs.

The Intrinsic Biological Complexity of Prion Diseases Prion diseases could be considered monogenic disorders since a central molecular event causes them: the conformational conversion of PrP^{C} into PrP^{Sc}. However, the pathological mechanisms underlying prion diseases are complex and only partially understood (Casey and Sleator 2025; Biasini et al. 2012). PrP^{C} itself exhibits an incredible number of structural variations (Bizingre et al. 2024). The PrP gene (*PRNP*) is located on chromosome 20 and encodes a polypeptide that undergoes extensive post-translational modifications during its synthesis and trafficking to the cell surface (Büeler et al. 1993; Linden et al. 2008). These modifications include glycosylation at specific asparagine residues, typically at N-linked glycosylation sites at asparagine residues N181 and N197 (human PrP numbering) (Rudd 2002; Otvos Jr and Cudic 2002; Lawson et al. 2005). The heterogeneous glycosylation of PrP^{C} results in various glycoforms that affect its folding, cellular trafficking, and susceptibility to conversion into PrP^{Sc} (Atkinson 2004; Ermonval 2003). Differences in glycosylation patterns also play a role in determining the strain characteristics and species barrier of prion diseases. A critical disulphide bond between cysteine residues C179 and C214 is essential for maintaining the structural integrity of PrP^{C}, and disruption of this bond can lead to structural instability, promoting misfolding and aggregation

(Maiti and Surewicz 2001; Turk et al. 1988). PrP^C is attached to the cell membrane via a glycosylphosphatidylinositol (GPI) anchor at its C-terminus, which is essential for the protein localization to lipid rafts, membrane microdomains involved in cell signalling, and protein trafficking (Hegde and Rane 2003). The presence of the GPI anchor also plays a crucial role in the endocytic recycling of PrP^C and may influence the conversion dynamics of PrP^{Sc} (Puig et al. 2014). In its mature form, PrP^C can be cleaved by various proteases at specific sites, producing several C- and N-terminal truncated forms (Chen et al. 1995; Haigh et al. 2009). Some of these proteolytic forms, such as the C1 and C2 fragments, are believed to have neuroprotective roles and may be involved in specific signalling pathways (Walmsley et al. 2009; Mangé et al. 2004). For example, the balance between full-length PrP^C and its cleaved counterparts may be critical in regulating its physiological functions and could impact prion disease pathogenesis (Vanni et al. 2023; Haigh et al. 2009). Another PrP cleaved form, shed-PrP, is generated by proteases that cleave PrP^C close to its GPI anchorage site at the C-terminus (Parizek et al. 2001; Taylor et al. 2009). The resulting soluble protein is found in the extracellular environment and can circulate in body fluids such as blood and cerebrospinal fluid (CSF). The proteolytic cleavage that leads to shed-PrP production is primarily mediated by a family of proteases known as ADAMs ("a disintegrin and metalloproteinase domain proteins"), with ADAM10 identified as a key enzyme (Taylor et al. 2009). This proteolytic activity can vary depending on the physiological and pathological state of the tissue, suggesting a regulated process influenced by cellular contexts and signals (Altmeppen et al. 2015). In the context of prion diseases, the role of shed-PrP is complex and still uncertain. While it lacks the GPI anchor necessary for certain interactions that promote prion replication, its soluble nature might influence the distribution and aggregation of pathological PrP^{Sc} (Puig et al. 2019; Linsenmeier et al. 2021). Understanding shed-PrP dynamics is therefore crucial for several reasons. Varying levels of shed-PrP across different prion disease states suggest that its detection in biological fluids could serve as a biomarker for diagnosing or monitoring prion disease progression. Additionally, manipulating the shedding process may offer new avenues for therapeutic intervention (Mohammadi et al. 2023; Linsenmeier et al. 2018). Further research is needed to delineate the precise mechanisms through which shed-PrP influences cellular functions and its exact role in prion pathogenesis. However, advances in these areas could pave the way for novel diagnostic and therapeutic strategies targeting shed-PrP production or function in prion diseases.

Another fundamental layer of complexity in prion disease research is represented by the pathological conversion of PrP^C into PrP^{Sc}, as the precise mechanism of such conversion remains elusive. This process represents the first and most well-established example of protein-mediated infectivity, with the misfolded PrP^{Sc} acting as a template and a catalyst for converting PrP^C into additional PrP^{Sc} molecules (Spagnolli et al. 2020). While the exact molecular details of the conversion process are not fully understood, it is known to involve a significant structural rearrangement that occurs through a direct interaction between PrP^C and PrP^{Sc}. The prevailing model, known as "template-assisted conversion," suggests that PrP^{Sc} serves as a template upon which the normal PrP^C is refolded into the pathogenic conformation

(Krauss and Vorberg 2013; Prusiner et al. 1998). This mechanism is thought to occur at the cell surface or within cellular compartments such as the endoplasmic reticulum or endosomes, where PrP^{C} and PrP^{Sc} can interact (Porto-Carreiro et al. 2005). Another, not alternative, model proposes that prion replication involves a nucleation phase, where a critical mass of PrP^{Sc} molecules forms a stable seed that then rapidly elongates by converting and incorporating PrP^{C} molecules into its structure (Gajdusek 1994; Carbonell et al. 2018). The growth of the prion fibril may occur through this repeated addition, which can eventually lead to the fibril breaking into smaller seeds, spreading the infection (Baxa 2008). Prion strains, which exhibit different biochemical and biological properties and cause different disease phenotypes, represent diverse conformational states of PrP^{Sc} (Weissmann 2009). These strains may also show varying abilities to infect different species, a phenomenon known as the species barrier (Priola 1999). The specific conformation of PrP^{Sc} is thus believed to dictate the disease phenotype and its ability to convert PrP^{C} from other species. Beyond the known complexity of PrP^{C} and PrP^{Sc} biology, prion diseases likely result from an intricate interplay of protein dynamics, physiological processes, and systemic interactions, most of which remain poorly understood. Understanding these complex aspects of prion diseases appears crucial for developing effective treatments.

Diagnostic Challenges The diagnosis of prion diseases is notoriously difficult, with a definitive diagnosis often possible only through post-mortem examination of brain tissue. Early in the disease, symptoms such as rapid-onset dementia, ataxia, and myoclonus are usually indistinguishable from those of other neurodegenerative disorders, complicating early and accurate diagnosis (Weber et al. 1997). The landscape of prion disease diagnostics has evolved substantially over the past decades, acquiring sophisticated biochemical and imaging techniques. Currently, diagnostic tools include magnetic resonance imaging (MRI) (Mattoli et al. 2024), which can reveal characteristic brain changes; CSF tests for markers like 14-3-3 and Tau proteins (Altuna et al. 2022; Satoh 2022), which are indicative of prion disease but not exclusive to it since they reflect neuronal damage rather than specific prion-induced effects; and the real-time quaking-induced conversion (RT-QuIC), which introduced an unprecedented level of specificity and sensitivity (Atarashi et al. 2011). Despite significant advancements in the clinical diagnosis of prion diseases, their complexity and rapid progression continue to challenge current diagnostic capabilities, underscoring the urgent need for novel biomarkers that can effectively support early diagnosis and therapeutic assessments. Recent progress has led to the identification of novel blood-based biomarkers, particularly brain-derived tau (BD-tau) and phospho-tau217 (p-tau217), which promise non-invasive CJD diagnosis and monitoring (Bentivenga et al. 2024), although their specificity still needs to be fully elucidated. Studies have also explored the diagnostic capabilities of 14-3-3, t-tau, and α-synuclein in differentiating genetic prion diseases from other neurological conditions (Schmitz et al. 2022). These biomarkers were evaluated for their seeding activity through the RT-QuIC conversion assay and compared with those obtained in healthy and neurological controls. Seeding activity assays typically involve introducing pre-formed amyloid fibrils or aggregates to a solution of monomeric proteins to accelerate fibril formation and mimic the nucleation-dependent polymerization process. BD-tau, reflective of neurodegeneration or neuronal injury, showed increased levels in CJD patients, correlating with disease progression and survival rates (Bentivenga et al. 2024). When combined with p-tau217, BD-tau plasma levels matched the diagnostic accuracy of traditional CSF markers like 14-3-3 for CJD, positioning it as a potent non-invasive diagnostic tool. In addition to diagnostic utility, plasma BD-tau levels correlate with patient survival, surpassing the performance of other plasma biomarkers like t-tau and NfL. These data suggest that BD-tau could serve as a diagnostic biomarker and a prognostic indicator in clinical settings, potentially guiding treatment decisions and monitoring disease progression. While traditional CSF markers such as t-tau and 14-3-3 proteins have good diagnostic accuracy for typical CJD subtypes, the requirement for CSF collection limits their widespread use. Conversely, the BD-tau and p-tau217 offer a less invasive method with comparable accuracy, making them particularly valuable in non-specialized medical settings and for preliminary screenings. Further research and validation are necessary to integrate these biomarkers into routine clinical practice, but their potential to improve patient outcomes remains promising.

Small Molecule Strategies to Halt Prion Pathogenesis

Strategies to halt prion pathogenesis using small molecules exploit various aspects of prion biology, replication cycle, and effects on the CNS (Uliassi et al. 2023). One key approach involves inhibiting prion replication by stopping the amplification of PrP^{Sc} after its formation. This strategy historically focused on developing compounds that can directly interfere with the replication process. For instance, one strategy uses small molecule inhibitors that bind to PrP^{Sc}, thereby preventing its interaction with PrP^{C} and its subsequent templating effect. Examples include compounds like **doxycycline** (Forloni et al. 2002a) and **suramin** (Ladogana et al. 1992), which have shown some efficacy in laboratory settings. Other strategies target the cellular machinery or environmental factors that facilitate prion replication, such as metal ions or pH, influencing prion stability and replication rates (Legname 2023). Enhancing the degradation and clearance of PrP^{Sc} from neural tissue is also a promising avenue. This includes drugs that stimulate autophagy or antibodies that specifically recognize and bind to PrP^{Sc}, acting as anti-prion agents by tagging the protein for degradation (Trevitt 2006). A primary preventive strategy to tackle prion diseases is stopping the disease at its earliest stages by ensuring that PrP^{C} does not undergo pathological conversion. Approaches include compounds that bind to PrP^{C} and stabilize its conformation (Mallucci and Collinge 2005; Iraci et al. 2015; Rigoli et al. 2019; Elezgarai and Biasini 2016). One example is molecule **GN8**, which has demonstrated potential by binding to PrP^{C} and stabilizing its native structure, reducing its susceptibility to misfolding (Kuwata et al. 2007). Most of these strategies, alone or in combination, represent the forefront of research in combating prion diseases. Each addresses different stages of the disease process, from the initial formation of PrP^{Sc} to its spread and the resulting neurodegeneration, highlighting the need for a multifaceted approach to tackle these complex diseases effectively (Table 5.1). Further details are provided in the following sections, which delve deeper into specific small molecules (Fig. 5.1).

Targeting Prion Replication One of the most direct approaches to tackle prion diseases is to inhibit the replication of the pathogenic form PrP^{Sc}. Research has identified various compounds that show potential to achieve this goal in lab settings. Pentosan polysulfate (PPS), a heparin-like compound, has demonstrated some ability to inhibit prion replication in various experimental models, although its mechanism is not fully understood (Diringer and Ehlers 1991; Caughey and Raymond 1993; Ladogana et al. 1992). Despite promising in vitro and in vivo results, PPS faces significant challenges in clinical applications (Todd et al. 2005). One major limitation is its inability to efficiently cross the BBB, primarily due to its large molecular size and negative charge, in addition to poor absorption rate from the gastrointestinal tract. Although invasive administration methods such as intraventricular or intrathecal delivery allow PPS to reach the CNS, these approaches are not ideal for long-term treatment. Moreover, although observational studies suggested extended survival in some cases, no direct symptomatic relief was observed, and the treatment was undermined by numerous complications (Tsuboi et al. 2009).

Another group of compounds studied for their anti-prion properties includes the antimalarial drugs **quinacrine** and **chloroquine**. These agents have been found to inhibit PrP^{Sc} formation in infected cell cultures, likely by interfering with the conformational change required to convert the normal PrP^{C} into the misfolded PrP^{Sc} (Doh-ura et al. 2000; Korth et al. 2001). Their ability to cross the BBB and exhibit anti-prion effects in vitro initially made them promising candidates for treating prion diseases. Quinacrine underwent clinical evaluation in two studies: the PRION1 open-label trial in the UK and a double-anonymized, placebo-controlled trial in the US (Collinge et al. 2009). Unfortunately, although some patients reported early symptomatic relief, neither trial demonstrated a significant survival benefit. Several factors may account for this failure, including inadequate CNS drug concentrations, limited ability to target prion reservoirs, or differences in prion strain susceptibility. Additionally, these drugs are associated with significant side effects, such as potential retinal toxicity and gastrointestinal disturbances, which further complicate their clinical utility, especially given the absence of proven therapeutic benefits (Mead and Tagliavini 2018).

The most remarkable example of drug discovery and development in prion diseases is represented by 2-aminothiazoles, a class of compounds identified through cell-based screens for their anti-prion properties. Notably, two molecules of this group, **IND24** and **IND81**, have shown promise by significantly extending the survival times of scrapie-infected mice in preclinical studies (Silber et al. 2013). However, their therapeutic potential was complicated by the emergence of drug-resistant prion strains during treatment. For instance, mice infected with Rocky Mountain Laboratory (RML) prions and treated with IND24 initially exhibited

Table 5.1 Therapeutic features of the discussed anti-prion compounds

Small molecule	Primary mechanism of action	Efficacy in models	Main limitations
Pentosan Polysulfate (PPS)	Heparin mimetic—competes with endogenous heparan sulfate proteoglycans, inhibiting prion aggregation and propagation	Effective in cells, prolonged survival in rodents, and mixed outcomes in humans	Does not cross BBB, requires invasive administration, and has inconsistent efficacy
Quinacrine	Antimalarian agent—binds to PrPC and inhibits conversion to PrPSc	High efficacy in cells, no effects in rodents, and no significant improvements in humans	Lack of clinical efficacy, pharmacokinetics issues, and significant side effects
2-Aminothiazoles (e.g., IND24)	Stabilizes PrPC and interferes with PrPSc aggregation	Effective in reducing PrPSc in cells, extends survival in some murine models, and not reported in humans	Development of drug-resistant strains and strain-dependent efficacy
Rapamycin	mTOR inhibitor: promotes PrPSc autophagic degradation	Reduces PrPSc in cells, extends survival in rodents, and not reported in humans	Long-term use concerns for immunosuppressive and metabolic side effects
ARN1468	SERPINA3 inhibitor: enables proteases to clear prion aggregates	Reduces prion accumulation in cells and not reported in humans	Poor bioavailability, low BBB penetration, and high plasma clearance
GN8	Binds and stabilizes PrPC's native conformation to prevent misfolding	Prevents new and reduces existing PrPSc aggregates in cells and animals and not reported in humans	Pharmacokinetics insufficiently investigated
SM875	Stabilizes PrPC folding intermediates to promote protein degradation	Lowers PrPC levels in cells, inhibits prion replication, promotes degradation, and not reported in humans	Not yet fully optimized for in vivo use
Zn(II)-BnPyP	Zinc complex: binds and destabilizes native PrPC, promoting its degradation	Anti-prion effects in cells and brain cultures and not reported in humans	Poor BBB permeability and potential toxicity due to metal-ion interactions
Flupirtine	Modulates glutamatergic signalling and affects PrPSc aggregation	Reduced aggregation and neurotoxicity in cells, extended survival in animals, and reduced cognitive decline in humans	Limited clinical evidence and potential for hepatotoxicity
Doxycycline	Tetracycline antibiotic: Binds to PrPSc, reducing its aggregation and protease-resistance	Reduces PrPSc accumulation and prevents aggregation in cells, prolongs survival in rodents, and has mixed results in humans	No clear survival benefit in trials and higher doses may be required for efficacy

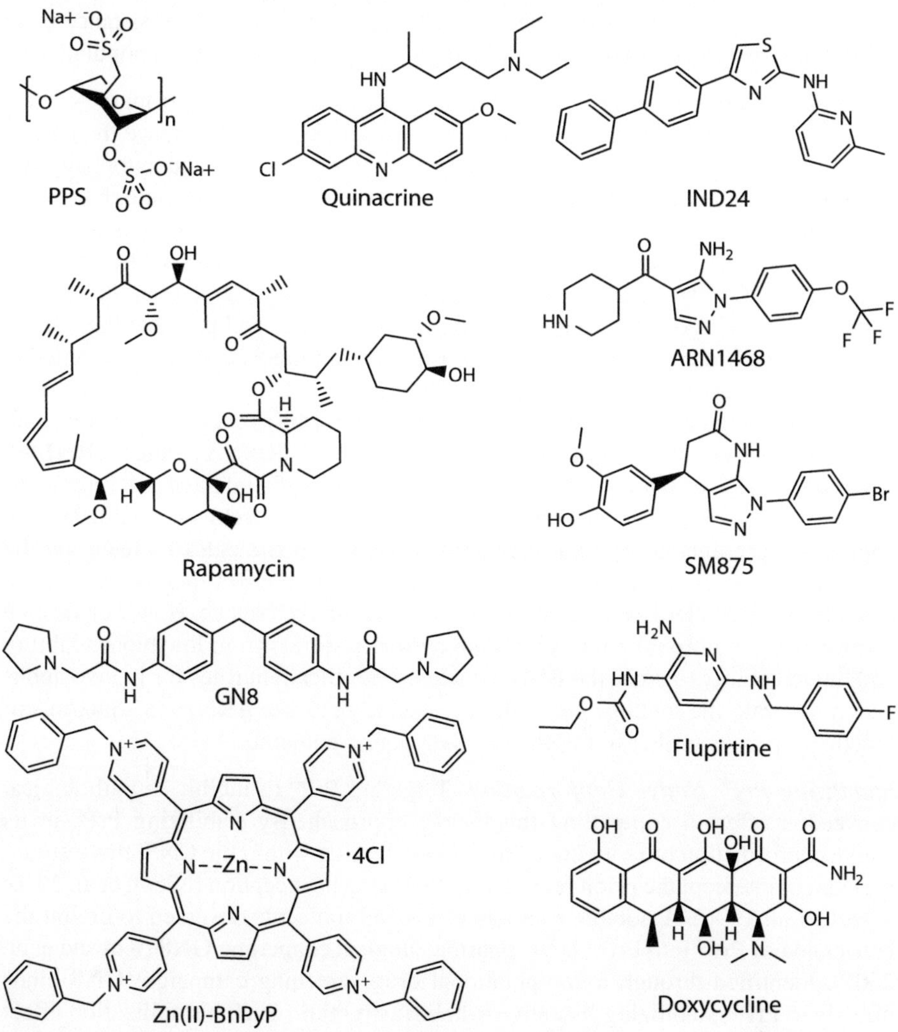

Fig. 5.1 Chemical structures of the discussed anti-prion compounds

prolonged survival but eventually developed neurological dysfunction over time due to a drug-resistant prion strain named RML[IND24] (Berry et al. 2013). The results underscore the complex interplay between prion strains and therapeutic agents, where resistance can significantly impede the effectiveness of treatments. The case of **IND24** offers valuable insights into the dynamics of therapeutic resistance and highlights the need for innovative treatment regimes. The findings suggest that monotherapies targeting PrP^{Sc} may be insufficient for treating prion diseases, as they impose selective pressures that drive the emergence of resistant strains. Instead, intermittent combination therapy with mixtures of different anti-prion compounds with a different mechanism of action might be required to manage the disease effectively (Burke et al. 2020). The approach could circumvent the resistance issue by altering the selection pressures exerted on the specific prion strain populations.

Destabilizing and Clearing PrP^{Sc} Clearing the accumulated PrP^{Sc} from the CNS is another viable option for treating prion diseases. Strategies that enhance the natural degradation pathways, such as the ubiquitin-proteasome system and autophagy, are under investigation. For instance, studies have shown that boosting autophagy with drugs like **rapamycin** can reduce PrP^{Sc} levels in infected neuronal cells and mice (Cortes et al. 2012). Another avenue of research involves using specific chemical chaperones that can refold misfolded proteins. Specific chaperones have been explored for their potential to destabilize PrP^{Sc} aggregates and promote their clearance from the brain (Tatzelt et al. 1996). Recently, an innovative strategy aimed at treating prion diseases by targeting the SERPINA3 protein, a protease previously found upregulated in several prion disease settings (Barbisin et al. 2014; Vanni et al. 2017). A newly developed small molecule inhibitor of SERPINA3 called **ARN1468** was reported to effectively reduce prion loads in chronically infected cell lines without interacting directly with PrP^{C} or PrP^{Sc} (Colini Baldeschi et al. 2022). The approach represents an original paradigm compared to strategies focusing on the two isoforms as molecular targets, potentially offering a novel therapeutic avenue that avoids the challenges of prion strain variability and resistance. However, despite its promising in vitro efficacy, **ARN1468** currently suffers from low bioavailability and limited ability to cross the BBB, which are significant hurdles for in vivo application. Overall, the findings highlight the need to optimize ARN1468's pharmacokinetic properties further to improve its therapeutic potential.

Stabilizing PrP^{C} Native Conformation Targeting PrP^{C} to inhibit its pathological conversion offers a compelling therapeutic approach. By stabilizing PrP^{C} in its native conformation or reducing its overall expression within the CNS, these strategies aim to intercept the prion replication cascade at its inception (Nicoll et al. 2010; Tatzelt et al. 1996). A notable example of a small molecule designed to inhibit the conversion of PrP^{C} into PrP^{Sc} is the pharmacological chaperone **GN8** (Kuwata et al. 2007). Identified through a computational drug screening campaign, **GN8** binds directly to PrP^{C}, stabilizing its native α-helical structure. Such a stabilization effect is believed to prevent the conformational change necessary for PrP^{C} to convert into PrP^{Sc}. The precise binding site of **GN8** on PrP^{C} has been a subject of study, as understanding this interaction at the molecular level is critical for optimizing its efficacy and specificity. Preclinical studies, primarily in cell culture and animal models, have shown that **GN8** can effectively reduce the levels of PrP^{Sc} (Kamatari et al. 2013). In these models, treatment with GN8 prevented the formation of new PrP^{Sc} and appeared to reduce the accumulation of existing PrP^{Sc} aggregates. This dual effect suggests that **GN8** may halt the progression of prion diseases and mitigate some of the pathological features associated with these disorders, although the latter claim requires further validation (Barreca et al. 2018). Despite its promising in vitro and in vivo efficacy, **GN8** faces several challenges that must be addressed before it can be considered a viable therapeutic agent. One major issue is its pharmacokinetics, which has not been thoroughly investigated. Similarly, the long-term effects and safety of **GN8** have not been fully established. Like any compound affecting protein folding, there is a risk of off-target effects or unintended interactions with other cellular proteins, necessitating extensive safety testing (Biasini 2019). Future research should focus on further refining the chemical scaffold of **GN8** to improve its efficacy, selectivity, and pharmacokinetic properties.

Promoting PrP^{C} Degradation **SM875** is another small molecule derived from computational efforts currently being explored for its potential as a therapeutic agent in prion diseases (Spagnolli et al. 2021). **SM875** operates through an unusual mechanism by interfering with the protein's folding process. The compound was identified through a novel drug discovery paradigm called Pharmacological Protein Inactivation by Folding Intermediate Targeting (PPI-FIT). The technique aims to reduce protein levels by selectively targeting intermediates appearing along the folding process. The rationale of PPI-FIT is that by stabilizing these intermediates pharmacologically, they can be marked for degradation by the cellular quality control machinery. Indeed, cells possess designed mechanisms to identify and degrade misfolded proteins or unsuccessful folding attempts, thus preventing the accumulation of dysfunctional or toxic proteins. Small molecules can make these intermediates resemble misfolded proteins by extending their lifespan and impeding the completion of the folding process. Such an artificial stabilization triggers the cellular quality control system to recognize the protein folding intermediate for degradation, thereby reducing the overall level of the target polypeptide. Implementing PPI-FIT involves identifying potential binding sites that are unique to the folding intermediates and not present in the native structure of the protein. The structure of an on-fold PrP intermediate was predicted using computational methods, and then a virtual screening campaign identified molecules binding selectively to this conformer. **SM875** emerged as the most effective compound among the positive hits in reducing PrP^{C} expression in different cells. Notably, the compound was also shown to selectively lower PrP loads without affecting the levels of other cellular proteins, highlighting its specificity. Additionally, **SM875** inhibited prion replication, making it a promising candidate for therapeutic development against prion diseases. The compound provides strong experimental support for targeting protein folding intermediates, which can be a novel, effective, and selective strategy for drug development, particularly for diseases involving protein misfolding and aggregation, like prion diseases.

Another mechanism for targeting PrP^{C} to degradation has been identified thanks to a tetracationic porphyrin compound called **Zn(II)-BnPyP** (Masone et al. 2023). The molecule shows a dual action against prion diseases by binding to different domains of PrP^{C}. Such a complex binding results in a strong destabilization of the native PrP^{C} fold, promoting its endocytosis and lysosomal degradation and thus reducing the substrate available for PrP^{Sc} formation. Zn(II)-BnPyP has shown robust anti-prion effects across various prion strains in neuronal cells and organotypic brain cultures. Like the other strategies targeting PrP^{C}, this approach could circumvent the issue of drug resistance arising from the prion strain phenomenon. While problems remain with moving Zn(II)-BnPyP to the clinical setting due to its pharmacokinetic properties, including poor BBB permeability, further optimization of this compound represents an interesting opportunity for treating prion diseases.

Additional Small Molecules Tested in Clinical Settings Two pharmacological interventions have been explored through clinical trials with varied outcomes. These include **flupirtine**, a triaminopyridine compound that demonstrated in vitro neuroprotection against damage from amyloid beta peptides and prion protein fragment-induced damage (Perovic et al. 1995). Clinically, **flupirtine** was evaluated for its potential to mitigate cognitive deterioration in CJD patients (Otto et al. 2004). In the small trial, 13 participants receiving **flupirtine** experienced a slower progression of dementia compared to 15 placebo recipients. However, the treatment failed to improve survival rates, limiting its clinical applicability.

Doxycycline, a tetracycline antibiotic known for its effective penetration of the BBB and its potential to inhibit PrP aggregation and reduce PrP^{Sc} protease resistance, showed promise in preclinical studies (Forloni et al. 2002b). Findings from a long-term preventive treatment trial in FFI patients are awaited in 2025, which may offer valuable insights into its therapeutic utility.

Immunotherapy for Prion Diseases

Immunotherapy is renowned for its targeted efficacy and minimal side effects, positioning it as a promising treatment for a variety of challenging, incurable conditions, including neurodegenerative diseases. Significant research efforts have been directed towards developing immunotherapies for prion-like disorders such as Alzheimer's and Parkinson's (Vroom and Dodart 2024). Although some clinical trials have yielded disappointing results and raised concerns about adverse effects, re-evaluated data from Aducanumab trials, a monoclonal antibody against Aβ aggregates, have shown consistent improvements in patient outcomes (Medel Sánchez et al. 2024). These data support continued exploration of immunotherapy as a feasible treatment for Alzheimer's disease.

Compared to other late-onset neurodegenerative disorders, prion diseases present distinct advantages for immunotherapy applications. The pathogenic misfolded PrP^{Sc} is a well-recognized causative agent, making it an appropriate target. Additionally, since the conversion of PrP^{C} into PrP^{Sc} occurs on cell surfaces or within the endocytic pathway, these proteins are readily accessible to therapeutic agents. Pharmacological strategies should ideally target PrP^{Sc} to prevent potential adverse effects associated with binding to PrP^{C} regions or interfering with essential cellular functions. Efforts to develop small molecules targeting PrP^{Sc} have shown promise, but this approach faces challenges due to subtle structural variations among different prion strains or species. It is known that minor amino acid differences, sometimes only one or two, can significantly hinder prion transmission, highlighting structural variations at crucial sites that facilitate seeding activity. Nonetheless, there is a consensus that all PrP^{Sc} molecules share a fundamental architecture (Manka et al. 2023). This is supported by the substantial similarity across mammalian PrP structures, the prevalent β-sheet composition in PrP^{Sc}, their strong resistance to protease at the C-terminal, and insights from structural analyses of PrP^{Sc}. Hence, an effective therapeutic agent with broad anti-prion capabilities would require engaging a substantial portion of PrP^{Sc} while accommodating minor structural differences across various prion strains. Compared to small molecules, antibodies are often regarded as more effective for treating prion diseases (Napper and Schatzl 2023a). Antibodies typically bind to a linear epitope encompassing several amino acids and can generally tolerate minor variations, such as one or two amino acids. When the epitope is conformational, the interaction between the antibody and the antigen spans a broader area, enhancing the resilience of the antibody-PrP^{Sc} interaction to structural discrepancies. This makes antibodies, especially in a vaccine formulation, potentially effective, as they can evoke a polyclonal response aimed at numerous epitopes on PrP^{Sc}, reducing the impact of minor structural differences on overall efficacy.

Passive Immunization Strategies The development of PrP knockout mice and the aspiration to create PrP^{Sc}-specific antibodies has produced numerous anti-PrP antibodies. However, none have yet robustly demonstrated specificity solely for PrP^{Sc}. Research has focused on the idea that antibodies binding to normal PrP^{C} can inhibit its conversion into PrP^{Sc}, potentially offering a strategy to combat prion replication and disease. The ability of antibodies to block PrP^{C} conversion has primarily been assessed in vitro with prion-infected cell cultures. Specifically, the **6H4 antibody**, targeting residues 144–152, successfully removed PrP^{Sc} from infected neuroblastoma (N2a) cells (Enari et al. 2001). PrP-specific fragment antigen-binding regions, such as **D13** (targeting residues 95–103) and **D18** (targeting residues 132–156), have efficiently cleared PrP^{Sc} from cells (Moroncini et al. 2004). Comprehensive screenings of extensive antibody panels have identified several antibodies effective in reducing PrP^{Sc} levels in these cell models, suggesting they can bind to various PrP epitopes and possibly work via multiple mechanisms (Biasini et al. 2008a, b). In vivo tests using prion-infected mice have further evaluated the effectiveness of these anti-PrP antibodies. Weekly intraperitoneal injections of antibodies **8B4** (targeting residues 34–52) or **8H4** (targeting residues 175–185) prolonged the lifespan of CD-1 mice inoculated with the 139A prion strain (Pan et al. 2004). Substantial therapeutic effects were achieved with the **ICSM18** (targeting residues 146–159) or **ICSM35** (targeting residues 91 to 110) antibodies (Reilly et al. 2022). Administering these macromolecules twice weekly intraperitoneally enabled mice inoculated with the RML prion strain to survive over 500 days without showing disease symptoms. However, these treatments were ineffective if initiated during the later stages of the infection or after intracerebral prion inoculation, underscoring the inability of these antibodies to traverse BBB and halt CNS progression of the disease. Short-term antibody treatments have also been explored. The **6D11** anti-PrP antibody (targeting residues 97–100) was tested in a 4-week or 8-week trial upon intraperitoneal injection, showing lifespan extension in mice infected with the 22L prion strain (Sadowski et al. 2009). Other antibodies were investigated for their effects upon intraventricular infusions starting at various intervals post-infection. Although only modestly effective, these treatments extended survival, demonstrating that anti-PrP antibodies can influence the course of prion disease within the CNS (Jeong et al. 2012; Song et al. 2008).

Active Immunization Attempts Specific PrP peptides, such as **PrP131–150** and **PrP211–230** containing MHC-I-binding motifs, have been used as immunogens to elicit strong immune responses in wild-type mice, reducing levels of protease-resistant PrP^{Sc} (Souan et al. 2001). Strategies like loading bone marrow-derived dendritic cells with specific PrP peptides have also proven effective in extending survival in prion-infected mice (Bachy et al. 2010). Other approaches include linking PrP or its peptides to immune-stimulatory molecules, such as the bacterial heat shock protein Hsp70 homolog DnaK (Koller et al. 2002) or keyhole limpet hemocyanin (KLH), to enhance immune response and survival in experimental models (Matsushita et al. 1998). Other strategies have included using CpG adjuvants combined with recombinant deer or mouse PrP as immunogens, demonstrating the ability to extend survival and break self-tolerance in models of chronic wasting disease (CWD) in transgenic mice (Abdelaziz et al. 2018). DNA vaccines enhancing immune responses by encoding heterologous PrP have also been explored, although their effectiveness in preventing prion infection remains unconfirmed (Fernandez-Borges et al. 2006). All these active immunization strategies generally target the normal host-encoded PrP^{C}, stabilizing it to prevent conversion to the pathogenic PrP^{Sc}. Nevertheless, ideal targets would be the structurally distinct neo-epitopes in PrP^{Sc}, as the immune system might recognize them as foreign, potentially leading to an effective immune response. Research has shown that specific PrP-specific antibodies are present late in prion disease, indicating an immune response against aberrantly folded PrP molecules (Senatore et al. 2020). Vaccines targeting specific PrP^{Sc} epitopes have been developed to induce sustained immune responses in animals (Napper and Schatzl 2023b; Ma and Ma 2020). However, the protective efficacy of these vaccines, particularly in larger animal models such as deer, remains uncertain, with some studies suggesting that vaccination could negatively alter disease susceptibility and progression (Bremer et al. 2009; Wood et al. 2018). These conclusions underscore the complexity and the need for further research in developing effective immunization strategies against prion diseases.

Potentials and Limits of Immunotherapy in Prion Diseases Advances in immunological strategies for tackling prion disease have been substantial. Nevertheless, creating a successful immunological approach is hindered by several factors, including the safety of treatments, the BBB's poor permeability, and the complexity of developing therapies that target PrP^{Sc} specifically. Delivering some anti-PrP antibodies to the CNS has been linked to neurotoxic effects (Reimann et al. 2016; Lefebvre-Roque et al. 2007). A previous comprehensive analysis revealed that specific antibodies targeting the globular domain of PrP induced neurotoxic effects, particularly at higher dosages (Reimann et al. 2016). Conversely, antibodies recognizing the octarepeat region showed no toxicity and were protective against neurotoxicity induced by other antibodies. These findings suggest that neurotoxicity from anti-PrP antibodies depends on the targeted epitope and dosage, necessitating careful evaluation in therapeutic applications. For passive immunotherapy to be effective, antibodies must be meticulously characterized to avoid adverse effects. Active immunization strategies typically produce a polyclonal antibody response, which might inadvertently target neurotoxic epitopes. One approach to circumvent this issue is using immunogens that differ conformationally from PrP^{C} but are similar to the pathogenic PrP^{Sc}, directing the immune response more specifically toward the pathogenic form and reducing the likelihood of neurotoxicity. Potential immunogens like recombinant PrP amyloid fibrils and multimeric PrP forms have shown promise in preclinical studies and should be explored further. The large size of conventional antibodies (~150 kDa) complicates their ability to cross the BBB, posing a significant challenge for therapeutic delivery to the CNS. Technologies like receptor-mediated transport and viral vector-based gene delivery are being developed to facilitate this process (Marciniuk et al. 2016). Studies have shown that targeted delivery of antibody fragments via vectors can prolong survival in prion-infected models and reduce PrP^{Sc} levels effectively (Wuertzer et al. 2008). Despite efforts, no antibodies have been confirmed specifically targeting PrP^{Sc} without reacting with PrP^{C} or other misfolded protein species (Stravalaci et al. 2016; Tapella et al. 2013). Advances in structural analysis of PrP^{Sc} might enable the development of such antibodies. Alternatively, nanobodies, which are small and preferentially bind conformational epitopes, could be engineered to recognize PrP^{Sc} specifically. These small, single-gene-encoded antibodies could be adapted via gene therapy vectors for CNS delivery. Advancing passive and active immunization strategies, refining delivery methods, and focusing on PrP^{Sc}-specific targets could potentially lead to successful immunotherapies for prion diseases (Rovis and Legname 2014). Additionally, given the involvement of PrP^{C} in other diseases (Biasini and Harris 2012; Manni et al. 2020), these immunotherapies might have broader applications, suggesting a promising future for anti-PrP immunological strategies.

Genetic Tools to Treat Prion Diseases

Techniques like RNAi, ASOs, and other cutting-edge biotechnological paradigms discussed below can effectively reduce the overall levels of PrP^{C} available for conversion. Lower expression levels of PrP^{C} directly translate into reduced substrate availability for conversion to PrP^{Sc}, thereby slowing or halting disease progression. CRISPR/Cas9 and other newly developed genetic technologies offer a long-term solution by potentially editing the PRNP gene in germ cells or somatic cells, thus providing resistance to prion diseases (Mehrabian et al. 2014; Castle et al. 2022). These strategies aim to address the fundamental biological mechanisms of prion diseases, offering potential routes to mitigate or reverse the pathological processes. Researchers have experimented with vectors such as adeno-associated viruses (AAVs) to deliver nucleic acids that either suppress PRNP expression or introduce mutations that halt the pathogenic conversion process. RNAi tools provide potent resources for reducing or modifying the expression of specific genes associated with disease states, including prion diseases. Several studies have directly shown that silencing the PRNP gene in infected mouse models leads to prolonged survival and delayed disease onset, highlighting the potential efficacy of this strategy (White et al. 2008; Lehmann et al. 2014).

Advanced Genetic Tools against Prion Diseases A recent report introduced an advanced therapeutic strategy to tackle prion diseases (Gentile et al. 2024) through divalent small interfering RNA (di-siRNA) molecules (Alterman et al. 2019). The uniqueness of this approach lies in developing "divalent" siRNA structures binding to two distinct sites on the PrP's mRNA. This dual-targeting feature is designed to enhance the efficiency and specificity of mRNA recognition and cleavage, ensuring more effective gene silencing with reduced risk of off-target effects. In vivo experiments in mice showed that treatment with divalent siRNAs significantly reduced PrP^C levels in the brain. Notably, the treated animals also showed a marked delay in the onset of prion disease symptoms and a substantial extension in survival times compared to control groups treated with conventional siRNA or placebo. Such an enhanced efficacy of di-siRNAs may result from more efficient recruitment of the RNA-induced silencing complex (RISC) and subsequent mRNA degradation. This could theoretically allow lower doses of siRNA to be used in therapeutic contexts, minimizing potential side effects associated with RNA therapies. Similar di-siRNA strategies could theoretically be developed to target mRNAs of other pathological proteins, potentially offering a new genetic tool in the fight against other neurodegenerative disorders.

Another recent study reported developing and applying a novel gene-editing technology called Coupled Histone tail for Autoinhibition Release of Methyltransferase (CHARM) to effectively reduce the expression of PrP^C across the brain (Neumann et al. 2024). The approach also uses a novel AAV vector to deliver the epigenetic editor directly into the brain. The editor is engineered to specifically recognize and bind to the PRNP gene, modifying the epigenetic marks that regulate its expression. The core innovation of this technology lies in the "compact" nature of the epigenetic editor. Traditional CRISPR-Cas systems, while powerful, are often too large to be efficiently packaged into AAV vectors, which are used for gene therapy due to their safety and ability to target nervous tissue. The engineered editor used in this study is small enough to fit within the AAV's packaging limits and has been optimized to avoid off-target effects that can lead to unintended gene modifications. The efficacy of this approach was assessed through comprehensive experiments showing a significant reduction of PrP^C expression throughout the brain. Importantly, the treatment was durable, with PrP^C silencing observed over extended periods post-administration. The ability to silence genes across the brain without requiring invasive procedures or systemic treatments that might cause widespread side effects represents a significant advancement in gene therapy.

The PrProfile Trial As this chapter was written, Ionis Pharmaceuticals announced the completion of PrProfile enrolment, its Phase 1/2a trial of **ION717**, a PrP-lowering ASO for treating prion diseases (NCT06153966). The PrProfile study is a global, multi-centre, early-phase clinical trial designed to evaluate the safety, tolerability, and pharmacokinetics of ION717 in individuals diagnosed with symptomatic prion disease. The trial, carried out by Ionis Pharmaceuticals in collaboration with leading neurological research institutions worldwide, represents a significant step forward to slow the disease progression and mitigate the devastating effects of prion disorders. ASOs are short, synthetic DNA or RNA molecules designed to bind to the target's mRNA. The binding effectively deactivates the mRNA, preventing the production of specific proteins. In the context of prion diseases, ASOs are engineered to reduce the production of PrP^{C} (Reidenbach et al. 2019). The trial is set up as a multi-centre, randomized, double-blind, placebo-controlled study to ensure rigorous evaluation of the ASO's effectiveness and safety. Participants undergo several spinal taps for the direct administration of ASOs into the CSF, aiming to maximize drug penetration into the CNS. The expected outcomes of the Ionis trial are multifaceted. Primarily, the trial seeks to demonstrate the safety profile of the ION717, evaluating side effects or potential complications arising from the therapy. Secondary objectives include assessing pharmacokinetics and exploring early indications of efficacy in slowing disease progression. The PrProfile trial is a landmark study in prion diseases, offering hope for a targeted therapeutic approach to a group of disorders that currently lack effective treatments. If successful, this trial could not only revolutionize the management of prion diseases but also potentially offer insights applicable to other protein misfolding diseases, such as Alzheimer's and Parkinson's diseases.

Future Directions in Human Prion Disease Therapeutics

The upcoming future of prion disease therapeutics presents several promising new strategies and a deeper understanding of prion biology that could transform the treatment landscape. Advances in molecular diagnostics and bioinformatics will likely drive early detection and intervention, key components that could change the course of these invariably fatal diseases (Rigoli et al. 2019). Moreover, innovative therapeutic approaches currently under exploration, such as targeted pharmacological therapies, immunomodulatory agents, and advanced gene-editing techniques, offer hope for effective interventions. A focal point of future research will be the development of treatments capable of effectively and safely crossing the BBB, a longstanding obstacle in neurodegenerative disease therapy. Nanotechnology and other drug delivery systems, such as encapsulated therapeutic agents and receptor-mediated transport mechanisms, are poised to play crucial roles in overcoming this obstacle. Furthermore, delving into the lifecycle and cellular interactions of prions may unveil new therapeutic targets. For instance, efforts to modulate the cellular pathways involved in the synthesis, trafficking, and degradation of PrP^{C} could provide innovative methods to prevent its pathological misfolding. Another promising direction involves harnessing the power of immunotherapy, which has shown some potential in selectively targeting disease-specific epitopes without affecting the function of the normal protein. Refining vaccine approaches and monoclonal antibodies to induce a robust and specific immune response against PrP^{Sc} could provide therapeutic and preventive benefits. Collaboration between genetic counselling and clinical interventions will be essential to effectively address the familial forms of prion disorders. Finally, as the research community continues to unravel the complex mechanisms underlying prion diseases, interdisciplinary approaches integrating neurobiology, molecular genetics, structural biology, and computational modelling will be essential. These efforts will enhance our understanding of prion pathogenesis and accelerate the translation of research findings into clinical applications.

In summary, the future of human prion disease therapeutics stands to benefit enormously from current scientific advancements, technological innovation, and ongoing trials. With a concerted effort towards elucidating the molecular underpinnings of prion diseases and developing innovative therapeutic strategies, there is strong hope that the coming years will bring transformative breakthroughs in managing and possibly curing these challenging neurodegenerative disorders.

References

Abdelaziz DH et al (2018) Recombinant prion protein vaccination of transgenic elk PrP mice and reindeer overcomes self-tolerance and protects mice against chronic wasting disease. J Biol Chem 293:19812–19822

Alterman JF et al (2019) A divalent siRNA chemical scaffold for potent and sustained modulation of gene expression throughout the central nervous system. Nat Biotechnol 37:884–894

Altmeppen HC et al (2015) The sheddase ADAM10 is a potent modulator of prion disease. elife 4:e04260

Altuna M, Ruiz I, Zelaya MV, Mendioroz M (2022) Role of biomarkers for the diagnosis of Prion diseases: a narrative review. Medicina (B Aires) 58:473

Atarashi R et al (2011) Ultrasensitive human prion detection in cerebrospinal fluid by real-time quaking-induced conversion. Nat Med 17:175–178

Atkinson P (2004) Glycosylation of prion strains in transmissible spongiform encephalopathies. Aust Vet J 82:292–299

Bachy V et al (2010) Mouse vaccination with dendritic cells loaded with prion protein peptides overcomes tolerance and delays scrapie. J Gen Virol 91:809–820

Barbisin M et al (2014) Gene expression profiling of brains from bovine spongiform encephalopathy (BSE)-infected cynomolgus macaques. BMC Genomics 15:434

Barreca M, Iraci N, Biggi S, Cecchetti V, Biasini E (2018) Pharmacological agents targeting the cellular Prion protein. Pathogens 7:27

Baxa U (2008) Structural basis of infectious and non-infectious amyloids. Curr Alzheimer Res 5:308–318

Bentivenga GM et al (2024) Clinical value of novel blood-based tau biomarkers in Creutzfeldt–Jakob disease. Alzheimers Dement 21. https://doi.org/10.1002/alz.14422

Berry DB et al (2013) Drug resistance confounding prion therapeutics. Proc Natl Acad Sci 110:E4160

Biasini E (2019) A designer chaperone against prion diseases. Nat Biomed Eng 3:167–168

Biasini E, Harris DA (2012) Targeting the cellular Prion protein to treat neurodegeneration. Future Med Chem 4:1655–1658

Biasini E et al (2008a) Non-infectious aggregates of the prion protein react with several PrP^{Sc}-directed antibodies. J Neurochem 105:2190–2204

Biasini E, Medrano AZ, Thellung S, Chiesa R, Harris DA (2008b) Multiple biochemical similarities between infectious and non-infectious aggregates of a prion protein carrying an octapeptide insertion. J Neurochem 104:1293–1308

Biasini E, Turnbaugh JA, Unterberger U, Harris DA (2012) Prion protein at the crossroads of physiology and disease. Trends Neurosci 35:92–103

Bizingre C et al (2024) Post-translational modifications in prion diseases. Front Mol Neurosci 17:1405415

Bremer J et al (2009) Repetitive immunization enhances the susceptibility of mice to peripherally administered Prions. PLoS One 4:e7160

Brown K, Mastrianni JA (2010) The Prion diseases. J Geriatr Psychiatry Neurol 23:277–298

Büeler H et al (1993) Mice devoid of PrP are resistant to scrapie. Cell 73:1339–1347

Burke CM, Mark KMK, Kun J, Beauchemin KS, Supattapone S (2020) Emergence of prions selectively resistant to combination drug therapy. PLoS Pathog 16:e1008581

Carbonell F, Iturria-Medina Y, Evans AC (2018) Mathematical modeling of protein misfolding mechanisms in neurological diseases: a historical overview. Front Neurol 9:37

Casey C, Sleator RD (2025) Prions: structure, function, evolution, and disease. Arch Microbiol 207:1

Castle AR, Wohlgemuth S, Arce L, Westaway D (2022) Investigating CRISPR/Cas9 gene drive for production of disease-preventing prion gene alleles. PLoS One 17:e0269342

Caughey B, Raymond GJ (1993) Sulfated polyanion inhibition of scrapie-associated PrP accumulation in cultured cells. J Virol 67:643–650

Chen SG et al (1995) Truncated forms of the human Prion protein in normal brain and in Prion diseases. J Biol Chem 270:19173–19180

Colini Baldeschi A et al (2022) Innovative non-PrP-targeted drug strategy designed to enhance Prion clearance. J Med Chem 65:8998–9010

Collinge J et al (2009) Safety and efficacy of quinacrine in human prion disease (PRION-1 study): a patient-preference trial. Lancet Neurol 8:334–344

Cortes CJ, Qin K, Cook J, Solanki A, Mastrianni JA (2012) Rapamycin delays disease onset and prevents PrP plaque deposition in a Mouse Model of Gerstmann–Sträussler–Scheinker disease. J Neurosci 32:12396–12405

Diringer H, Ehlers B (1991) Chemoprophylaxis of scrapie in mice. J Gen Virol 72:457–460

Doh-ura K, Iwaki T, Caughey B (2000) Lysosomotropic agents and cysteine protease inhibitors inhibit scrapie-associated prion protein accumulation. J Virol 74:4894–4897

Elezgarai SR, Biasini E (2016) Common therapeutic strategies for prion and Alzheimer's diseases. Biol Chem 397:1115–1124

Enari M, Flechsig E, Weissmann C (2001) Scrapie prion protein accumulation by scrapie-infected neuroblastoma cells abrogated by exposure to a prion protein antibody. Proc Natl Acad Sci 98:9295–9299

Ermonval M (2003) Evolving views in prion glycosylation: functional and pathological implications. Biochimie 85:33–45

Fernandez-Borges N et al (2006) DNA vaccination can break immunological tolerance to PrP in wild-type mice and attenuates Prion disease after intracerebral challenge. J Virol 80:9970–9976

Forloni G et al (2002a) Tetracyclines affect prion infectivity. Proc Natl Acad Sci 99:10849–10854

Forloni G et al (2002b) Tetracyclines affect prion infectivity. Proc Natl Acad Sci USA 99:10849–10854

Gajdusek DC (1994) Spontaneous generation of infectious nucleating amyloids in the transmissible and nontransmissible cerebral amyloidoses. Mol Neurobiol 8:1–13

Gentile JE et al (2024) Divalent siRNA for prion disease. Preprint at https://doi.org/10.1101/2024.12.05.627039

Geschwind MD (2015) Prion diseases. Continuum: Lifelong Learning in Neurology 21:1612–1638

Haigh CL et al (2009) PrPC-related signal transduction is influenced by copper, membrane integrity and the alpha cleavage site. Cell Res 19:1062–1078

Hegde RS, Rane NS (2003) Prion protein trafficking and the development of neurodegeneration. Trends Neurosci 26:337–339

Iraci N, Stincardini C, Barreca ML, Biasini E (2015) Decoding the function of the N-terminal tail of the cellular prion protein to inspire novel therapeutic avenues for neurodegenerative diseases. Virus Res 207:62–68

Jeong H-J et al (2012) Development of monoclonal antibodies against the abnormal prion protein isoform (PrP^{res}) associated with chronic wasting disease (CWD). J Vet Sci 13:429

Jurcau MC, Jurcau A, Diaconu RG, Hogea VO, Nunkoo VS (2024) A systematic review of Sporadic Creutzfeldt-Jakob disease: pathogenesis, diagnosis, and therapeutic attempts. Neurol Int 16:1039–1065

Kamatari YO, Hayano Y, Yamaguchi K, Hosokawa-Muto J, Kuwata K (2013) Characterizing anti-prion compounds based on their binding properties to prion proteins: implications as medical chaperones. Protein Sci 22:22–34

Kishida H, Ueda N, Tanaka F (2023) The advances in the early and accurate diagnosis of Creutzfeldt–Jakob disease and other prion diseases: where are we today? Expert Rev Neurother 23:803–817

Koller MF, Grau T, Christen P (2002) Induction of antibodies against murine full-length prion protein in wild-type mice. J Neuroimmunol 132:113–116

Korth C, May BCH, Cohen FE, Prusiner SB (2001) Acridine and phenothiazine derivatives as pharmacotherapeutics for prion disease. Proc Natl Acad Sci 98:9836–9841

Krauss S, Vorberg I (2013) Prions *ex vivo*: what cell culture models tell us about infectious proteins. Int J Cell Biol 2013:1–14

Kuwata K et al (2007) Hot spots in prion protein for pathogenic conversion. Proc Natl Acad Sci 104:11921–11926

Ladogana A et al (1992) Sulphate polyanions prolong the incubation period of scrapie-infected hamsters. J Gen Virol 73:661–665

Lawson VA, Collins SJ, Masters CL, Hill AF (2005) Prion protein glycosylation. J Neurochem 93:793–801

Lefebvre-Roque M et al (2007) Toxic effects of intracerebral PrP antibody administration during the course of BSE infection in mice. Prion 1:198–206

Legname G (2023) Copper coordination modulates prion conversion and infectivity in mammalian prion proteins. Prion 17:1–6

Lehmann S et al (2014) Systemic delivery of siRNA down regulates brain Prion protein and ameliorates neuropathology in Prion disorder. PLoS One 9:e88797

Linden R et al (2008) Physiology of the Prion protein. Physiol Rev 88:673–728

Linsenmeier L et al (2018) Structural and mechanistic aspects influencing the ADAM10-mediated shedding of the prion protein. Mol Neurodegener 13:18

Linsenmeier L et al (2021) Ligands binding to the prion protein induce its proteolytic release with therapeutic potential in neurodegenerative proteinopathies. Sci Adv 7:eabj1826

Llorens F, Zarranz J-J, Fischer A, Zerr I, Ferrer I (2017) Fatal familial insomnia: clinical aspects and molecular alterations. Curr Neurol Neurosci Rep 17:30

Ma Y, Ma J (2020) Immunotherapy against Prion disease. Pathogens 9:216

Maiti NR, Surewicz WK (2001) The role of disulfide bridge in the folding and stability of the recombinant human Prion protein. J Biol Chem 276:2427–2431

Mallucci G, Collinge J (2005) Rational targeting for prion therapeutics. Nat Rev Neurosci 6:23–34

Mangé A et al (2004) Alpha- and beta- cleavages of the amino-terminus of the cellular prion protein. Biol Cell 96:125–132

Manka SW, Wenborn A, Collinge J, Wadsworth JDF (2023) Prion strains viewed through the lens of cryo-EM. Cell Tissue Res 392:167–178

Manni G et al (2020) The cellular prion protein beyond prion diseases. Swiss Med Wkly 150:w20222

Marciniuk K, Taschuk R, Napper S (2016) Methods and protocols for developing Prion vaccines, pp 657–680. https://doi.org/10.1007/978-1-4939-3387-7_37

Masone A et al (2023) A tetracationic porphyrin with dual anti-prion activity. iScience 26:107480

Matsushita K et al (1998) Chicken monoclonal antibodies against synthetic Bovine Prion protein peptide. J Vet Med Sci 60:777–779

Mattoli MV, Giancipoli RG, Cocciolillo F, Calcagni ML, Taralli S (2024) The role of PET imaging in patients with Prion disease: a literature review. Mol Imaging Biol 26:195–212

Mead S, Tagliavini F (2018) Clinical trials, pp 431–444. https://doi.org/10.1016/B978-0-444-63945-5.00024-6

Medel Sánchez A et al (2024) Aducanumab in Alzheimer's disease: a comparative study of its effects on dementia and mild cognitive impairment. Cureus 16:e75907. https://doi.org/10.7759/cureus.75907

Mehrabian M et al (2014) CRISPR-Cas9-based knockout of the Prion protein and its effect on the proteome. PLoS One 9:e114594

Mohammadi B et al (2023) Anchorless risk or released benefit? An updated view on the ADAM10-mediated shedding of the prion protein. Cell Tissue Res 392:215–234

Moroncini G et al (2004) Motif-grafted antibodies containing the replicative interface of cellular PrP are specific for PrP ^{Sc}. Proc Natl Acad Sci 101:10404–10409

Napper S, Schatzl HM (2023a) Vaccines for prion diseases: a realistic goal? Cell Tissue Res 392:367–392

Napper S, Schatzl HM (2023b) Oral vaccination as a potential strategy to manage chronic wasting disease in wild cervid populations. Front Immunol 14:1156451

Neumann EN et al (2024) Brainwide silencing of prion protein by AAV-mediated delivery of an engineered compact epigenetic editor. Science (1979) 384:ado7082

Nicoll AJ et al (2010) Pharmacological chaperone for the structured domain of human prion protein. Proc Natl Acad Sci 107:17610–17615

Otto M et al (2004) Efficacy of flupirtine on cognitive function in patients with CJD. Neurology 62:714–718

Otvos L Jr, Cudic M (2002) Post-translational modifications in Prion proteins. Curr Protein Pept Sci 3:643–652

Pan T et al (2004) Epitope scanning reveals gain and loss of strain specific antibody binding epitopes associated with the conversion of normal cellular prion to scrapie prion. J Neurochem 90:1205–1217

Parizek P et al (2001) Similar turnover and shedding of the cellular Prion protein in primary lymphoid and neuronal cells. J Biol Chem 276:44627–44632

Perovic S et al (1995) Flupirtine partially prevents neuronal injury induced by Prion protein fragment and lead acetate. Neurodegeneration 4:369–374

Porto-Carreiro I, Février B, Paquet S, Vilette D, Raposo G (2005) Prions and exosomes: from PrPc trafficking to PrPsc propagation. Blood Cells Mol Dis 35:143–148

Priola SA (1999) Prion protein and species barriers in the transmissible spongiform encephalopathies. Biomed Pharmacother 53:27–33

Prusiner SB (1998) Prions. Proc Natl Acad Sci 95:13363–13383

Prusiner SB, Scott MR, DeArmond SJ, Cohen FE (1998) Prion protein biology. Cell 93:337–348

Puig B, Altmeppen H, Glatzel M (2014) The GPI-anchoring of PrP. Prion 8:11–18

Puig B et al (2019) GPI-anchor signal sequence influences PrPC sorting, shedding and signalling, and impacts on different pathomechanistic aspects of prion disease in mice. PLoS Pathog 15:e1007520

Reidenbach AG et al (2019) Characterization of the Prion protein binding properties of antisense Oligonucleotides. Biomol Ther 10:1

Reilly M et al (2022) A high-content neuron imaging assay demonstrates inhibition of prion disease-associated neurotoxicity by an anti-prion protein antibody. Sci Rep 12:9493

Reimann RR et al (2016) Differential toxicity of antibodies to the Prion protein. PLoS Pathog 12:e1005401

Rigoli M, Spagnolli G, Faccioli P, Requena JR, Biasini E (2019) Ok Google, how could I design therapeutics against prion diseases? Curr Opin Pharmacol 44:39–45

Rovis T, Legname G (2014) Prion protein-specific antibodies-development, modes of action and therapeutics application. Viruses 6:3719–3737

Rudd P (2002) Glycosylation and prion protein. Curr Opin Struct Biol 12:578–586

Sadowski MJ et al (2009) Anti-PrP Mab 6D11 suppresses PrPSc replication in prion infected myeloid precursor line FDC-P1/22L and in the lymphoreticular system in vivo. Neurobiol Dis 34:267–278

Satoh K (2022) CSF biomarkers for prion diseases. Neurochem Int 155:105306

Schmitz M et al (2022) Diagnostic accuracy of cerebrospinal fluid biomarkers in genetic prion diseases. Brain 145:700–712

Senatore A et al (2020) Protective anti-prion antibodies in human immunoglobulin repertoires. EMBO Mol Med 12:e12739

Silber BM et al (2013) Pharmacokinetics and metabolism of 2-Aminothiazoles with Antiprion activity in mice. Pharm Res 30:932–950

Song C-H et al (2008) Effect of intraventricular infusion of anti-prion protein monoclonal antibodies on disease progression in prion-infected mice. J Gen Virol 89:1533–1544

Souan L et al (2001) Modulation of proteinase-K resistant prion protein by prion peptide immunization. Eur J Immunol 31:2338–2346

Spagnolli G, Requena JR, Biasini E (2020) Understanding prion structure and conversion, pp 19–30. https://doi.org/10.1016/bs.pmbts.2020.07.005

Spagnolli G et al (2021) Pharmacological inactivation of the prion protein by targeting a folding intermediate. Commun Biol 4:62

Stravalaci M et al (2016) The anti-Prion antibody 15B3 detects toxic amyloid-β Oligomers. J Alzheimers Dis 53:1485–1497

Tapella L et al (2013) Epitope scanning indicates structural differences in brain-derived monomeric and aggregated mutant prion proteins related to genetic prion diseases. Biochem J 454:417–425

Tatzelt J, Prusiner SB, Welch WJ (1996) Chemical chaperones interfere with the formation of scrapie prion protein. EMBO J 15(23):6363–6373

Taylor DR et al (2009) Role of ADAMs in the Ectodomain shedding and conformational conversion of the Prion protein. J Biol Chem 284:22590–22600

Todd NV et al (2005) Cerebroventricular infusion of pentosan polysulphate in human variant Creutzfeldt–Jakob disease. J Infect 50:394–396

Trevitt CR (2006) A systematic review of prion therapeutics in experimental models. Brain 129:2241–2265

Tsuboi Y, Doh-ura K, Yamada T (2009) Continuous intraventricular infusion of pentosan polysulfate: clinical trial against prion diseases. Neuropathology 29:632–636

Turk E, Teplow DB, Hood LE, Prusiner SB (1988) Purification and properties of the cellular and scrapie hamster prion proteins. Eur J Biochem 176:21–30

Uliassi E, Nikolic L, Bolognesi ML, Legname G (2023) Therapeutic strategies for identifying small molecules against prion diseases. Cell Tissue Res 392:337–347

Vanni S et al (2017) Differential overexpression of SERPINA3 in human prion diseases. Sci Rep 7:15637

Vanni I et al (2023) An optimized Western blot assay provides a comprehensive assessment of the physiological endoproteolytic processing of the prion protein. J Biol Chem 299:102823

Vroom MM, Dodart J-C (2024) Active immunotherapy for the prevention of Alzheimer's and Parkinson's disease. Vaccines (Basel) 12:973

Walmsley AR, Watt NT, Taylor DR, Perera WSS, Hooper NM (2009) α-cleavage of the prion protein occurs in a late compartment of the secretory pathway and is independent of lipid rafts. Mol Cell Neurosci 40:242–248

Weber T, Otto M, Bodemer M, Zerr I (1997) Diagnosis of Creutzfeldt-Jakob disease and related human spongiform encephalopathies. Biomed Pharmacother 51:381–387

Weissmann C (2009) Thoughts on mammalian prion strains. Folia Neuropathol 47:104–113

White MD et al (2008) Single treatment with RNAi against prion protein rescues early neuronal dysfunction and prolongs survival in mice with prion disease. Proc Natl Acad Sci 105:10238–10243

Wood ME et al (2018) Accelerated onset of chronic wasting disease in elk (Cervus canadensis) vaccinated with a PrPSc-specific vaccine and housed in a prion contaminated environment. Vaccine 36:7737–7743

Wuertzer CA, Sullivan MA, Qiu X, Federoff HJ (2008) CNS delivery of Vectored Prion-specific single-chain antibodies delays disease onset. Mol Ther 16:481–486

Zerr I et al (2024) Creutzfeldt–Jakob disease and other prion diseases. Nat Rev Dis Primers 10:14

Chapter 6
Prion and "Prion-Like" Detection: From Conventional Methods to Microfluidics or Lab-on-Chip Platforms to Monitor Seeding and Spreading of Misfolded Proteins

José A. del Río, Laia Lidón, and Rosalina Gavín

Abstract Misfolded protein neurodegeneration includes several pathologies characterized by the accumulation of a group of proteins that can modify their folding due to intrinsic or extrinsic factors, leading to the generation of aberrant forms characterized by their high insolubility, cytotoxicity, and the ability to propagate among various cell types and regions in affected brains. Due to this capacity and based on the properties of bona fide prions, a large number of "prion-like" or "prionoid" proteins with this ability have been described in recent years. Their study presents challenges, including the development of a detailed understanding of the processes involved in the formation of these insoluble aggregates and in establishing the cellular and molecular bases underlying the process of intercellular propagation. To address these processes, various laboratories have developed techniques to detect their presence in brain or peripheral samples. The detection of these molecules is, as

J. A. del Río (✉) · R. Gavín
Molecular and Cellular Neurobiotechnology, Institute for Bioengineering of Catalonia (IBEC), Barcelona Institute of Science and Technology (BIST), Science Park of Barcelona, Barcelona, Spain

Department of Cell Biology, Physiology and Immunology, Faculty of Biology, University of Barcelona, Barcelona, Spain

Network Center for Biomedical Research of Neurodegenerative Diseases, Institute Carlos III, Ministry of Health, Barcelona, Spain

Institute of Neuroscience, University of Barcelona, Barcelona, Spain
e-mail: jadelrio@ibecbarcelona.eu

L. Lidón
Network Center for Biomedical Research of Neurodegenerative Diseases, Institute Carlos III, Ministry of Health, Barcelona, Spain

Sant Pau Memory Unit, Neurology Department and IIB-Sant Pau, Hospital de La Santa Creu I Sant Pau, Universitat Autonoma de Barcelona, Barcelona, Spain

G. Legname, F. Moda (eds.), *Biomarkers and Therapeutical Targets for Prion Diseases*, Subcellular Biochemistry 112, https://doi.org/10.1007/978-3-031-97055-9_6

of today, very effective and selective. However, the processes of transmission and propagation are not fully characterized. Indeed, various classical detection techniques have been developed, generally based on controlled polymerization processes and effective detection methods. Nevertheless, these conventional techniques have now incorporated various methodologies employed in other disciplines, such as nanotechnology, which have increased our understanding of these processes and are useful in the development of future therapies and drug discovery. In this chapter, we summarize the current state of the art of these conventional methods, their limitations, and the use of new platforms to deepen our understanding of these processes.

Keywords Prion protein · PCMA · RT-QuIC · Microfluidic devices · Organotypic slices · Brain-on-chip · Amyloid seeding · Amyloid spreading · "prion-like" and prionoids

Introduction

Prion diseases are neurodegenerative diseases characterized by neuronal loss, vacuolation of brain parenchyma, deposition of protein aggregates, inflammatory responses, and a fatal course. They are caused by the posttranslational conversion of the cellular prion protein (PrP^{C}), a glycoprotein physiologically relevant in mammals, into a misfolded and aggregated pathogenic isoform (PrP^{res} or PrP^{Sc}) (Prusiner 1982). Following this conformational change, PrP^{Sc} acquires partial resistance to proteinase K digestion, insolubility in nonionic detergents, and great resistance to both physical and chemical sterilization. Examples of these processes in animals are scrapie in sheep and goats, bovine spongiform encephalopathy (BSE), and chronic wasting disease (CWD) in cervids, among others. In humans, the most common is Creutzfeldt–Jakob disease (CJD), in sporadic (sCJD), iatrogenic (iCJD), and familial (fCJD) forms, and a variant of CJD (vCJD). In addition to genetic forms, there are fatal familial insomnia (FFI), Gerstmann–Sträussler–Scheinker syndrome (GSS), and variable protease-sensitive prionopathy (VPSPr). However, other neurodegenerative diseases are characterized by protein misfolding aggregation (PMA), such as Alzheimer's and Parkinson's diseases (AD and PD, respectively), multiple system atrophy (MSA), and numerous tauopathies (Scheckel and Aguzzi 2018; Kovacs 2019). Although with relevant differences, recent evidence suggests that many of these misfolded/amyloid proteins can exhibit behavior similar to prions, propagating between neural cells in a "prion-like" or prionoid manner (see (Costanzo and Zurzolo 2013; Goedert et al. 2017; Holmes and Diamond 2017; Del Rio et al. 2018; Kara et al. 2018; Vilette et al. 2018; Scialo et al. 2019; Meisl et al. 2020; Peng et al. 2020; Willbold et al. 2021; Joshi and Ahuja 2023; Zerr et al. 2024)). Indeed, in contrast to infective prions (i.e., (Blattler et al. 1997)), the seeding and propagation of the prionoid can take place in some experiments in the absence of the endogenous protein counterpart (e.g., for tau, see (Wegmann et al. 2015); see also (Kara et al. 2018) for review), which reinforces the intrinsic differences between prions and "prion-like" proteins (Hall and Patuto 2012; Erana 2019; Jellinger et al. 2021). The seeding and progression of these bona fide prions and "prion-like" or prionoids can be determined using established detection methods in parallel to biochemical or histopathological examination. To enhance understanding of the behavior of misfolded amyloids in protein misfolding diseases (PMDs), microfluidics and brain-on-chip approaches, including 3D organotypic slices and brain organoids, have been developed as valuable tools (see (Del Rio and Ferrer 2020; Pineau and Sim 2020; Chia et al. 2022; Sala-Jarque et al. 2022; Walters and Haigh 2023) for reviews). These techniques enable the monitoring of changes in specific cellular and molecular processes responsible for amyloid seeding and cell spreading, along with their effects on neuronal physiology. Furthermore, these platforms offer enhanced reproducibility and represent a potential alternative to conventional approaches for better understanding neurodegeneration. In this chapter, we recapitulate classical/conventional and recent progress in neurobiological research on prion and "prion-like" proteins using microfluidic lab-on-chip (LOC) approaches as well as cellular models (from cells to brain organoids). These approaches, propelled by various fields including biochemistry, nanotechnology, and cell biology, aim to facilitate the development of more effective detection methods and precise models for basic mechanistic studies in protein–protein interactions and for fast and high-throughput drug screening for these devastating diseases.

Classical Prion-Seed Detection and Aggregation Methods and Their Use in PMDs

Protein Misfolding Cyclic Amplification (PMCA)

Some studies have demonstrated that the formation of amyloids might occur via seeded nucleation (Jarrett and Lansbury 1993; Eigen 1996), displaying sigmoidal growth kinetics (Ferrone et al. 1985; Knowles et al. 2009; Morris et al. 2009; Prusiner 2017). The limiting step in the development of the process lies in the formation of small aggregates of misfolded protein in the so-called "lag phase" (Morris et al. 2009; Arosio et al. 2015). The aggregates, commonly termed "nuclei," recruit and convert properly folded protein, forming larger fibrillar aggregates. The full process includes the indicated "lag phase" followed by a fast-growing phase and the last phase, or "steady state" equilibrium, as described in several other proteins (Ferrone et al. 1985; Knowles et al. 2009; Morris et al. 2009; Prusiner 2017). Techniques allowing in vitro detection and amplification of misfolded proteins are of great interest to the bigger field of PMDs. One of the most widely used techniques is protein misfolding cyclic amplification (PMCA), developed in Claudio Soto's lab around 2000 ((Saborio et al. 2001); see also (Wang et al. 2023a) for a recent review). Similar to DNA amplification by polymerase chain reaction (PCR), the amplification of PrP^{Sc} by PMCA is achieved by exposing the PrP^{Sc} seeds present in the analyzed sample and the PrP^{C} substrates to a cyclical process of incubation/elongation and fragmentation (Soto et al. 2002; Castilla et al. 2006; Morales et al. 2012). The method is based on the capacity of PrP^{Sc} to induce a conformational change of PrP^{C} leading to the amplification of very small amounts of PrP^{Sc} in the sample to biochemically measurable amounts of generated fibrillar PrP^{Sc}. Thus, after the processes, samples are usually followed by the detection of generated prions, usually by Western blotting (Saa et al. 2006). The PCMA process includes an incubation step characterized by the role of PrP^{Sc} as a template for the added PrP^{C} protein and a second fragmentation step that breaks the preformed PrP^{Sc} fibrils into shorter portions that, in a second PCMA cycle, play the role of new templates for increased elongation (Saborio et al. 2001). Thus, cyclic alternation of sonication and incubation phases enables exponential amplification of PrP^{Sc} that can be finally detected. The original method was improved by Joaquin Castilla in Claudio Soto's lab utilizing a programmable microplate horn sonicating system (Castilla et al. 2005) used in several studies (see (Wang et al. 2023a) for review). Concerning detection levels, it has been considered that PCMA can detect a single molecule of PrP^{Sc} present in an affected brain extract. This is a $>10^{12}$ range of higher sensibility compared to classical Western blotting techniques (Bieschke et al. 2004; Saa et al. 2006; Wang et al. 2023a). Since its development, PMCA has been improved by several modifications that include the addition of Teflon beads (Gonzalez-Montalban et al. 2011; Johnson et al. 2012) and EDTA or digitonin (Moda et al. 2014) in the reaction mixture, leading to increased fragmentation, elongation, and further detection. Considering the great potential use of this technique in prion detection in biological samples as well as other samples or materials, several studies have focused on applying this technique to other PMDs. Indeed, PCMA has been reported to be useful for β-amyloid (Salvadores et al. 2014), α-synuclein (Herva et al. 2014; Shahnawaz et al. 2017), and tau (Meyer et al. 2014) seed detection. However, real-time quaking-induced conversion (RT-QuIC) seems to be more useful for these "prion-like" proteins due to its intrinsic advantages and broader use (see below).

Real-Time Quaking-Induced Conversion (RT-QuIC)

As indicated, protein misfolding cyclic amplification assay (PMCA) and RT-QuIC are the two most widely used PrP^{Sc} amplification techniques. Around ten years after the development of the PMCA technique, a new technique was developed: the "quaking-induced conversion" (QuIC), which improves speed and practicality by using recombinant bacterial-expressed PrP, which is easy to prepare in large quantities, as a substrate for the conversion reaction, and the reaction is promoted by intermittent strong agitation replacing sonication (Wilham et al. 2010; Atarashi et al. 2011b). With a similar detection range of PMCA for most samples, the original method and its modifications (e.g., e-QuIC) improved their use in different samples and amyloids, increasing sensitivity (Orru et al. 2011), although the e-QuIC assay is more difficult to standardize and less reproducible than the standard RT-QuIC (Atarashi et al. 2011b). In the currently used method, RT-QuIC uses different forms of recombinant PrP as a substrate to amplify small amounts of PrP^{Sc} in the analyzed sample. The generated fibrils are linked to amyloidophilic dyes such as Thioflavin T (Saeed and Fine 1967), Congo red (Yakupova et al. 2019), and benzofuranone (Lengyel-Zhand et al. 2020) monitored in real-time using a fluorescence multiplate reader (Atarashi et al. 2011a). Alternatively, gold nanoparticles (Zhou et al. 2015) or CdTe quantum dots (Xia et al. 2016) have been used as an alternative to Thioflavin T. However, a modification of the original RT-QuIC was described using the 90-231 aa of hamster PrP instead of recombinant *E. coli*-generated PrP, 0.002% SDS, and increased temperature largely reduced the "lag phase" of the reaction (Orru et al. 2015). In fact, this modification is considered a second generation of RT-QuIC by some authors (Atarashi 2023). The method was used to detect pathogenic prions in cerebrospinal fluid (CSF) from CJD patients (Orru et al. 2015), since one of the disadvantages of the first-generation technique is its unspecific results in the presence of other proteins in the sample in addition to the pathogenic prion (e.g., blood cells in CSF). Current detection ranges of RT-QuIC are close to the attogram (10^{-18} g).

As was previously the case with PMCA, RT-QuIC was also used to detect amyloid, "prion-like" proteins of other PMDs in affected brains. Thus, α-syn RT-QuIC has been developed to be applied in α-synuclein detection in PD, MSA, dementia with Lewy bodies (DLB), or Lewy body dementia (LBD) (see (Soto 2024) for a recent review). Numerous studies have demonstrated that the α-syn RT-QuIC method is useful in detecting α-synuclein seed presence in CSF and serum with a sensitivity of ≈ 92–95% and a specificity of 100% (e.g., (Fairfoul et al. 2016; Shahnawaz et al. 2017; Groveman et al. 2018; Sano et al. 2018; Huang et al. 2024)). In addition, modifications of the technique have been used to detect α-synuclein of p-α-synuclein in peripheral tissues such as submandibular gland, biopsies of skin, and olfactory mucosa, aiming at noninvasive testing, although with lower sensitivity (≈ 85–90%) (e.g., (Manne et al. 2020a, b, Kuzkina et al. 2021; Stefani et al. 2021). In parallel, although some reports point to the potential use of the RT-QuIC for AD (e.g., (Jack et al. 2018)), most of the published studies used this technique in pure tauopathies since some of them are adapted to determine the presence of the 3R and 4Rtau isoforms in samples from Pick's disease (PID) (Saijo et al. 2017; Metrick et al. 2020), progressive supranuclear palsy (PSP) (Metrick et al. 2020), corticobasal degeneration (CBD), and frontotemporal lobar degeneration linked to chromosome 17 (FTDP17 *MAPT*) (Saijo et al. 2020). However, Frey and coworkers recently developed a novel tau seeding assay based on RT-QuIC but using the full-length 0N3R tau as a template. This method was able to distinguish between different tauopathies, including AD, based on the 3R/4R tau ratio (Frey et al. 2023).

Semi-Denaturing Amyloid Seeding Assays and Native Aggregation Assays

As an alternative to RT-QuIC, semi-denaturing amyloid seeding assay (SDASA) uses similar agitation of recombinant PrP in a solution that contains zirconium beads to accelerate fibril growth (Sabareesan and Udgaonkar 2017). The process is also monitored in real time by different fluorochromes. The main advantage is that SDASA does not involve elevated temperatures like RT-QuIC but instead uses chaotropes in the buffer mixture to unfold the PrP substrate, destabilize hydrophobic aggregates, and increase the solubility of hydrophobes, thereby enhancing fibril formation. The method, like RT-QuIC, can use full-length recombinant PrP or fragments of the protein (Sabareesan and Udgaonkar 2017). In contrast, as an alternative, a recent study used a native aggregation method that largely avoids denaturing conditions (Sangar et al., BioRxiv 2022, DOI: 10.1101/2022.08.25.505283). However, these two methods require extensive setup and are still under development.

New Methods for Prion-Seeded Aggregation and Propagation Studies and Their Use in Other Neurodegenerative Diseases

An Overview of New Methods to Detect Prion and "Prion-Like" Proteins

As an alternative to the abovementioned methods (PMCA and RT-QuIC) and their modifications, several studies developed new methods of prion and "prion-like" detection in biological samples. Most of them are based on the ELISA protocol and antibody-mediated reactions, although other electrochemical and nanotechnological methods have also been developed. Most of the methods largely increase sensitivity by several orders of magnitude. Some of these examples are the single-molecule array technology (SIMOA) that is able to handle very minute samples (≈ fg range) with a relevant detection range of ($\approx 10^{-16}$ M) of several molecules or biomarkers (Rissin et al. 2010). In fact, their combined use with RT-QuIC has recently been valuable in determining risk factors and seed formation in larger prion at-risk cohorts (Mok et al. 2023). Other alternatives, such as modified chemiluminescence (CLEIA) and proximity extension assays mainly used for biomarker detection, such as SIMOA, displayed detection ranges of ≈100 pg/ml for some analytes (Hirose et al. 2015). In addition to these approaches, researchers have used surface-enhanced Raman scattering (SERS) (Manno et al. 2010), electrochemical detection (Yoon et al. 2021), improved fluorescence detection (Hu et al. 2013), fluorescence method with dextran-labeled probes (Azam et al. 2011), and plasmon resonance (Hianik et al. 2009; Jiayu et al. 2009) to detect prions. The sensitivity of these methods ranged from 0.1 to 10^{-9} nM. However, these methods are unfortunately not available for all laboratories and are not yet a powerful alternative to PCMA or RT-QuIC for prion detection, although, as noted, both conventional techniques still have some limitations (Gough and Maddison 2010).

Cellular-Based Biosensors for Prion and Prion-Like Seed Detection

Several groups have developed in vitro cellular models to monitor the seeding properties of some "prion-like" proteins and their effects on cell physiology (e.g., cytotoxicity). Cell lines overexpressing mutated proteins fused with a flag sequence (e.g., for tau, P301L-V5 (Xu et al. 2016)) have been used for high-throughput assays of tau seeding. Some of these approaches have also included Förster resonance energy transfer (FRET) detection methods (Chun and Johnson 2007; Kfoury et al. 2012; Lo et al. 2019; Shin et al. 2019). These "biosensor" cell lines are highly specific, generating fluorescence FRET signals only when the appropriate tau seed is transferred into the cell line, mainly by lipofection (see Sala-Jarque et al., bioRxiv, DOI: 10.1101/2024.01.20.576414). A new biosensor cell line has recently been developed by using nanoluciferase (Nluc) binary technology (NanoBiT) (Sadeghzadeh et al. 2023). These cellular biosensor approaches, mainly developed in HEK293 or H4 cells, have been adapted for α-synuclein (Prusiner et al. 2015; Holmes and Diamond 2017; Braun et al. 2023). The reader may find more information about these cellular biosensors and their putative potential in the following reviews (Kara et al. 2018; Ferreira and Caughey 2020; Sala-Jarque et al. 2022; Braun et al. 2023). Although these methods have revealed relevant information about the "prion-like" seeding and strain-dependent seeding capacity of aggregates (e.g., tau), their use for routine clinical diagnostic purposes in clinical practice is limited by their needs: tissue cultures, immunohistochemical, and advanced microscopy or flow cytometry facilities.

For pathogenic prions, bioassays in mice or hamsters are the gold standard for studying prion seeding and propagation, since only PCMA but not RT-QuIC can generate bona fide infective prions (Kara et al. 2018). In recent years, many different cell lines that can become chronically infected with prions have been developed. These cell lines are useful to mimic the intracellular machinery or processes involved in prion seeding, and in addition, are powerful tools for drug screening. The first cell line able to replicate prion, developed by Chandler in 1961, was termed the scrapie mouse brain (SMB) cell model (Chandler 1961). From this model, researchers developed other cell lines suitable to be infected with prions, mainly derived from neuroblastoma cell lines (e.g., (Race et al. 1987)). However, these neuroblastoma cell lines, especially N2a, generate clones with different susceptibility to being infected with prions (Marbiah et al. 2014) (J.M. Torres (INIA, Madrid, Spain) personal communication). This has been partially solved by developing a cellular bioassay for pathogenic prions. One example is the standard scrapie cell assay (SSCA) that is well suited for mouse-adapted prions such as RML (Klohn et al. 2003). Today, the ScN2a, GT1, SN56, 1C11, and CAD5 neural-derived cells from rodents are examples of cell lines infected with prions that have been used for several years in numerous studies (see (Krance et al. 2020) for review). In addition to these, nonneuronal cell lines such as NIH-3T3 and L929 (LD9) and genetically engineered cells expressing mouse PrP^C, such as Npl2, Hpl3-4, or RK13, are used (Krance et al. 2020). However, although other cell lines can be genetically engineered to express non-murine prions for specific purposes, there is a lack of reliable 2D cell lines or protocols that support infectivity and propagation of human prions, and this can only be achieved by using 3D human-derived organoids (Groveman et al. 2019, 2023; Foliaki et al. 2020) or organotypic slice preparations (see below). The use of brain organoids derived from pluripotent embryonic stem cells (e.g., H9 cells, WiCell Research Institute) or induced-pluripotent stem cells (iPSCs) is a plausible strategy to study these processes for human prions (Groveman et al. 2020, 2021; Pineau and Sim 2021; Walters and Haigh 2023). 3D organization, but also neural development of the organoid, seems to be mandatory to efficiently infect cells with human prions since undifferentiated mouse neurospheres are not infected with prions (Herva et al. 2010; Iwamaru et al. 2017), and neurons derived from iPSCs obtained from a GSS patient growing in a 2D substrate do not generate prions after treatment (Matamoros-Angles et al. 2018). In contrast, iPSC-derived astrocytes seemed to reproduce prion infectivity (Krejciova et al. 2017); see (Aguzzi and Liu 2017) for comments on this interesting result. This appears to be specific for bona fide prions since other "prion-like" proteins can be generated in cultured iPSCs and brain organoids (see (Foliaki et al. 2021) and (Walters and Haigh 2023) as examples) using mainly genetic approaches (see also (Gonzalez et al. 2018)). Examples of organoids include ptau (Bowles et al. 2021; Shimada et al. 2022), ptau and β-amyloid (Raja et al. 2016; Chen et al. 2021), α-synuclein (Raja et al. 2022; Becerra-Calixto et al. 2023; Ra et al. 2023), and TDP-43 (Tamaki et al. 2023).

Organotypic Slice Cultures for Prion and "Prion-Like" Seed Detection and Spreading

For most purposes, cultured slices of the developing brain, termed organotypic slice cultures (OSCs), preserve a high degree of cellular differentiation and tissue organization, offering unique advantages over other in vitro methods, as they mimic numerous in vivo aspects such as the 3D organization of the cultured region with preserved cytoarchitectonics (Gahwiler 1988; Gahwiler et al. 1997). OSCs have been prepared from various brain regions, including the hippocampus, neocortex, striatum, spinal cord, hypothalamus, and cerebellum (e.g., (Gahwiler 1981; Zimmer and Gahwiler 1984; Sofroniew et al. 1988; Stoppini et al. 1991; Lonchamp et al. 2006; Del Turco and Deller 2007)). The current protocol for OSCs used in several laboratories was developed by Stoppini and coworkers (Stoppini et al. 1991) by culturing slices from perinatal rodent brains on transwells (Stoppini et al. 1991; del

Rio and Soriano 2010). One of the advantages of OSCs is the maintenance of the main neural types in the preparation (Gahwiler et al. 1997; Mingorance et al. 2006). Another advantage is that the culture can be prolonged for >2–3 months, being a useful method to analyze developmental changes, axonal plasticity, and neurodegeneration.

Concerning the use of OSCs in prion biology, Aguzzi and coworkers developed an infectivity assay of pathogenic prions by using OSCs derived from the cerebellum of wild-type mice, since the endogenous expression of mouse PrP^C is greater in this region (Falsig and Aguzzi 2008; Falsig et al. 2008). This method was termed prion organotypic slice culture assay (POSCA) (see also (Pineau and Sim 2020) for review). In the original study, Falsig and Aguzzi were able to demonstrate that PrP^{Sc} accumulation occurs faster in POSCA preparations compared to in vivo infection (Falsig and Aguzzi 2008; Falsig et al. 2008; Kondru et al. 2017). This might be the result of the observed accelerated aging in the OSCs in contrast to other cell culture methods (Liu et al. 2022). In fact, the aging process analyzed by means of genome-wide gene expression analyses (GEWAS) revealed that prions accelerated the development of age-related molecular signatures and senescence markers (Liu et al. 2022). The POSCA method was combined with RT-QuIC, and PrP^{Sc} in the slice was detected 7 days after prion infection (Kondru et al. 2017). This combination was termed the OSCAR assay by the authors (Kondru et al. 2017). Relevantly, POSCA cultures can be infected by prions that fail to infect other cell lines (e.g., mouse-adapted scrapie strain ME7 and the BSE-derived strain 301C). Most of the studies used wild-type mice as the source of the cerebellar slices. However, Kondru and coworkers extended their use to analyze the infection properties of CWD-derived prions by using cerebellar slices from mice overexpressing elk PrP^C (Kondru et al. 2020), and Pino and Sim were able to culture complete serial coronal sections of mouse brain as OSCs to infect them with different scrapie strains RML, 22L, and ME7 (Pineau and Sim 2020). This is of relevance since the infective behavior of different prion strains in the same recipient PrP^C background slice can be monitored and compared with different techniques (from biochemical to histological), also reducing animal experimentation and repetitive experiments. Thus, the pathological effects derived from prion infections can be recapitulated in POSCA, such as astrogliosis, spongiform vacuolization, microglia activation, neuronal death, and synapse degeneration (e.g., (Falsig et al. 2008; Campeau et al. 2013; Wolf et al. 2015)). In fact, the roles of particular domains (e.g., the globular domain) of the prion protein in their associated neurotoxicity have recently been analyzed by using POSCA (Reimann et al. 2023). Finally, POSCA is also used to develop anti-prion drugs (e.g., (Margalith et al. 2012; Cortez et al. 2015; Bamia et al. 2021; Masone et al. 2023)) with varying effectiveness. This anti-prion test has also been developed using organotypic slices from the hippocampus (Goniotaki et al. 2017).

Concerning "prion-like" proteins observed in other PMDs, OSCs have been used to monitor seeding as well as the progression of, among others, P301S tau or ptau (Croft et al. 2017; Miller et al. 2021; Korde and Humpel 2022), β-amyloid (Harwell and Coleman 2016; Novotny et al. 2016; Croft and Noble 2018; Moelgg et al. 2021), α-synuclein (Elfarrash et al. 2019; Croft et al. 2022; Ucar et al. 2022), and TDP-43

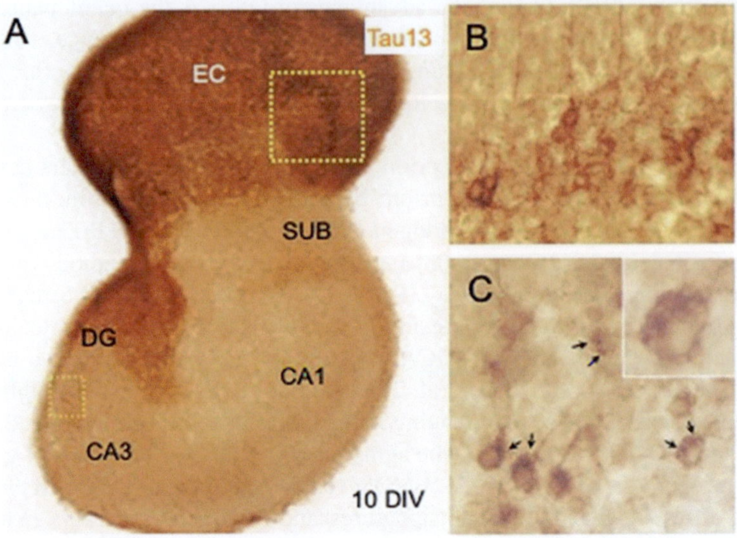

Fig. 6.1 Example of the seeding and spreading of P301L human tau in OSCs of the entorhinal-hippocampal formation of wild-type mice. The entorhinal cortex (EC) was infected with AAV expressing mutated human tau, and the appearance of human tau in infected and recipient cells can be detected by using human-specific tau antibodies (Tau13). The boxed areas in (**a**) can be seen at high magnification in (**b**) and (**c**). Infected neurons and recipient cells labeled by the Tau13 antibody showed dystrophic neurites and relevant presence of human tau in their vacuolized cytoplasm. These experiments demonstrate the axonal transport and intercellular transfer of pathogenic tau species between infected and recipient neurons in OSCs shortly after 10 days of culture. Abbreviations: EC entorhinal cortex, CA1–3 hippocampal fields 1 and 3, DG dentate gyrus, SUB subicular region

(Leggett et al. 2012). In most cases, OSCs, as with brain organoids, were prepared from mice carrying mutation/s in the protein of interest (e.g., 3xTgAD (Croft et al. 2017), huAPP mice (Harwell and Coleman 2016), or mice overexpressing mutated tau ΔK280 (Messing et al. 2013)). Alternatively, the seeding properties can be analyzed by incubating or injecting the OSCs with preformed fibrils or seeds (e.g., TDP-43 (Leal-Lasarte et al. 2017), tau (Suttkus et al. 2016), α-synuclein oligomers (Xu et al. 2013)), or, as an alternative, by viral delivery of mutated forms of the amyloid (e.g., using recombinant adeno-associated virus (AAV) expressing A53T (Croft et al. 2019), or AAV expressing P301L (Fig. 6.1)). As with infective prions, the accumulation of the misfolded amyloids takes place faster than in in vivo rodent models, and for some experiments it appeared after 7 days (β-amyloid (Novotny et al. 2016), 15 days (α-synuclein (Elfarrash et al. 2019), and 28 days (ptau (Croft et al. 2019)), leading to fast monitoring of these processes in a complex culture ex vivo model. Our current experiments illustrate how human tau with P301L mutation can be propagated in the entorhino-hippocampal connection of wild-type mice shortly after 10 days of in vitro after treatment in the entorhinal cortex (EC) (Fig. 6.1).

Lab-on-Chip Microfluidic Devices to Prion and "Prion-Like" Seeding and Spreading

An Overview of Lab-on-Chip Devices

Microfluidics and lab-on-chip (LOC) technologies emerged in the last 20 years as part of a plausible strategy for monitoring amyloid seeding and propagation and amyloid-biological interactions, as well as being a valuable, reproducible tool to analyze cell-to-cell seeding and spreading of pathogenic seeds from "prion-like" or prionoids (Aguzzi and Rajendran 2009; Scheckel and Aguzzi 2018). Microfluidics platforms are devices containing microchannels interconnecting different cell culture reservoirs with a height ranging from nanometers to micrometers (Tabeling 2005; Folch i Folch 2013; Lagally 2014; Li 2015). Although pioneering microfluidic studies were developed by Champenot et al. for neuroscience ((Campenot 1977; Campenot 1982); see also (Taylor and Jeon 2010; Neto et al. 2016) for reviews), their origins were parallel to micromanufacturing techniques from the semiconductor industry in the '70s–'80s (please see (Tabeling 2005; Li 2015) for reviews). After this, the concept of μTAS (miniaturized total chemical analysis systems) was developed to describe a microfluidic device/platform that could carry out all the functions required for the analysis of an analyte. From this, manufacturing terms like MEMs (microelectromechanical systems) and LOC devices were extended to biomedical fields (Folch i Folch 2013; Song et al. 2018).

Most microfluidic LOC devices used in biology are generated using the silicone elastomer Polydimethylsiloxane (PDMS) (McDonald et al. 2000; McDonald and Whitesides 2002; Ng et al. 2002; Kuncova-Kallio and Kallio 2006; Dixit and Kaushik 2016). PDMS is a cheap, biocompatible, soft, and easy-to-handle elastomer with an optical refraction index like a glass coverslip. This PDMS-based microfluidic application using soft-lithography micromanufacturing technologies for neurobiology was developed around 2005 by Jeon's lab ((Taylor et al. 2003, 2005; Rhee et al. 2005; Park et al. 2006) with great evolution in recent years (see (Neto et al. 2016; Choi et al. 2017; Habibey et al. 2022) for reviews). These studies follow the pioneer platform (Campenot 1977, 1982) to generate a simple and reproducible culture device for compartmentalized neural growth and differentiation. Most of the current experiments designed to explore the cell dynamics of different amyloids include experiments developed using microfluidic devices. Although marketed by different companies (e.g., Xona™ microfluidics), ad-hoc PDMS-based manufacture of LOC devices is mainly based on soft-lithography protocols (McDonald et al. 2000; Whitesides et al. 2001). Readers may obtain more information about LOC platform manufacturing strategies in reference books (Minteer 2006; Herold and Rasooly 2009; Lee and Sundararajan 2010; Folch i Folch 2013). In this chapter, we will focus on the PDMS-derived LOC devices with the greatest impact on the study of amyloids associated with PMDs.

Microfluidics and LOC devices hold a number of advantages for amyloid-related research at different levels: (i) small reaction volumes; (ii) putative large number of independent but repetitive compartments allowing protein–protein interactions (e.g., droplet microfluidics); (iii) almost complete control over spatial and temporal parameters of the reaction; (iv) compatibility with several detection methods (from optical to electrical sensors); and finally (v) compartmentalized LOC devices allow for the specific and safe culture of different cell types (e.g., neurons, astroglia, oligodendroglia, microglia, etc.) and nonneuronal cells in the same device in an interactive manner.

Although most amyloid species were initially identified within the context of prion diseases and several PMDs (see (Peng et al. 2020) for review), several nonneural proteins also form misfolded fibrils (e.g., amylin (Zheng et al. 2020) or insulin (Zheng et al. 2020)). The monitoring of this process can be developed in LOC (e.g., in hydrodynamic focusing systems (e.g., (Fitzpatrick et al. 2013; Arosio et al. 2016)), or using electrophoretic approaches in these systems (e.g., (Saar et al. 2018)). In addition, the proteinase K resistance of generated prions can be analyzed more quickly in microfluidic devices (Le Nel et al. 2008).

However, it is well known that the amyloid aggregation process is largely dependent on the interaction of the protein with several ions (e.g., (Kim et al. 2018)), membranes (e.g., (Terakawa et al. 2018; Alghrably et al. 2019)), or other surfaces (e.g., water (Schladitz et al. 1999). In fact, these are some of the disadvantages of PMCA and RT-QuIC. Thus, researchers use different methods to avoid or control these sometimes unwanted interactions. One of the most widely used methods is the aggregation of different proteins inside micelles or micro/nanodroplets (water-oil) (e.g., (Shim et al. 2007; Teh et al. 2008; Casadevall i Solvas et al. 2012; Shembekar et al. 2016; Wang et al. 2023b)). Using microdroplets, several independent reactions can be monitored instantaneously in identical droplet sizes, which are several orders of magnitude smaller than what has been common in biochemical assays. On a practical basis, the formation of the droplets, mainly by microfluidic focusing devices, is followed by their harvesting and maintenance, trapped or retained in reaction chambers to be further analyzed (e.g., (Shim et al. 2007; Teh et al. 2008; Casadevall i Solvas et al. 2012; Shembekar et al. 2016; Wang et al. 2023b)). In these cases (especially for droplets exposed to air) (Casadevall i Solvas et al. 2012), the decrease in droplet volume over time due to the evaporation of water molecules increases the analyte/protein concentration inside the generated droplet, thereby increasing protein–protein interaction and, more relevantly, amplifying signal(s) associated with the aggregation that can be monitored with several methods, including FRET. Other alternatives to enhance the process of reduction of microdroplet volume by using PDMS-derived microflows to enhance the reactions were also recently reported (e.g., DroMiCo, (Kopp et al. 2020)). This published method generates and traps the microdroplets to accelerate droplet shrinking and prevent entry of air into the device, thereby furthering analysis.

Indeed, changes in the aggregation of fibrils during the three-phase curve (see introduction) can be monitored easily using Thioflavin T fluorescence or, depending on the protein sequence with some amino acid (e.g., tryptophan), with specific fluorescence features after aggregation (e.g., (Knowles et al. 2009; Meisl et al. 2016;

Toprakcioglu et al. 2019). In fact, the time evolution of the "lag phase" is also of interest for the characterization of some amyloids since in most studies the molecular events that occur in this "lag phase" of the amyloid formation cannot be fully ascertained using classical methods. However, LOC devices and microdroplet formation were very useful in describing in greater detail the sequential events during this fibril formation for different misfolded proteins of neural and nonneural origin (e.g., B-amyloid (Arosio et al. 2014), fibroin, β-lactoglobulin (Toprakcioglu et al. 2019), human islet amyloid polypeptide (Marek et al. 2010), or hemoglobin (Ferrone et al. 1985).

In neurodegeneration, researchers aim to determine the properties of the "propagative seeds" or "propagons" for particular amyloids. This is a challenging question, especially for "prion-like" proteins (Aguzzi and Lakkaraju 2016). Over time, classical bulk methods have been modified to solve these challenges by adding an amplification step. For example, using a new method, Arosio et al. detected β-amyloid-"propagons" in a "lag phase" of aggregation by sample filtration during the phase, followed by an amplification method with fresh β−amyloid monomer. This was followed by the quantification of the original "propagon" concentration using a calibration curve based on controlled seed concentration (Arosio et al. 2014). Using this method, the authors improved by two orders of magnitude the bulk technical approaches to allow the concentration of fibrillar β−amyloid (Arosio et al. 2014). However, the development of "digital microfluidics," which combines the use of microfluidics and high-throughput biological assays (Guo et al. 2012), has helped researchers to develop a digital amyloid quantitative assay (d-AQuA) aimed at allowing the absolute quantification of single replicative units, the "propagons" of insulin (Pfammatter et al. 2017). In fact, the authors used a dilution method of nanodroplets with picoliters of volume containing (or not) "propagon" species being further evaluated (Pfammatter et al. 2017). This method, although not developed for other amyloids, is faster than currently available methods (e.g., microplate assays) and will be of relevance for fast diagnosis of the presence of "pathological seeds" for different PMDs. Parallel to these approaches, other groups have developed more automatic methods, such as a microchannel-connected multi-well plate (μCHAMP) device (Park et al. 2016) that uses microdroplet formation and microfluidic transport to 96-well plates. The amount of Aβ (at a range of ≈10 pg/mL) is detected using a droplet-based magnetic bead immunoassay (Park et al. 2016). Of relevance, as several PMDs show the presence of different amyloids, micro/nanodroplet assays are useful to mimic these putative interactions at the nanoscale level (Agarwal et al. 2022). Thus, LOC devices are of relevance not only for the reduced sample volume required but also for the automatization and high-throughput assays of amyloid detection. As an example, microfluidic devices have been used to determine the prion size in brain extracts using sample diffusion in H-shaped devices and further FRET detection (Meisl et al. 2021). Lastly, a recent example may be seen in a study that developed a specific microfluidic biosensor for CWD (Muhsin et al. 2023).

LOC as a Tool to Monitor Cell-to-Cell Transport of Prion and "Prion-Like" Proteins

However, as indicated, researchers have tried to determine cell-to-cell mechanisms implicated in amyloid seeding and spreading that, for some "prionoids," are still not fully clarified (see introduction). In this respect, several studies reported using LOC devices, with two or more consecutive chambers: β-amyloid (e.g., (Deleglise et al. 2014; Song et al. 2014)), α-synuclein (e.g., (Volpicelli-Daley et al. 2011; Freundt et al. 2012; Brahic et al. 2016; Urrea et al. 2018; Wang et al. 2018; Gribaudo et al. 2019, 2023)), tau (e.g., (Wu et al. 2013; Dujardin et al. 2014; Calafate et al. 2015; Takeda et al. 2015; Usenovic et al. 2015; Congdon et al. 2016; Polanco et al. 2018; Hallinan et al. 2019, 2020)), TDP-43 (i.e., (Feiler et al. 2015)) or dipeptide repeat proteins (DPRs) of the *C9orf72* gene product associated with ALS and frontotemporal dementia (FTD) (Westergard et al. 2016) were able to perform cell-to-cell transmission by navigating intracellularly along axons allowing seeding and propagation of the amyloid (Fig. 6.2) (reviewed in (Del Rio et al. 2018; Urrea et al. 2018; Peng et al. 2020; Uemura et al. 2020)).

These studies reported the sequential transport of the "pathogenic seeds" as free seeds or in exosomes between different cell populations cultured in hydrodynamically isolated microfluidic chambers. In these experiments, a differential volume between reservoirs is established to avoid diffusion transfer by the media between reservoirs. More relevantly, in several studies, neurons derived from

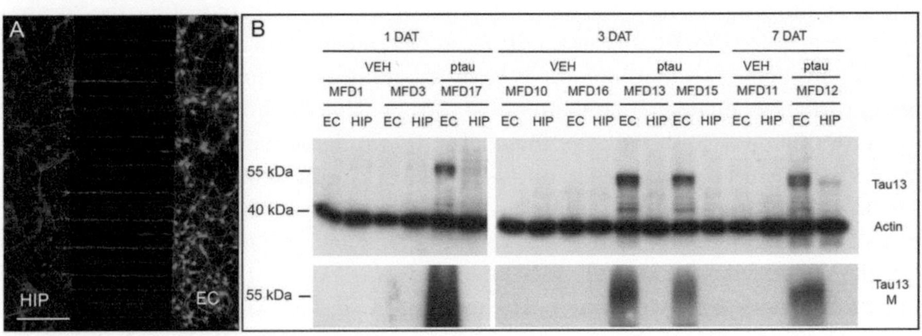

Fig. 6.2 Uptake and transport between EC and HIP neurons of different forms of tau in microfluidic devices (MFDs). (**a**) Example of an MFD used in the experiments. EC and HIP were seeded in the different chambers; 7 days after seeding, EC neurons were labeled with AAV9.h*Syn*.GCaMP6s, and HIP neurons with AAV9.*Syn*.JRCaMP1b. Neurons interconnect by crossing their axons through the microchannels of 950 μm length. (**b**) Western blots of MFDs treated with vehicle (VEH), fibrillar tau, and fibrillar ptau. The number of days after treatment (DAT) is indicated in each panel. For each MFD, the result of the Western blot in the cells (up) and the culture media (down, i.e., Tau13 M in (**b**)) is illustrated. In addition, the number of the MFD used is included, and the bottom line of each MFD links the EC and HIP chambers of the device. Cells and media extract were immunoblotted using Tau13 and actin (**b**). As observed in (**b**), labeling of Tau13 appeared in the HIP cells after ptau incubation of the EC chamber only after 7 DAT. Note that in all cases the observed transport is a cell-mediated transport, and no diffusion in the media of the tau species may be detected. Abbreviations as in Fig. 6.1. Scale bar: A = 450 μm

induced-pluripotent stem cells or neuronal progenitors are included to mimic specific neurodegenerative diseases (e.g., (Choi et al. 2014; Ruiz et al. 2014; Park et al. 2018)). These LOC approaches help researchers to ascertain the role of nonneuronal cells in the seeding and propagation process for particular amyloids (e.g., β−amyloid roles of microglia reactivity and migration (Cho et al. 2013; Park et al. 2018) and the role of astrocytes in α-synuclein seeding (Cavaliere et al. 2017). However, the emerging role of oligodendrocytes during some amyloid transmission (e.g., α-synuclein (Tu et al. 1998; Uemura et al. 2019), tau (Ferrer et al. 2019, 2020); see also (Ferrer 2018) for a recent review) has yet to be fully described in LOC devices. Interestingly, the role of PrP^C, LRP1, and other putative receptors for these "prionoids" has been evaluated in these microfluidic compartmentalized platforms (e.g., (Urrea et al. 2018; Chen et al. 2022; Thom et al. 2022; Courte et al. 2023; Rivas-Santisteban et al. 2023; Chen et al. 2024), establishing platforms for drug screening.

Conclusions

The clinical presentation of human prion diseases bears some resemblance to the pathological hallmarks of many other commonly known PMDs. They have been largely studied, but the mechanisms by which they cause neurodegeneration have not been fully uncovered. The presence and detection of prions and "prion-like" proteins in different samples presents a challenging and evolving problem. Thus, the development of more precise and efficient detection methods has been the focus of research in recent years. Other challenging processes, such as intercellular propagation and generation in more complex tissue models, have been addressed in recent years with the use of humanized models based on iPSCs, organoids, and organotypic slice preparations. These new approaches complement animal experimentation to ascertain basic concepts, such as strain-derived effects. In parallel, microfluidics and LOC platforms have emerged as plausible platforms for characterizing amyloid generation in more controlled environments. Significantly, these newly developed techniques are useful for analyzing cell-to-cell transfers of misfolded proteins and determining the molecular factors involved in seeding and spreading. Moreover, they are particularly valuable for initial drug screening aimed at developing anti-prion treatments.

Funding J.A. del Río was supported by PRPCDEVTAU PID2021-123714OB-I00, ALTERNed PLEC2022-009401, PDC2022-133268-I00, and THRIVE PID2024-162521OB-I00 funded by *MCIN/AEI/10.13039/501100011033 and by "ERDF A way of making Europe,"* the CERCA Programme, and by the Commission for Universities and Research of the Department of Innovation, Universities, and Enterprise of the Generalitat de Catalunya (SGR2021-00453). L. Lidón is supported by the Margarita Salas program (MICINN).

Compliance with Ethical Standards The authors declare no competing financial or nonfinancial interests.

References

Agarwal A, Arora L, Rai SK, Avni A, Mukhopadhyay S (2022) Spatiotemporal modulations in heterotypic condensates of prion and alpha-synuclein control phase transitions and amyloid conversion. Nat Commun 13(1):1154

Aguzzi A, Lakkaraju AKK (2016) Cell biology of prions and prionoids: a status report. Trends Cell Biol 26(1):40–51

Aguzzi A, Liu Y (2017) A role for astroglia in prion diseases. J Exp Med 214(12):3477–3479

Aguzzi A, Rajendran L (2009) The transcellular spread of cytosolic amyloids, prions, and prionoids. Neuron 64(6):783–790

Alghrably M, Czaban I, Jaremko L, Jaremko M (2019) Interaction of amylin species with transition metals and membranes. J Inorg Biochem 191:69–76

Arosio P, Cukalevski R, Frohm B, Knowles TP, Linse S (2014) Quantification of the concentration of Abeta42 propagons during the lag phase by an amyloid chain reaction assay. J Am Chem Soc 136(1):219–225

Arosio P, Knowles TP, Linse S (2015) On the lag phase in amyloid fibril formation. Phys Chem Phys 17(12):7606–7618

Arosio P, Muller T, Rajah L, Yates EV, Aprile FA, Zhang Y, Cohen SI, White DA, Herling TW, De Genst EJ, Linse S, Vendruscolo M, Dobson CM, Knowles TP (2016) Microfluidic diffusion analysis of the sizes and interactions of proteins under native solution conditions. ACS Nano 10(1):333–341

Atarashi R (2023) RT-QuIC as ultrasensitive method for prion detection. Cell Tissue Res 392(1):295–300

Atarashi R, Sano K, Satoh K, Nishida N (2011a) Real-time quaking-induced conversion: a highly sensitive assay for prion detection. Prion 5(3):150–153

Atarashi R, Satoh K, Sano K, Fuse T, Yamaguchi N, Ishibashi D, Matsubara T, Nakagaki T, Yamanaka H, Shirabe S, Yamada M, Mizusawa H, Kitamoto T, Klug G, McGlade A, Collins SJ, Nishida N (2011b) Ultrasensitive human prion detection in cerebrospinal fluid by real-time quaking-induced conversion. Nat Med 17(2):175–178

Azam G, Shibata T, Kabashima T, Kai M (2011) Sensitive chemiluminescence detection of prion protein on a membrane by using a peroxidase-labeled dextran probe. Anal Sci 27(7):715

Bamia A, Sinane M, Nait-Saidi R, Dhiab J, Keruzore M, Nguyen PH, Bertho A, Soubigou F, Halliez S, Blondel M, Trollet C, Simonelig M, Friocourt G, Beringue V, Bihel F, Voisset C (2021) Anti-prion drugs targeting the protein folding activity of the ribosome reduce PABPN1 aggregation. Neurotherapeutics 18(2):1137–1150

Becerra-Calixto A, Mukherjee A, Ramirez S, Sepulveda S, Sinha T, Al-Lahham R, De Gregorio N, Gherardelli C, Soto C (2023) Lewy body-like pathology and loss of dopaminergic neurons in midbrain organoids derived from familial Parkinson's disease patient. Cells 12(4):625

Bieschke J, Weber P, Sarafoff N, Beekes M, Giese A, Kretzschmar H (2004) Autocatalytic self-propagation of misfolded prion protein. Proc Natl Acad Sci USA 101(33):12207–12211

Blattler T, Brandner S, Raeber AJ, Klein MA, Voigtlander T, Weissmann C, Aguzzi A (1997) PrP-expressing tissue required for transfer of scrapie infectivity from spleen to brain. Nature 389(6646):69–73

Bowles KR, Silva MC, Whitney K, Bertucci T, Berlind JE, Lai JD, Garza JC, Boles NC, Mahali S, Strang KH, Marsh JA, Chen C, Pugh DA, Liu Y, Gordon RE, Goderie SK, Chowdhury R, Lotz S, Lane K, Crary JF, Haggarty SJ, Karch CM, Ichida JK, Goate AM, Temple S (2021) ELAVL4, splicing, and glutamatergic dysfunction precede neuron loss in MAPT mutation cerebral organoids. Cell 184(17):4547–4563 e4517

Brahic M, Bousset L, Bieri G, Melki R, Gitler AD (2016) Axonal transport and secretion of fibrillar forms of alpha-synuclein, Abeta42 peptide and HTTExon 1. Acta Neuropathol 131(4):539–548

Braun AR, Kochen NN, Yuen SL, Liao EE, Cornea RL, Thomas DD, Sachs JN (2023) Advancements in a FRET biosensor for live-cell fluorescence-lifetime high-throughput screening of alpha-synuclein. ASN Neuro 15:17590914231184086

Calafate S, Buist A, Miskiewicz K, Vijayan V, Daneels G, de Strooper B, de Wit J, Verstreken P, Moechars D (2015) Synaptic contacts enhance cell-to-cell tau pathology propagation. Cell Rep 11(8):1176–1183

Campeau JL, Wu G, Bell JR, Rasmussen J, Sim VL (2013) Early increase and late decrease of purkinje cell dendritic spine density in prion-infected organotypic mouse cerebellar cultures. PLoS One 8(12):e81776

Campenot RB (1977) Local control of neurite development by nerve growth factor. Proc Natl Acad Sci USA 74(10):4516–4519

Campenot RB (1982) Development of sympathetic neurons in compartmentalized cultures. Il Local control of neurite growth by nerve growth factor. Dev Biol 93(1):1–12

Casadevall i Solvas X, Turek V, Prodromakis T, Edel JB (2012) Microfluidic evaporator for on-chip sample concentration. Lab Chip 12(20):4049–4054

Castilla J, Saa P, Hetz C, Soto C (2005) In vitro generation of infectious scrapie prions. Cell 121(2):195–206

Castilla J, Saa P, Morales R, Abid K, Maundrell K, Soto C (2006) Protein misfolding cyclic amplification for diagnosis and prion propagation studies. Methods Enzymol 412:3–21

Cavaliere F, Cerf L, Dehay B, Ramos-Gonzalez P, De Giorgi F, Bourdenx M, Bessede A, Obeso JA, Matute C, Ichas F, Bezard E (2017) In vitro alpha-synuclein neurotoxicity and spreading among neurons and astrocytes using Lewy body extracts from Parkinson disease brains. Neurobiol Dis 103:101–112

Chandler RL (1961) Encephalopathy in mice produced by inoculation with scrapie brain material. Lancet 1(7191):1378–1379

Chen X, Sun G, Tian E, Zhang M, Davtyan H, Beach TG, Reiman EM, Blurton-Jones M, Holtzman DM, Shi Y (2021) Modeling sporadic Alzheimer's disease in human brain organoids under serum exposure. Adv Sci (Weinh) 8(18):e2101462

Chen K, Martens YA, Meneses A, Ryu DH, Lu W, Raulin AC, Li F, Zhao J, Chen Y, Jin Y, Linares C, Goodwin M, Li Y, Liu CC, Kanekiyo T, Holtzman DM, Golde TE, Bu G, Zhao N (2022) LRP1 is a neuronal receptor for alpha-synuclein uptake and spread. Mol Neurodegener 17(1):57

Chen C, Kumbhar R, Wang H, Yang X, Gadhave K, Rastegar C, Kimura Y, Behensky A, Kotha S, Kuo G, Katakam S, Jeong D, Wang L, Wang A, Chen R, Zhang S, Jin L, Workman CJ, Vignali DAA, Pletinkova O, Jia H, Peng W, Nauen DW, Wong PC, Redding-Ochoa J, Troncoso JC, Ying M, Dawson VL, Dawson TM, Mao X (2024) Lymphocyte-activation gene 3 facilitates pathological tau neuron-to-neuron transmission. Adv Sci (Weinh) 11:e2303775

Chia SPS, Kong SLY, Pang JKS, Soh BS (2022) 3D human organoids: the next "viral" model for the molecular basis of infectious diseases. Biomedicines 10(7):1541

Cho H, Hashimoto T, Wong E, Hori Y, Wood LB, Zhao L, Haigis KM, Hyman BT, Irimia D (2013) Microfluidic chemotaxis platform for differentiating the roles of soluble and bound amyloid-beta on microglial accumulation. Sci Rep 3:1823

Choi SH, Kim YH, Hebisch M, Sliwinski C, Lee S, D'Avanzo C, Chen H, Hooli B, Asselin C, Muffat J, Klee JB, Zhang C, Wainger BJ, Peitz M, Kovacs DM, Woolf CJ, Wagner SL, Tanzi RE, Kim DY (2014) A three-dimensional human neural cell culture model of Alzheimer's disease. Nature 515(7526):274–278

Choi JH, Cho HY, Choi JW (2017) Microdevice platform for in vitro nervous system and its disease model. Bioengineering (Basel) 4(3):77

Chun W, Johnson GV (2007) Activation of glycogen synthase kinase 3beta promotes the intermolecular association of tau. The use of fluorescence resonance energy transfer microscopy. J Biol Chem 282(32):23410–23417

Congdon EE, Lin Y, Rajamohamedsait HB, Shamir DB, Krishnaswamy S, Rajamohamedsait WJ, Rasool S, Gonzalez V, Levenga J, Gu J, Hoeffer C, Sigurdsson EM (2016) Affinity of Tau antibodies for solubilized pathological Tau species but not their immunogen or insoluble Tau aggregates predicts in vivo and ex vivo efficacy. Mol Neurodegener 11(1):62

Cortez LM, Campeau J, Norman G, Kalayil M, Van der Merwe J, McKenzie D, Sim VL (2015) Bile acids reduce prion conversion, reduce neuronal loss, and prolong male survival in models of prion disease. J Virol 89(15):7660–7672

Costanzo M, Zurzolo C (2013) The cell biology of prion-like spread of protein aggregates: mechanisms and implication in neurodegeneration. Biochem J 452(1):1–17

Courte J, Le NA, Pan T, Bousset L, Melki R, Villard C, Peyrin JM (2023) Synapses do not facilitate prion-like transfer of alpha-synuclein: a quantitative study in reconstructed unidirectional neural networks. Cell Mol Life Sci 80(10):284

Croft CL, Noble W (2018) Preparation of organotypic brain slice cultures for the study of Alzheimer's disease. F1000Res 7:592

Croft CL, Wade MA, Kurbatskaya K, Mastrandreas P, Hughes MM, Phillips EC, Pooler AM, Perkinton MS, Hanger DP, Noble W (2017) Membrane association and release of wild-type and pathological tau from organotypic brain slice cultures. Cell Death Dis 8(3):e2671

Croft CL, Cruz PE, Ryu DH, Ceballos-Diaz C, Strang KH, Woody BM, Lin WL, Deture M, Rodriguez-Lebron E, Dickson DW, Chakrabarty P, Levites Y, Giasson BI, Golde TE (2019) rAAV-based brain slice culture models of Alzheimer's and Parkinson's disease inclusion pathologies. J Exp Med 216(3):539–555

Croft CL, Paterno G, Vause AR, Rowe LA, Ryu DH, Goodwin MS, Moran CA, Cruz PE, Giasson BI, Golde TE (2022) Optical pulse labeling studies reveal exogenous seeding slows alpha-synuclein clearance. NPJ Parkinsons Dis 8(1):173

Del Rio JA, Ferrer I (2020) Potential of microfluidics and lab-on-chip platforms to improve understanding of "prion-like" protein assembly and behavior. Front Bioeng Biotechnol 8:570692

del Rio JA, Soriano E (2010) Regenerating cortical connections in a dish: the entorhino-hippocampal organotypic slice co-culture as tool for pharmacological screening of molecules promoting axon regeneration. Nat Protoc 5(2):217–226

Del Rio JA, Ferrer I, Gavin R (2018) Role of cellular prion protein in interneuronal amyloid transmission. Prog Neurobiol 165–167:87–102

Del Turco D, Deller T (2007) Organotypic entorhino-hippocampal slice cultures—a tool to study the molecular and cellular regulation of axonal regeneration and collateral sprouting in vitro. Methods Mol Biol 399:55–66

Deleglise B, Magnifico S, Duplus E, Vaur P, Soubeyre V, Belle M, Vignes M, Viovy JL, Jacotot E, Peyrin JM, Brugg B (2014) Beta-amyloid induces a dying-back process and remote trans-synaptic alterations in a microfluidic-based reconstructed neuronal network. Acta Neuropathol Commun 2:145

Dixit CK, Kaushik AK (2016) Microfluidics for biologists: fundamentals and applications. Springer International Publishing: Imprint: Springer: 1 online resource (VII, 252 pages 112 illustrations, 298 illustrations in color, Cham

Dujardin S, Lecolle K, Caillierez R, Begard S, Zommer N, Lachaud C, Carrier S, Dufour N, Auregan G, Winderickx J, Hantraye P, Deglon N, Colin M, Buee L (2014) Neuron-to-neuron wild-type Tau protein transfer through a trans-synaptic mechanism: relevance to sporadic tauopathies. Acta Neuropathol Commun 2:14

Eigen M (1996) Prionics or the kinetic basis of prion diseases. Biophys Chem 63(1):A1–18

Elfarrash S, Jensen NM, Ferreira N, Betzer C, Thevathasan JV, Diekmann R, Adel M, Omar NM, Boraie MZ, Gad S, Ries J, Kirik D, Nabavi S, Jensen PH (2019) Organotypic slice culture model demonstrates inter-neuronal spreading of alpha-synuclein aggregates. Acta Neuropathol Commun 7(1):213

Erana H (2019) The Prion 2018 round tables (II): Abeta, tau, alpha-synuclein... are they prions, prion-like proteins, or what? Prion 13(1):41–45

Fairfoul G, McGuire LI, Pal S, Ironside JW, Neumann J, Christie S, Joachim C, Esiri M, Evetts SG, Rolinski M, Baig F, Ruffmann C, Wade-Martins R, Hu MT, Parkkinen L, Green AJ (2016) Alpha-synuclein RT-QuIC in the CSF of patients with alpha-synucleinopathies. Ann Clin Transl Neurol 3(10):812–818

Falsig J, Aguzzi A (2008) The prion organotypic slice culture assay—POSCA. Nat Protoc 3(4):555–562

Falsig J, Julius C, Margalith I, Schwarz P, Heppner FL, Aguzzi A (2008) A versatile prion replication assay in organotypic brain slices. Nat Neurosci 11(1):109–117

Feiler MS, Strobel B, Freischmidt A, Helferich AM, Kappel J, Brewer BM, Li D, Thal DR, Walther P, Ludolph AC, Danzer KM, Weishaupt JH (2015) TDP-43 is intercellularly transmitted across axon terminals. J Cell Biol 211(4):897–911

Ferreira NDC, Caughey B (2020) Proteopathic seed amplification assays for neurodegenerative disorders. Clin Lab Med 40(3):257–270

Ferrer I (2018) Oligodendrogliopathy in neurodegenerative diseases with abnormal protein aggregates: the forgotten partner. Prog Neurobiol 169:24–54

Ferrer I, Aguilo Garcia M, Carmona M, Andres-Benito P, Torrejon-Escribano B, Garcia-Esparcia P, Del Rio JA (2019) Involvement of oligodendrocytes in tau seeding and spreading in tauopathies. Front Aging Neurosci 11:112

Ferrer I, Andres-Benito P, Zelaya MV, Aguirre MEE, Carmona M, Ausin K, Lachen-Montes M, Fernandez-Irigoyen J, Santamaria E, Del Rio JA (2020) Familial globular glial tauopathy linked to MAPT mutations: molecular neuropathology and seeding capacity of a prototypical mixed neuronal and glial tauopathy. Acta Neuropathol 139(4):735–771

Ferrone FA, Hofrichter J, Eaton WA (1985) Kinetics of sickle hemoglobin polymerization. I. Studies using temperature-jump and laser photolysis techniques. J Mol Biol 183(4):591–610

Fitzpatrick AW, Debelouchina GT, Bayro MJ, Clare DK, Caporini MA, Bajaj VS, Jaroniec CP, Wang L, Ladizhansky V, Muller SA, MacPhee CE, Waudby CA, Mott HR, De Simone A, Knowles TP, Saibil HR, Vendruscolo M, Orlova EV, Griffin RG, Dobson CM (2013) Atomic structure and hierarchical assembly of a cross-beta amyloid fibril. Proc Natl Acad Sci USA 110(14):5468–5473

Folch i Folch A (2013) Introduction to bioMEMS. CRC Press, Boca Raton

Foliaki ST, Groveman BR, Yuan J, Walters R, Zhang S, Tesar P, Zou W, Haigh CL (2020) Pathogenic prion protein isoforms are not present in cerebral organoids generated from asymptomatic donors carrying the e200k mutation associated with familial prion disease. Pathogens 9(6):482

Foliaki ST, Schwarz B, Groveman BR, Walters RO, Ferreira NC, Orru CD, Smith A, Wood A, Schmit OM, Freitag P, Yuan J, Zou W, Bosio CM, Carroll JA, Haigh CL (2021) Neuronal excitatory-to-inhibitory balance is altered in cerebral organoid models of genetic neurological diseases. Mol Brain 14(1):156

Freundt EC, Maynard N, Clancy EK, Roy S, Bousset L, Sourigues Y, Covert M, Melki R, Kirkegaard K, Brahic M (2012) Neuron-to-neuron transmission of alpha-synuclein fibrils through axonal transport. Ann Neurol 72(4):517–524

Frey B, Holzinger D, Taylor K, Ehrnhoefer DE, Striebinger A, Biesinger S, Gasparini L, O'Neill MJ, Wegner F, Barghorn S, Hoglinger GU, Heym RG (2023) Tau seed amplification assay reveals relationship between seeding and pathological forms of tau in Alzheimer's disease brain. Acta Neuropathol Commun 11(1):181

Gahwiler BH (1981) Morphological differentiation of nerve cells in thin organotypic cultures derived from rat hippocampus and cerebellum. Proc R Soc Lond B Biol Sci 211(1184):287–290

Gahwiler BH (1988) Organotypic cultures of neural tissue. Trends Neurosci 11(11):484–489

Gahwiler BH, Capogna M, Debanne D, McKinney RA, Thompson SM (1997) Organotypic slice cultures: a technique has come of age. Trends Neurosci 20(10):471–477

Goedert M, Masuda-Suzukake M, Falcon B (2017) Like prions: the propagation of aggregated tau and alpha-synuclein in neurodegeneration. Brain 140(Pt 2):266–278

Goniotaki D, Lakkaraju AKK, Shrivastava AN, Bakirci P, Sorce S, Senatore A, Marpakwar R, Hornemann S, Gasparini F, Triller A, Aguzzi A (2017) Inhibition of group-I metabotropic glutamate receptors protects against prion toxicity. PLoS Pathog 13(11):e1006733

Gonzalez C, Armijo E, Bravo-Alegria J, Becerra-Calixto A, Mays CE, Soto C (2018) Modeling amyloid beta and tau pathology in human cerebral organoids. Mol Psychiatry 23(12):2363–2374

Gonzalez-Montalban N, Makarava N, Ostapchenko VG, Savtchenk R, Alexeeva I, Rohwer RG, Baskakov IV (2011) Highly efficient protein misfolding cyclic amplification. PLoS Pathog 7(2):e1001277

Gough KC, Maddison BC (2010) Prion transmission: prion excretion and occurrence in the environment. Prion 4(4):275–282

Gribaudo S, Tixador P, Bousset L, Fenyi A, Lino P, Melki R, Peyrin JM, Perrier AL (2019) Propagation of alpha-synuclein strains within human reconstructed neuronal network. Stem Cell Rep 12(2):230–244

Gribaudo S, Bousset L, Courte J, Melki R, Peyrin JM, Perrier AL (2023) Propagation of distinct alpha-synuclein strains within human reconstructed neuronal network and associated neuronal dysfunctions. Methods Mol Biol 2551:357–378

Groveman BR, Orru CD, Hughson AG, Raymond LD, Zanusso G, Ghetti B, Campbell KJ, Safar J, Galasko D, Caughey B (2018) Rapid and ultra-sensitive quantitation of disease-associated alpha-synuclein seeds in brain and cerebrospinal fluid by alphaSyn RT-QuIC. Acta Neuropathol Commun 6(1):7

Groveman BR, Foliaki ST, Orru CD, Zanusso G, Carroll JA, Race B, Haigh CL (2019) Sporadic Creutzfeldt-Jakob disease prion infection of human cerebral organoids. Acta Neuropathol Commun 7(1):90

Groveman BR, Walters R, Haigh CL (2020) Using our mini-brains: cerebral organoids as an improved cellular model for human prion disease. Neural Regen Res 15(6):1019–1020

Groveman BR, Smith A, Williams K, Haigh CL (2021) Cerebral organoids as a new model for prion disease. PLoS Pathog 17(7):e1009747

Groveman BR, Race B, Foliaki ST, Williams K, Hughson AG, Baune C, Zanusso G, Haigh CL (2023) Sporadic Creutzfeldt-Jakob disease infected human cerebral organoids retain the original human brain subtype features following transmission to humanized transgenic mice. Acta Neuropathol Commun 11(1):28

Guo MT, Rotem A, Heyman JA, Weitz DA (2012) Droplet microfluidics for high-throughput biological assays. Lab Chip 12(12):2146–2155

Habibey R, Rojo Arias JE, Striebel J, Busskamp V (2022) Microfluidics for neuronal cell and circuit engineering. Chem Rev 122(18):14842–14880

Hall GF, Patuto BA (2012) Is tau ready for admission to the prion club? Prion 6(3):223–233

Hallinan GI, Pitera AP, Patel P, West J, Deinhardt K (2019) Minimalistic in vitro systems for investigating tau pathology. J Neurosci Methods 319:69–76

Hallinan GI, Lopez DM, Vargas-Caballero M, West J, Deinhardt K (2020) Co-culture of murine neurons using a microfluidic device for the study of tau misfolding propagation. Bio Protoc 10(16):e3718

Harwell CS, Coleman MP (2016) Synaptophysin depletion and intraneuronal Abeta in organotypic hippocampal slice cultures from huAPP transgenic mice. Mol Neurodegener 11(1):44

Herold KE, Rasooly A (2009) Lab on a chip technology. Caister Academic Press, Norfolk

Herva ME, Relano-Gines A, Villa A, Torres JM (2010) Prion infection of differentiated neurospheres. J Neurosci Methods 188(2):270–275

Herva ME, Zibaee S, Fraser G, Barker RA, Goedert M, Spillantini MG (2014) Anti-amyloid compounds inhibit alpha-synuclein aggregation induced by protein misfolding cyclic amplification (PMCA). J Biol Chem 289(17):11897–11905

Hianik T, Porfireva A, Grman I, Evtugyn G (2009) EQCM biosensors based on DNA aptamers and antibodies for rapid detection of prions. Protein Pept Lett 16(4):363–367

Hirose O, Itabashi M, Takei T, Nitta K (2015) Comparison of a novel chemiluminescence enzyme immunoassay (CLEIA) with enzyme-linked immunosorbent assay (ELISA) for the determination of MPO-ANCA in patients with ANCA-associated vasculitis. Mod Rheumatol 25(2):230–234

Holmes BB, Diamond MI (2017) Cellular models for the study of Prions. Cold Spring Harb Perspect Med 7(2):a024026

Hu PP, Zheng LL, Zhan L, Li JY, Zhen SJ, Liu H, Luo LF, Xiao GF, Huang CZ (2013) Metal-enhanced fluorescence of nano-core-shell structure used for sensitive detection of prion protein with a dual-aptamer strategy. Anal Chim Acta 787:239–245

Huang J, Yuan X, Chen L, Hu B, Wang H, Wang Y, Huang W (2024) Pathological alpha-synuclein detected by real-time quaking-induced conversion in synucleinopathies. Exp Gerontol 187:112366

Iwamaru Y, Mathiason CK, Telling GC, Hoover EA (2017) Chronic wasting disease prion infection of differentiated neurospheres. Prion 11(4):277–283

Jack CR Jr, Bennett DA, Blennow K, Carrillo MC, Dunn B, Haeberlein SB, Holtzman DM, Jagust W, Jessen F, Karlawish J, Liu E, Molinuevo JL, Montine T, Phelps C, Rankin KP, Rowe CC, Scheltens P, Siemers E, Snyder HM, Sperling R, Contributors (2018) NIA-AA Research Framework: toward a biological definition of Alzheimer's disease. Alzheimers Dement 14(4):535–562

Jarrett JT, Lansbury PT Jr (1993) Seeding "one-dimensional crystallization" of amyloid: a pathogenic mechanism in Alzheimer's disease and scrapie? Cell 73(6):1055–1058

Jellinger KA, Wenning GK, Stefanova N (2021) Is multiple system atrophy a prion-like disorder? Int J Mol Sci 22(18):10093

Jiayu W, Xiong W, Jiping L, Wensen L, Ming X, Linna L, Jing X, Haiying W, Hongwei G (2009) A rapid method for detection of PrP by surface plasmon resonance (SPR). Arch Virol 154(12):1901–1908

Johnson CJ, Aiken JM, McKenzie D, Samuel MD, Pedersen JA (2012) Highly efficient amplification of chronic wasting disease agent by protein misfolding cyclic amplification with beads (PMCAb). PLoS One 7(4):e35383

Joshi T, Ahuja N (2023) The prion basis of progressive neurodegenerative disorders. Interdiscip Perspect Infect Dis 2023:6687264

Kara E, Marks JD, Aguzzi A (2018) Toxic protein spread in neurodegeneration: reality versus fantasy. Trends Mol Med 24(12):1007–1020

Kfoury N, Holmes BB, Jiang H, Holtzman DM, Diamond MI (2012) Trans-cellular propagation of Tau aggregation by fibrillar species. J Biol Chem 287(23):19440–19451

Kim AC, Lim S, Kim YK (2018) Metal ion effects on abeta and tau aggregation. Int J Mol Sci 19(1):128

Klohn PC, Stoltze L, Flechsig E, Enari M, Weissmann C (2003) A quantitative, highly sensitive cell-based infectivity assay for mouse scrapie prions. Proc Natl Acad Sci USA 100(20):11666–11671

Knowles TP, Waudby CA, Devlin GL, Cohen SI, Aguzzi A, Vendruscolo M, Terentjev EM, Welland ME, Dobson CM (2009) An analytical solution to the kinetics of breakable filament assembly. Science 326(5959):1533–1537

Kondru N, Manne S, Greenlee J, West Greenlee H, Anantharam V, Halbur P, Kanthasamy A, Kanthasamy A (2017) Integrated organotypic slice cultures and RT-QuIC (OSCAR) assay: implications for translational discovery in protein misfolding diseases. Sci Rep 7:43155

Kondru N, Manne S, Kokemuller R, Greenlee J, Greenlee MHW, Nichols T, Kong Q, Anantharam V, Kanthasamy A, Halbur P, Kanthasamy AG (2020) An ex vivo brain slice culture model of chronic wasting disease: implications for disease pathogenesis and therapeutic development. Sci Rep 10(1):7640

Kopp MRG, Linsenmeier M, Hettich B, Prantl S, Stavrakis S, Leroux JC, Arosio P (2020) Microfluidic shrinking droplet concentrator for analyte detection and phase separation of protein solutions. Anal Chem 92(8):5803–5812

Korde DS, Humpel C (2022) Spreading of P301S aggregated tau investigated in organotypic mouse brain slice cultures. Biomol Ther 12(9):1164

Kovacs GG (2019) Molecular pathology of neurodegenerative diseases: principles and practice. J Clin Pathol 72(11):725–735

Krance SH, Luke R, Shenouda M, Israwi AR, Colpitts SJ, Darwish L, Strauss M, Watts JC (2020) Cellular models for discovering prion disease therapeutics: Progress and challenges. J Neurochem 153(2):150–172

Krejciova Z, Alibhai J, Zhao C, Krencik R, Rzechorzek NM, Ullian EM, Manson J, Ironside JW, Head MW, Chandran S (2017) Human stem cell-derived astrocytes replicate human prions in a PRNP genotype-dependent manner. J Exp Med 214(12):3481–3495

Kuncova-Kallio J, Kallio PJ (2006) PDMS and its suitability for analytical microfluidic devices. Conf Proc IEEE Eng Med Biol Soc 2006:2486–2489

Kuzkina A, Bargar C, Schmitt D, Rossle J, Wang W, Schubert AL, Tatsuoka C, Gunzler SA, Zou WQ, Volkmann J, Sommer C, Doppler K, Chen SG (2021) Diagnostic value of skin RT-QuIC in Parkinson's disease: a two-laboratory study. NPJ Parkinsons Dis 7(1):99

Lagally E (2014) Microfluidics and nanotechnology : biosensing to the single molecule limit. CRC Press/Taylor & Francis Group, Boca Raton

Le Nel A, Minc N, Smadja C, Slovakova M, Bilkova Z, Peyrin JM, Viovy JL, Taverna M (2008) Controlled proteolysis of normal and pathological prion protein in a microfluidic chip. Lab Chip 8(2):294–301

Leal-Lasarte MM, Franco JM, Labrador-Garrido A, Pozo D, Roodveldt C (2017) Extracellular TDP-43 aggregates target MAPK/MAK/MRK overlapping kinase (MOK) and trigger caspase-3/IL-18 signaling in microglia. FASEB J 31(7):2797–2816

Lee S-JJ, Sundararajan N (2010) Microfabrication for microfluidics. Artech House, Boston

Leggett C, McGehee DS, Mastrianni J, Yang W, Bai T, Brorson JR (2012) Tunicamycin produces TDP-43 cytoplasmic inclusions in cultured brain organotypic slices. J Neurol Sci 317(1–2):66–73

Lengyel-Zhand Z, Ferrie JJ, Janssen B, Hsieh CJ, Graham T, Xu KY, Haney CM, Lee VM, Trojanowski JQ, Petersson EJ, Mach RH (2020) Synthesis and characterization of high affinity fluorogenic alpha-synuclein probes. Chem Commun (Camb) 56(24):3567–3570

Li D (2015) Encyclopedia of microfluidics and nanofluidics. Springer, New York

Liu Y, Senatore A, Sorce S, Nuvolone M, Guo J, Gumus ZH, Aguzzi A (2022) Brain aging is faithfully modelled in organotypic brain slices and accelerated by prions. Commun Biol 5(1):557

Lo CH, Lim CK, Ding Z, Wickramasinghe SP, Braun AR, Ashe KH, Rhoades E, Thomas DD, Sachs JN (2019) Targeting the ensemble of heterogeneous tau oligomers in cells: a novel small molecule screening platform for tauopathies. Alzheimers Dement 15(11):1489–1502

Lonchamp E, Dupont JL, Beekenkamp H, Poulain B, Bossu JL (2006) The mouse cerebellar cortex in organotypic slice cultures: an in vitro model to analyze the consequences of mutations and pathologies on neuronal survival, development, and function. Crit Rev Neurobiol 18(1–2):179–186

Manne S, Kondru N, Jin H, Anantharam V, Huang X, Kanthasamy A, Kanthasamy AG (2020a) alpha-Synuclein real-time quaking-induced conversion in the submandibular glands of Parkinson's disease patients. Mov Disord 35(2):268–278

Manne S, Kondru N, Jin H, Serrano GE, Anantharam V, Kanthasamy A, Adler CH, Beach TG, Kanthasamy AG (2020b) Blinded RT-QuIC analysis of alpha-synuclein biomarker in skin tissue from Parkinson's disease patients. Mov Disord 35(12):2230–2239

Manno D, Filippo E, Fiore R, Serra A, Urso E, Rizzello A, Maffia M (2010) Monitoring prion protein expression in complex biological samples by SERS for diagnostic applications. Nanotechnology 21(16):165502

Marbiah MM, Harvey A, West BT, Louzolo A, Banerjee P, Alden J, Grigoriadis A, Hummerich H, Kan HM, Cai Y, Bloom GS, Jat P, Collinge J, Klohn PC (2014) Identification of a gene regulatory network associated with prion replication. EMBO J 33(14):1527–1547

Marek P, Mukherjee S, Zanni MT, Raleigh DP (2010) Residue-specific, real-time characterization of lag-phase species and fibril growth during amyloid formation: a combined fluorescence and IR study of p-cyanophenylalanine analogs of islet amyloid polypeptide. J Mol Biol 400(4):878–888

Margalith I, Suter C, Ballmer B, Schwarz P, Tiberi C, Sonati T, Falsig J, Nystrom S, Hammarstrom P, Aslund A, Nilsson KP, Yam A, Whitters E, Hornemann S, Aguzzi A (2012) Polythiophenes inhibit prion propagation by stabilizing prion protein (PrP) aggregates. J Biol Chem 287(23):18872–18887

Masone A, Zucchelli C, Caruso E, Lavigna G, Erana H, Giachin G, Tapella L, Comerio L, Restelli E, Raimondi I, Elezgarai SR, De Leo F, Quilici G, Taiarol L, Oldrati M, Lorenzo NL, Garcia-Martinez S, Cagnotto A, Lucchetti J, Gobbi M, Vanni I, Nonno R, Di Bari MA, Tully MD, Cecatiello V, Ciossani G, Pasqualato S, Van Anken E, Salmona M, Castilla J, Requena JR, Banfi S, Musco G, Chiesa R (2023) A tetracationic porphyrin with dual anti-prion activity. iScience 26(9):107480

Matamoros-Angles A, Gayosso LM, Richaud-Patin Y, di Domenico A, Vergara C, Hervera A, Sousa A, Fernandez-Borges N, Consiglio A, Gavin R, Lopez de Maturana R, Ferrer I, Lopez de Munain A, Raya A, Castilla J, Sanchez-Pernaute R, Del Rio JA (2018) iPS cell cultures from a Gerstmann-Straussler-Scheinker patient with the Y218N PRNP mutation recapitulate tau pathology. Mol Neurobiol 55(4):3033–3048

McDonald JC, Whitesides GM (2002) Poly(dimethylsiloxane) as a material for fabricating microfluidic devices. Acc Chem Res 35(7):491–499

McDonald JC, Duffy DC, Anderson JR, Chiu DT, Wu H, Schueller OJ, Whitesides GM (2000) Fabrication of microfluidic systems in poly(dimethylsiloxane). Electrophoresis 21(1):27–40

Meisl G, Kirkegaard JB, Arosio P, Michaels TC, Vendruscolo M, Dobson CM, Linse S, Knowles TP (2016) Molecular mechanisms of protein aggregation from global fitting of kinetic models. Nat Protoc 11(2):252–272

Meisl G, Knowles TP, Klenerman D (2020) The molecular processes underpinning prion-like spreading and seed amplification in protein aggregation. Curr Opin Neurobiol 61:58–64

Meisl G, Kurt T, Condado-Morales I, Bett C, Sorce S, Nuvolone M, Michaels TCT, Heinzer D, Avar M, Cohen SIA, Hornemann S, Aguzzi A, Dobson CM, Sigurdson CJ, Knowles TPJ (2021) Scaling analysis reveals the mechanism and rates of prion replication in vivo. Nat Struct Mol Biol 28(4):365–372

Messing L, Decker JM, Joseph M, Mandelkow E, Mandelkow EM (2013) Cascade of tau toxicity in inducible hippocampal brain slices and prevention by aggregation inhibitors. Neurobiol Aging 34(5):1343–1354

Metrick MA 2nd, Ferreira NDC, Saijo E, Kraus A, Newell K, Zanusso G, Vendruscolo M, Ghetti B, Caughey B (2020) A single ultrasensitive assay for detection and discrimination of tau aggregates of Alzheimer and Pick diseases. Acta Neuropathol Commun 8(1):22

Meyer V, Dinkel PD, Rickman Hager E, Margittai M (2014) Amplification of Tau fibrils from minute quantities of seeds. Biochemistry 53(36):5804–5809

Miller LVC, Mukadam AS, Durrant CS, Vaysburd MJ, Katsinelos T, Tuck BJ, Sanford S, Sheppard O, Knox C, Cheng S, James LC, Coleman MP, McEwan WA (2021) Tau assemblies do not behave like independently acting prion-like particles in mouse neural tissue. Acta Neuropathol Commun 9(1):41

Mingorance A, Sole M, Muneton V, Martinez A, Nieto-Sampedro M, Soriano E, del Rio JA (2006) Regeneration of lesioned entorhino-hippocampal axons in vitro by combined degradation of inhibitory proteoglycans and blockade of Nogo-66/NgR signaling. FASEB J 20(3):491–493

Minteer SD (2006) Microfluidic techniques: reviews and protocols. Humana Press, Totowa

Moda F, Gambetti P, Notari S, Concha-Marambio L, Catania M, Park KW, Maderna E, Suardi S, Haik S, Brandel JP, Ironside J, Knight R, Tagliavini F, Soto C (2014) Prions in the urine of patients with variant Creutzfeldt-Jakob disease. N Engl J Med 371(6):530–539

Moelgg K, Jummun F, Humpel C (2021) Spreading of beta-amyloid in organotypic mouse brain slices and microglial elimination and effects on cholinergic neurons. Biomol Ther 11(3):434

Mok TH, Nihat A, Majbour N, Sequeira D, Holm-Mercer L, Coysh T, Darwent L, Batchelor M, Groveman BR, Orr CD, Hughson AG, Heslegrave A, Laban R, Veleva E, Paterson RW, Keshavan A, Schott JM, Swift IJ, Heller C, Rohrer JD, Gerhard A, Butler C, Rowe JB, Masellis M, Chapman M, Lunn MP, Bieschke J, Jackson GS, Zetterberg H, Caughey B, Rudge P,

Collinge J, Mead S (2023) Seed amplification and neurodegeneration marker trajectories in individuals at risk of prion disease. Brain 146(6):2570–2583

Morales R, Duran-Aniotz C, Diaz-Espinoza R, Camacho MV, Soto C (2012) Protein misfolding cyclic amplification of infectious prions. Nat Protoc 7(7):1397–1409

Morris AM, Watzky MA, Finke RG (2009) Protein aggregation kinetics, mechanism, and curve-fitting: a review of the literature. Biochim Biophys Acta 1794(3):375–397

Muhsin SA, Abdullah A, Kobashigawa E, Al-Amidie M, Russell S, Zhang MZ, Zhang S, Almasri M (2023) A microfluidic biosensor for the diagnosis of chronic wasting disease. Microsyst Nanoeng 9:104

Neto E, Leitao L, Sousa DM, Alves CJ, Alencastre IS, Aguiar P, Lamghari M (2016) Compartmentalized microfluidic platforms: the unrivaled breakthrough of in vitro tools for neurobiological research. J Neurosci 36(46):11573–11584

Ng JM, Gitlin I, Stroock AD, Whitesides GM (2002) Components for integrated poly(dimethylsiloxane) microfluidic systems. Electrophoresis 23(20):3461–3473

Novotny R, Langer F, Mahler J, Skodras A, Vlachos A, Wegenast-Braun BM, Kaeser SA, Neher JJ, Eisele YS, Pietrowski MJ, Nilsson KP, Deller T, Staufenbiel M, Heimrich B, Jucker M (2016) Conversion of synthetic abeta to in vivo active seeds and amyloid plaque formation in a hippocampal slice culture model. J Neurosci 36(18):5084–5093

Orru CD, Wilham JM, Raymond LD, Kuhn F, Schroeder B, Raeber AJ, Caughey B (2011) Prion disease blood test using immunoprecipitation and improved quaking-induced conversion. MBio 2(3):e00078–e00011

Orru CD, Groveman BR, Hughson AG, Zanusso G, Coulthart MB, Caughey B (2015) Rapid and sensitive RT-QuIC detection of human Creutzfeldt-Jakob disease using cerebrospinal fluid. MBio 6(1):e02451

Park JW, Vahidi B, Taylor AM, Rhee SW, Jeon NL (2006) Microfluidic culture platform for neuroscience research. Nat Protoc 1(4):2128–2136

Park MC, Kim M, Lim GT, Kang SM, An SS, Kim TS, Kang JY (2016) Droplet-based magnetic bead immunoassay using microchannel-connected multiwell plates (muCHAMPs) for the detection of amyloid beta oligomers. Lab Chip 16(12):2245–2253

Park J, Wetzel I, Marriott I, Dreau D, D'Avanzo C, Kim DY, Tanzi RE, Cho H (2018) A 3D human triculture system modeling neurodegeneration and neuroinflammation in Alzheimer's disease. Nat Neurosci 21(7):941–951

Peng C, Trojanowski JQ, Lee VM (2020) Protein transmission in neurodegenerative disease. Nat Rev Neurol 16(4):199–212

Pfammatter M, Andreasen M, Meisl G, Taylor CG, Adamcik J, Bolisetty S, Sanchez-Ferrer A, Klenerman D, Dobson CM, Mezzenga R, Knowles TPJ, Aguzzi A, Hornemann S (2017) Absolute quantification of amyloid propagons by digital microfluidics. Anal Chem 89(22):12306–12313

Pineau H, Sim V (2020) POSCAbilities: the application of the prion organotypic slice culture assay to neurodegenerative disease research. Biomol Ther 10(7):1079

Pineau H, Sim VL (2021) From cell culture to organoids-model systems for investigating prion strain characteristics. Biomol Ther 11(1):106

Polanco JC, Li C, Durisic N, Sullivan R, Gotz J (2018) Exosomes taken up by neurons hijack the endosomal pathway to spread to interconnected neurons. Acta Neuropathol Commun 6(1):10

Prusiner SB (1982) Novel proteinaceous infectious particles cause scrapie. Science 216(4542):136–144

Prusiner SB (2017) Prion biology: a subject collection from Cold Spring Harbor perspectives in biology. Cold Spring Harbor Laboratory Press, Cold Spring Harbor/New York

Prusiner SB, Woerman AL, Mordes DA, Watts JC, Rampersaud R, Berry DB, Patel S, Oehler A, Lowe JK, Kravitz SN, Geschwind DH, Glidden DV, Halliday GM, Middleton LT, Gentleman SM, Grinberg LT, Giles K (2015) Evidence for alpha-synuclein prions causing multiple system atrophy in humans with parkinsonism. Proc Natl Acad Sci USA 112(38):E5308–E5317

Ra EA, Kim MS, Lee G (2023) Optogenetic induction of alpha-synuclein aggregation in human dopaminergic neurons to model Parkinson's disease pathology. STAR Protoc 4(4):102609

Race RE, Fadness LH, Chesebro B (1987) Characterization of scrapie infection in mouse neuroblastoma cells. J Gen Virol 68(Pt 5):1391–1399

Raja WK, Mungenast AE, Lin YT, Ko T, Abdurrob F, Seo J, Tsai LH (2016) Self-Organizing 3D human neural tissue derived from induced pluripotent stem cells recapitulate Alzheimer's disease phenotypes. PLoS One 11(9):e0161969

Raja WK, Neves E, Burke C, Jiang X, Xu P, Rhodes KJ, Khurana V, Scannevin RH, Chung CY (2022) Patient-derived three-dimensional cortical neurospheres to model Parkinson's disease. PLoS One 17(12):e0277532

Reimann RR, Puzio M, Rosati A, Emmenegger M, Schneider BL, Valdes P, Huang D, Caflisch A, Aguzzi A (2023) Rapid ex vivo reverse genetics identifies the essential determinants of prion protein toxicity. Brain Pathol 33(2):e13130

Rhee SW, Taylor AM, Tu CH, Cribbs DH, Cotman CW, Jeon NL (2005) Patterned cell culture inside microfluidic devices. Lab Chip 5(1):102–107

Rissin DM, Kan CW, Campbell TG, Howes SC, Fournier DR, Song L, Piech T, Patel PP, Chang L, Rivnak AJ, Ferrell EP, Randall JD, Provuncher GK, Walt DR, Duffy DC (2010) Single-molecule enzyme-linked immunosorbent assay detects serum proteins at subfemtomolar concentrations. Nat Biotechnol 28(6):595–599

Rivas-Santisteban R, Raich I, Aguinaga D, Saura CA, Franco R, Navarro G (2023) The expression of cellular prion protein, PrPC, favors pTau propagation and blocks NMDAR signaling in primary cortical neurons. Cells 12(2):283

Ruiz A, Joshi P, Mastrangelo R, Francolini M, Verderio C, Matteoli M (2014) Testing Abeta toxicity on primary CNS cultures using drug-screening microfluidic chips. Lab Chip 14(15):2860–2866

Saa P, Castilla J, Soto C (2006) Ultra-efficient replication of infectious prions by automated protein misfolding cyclic amplification. J Biol Chem 281(46):35245–35252

Saar KL, Muller T, Charmet J, Challa PK, Knowles TPJ (2018) Enhancing the resolution of micro free flow electrophoresis through spatially controlled sample injection. Anal Chem 90(15):8998–9005

Sabareesan AT, Udgaonkar JB (2017) The G126V mutation in the mouse prion protein hinders nucleation-dependent fibril formation by slowing initial fibril growth and by increasing the critical concentration. Biochemistry 56(44):5931–5942

Saborio GP, Permanne B, Soto C (2001) Sensitive detection of pathological prion protein by cyclic amplification of protein misfolding. Nature 411(6839):810–813

Sadeghzadeh J, Shahabi P, Farhoudi M, Ebrahimi-Kalan A, Mobed A, Shahpasand K (2023) Tau protein biosensors in the diagnosis of neurodegenerative diseases. Adv Pharm Bull 13(3):502–511

Saeed SM, Fine G (1967) Thioflavin-T for amyloid detection. Am J Clin Pathol 47(5):588–593

Saijo E, Ghetti B, Zanusso G, Oblak A, Furman JL, Diamond MI, Kraus A, Caughey B (2017) Ultrasensitive and selective detection of 3-repeat tau seeding activity in Pick disease brain and cerebrospinal fluid. Acta Neuropathol 133(5):751–765

Saijo E, Metrick MA 2nd, Koga S, Parchi P, Litvan I, Spina S, Boxer A, Rojas JC, Galasko D, Kraus A, Rossi M, Newell K, Zanusso G, Grinberg LT, Seeley WW, Ghetti B, Dickson DW, Caughey B (2020) 4-Repeat tau seeds and templating subtypes as brain and CSF biomarkers of frontotemporal lobar degeneration. Acta Neuropathol 139(1):63–77

Sala-Jarque J, Zimkowska K, Avila J, Ferrer I, Del Rio JA (2022) Towards a mechanistic model of tau-mediated pathology in tauopathies: what can we learn from cell-based in vitro assays? Int J Mol Sci 23(19):11527

Salvadores N, Shahnawaz M, Scarpini E, Tagliavini F, Soto C (2014) Detection of misfolded Abeta oligomers for sensitive biochemical diagnosis of Alzheimer's disease. Cell Rep 7(1):261–268

Sano K, Atarashi R, Satoh K, Ishibashi D, Nakagaki T, Iwasaki Y, Yoshida M, Murayama S, Mishima K, Nishida N (2018) Prion-like seeding of misfolded alpha-synuclein in the brains of dementia with Lewy body patients in RT-QUIC. Mol Neurobiol 55(5):3916–3930

Scheckel C, Aguzzi A (2018) Prions, prionoids and protein misfolding disorders. Nat Rev Genet 19(7):405–418
Schladitz C, Vieira EP, Hermel H, Mohwald H (1999) Amyloid-beta-sheet formation at the air-water interface. Biophys J 77(6):3305–3310
Scialo C, De Cecco E, Manganotti P, Legname G (2019) Prion and prion-like protein strains: deciphering the molecular basis of heterogeneity in neurodegeneration. Viruses 11(3):261
Shahnawaz M, Tokuda T, Waragai M, Mendez N, Ishii R, Trenkwalder C, Mollenhauer B, Soto C (2017) Development of a biochemical diagnosis of parkinson disease by detection of alpha-synuclein misfolded aggregates in cerebrospinal fluid. JAMA Neurol 74(2):163–172
Shembekar N, Chaipan C, Utharala R, Merten CA (2016) Droplet-based microfluidics in drug discovery, transcriptomics and high-throughput molecular genetics. Lab Chip 16(8):1314–1331
Shim JU, Cristobal G, Link DR, Thorsen T, Jia Y, Piattelli K, Fraden S (2007) Control and measurement of the phase behavior of aqueous solutions using microfluidics. J Am Chem Soc 129(28):8825–8835
Shimada H, Sato Y, Sasaki T, Shimozawa A, Imaizumi K, Shindo T, Miyao S, Kiyama K, Kondo T, Shibata S, Ishii S, Kuromitsu J, Aoyagi H, Ito D, Okano H (2022) A next-generation iPSC-derived forebrain organoid model of tauopathy with tau fibrils by AAV-mediated gene transfer. Cell Rep Methods 2(9):100289
Shin WS, Di J, Murray KA, Sun C, Li B, Bitan G, Jiang L (2019) Different amyloid-beta self-assemblies have distinct effects on intracellular tau aggregation. Front Mol Neurosci 12:268
Sofroniew MV, Dreifuss JJ, Gahwiler BH (1988) Slice cultures of rat hypothalamus examined by immunohistochemical staining for neurohypophyseal peptides and GFAP. Brain Res Bull 20(6):669–674
Song HL, Shim S, Kim DH, Won SH, Joo S, Kim S, Jeon NL, Yoon SY (2014) Beta-Amyloid is transmitted via neuronal connections along axonal membranes. Ann Neurol 75(1):88–97
Song Y, Cheng D, Zhao L (2018) Microfluidics: fundamental, devices and applications. Wiley-VCH, Weinheim
Soto C (2024) alpha-Synuclein seed amplification technology for Parkinson's disease and related synucleinopathies. Trends Biotechnol 42:829
Soto C, Saborio GP, Anderes L (2002) Cyclic amplification of protein misfolding: application to prion-related disorders and beyond. Trends Neurosci 25(8):390–394
Stefani A, Iranzo A, Holzknecht E, Perra D, Bongianni M, Gaig C, Heim B, Serradell M, Sacchetto L, Garrido A, Capaldi S, Sanchez-Gomez A, Cecchini MP, Mariotto S, Ferrari S, Fiorini M, Schmutzhard J, Cocchiara P, Vilaseca I, Brozzetti L, Monaco S, Jose Marti M, Seppi K, Tolosa E, Santamaria J, Hogl B, Poewe W, Zanusso G, S. group (2021) Alpha-synuclein seeds in olfactory mucosa of patients with isolated REM sleep behaviour disorder. Brain 144(4):1118–1126
Stoppini L, Buchs PA, Muller D (1991) A simple method for organotypic cultures of nervous tissue. J Neurosci Methods 37(2):173–182
Suttkus A, Holzer M, Morawski M, Arendt T (2016) The neuronal extracellular matrix restricts distribution and internalization of aggregated Tau-protein. Neuroscience 313:225–235
Tabeling P (2005) Introduction to microfluidics. Oxford University Press, Oxford, UK/New York
Takeda S, Wegmann S, Cho H, DeVos SL, Commins C, Roe AD, Nicholls SB, Carlson GA, Pitstick R, Nobuhara CK, Costantino I, Frosch MP, Muller DJ, Irimia D, Hyman BT (2015) Neuronal uptake and propagation of a rare phosphorylated high-molecular-weight tau derived from Alzheimer's disease brain. Nat Commun 6:8490
Tamaki Y, Ross JP, Alipour P, Castonguay CE, Li B, Catoire H, Rochefort D, Urushitani M, Takahashi R, Sonnen JA, Stifani S, Dion PA, Rouleau GA (2023) Spinal cord extracts of amyotrophic lateral sclerosis spread TDP-43 pathology in cerebral organoids. PLoS Genet 19(2):e1010606
Taylor AM, Jeon NL (2010) Micro-scale and microfluidic devices for neurobiology. Curr Opin Neurobiol 20(5):640–647
Taylor AM, Rhee SW, Tu CH, Cribbs DH, Cotman CW, Jeon NL (2003) Microfluidic multicompartment device for neuroscience research. Langmuir 19(5):1551–1556

Taylor AM, Blurton-Jones M, Rhee SW, Cribbs DH, Cotman CW, Jeon NL (2005) A microfluidic culture platform for CNS axonal injury, regeneration and transport. Nat Methods 2(8):599–605

Teh SY, Lin R, Hung LH, Lee AP (2008) Droplet microfluidics. Lab Chip 8(2):198–220

Terakawa MS, Lin Y, Kinoshita M, Kanemura S, Itoh D, Sugiki T, Okumura M, Ramamoorthy A, Lee YH (2018) Impact of membrane curvature on amyloid aggregation. Biochim Biophys Acta Biomembr 1860:1741

Thom T, Schmitz M, Fischer AL, Correia A, Correia S, Llorens F, Pique AV, Mobius W, Domingues R, Zafar S, Stoops E, Silva CJ, Fischer A, Outeiro TF, Zerr I (2022) Cellular prion protein mediates alpha-synuclein uptake, localization, and toxicity in vitro and in vivo. Mov Disord 37(1):39–51

Toprakcioglu Z, Challa P, Xu C, Knowles TPJ (2019) Label-free analysis of protein aggregation and phase behavior. ACS Nano 13(12):13940–13948

Tu PH, Galvin JE, Baba M, Giasson B, Tomita T, Leight S, Nakajo S, Iwatsubo T, Trojanowski JQ, Lee VM (1998) Glial cytoplasmic inclusions in white matter oligodendrocytes of multiple system atrophy brains contain insoluble alpha-synuclein. Ann Neurol 44(3):415–422

Ucar B, Stefanova N, Humpel C (2022) Spreading of aggregated alpha-synuclein in sagittal organotypic mouse brain slices. Biomol Ther 12(2):163

Uemura N, Uemura MT, Lo A, Bassil F, Zhang B, Luk KC, Lee VM, Takahashi R, Trojanowski JQ (2019) Slow progressive accumulation of oligodendroglial alpha-synuclein (alpha-syn) pathology in synthetic alpha-syn fibril-induced mouse models of synucleinopathy. J Neuropathol Exp Neurol 78(10):877–890

Uemura N, Uemura MT, Luk KC, Lee VM, Trojanowski JQ (2020) Cell-to-cell transmission of tau and alpha-synuclein. Trends Mol Med 26:936

Urrea L, Segura-Feliu M, Masuda-Suzukake M, Hervera A, Pedraz L, Garcia Aznar JM, Vila M, Samitier J, Torrents E, Ferrer I, Gavin R, Hagesawa M, Del Rio JA (2018) Involvement of cellular prion protein in alpha-synuclein transport in neurons. Mol Neurobiol 55(3):1847–1860

Usenovic M, Niroomand S, Drolet RE, Yao L, Gaspar RC, Hatcher NG, Schachter J, Renger JJ, Parmentier-Batteur S (2015) Internalized tau oligomers cause neurodegeneration by inducing accumulation of pathogenic tau in human neurons derived from induced pluripotent stem cells. J Neurosci 35(42):14234–14250

Vilette D, Courte J, Peyrin JM, Coudert L, Schaeffer L, Andreoletti O, Leblanc P (2018) Cellular mechanisms responsible for cell-to-cell spreading of prions. Cell Mol Life Sci 75(14):2557–2574

Volpicelli-Daley LA, Luk KC, Patel TP, Tanik SA, Riddle DM, Stieber A, Meaney DF, Trojanowski JQ, Lee VM (2011) Exogenous alpha-synuclein fibrils induce Lewy body pathology leading to synaptic dysfunction and neuron death. Neuron 72(1):57–71

Walters RO, Haigh CL (2023) Organoids for modeling prion diseases. Cell Tissue Res 392(1):97–111

Wang B, Underwood R, Kamath A, Britain C, McFerrin MB, McLean PJ, Volpicelli-Daley LA, Whitaker RH, Placzek WJ, Becker K, Ma J, Yacoubian TA (2018) 14-3-3 proteins reduce cell-to-cell transfer and propagation of pathogenic alpha-synuclein. J Neurosci 38(38):8211–8232

Wang F, Pritzkow S, Soto C (2023a) PMCA for ultrasensitive detection of prions and to study disease biology. Cell Tissue Res 392(1):307–321

Wang J, Devarajan DS, Nikoubashman A, Mittal J (2023b) Conformational properties of polymers at droplet interfaces as model systems for disordered proteins. ACS Macro Lett 12(11):1472–1478

Wegmann S, Maury EA, Kirk MJ, Saqran L, Roe A, DeVos SL, Nicholls S, Fan Z, Takeda S, Cagsal-Getkin O, William CM, Spires-Jones TL, Pitstick R, Carlson GA, Pooler AM, Hyman BT (2015) Removing endogenous tau does not prevent tau propagation yet reduces its neurotoxicity. EMBO J 34(24):3028–3041

Westergard T, Jensen BK, Wen X, Cai J, Kropf E, Iacovitti L, Pasinelli P, Trotti D (2016) Cell-to-cell transmission of dipeptide repeat proteins linked to C9orf72-ALS/FTD. Cell Rep 17(3):645–652

Whitesides GM, Ostuni E, Takayama S, Jiang X, Ingber DE (2001) Soft lithography in biology and biochemistry. Annu Rev Biomed Eng 3:335–373

Wilham JM, Orru CD, Bessen RA, Atarashi R, Sano K, Race B, Meade-White KD, Taubner LM, Timmes A, Caughey B (2010) Rapid end-point quantitation of prion seeding activity with sensitivity comparable to bioassays. PLoS Pathog 6(12):e1001217

Willbold D, Strodel B, Schroder GF, Hoyer W, Heise H (2021) Amyloid-type protein aggregation and prion-like properties of amyloids. Chem Rev 121(13):8285–8307

Wolf H, Hossinger A, Fehlinger A, Buttner S, Sim V, McKenzie D, Vorberg IM (2015) Deposition pattern and subcellular distribution of disease-associated prion protein in cerebellar organotypic slice cultures infected with scrapie. Front Neurosci 9:410

Wu JW, Herman M, Liu L, Simoes S, Acker CM, Figueroa H, Steinberg JI, Margittai M, Kayed R, Zurzolo C, Di Paolo G, Duff KE (2013) Small misfolded Tau species are internalized via bulk endocytosis and anterogradely and retrogradely transported in neurons. J Biol Chem 288(3):1856–1870

Xia N, Zhou B, Huang N, Jiang M, Zhang J, Liu L (2016) Visual and fluorescent assays for selective detection of beta-amyloid oligomers based on the inner filter effect of gold nanoparticles on the fluorescence of CdTe quantum dots. Biosens Bioelectron 85:625–632

Xu B, Wu SW, Lu CW, Deng Y, Liu W, Wei YG, Yang TY, Xu ZF (2013) Oxidative stress involvement in manganese-induced alpha-synuclein oligomerization in organotypic brain slice cultures. Toxicology 305:71–78

Xu Y, Martini-Stoica H, Zheng H (2016) A seeding based cellular assay of tauopathy. Mol Neurodegener 11:32

Yakupova EI, Bobyleva LG, Vikhlyantsev IM, Bobylev AG (2019) Congo Red and amyloids: history and relationship. Biosci Rep 39(1):BSR20181415

Yoon S, Go G, Yoon Y, Lim J, Lee G, Lee S (2021) Harnessing the physiological functions of cellular prion protein in the kidneys: applications for treating renal diseases. Biomol Ther 11(6):784

Zerr I, Ladogana A, Mead S, Hermann P, Forloni G, Appleby BS (2024) Creutzfeldt-Jakob disease and other prion diseases. Nat Rev Dis Primers 10(1):14

Zheng X, Liu P, Yang C, Wu X (2020) Amyloid protein aggregation in diabetes mellitus accelerate intervertebral disc degeneration. Med Hypotheses 141:109739

Zhou Y, Dong H, Liu L, Xu M (2015) Simple colorimetric detection of amyloid beta-peptide (1-40) based on aggregation of gold nanoparticles in the presence of copper ions. Small 11(18):2144–2149

Zimmer J, Gahwiler BH (1984) Cellular and connective organization of slice cultures of the rat hippocampus and fascia dentata. J Comp Neurol 228(3):432–446

Chapter 7
Tracking Prions by RT-QuIC: An Update

Christina D. Orrù, Bradley R. Groveman, and Byron Caughey

Abstract Prion seed amplification assays, such as real-time quaking-induced conversion (RT-QuIC), are providing ultrasensitive, specific, and increasingly practical means of detecting prions and diagnosing prion diseases in humans and other animals. Widespread efforts have greatly increased the breadth of RT-QuIC applications for research, surveillance, and clinical purposes. Here, we summarize recent progress in several key areas in which RT-QuIC assays are being applied and further developed.

Keywords Prions · RT-QuIC · Skin · Diagnostics · Ear punches · Preclinical detection · sCJD subtypes · Pregnancy · Solid surfaces · Quantitation · Seed amplification assay

Prion protein (PrP) prion diseases are mammalian neurodegenerative disorders involving the accumulation of abnormal deposits of the host's prion protein in various tissues, most importantly the central nervous system (Sigurdson et al. 2019; Caughey and Sim 2024; Zerr et al. 2024). Prion disease pathogenesis involves the conversion of the normal "cellular" form of PrP (PrP^{C}) into pathological forms, which, when infectious, are generically called PrP^{Sc} (Meyer et al. 1986; Artikis et al. 2022; Manka 2025). In humans, prion diseases are most often sporadic, that is, of no known origin other than the spontaneous misfolding and aggregation of PrP. The most common of those is sporadic Creutzfeldt–Jakob disease (sCJD) and its multiple subtypes (Zerr and Parchi 2018; Zerr et al. 2024). Sporadic prion diseases are also thought to occur in other animal species, most notably the so-called atypical forms of bovine spongiform encephalopathy (BSE) (Houston and Andreoletti 2019; Caughey and Sim 2024) and sheep scrapie (Benestad et al. 2008). Much rarer in

C. D. Orrù (✉) · B. R. Groveman · B. Caughey (✉)
Laboratory of Neurological Infections and Immunity, Rocky Mountain Laboratories, Division of Intramural Research, National Institute of Allergy and Infectious Disease, National Institutes of Health, Hamilton, MT, USA
e-mail: christina.orru@nih.gov; bcaughey@niaid.nih.gov

G. Legname, F. Moda (eds.), *Biomarkers and Therapeutical Targets for Prion Diseases*, Subcellular Biochemistry 112, https://doi.org/10.1007/978-3-031-97055-9_7

humans but more common in sheep (classical scrapie) (Cassmann and Greenlee 2020; Sola et al. 2023), cervids (chronic wasting disease (CWD)) (Benestad and Telling 2018), and cattle (classical BSE, during its epidemic) are acquired prion diseases initiated by infections from exogenous sources (Houston and Andreoletti 2019). In humans, such infections are usually iatrogenic, i.e., caused inadvertently via the use of prion-contaminated medical instruments, transplanted tissues, or extracts thereof (Brown et al. 2012; Brown and Farrell 2015). A third etiology, particularly well described in humans, is genetic, i.e., due to specific pathogenic mutations in the host's PrP gene (*PRNP*) (Zerr and Parchi 2018; Mead et al. 2019; Caughey and Sim 2024; Zerr et al. 2024). At least 35 such mutations have been identified, accounting for genetic forms of CJD, Gerstmann–Straussler–Scheinker (GSS) syndrome, and fatal familial insomnia.

Among the challenges of coping with neurodegenerative proteinopathies, including PrP prion diseases, has been the accurate and sensitive detection of the causative prions or prion-like self-propagating protein aggregates. These aggregates can serve as diagnostic and prognostic biomarkers as well as indications of sources of infection. The structures of such assemblies that have been determined at near-atomic resolution have so far taken the form of amyloid fibrils [e.g., (Scheres et al. 2020; Caughey et al. 2022, 2023; Artikis et al. 2022)]. In the case of brain-derived fibrils of PrP, the PrP monomers are stacked parallel and in-register along the fibril axis (Alam et al. 2024; Artikis et al. 2022; Hallinan et al. 2022; Hoyt et al. 2022a, b; Kraus et al. 2021; Manka 2025; Manka et al. 2022, 2023). The fibrils can grow with the addition and complete refolding of PrP^C monomers at the templating surfaces at the ends of the fibrils through seeded polymerization (Kraus et al. 2021).

This inherent seeding and conformational templating activity of prions was first demonstrated in cell-free conversion reactions in which proteinase K (PK)-resistant PrP^{Sc} (PrP^{Res}) was shown to convert PrP^C into PrP^{Res} (Kocisko et al. 1994) with strain- and species-specificities (Raymond et al. 1997; Kocisko et al. 1995; Bessen et al. 1995; Bossers et al. 2000). The subsequent demonstration of continuous conversion in sonicated, brain-homogenate-based protein misfolding cyclic amplification (PMCA) reactions allowed the development of ultrasensitive assays that propagated prion infectivity (Castilla et al. 2005; Deleault et al. 2003, 2007, 2012; Saa et al. 2006b; Saborio et al. 2001). Other studies indicated that purified recombinant PrP^C (rPrP) can replace brain homogenate as a source of PrP^C substrate without diminishing prion detection in sonicated reactions called rPrP-PMCA (Atarashi et al. 2007). Then the often difficult-to-control sonication of assay tubes was replaced with shaking to give rise to tube-based quaking-induced conversion (QuIC) assays (Atarashi et al. 2008; Orrú et al. 2009). Concurrently, a shaken, multi-well plate-based conversion reaction was introduced, i.e., the amyloid seeding assay (ASA) (Colby et al. 2007). Improvements in conditions of the plate-based assays using $rPrP^C$ substrates to markedly increase kinetic differences between prion-seeded and spontaneous (unseeded) fibrillization led to the real-time QuIC (RT-QuIC) assays (Atarashi et al. 2011b; Wilham et al. 2010). In contrast to PMCA conversion products (Castilla et al. 2005), the products of at least human sporadic Creutzfeldt–Jakob disease (sCJD)-seeded RT-QuIC assays have been found to be

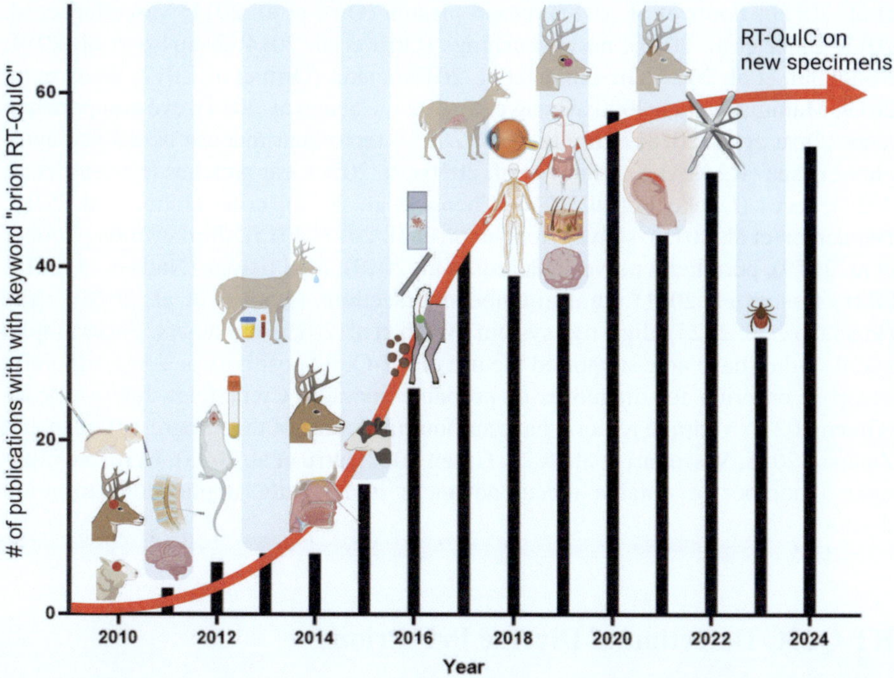

Fig. 7.1 Number of yearly peer-reviewed scientific publications with prion RT-QuIC applications. Graphics highlight when specimens were initially tested (to our knowledge) by prion RT-QuIC in representative publications listed in the main text. The X-axis indicates the year of publication, and the y-axis is the number (#) of papers identified via a PubMed search using the keywords "prion RT-QuIC." The red line highlights both the increased use of RT-QuIC over time as well as the rapid surge in new applications of the assay over the last decade. Dots indicate the anatomical region tested: brain (red), retropharyngeal lymph node (orange), RAMALT (green), and third eyelid (purple). The scissors and scalpels represent surgical instruments and other solid surfaces on which prions are now readily detectable with RT-QuIC assays (see main text)

noninfectious (Raymond et al. 2020). Collectively, these various types of nucleated polymerization-based assay formats, i.e., PMCA, ASA, and RT-QuIC, have been dubbed seed amplification assays (SAAs) (Russo et al. 2021). With each of these SAAs, strong correlations have been established between seeding activity and infectivity in tissues of prion-infected hosts [e.g., (Saa et al. 2006a; Colby et al. 2007; Wilham et al. 2010)].

Since the initial descriptions of RT-QuIC assays (Atarashi et al. 2011b; Wilham et al. 2010), there has been an explosion of applications to human and animal prion diseases [reviewed in (Green and Zanusso 2018; Hermann et al. 2021; Vascellari et al. 2022; Green 2022; Orru et al. 2023)] (Fig. 7.1). RT-QuIC assays are usually much more sensitive than more conventional assays such as ELISA, immunohistochemistry, and western blot (McNulty et al. 2019). This greater sensitivity has allowed prion detection, and associated diagnoses, using accessible biospecimens such as cerebrospinal fluid (CSF) (Atarashi et al. 2007; McGuire et al. 2012; Orru et al. 2015b; Foutz et al. 2017), blood plasma (Orru et al. 2011; Vascellari et al. 2012; Elder et al. 2013), nasal brushings (Orru et al. 2014; Zanusso et al. 2014; Bongianni et al. 2017; Groveman et al. 2017a), skin (Orru et al. 2017; Wang et al. 2019; Mammana et al. 2020; Zhang et al. 2024; Chen et al. 2024), eye components/tears (Orru et al. 2018; Schmitz et al. 2023), recto-anal mucosa associated lymphoid tissue (RAMALT) (Haley et al. 2016a, b, 2020), ear punches (Ferreira et al. 2021), feces (Denkers et al. 2016; Cheng et al. 2016), urine (John et al. 2013; Henderson et al. 2015), saliva (Henderson et al. 2013, 2015), third eyelids (Cooper et al. 2019), peripheral nerves (Baiardi et al. 2019), fetal tissues (Nalls et al. 2017, 2021; Luk et al. 2021) paraffin-embedded sections (Hoover et al. 2016), ticks (Inzalaco et al. 2023), digestive system (Satoh et al. 2019) and others. For example, many studies have now supported the use of RT-QuIC positivity as a key, officially accepted criterion for diagnoses of probable sporadic Creutzfeldt–Jakob disease (Green 2022). Multiple reviews have summarized most of this progress (Green and Zanusso 2018; Vascellari et al. 2022; Green 2022; Orru et al. 2023). Here, we highlight a number of notable recent advances in RT-QuIC applications to prion diseases.

RT-QuIC Detection of Diverse PrP Prions

To our knowledge, RT-QuIC seeding activity has been detected at some level in brain tissue from humans and animals with all forms/strains of prion disease, especially using bank vole rPrP as a substrate (Orru et al. 2015c). The list of detectable human prions has recently been extended to novel subtypes of sporadic Creutzfeldt–Jakob disease with PRNP codon 129MM genotype and distinct PrP plaque phenotypes affecting the gray matter (pGM) or the white matter (pWM) of sCJD cases (Bayazid et al. 2023). Overall, depending upon rPrP substrates and reaction conditions, prion RT-QuIC can be broadly reactive to mammalian prions or more selective, allowing for strain discrimination [e.g., (Masujin et al. 2016; Orru et al. 2015a, c)]. A long list of variables has been found to affect prion RT-QuIC sensitivity and specificity, including sample handling, sample matrices, rPrP substrate, salts, pH, detergents, temperature, shaking speed, shaking motion, shake-rest intervals, and beads. [e.g., (Atarashi et al. 2011a, b; Cramm et al. 2016; Elder et al. 2013; Green 2019; McGuire et al. 2016; Metrick 2nd et al. 2019; Orru et al. 2011, 2016; Peden et al. 2012; Wilham et al. 2010)]. Importantly, it remains possible that new types and strains of prions, such as those that might be generated upon prion transmission from one species or host genotype into another, could escape detection by existing RT-QuIC assays. This scenario is of particular concern in testing for any new zoonotic transmissions of animal prion strains into humans.

Detecting Prions in Skin Specimens

For the purposes of diagnostics and surveillance in living hosts, it is important to identify accessible biospecimens in which prion seeding activity can be reliably detected. In humans, the best validated and most widely used diagnostic specimen is cerebrospinal fluid (CSF) [(Atarashi et al. 2011b; McGuire et al. 2012) and many subsequent papers listed above]. However, the necessary lumbar punctures are invasive and can be contraindicated in certain patients. Skin represents another readily accessible diagnostic specimen. Our initial studies found RT-QuIC seeding activity in the skin of both living ($n = 3$) and deceased ($n = 35$) sCJD patients (Orru et al. 2017) and in experimental rodents early in the course of infection (Wang et al. 2019). Further studies showed that treatment of mice with an anti-prion compound reduced skin seeding activity (Ding et al. 2021). Analyses of a larger panel of human skin biopsies from CJD cases yielded 89% diagnostic sensitivity and 100% specificity (Mammana et al. 2020). A recent, and still larger, study compared RT-QuIC analyses of skin biopsies and CSF from live human prion disease cases (Chen et al. 2024), finding that testing of analysis of multiple skin biopsies provided slightly better sensitivity than analysis of CSF. Also recently, we reported an extensive study of skin samples that were collected at autopsy from two to three body areas (next to the ear, lower back, and/or apex of the head) of 339 cases with various types of neuropathologically confirmed prion diseases and non-sCJD controls (Zhang et al. 2024). Analysis of a retrospective cohort by two independent laboratories gave 87.3% or 91.3% sensitivity and 94.7% or 100% specificity, respectively, while a prospective cohort showed a sensitivity of 89.4% and specificity of 95.5%. CSF from 212 cases gave 89.7% sensitivity and 94.1% specificity.

In addition to these overall sensitivity and specificity parameters are the relative RT-QuIC sensitivities in detecting the multiple subtypes of sCJD and other human prion diseases that have been described (Zerr and Parchi 2018; Mead et al. 2019; Caughey and Sim 2024; Zerr et al. 2024). As also seen in previous studies (Orru et al. 2017; Mammana et al. 2020; Chen et al. 2024), the performance of the skin RT-QuIC analysis was sCJD subtype-dependent, in this case with sensitivity being highest in sCJDVV1–2 subtype, followed by VV2, MV1–2, MV1, MV2, MM1, MM1–2, MM2, and VV1 (Zhang et al. 2024). The skin next to the ear gave the highest sensitivity individually, but again, greater sensitivity was obtained by analyzing more than one site. In comparing cases with false negative versus true positive skin RT-QuIC results, the disease duration was significantly longer with the false negatives [12.0 ± 13.3 (months, SD) vs. 6.5 ± 6.4, $p < 0.001$]. Finally, Baranova et al. directly compared matching skin and CSF samples from 38 prion disease cases and 30 neurological controls (Baranova et al. 2024). Although positive skin samples had ~ten-fold higher median seeding activity, the CSF analyses gave higher diagnostic sensitivity (100% versus 89.5% for skin). Collectively, these studies are building validation of RT-QuIC analysis of skin as a valuable biomarker for the detection and diagnosis of prion diseases.

Detection of CWD in Deer Using Ear Punches

An extension of RT-QuIC applications to skin specimens is the use of ear punches from white-tailed and mule deer to detect CWD infections (Ferreira et al. 2021). With the aid of an iron oxide magnetic extraction step introduced by the Hoover and Mathiason labs (Denkers et al. 2016), we analyzed punches from various parts of the ear pinna from CWD-infected and control deer and compared the results to those obtained from gold-standard retropharyngeal lymph node specimens [e.g., (Darish et al. 2024)]. On this basis, the ear analyses provided an apparent diagnostic sensitivity of 81% and specificity of 91%. This performance was comparable to, or even improved upon, that achieved previously using RAMALT biopsies (Haley et al. 2016b), suggesting that RT-QuIC analysis of ear pinna punches may be helpful in detecting CWD infections in cervids.

Preclinical Detection of Prion Seeds in Individuals at Risk of Prion Disease

In humans, inherited prion diseases account for 10–15% of cases, while acquired, e.g., iatrogenic, infections constitute less than 1% of cases (Zerr and Parchi 2018; Mead et al. 2019; Caughey and Sim 2024; Zerr et al. 2024). The age of disease onset is difficult to predict in carriers of disease-associated *PRNP* mutations or iatrogenically infected individuals. The iatrogenic cases were recipients of implicated batches of cadaver-sourced human growth hormone. As treatments become available, as currently exemplified by Ionis' Phase 1/2a clinical trial of ION717 called PrProfile, it will be important to be able to assess when such treatments, with their inevitable costs and likely side effects, should be initiated in individual mutation carriers. In an effort to determine whether incipient prion disease pathogenesis can be detected prior to the onset of overt clinical signs, Mok and colleagues assembled and analyzed a longitudinal biofluid collection of CSF ($n = 72$) and plasma samples ($n = 220$) from individuals at risk of inherited prion disease and iatrogenic CJD as well as symptomatic cohorts over 13 years (CSF $n = 39$; plasma $n = 102$) (Mok et al. 2023). This collection included samples from 16 IPD converters ranging from 9.9 prior to and 7.4 years after clinical onset, as well as non-prion and healthy control samples (CSF $n = 83$; plasma $n = 132$). RT-QuIC was 100% sensitive and specific for sporadic CJD, iCJD, and familial CJD phenotypes (E200K + 6-OPRI-CJD). Prion seeds were also detected in CSF samples from three E200K carriers in the presymptomatic phase. One of the latter cases converted shortly after testing, but the other two remained asymptomatic for another 2–3 years. Partial sensitivity was seen for P102L disease (sensitivity 44.4%; specificity 98.2%) with a specially adapted RT-QuIC assay, which was positive for a CSF sample from a P102L *PRNP* mutation carrier. No RT-QuIC assay conditions were found to work well for CSF samples from classical 6-OPRI, A117V, and D178N carriers, despite our previous studies showing seed detection in brain tissue from these types of cases at autopsy, albeit with slower kinetics than is typical for many other human prion diseases (Orru et al. 2015c). In slowly progressive forms of inherited prion disease, plasma glial acidic fibrillary protein (GFAP) and neurofilament light (NfL) chain, and CSF NfL levels measured by Simoa served as proximity markers of neurodegeneration, distinguishing normal control (together with inherited prion disease >2 years to onset) from inherited prion disease cases with <2 years to onset and symptomatic cohorts. Based on these results, Mok proposed a presymptomatic staging system based on clinical, seeding, and neurodegeneration features.

Vertical Transmission and RT-QuIC Detection of sCJD Infectivity in Human Placental Tissues

Vertical transmission of prion disease has been demonstrated in cervids (Nalls et al. 2013) and sheep (Ligios et al. 2011; Konold et al. 2013; Foster et al. 2013; Spiropoulos et al. 2014). Although sCJD only rarely affects pregnant women and there is no evidence of vertical transmission in humans, cases have arisen that have raised issues of biosafety associated with performing Cesarean sections. Fortunately, RT-QuIC analyses of amniotic fluid, cord blood, and placental tissue from a woman with sCJD of the MM2 subtype indicated little to no detectable prion seeding activity, and hence infectivity, in such specimens (Luk et al. 2021). Nonetheless, precautions were advised with products of gestation.

RT-QuIC Detection of Prion Seeds in Human Cerebral Organoids

Cell culture models of prion infection have been important tools in the field. However, until recently these models had only been applicable to nonhuman prions. To address the lack of human models, Haigh and colleagues have developed a three-dimensional self-organizing culture of cerebral brain tissue (cerebral organoids), derived from human induced pluripotent stem cells (iPSCs), that takes up and propagates sCJD prions while maintaining subtype specificity (Groveman et al. 2019, 2023). Clearance of the initial sCJD brain inoculum and the associated prions from the organoids and the subsequent de novo accumulation of prions were monitored by RT-QuIC. Interestingly, cerebral organoids generated from asymptomatic carriers of human genetic prion disease mutations E200K CJD and D178N fatal familial insomnia showed metabolic disturbances but did not develop RT-QuIC-detectable seeding activity, indicating that the mutation itself is insufficient to cause prion seed formation (Foliaki et al. 2020, 2023), even under cellular stress (Smith et al. 2022). RT-QuIC and immunohistochemistry were also used on cerebral organoids that had been exposed to CWD to demonstrate that even direct exposure of human brain-like tissue to CWD prions was insufficient for zoonotic transmission (Groveman et al. 2024). Finally, CJD-infected cerebral organoids were used to test neural cell engraftment therapy (Williams et al. 2023) or anti-prion compounds (Groveman et al. 2021). In the latter case, RT-QuIC was used to quantitatively assess the impact of pentosan polysulfate treatment on prion propagation. Early prophylactic treatment with the drug delayed prion propagation and even a much-delayed, transient postinfection treatment reduced prion seed accumulation.

Detecting Prions on Solid Surfaces

Besides the development of prion disease therapeutics, it remains important to improve measures to prevent transmissions in the first place. As evidenced by the iatrogenic transmission of human prion disease via contaminated medical instruments (Brown and Farrell 2015; Brown et al. 2012) and the environment-mediated transmission of scrapie in sheep and CWD in cervids, the presence of prions on solid surfaces can be biohazardous. Indeed, many experimental studies have clearly documented transmission via environmental materials [e.g., see (Bartlet-Hunt et al. 2013; Zabel and Ortega 2017; Sakudo 2020; Pritzkow et al. 2018)]. Such concerns have driven the development of RT-QuIC assays for detecting prions bound to environmental materials, machinery, tools, and surgical instruments to help limit fomite-borne disease dissemination. Although some solids can be immersed directly in RT-QuIC reaction wells for highly sensitive detection (Hughson et al. 2016; Mori et al. 2016; Belondrade et al. 2016; Yuan et al. 2018; Moudjou et al. 2020; Belondrade et al. 2020), many objects are too large to fit. In addressing this issue, two studies reported the detection of CWD prions from flat surfaces by sampling with foam swabs that are sonicated to elute prion seeds that are then concentrated and analyzed by RT-QuIC and PMCA (Yuan et al. 2022; Soto et al. 2023). A recent study used this swabbing approach to show that water and ethanol treatments of several types of surfaces relevant to laboratories and clinics left readily detectable residual prions, whereas bleach eliminated such seeds (Simmons et al. 2024). However, this study also showed that certain soil minerals can inhibit RT-QuIC assays, advising that caution be used in assays of new materials. In contrast, as had been demonstrated earlier (Christenson et al. 2023, 2024), the addition of a specific type of nanoparticle can enhance RT-QuIC detection of CWD prions.

We have reported a somewhat different sampling method in which the surfaces were simply exposed to a sampling medium without the need for swabbing and subsequent elution steps (Orru et al. 2024). Using that approach, called surface RT-QuIC (sfRT-QuIC), we also showed that hamster 263K prion seeds can remain detectable on steel wires for at least a year, even after standard enzymatic cleaning and sterilization treatments. Moreover, prions on surgical instruments and plates exposed to even extreme brain tissue dilutions from prion-affected humans, sheep, cattle, and cervids could be detected using sfRT-QuIC. Similarly, surfaces exposed to comparable dilutions of brain tissue from human cases of Alzheimer's and Parkinson's disease were positive using adaptations of tau- and α-synuclein-RT-QuIC assays even after cleaning/sterilization. Altogether, these studies have demonstrated ultrasensitive methods of testing for prions and other prion-like disease-associated seeds bound to a variety of solid materials. PMCA assays and PMCA coupled to RT-QuIC assays (Kuznetsova et al. 2014) have also been successfully applied to prion detection on various solids such as soils [e.g., (Saunders et al. 2011; Xu et al. 2014; Wyckoff et al. 2016)] or steel wires (Belondrade et al. 2016) that can be immersed in an assay microtube.

Enhancing Quantitative Accuracy

Despite the common misconception that RT-QuIC assays provide only binary positive vs. negative assessments of the presence of proteopathic seeds, these assays have in fact been used quantitatively in many studies [e.g., (Favole et al. 2019; Groveman et al. 2017b, 2019; Metrick 2nd et al. 2019; Orru et al. 2011, 2012, 2014, 2015b, 2017, 2018; Peden et al. 2012; Vascellari et al. 2012; Wang et al. 2019)], including the first paper to describe RT-QuIC assays (Wilham et al. 2010). These and many other papers have demonstrated that prion RT-QuIC assays are usually more sensitive than prion bioassays in animals. However, the approaches that have typically been used are only able to reliably discriminate differences in proteopathic seed concentrations of at least five to tenfold. We have recently reported ways to improve the accuracy and inter-assay reproducibility of end-point dilution RT-QuIC assays to seeding activity differences of ~two fold (Srivastava et al. 2024). For this study, we used as examples α-synuclein-RT-QuIC (SAA) assays of a variety of biospecimens from Parkinson's disease and dementia with Lewy bodies cases, but most of the principles learned should also be similarly applicable to prion RT-QuIC assays. A combination of factors proved to be influential: sample handling; increasing the number of replicate reactions per sample dilution; decreasing the interval of serial dilutions; and data processing methods such as the midSIN algorithm developed by Catherine Beauchemin (Cresta et al. 2021) in place of the commonly used Spearman–Kärber analysis (Dougherty 1964). Better accuracy and reproducibility in RT-QuIC types of SAAs should improve comparisons of seeding activities with other disease parameters to improve diagnostics, prognostics, patient cohort selection for drug trials, and assessments of pharmacodynamics and target engagement in trials of drugs aimed at pathological protein aggregates such as prions.

Outlook

Many labs continue to make significant advances in the applications of RT-QuIC and other SAAs to important challenges in prion disease management, including fundamental research [e.g., (Aguilar-Calvo et al. 2023)], detection, diagnostics, biohazard mitigation, and the development of therapeutics. Clearly, work from many labs has collectively revolutionized the routine diagnosis of prion diseases in humans, at least. Still, it remains important to continue improving the practicality of these assays so that they can be more readily applied in the clinic and the field. One attractive example of such efforts is MN-QuIC, which is a visually based, field-deployable diagnostic assay for prions, with demonstrated application to chronic wasting disease in cervids (Christenson et al. 2022). It also seems likely that miniaturization of these assays into the realm of microfluidics should be both possible and beneficial in reducing costs and increasing the quantitative accuracy of RT-QuIC SAAs (Lee et al. 2023). In any case, a key advantage of RT-QuIC SAAs is that the prion seeds that they detect so sensitively and specifically are not indirect manifestations of infections or pathology but primary etiological biomarkers of prion diseases.

Acknowledgments This work was supported by the Division of Intramural Research of the NIAID, NIH and by generous gifts from Mary Hilderman Smith, Zoë Smith Jaye, and Jenny Smith Unruh in memory of Jeffrey Smith.

References

Aguilar-Calvo P, Malik A, Sandoval DR, Barback C, Orru CD, Standke HG, Thomas OR, Dwyer CA, Pizzo DP, Bapat J, Soldau K, Ogawa R, Riley MB, Nilsson KPR, Kraus A, Caughey B, Iliff JJ, Vera DR, Esko JD, Sigurdson CJ (2023) Neuronal Ndst1 depletion accelerates prion protein clearance and slows neurodegeneration in prion infection. PLoS Pathog 19:e1011487

Alam P, Hoyt F, Artikis E, Soukup J, Hughson AG, Schwartz CL, Barbian K, Miller MW, Race B, Caughey B (2024) Cryo-EM structure of a natural prion: chronic wasting disease fibrils from deer. Acta Neuropathol 148:56

Artikis E, Kraus A, Caughey B (2022) Structural biology of ex vivo mammalian prions. J Biol Chem 298:102181

Atarashi R, Moore RA, Sim VL, Hughson AG, Dorward DW, Onwubiko HA, Priola SA, Caughey B (2007) Ultrasensitive detection of scrapie prion protein using seeded conversion of recombinant prion protein. Nat Methods 4:645–650

Atarashi R, Wilham JM, Christensen L, Hughson AG, Moore RA, Johnson LM, Onwubiko HA, Priola SA, Caughey B (2008) Simplified ultrasensitive prion detection by recombinant PrP conversion with shaking. Nat Methods 5:211–212

Atarashi R, Sano K, Satoh K, Nishida N (2011a) Real-time quaking-induced conversion: a highly sensitive assay for prion detection. Prion 5:150–153

Atarashi R, Satoh K, Sano K, Fuse T, Yamaguchi N, Ishibashi D, Matsubara T, Nakagaki T, Yamanaka H, Shirabe S, Yamada M, Mizusawa H, Kitamoto T, Klug G, Mcglade A, Collins SJ, Nishida N (2011b) Ultrasensitive human prion detection in cerebrospinal fluid by real-time quaking-induced conversion. Nat Med 17:175–178

Baiardi S, Redaelli V, Ripellino P, Rossi M, Franceschini A, Moggio M, Sola P, Ladogana A, Fociani P, Magherini A, Capellari S, Giese A, Caughey B, Caroppo P, Parchi P (2019) Prion-related peripheral neuropathy in sporadic Creutzfeldt-Jakob disease. J Neurol Neurosurg Psychiatry 90:424–427

Baranova S, Mosko T, Bruzova M, Haldiman T, Kim C, Safar JG, Matej R, Holada K (2024) Detection of prions in matching post-mortem skin and cerebrospinal fluid samples using second-generation real-time quaking-induced conversion assay. Sci Rep 14:6294

Bartlet-Hunt SL, Bartz JC, Saunders SE (2013) Prions in the environment. Springer, New York

Bayazid R, Orru C, Aslam R, Cohen Y, Silva-Rohwer A, Lee SK, Occhipinti R, Kong Q, Shetty S, Cohen ML, Caughey B, Schonberger LB, Appleby BS, Cali I (2023) A novel subtype of sporadic Creutzfeldt-Jakob disease with Prnp codon 129mm genotype and PrP plaques. Acta Neuropathol 146:121–143

Belondrade M, Nicot S, Beringue V, Coste J, Lehmann S, Bougard D (2016) Rapid and highly sensitive detection of variant creutzfeldt-Jakob disease abnormal prion protein on steel surfaces by protein Misfolding cyclic amplification: application to prion decontamination studies. PLoS One 11:e0146833

Belondrade M, Jas-Duval C, Nicot S, Bruyere-Ostells L, Mayran C, Herzog L, Reine F, Torres JM, Fournier-Wirth C, Beringue V, Lehmann S, Bougard D (2020) Correlation between bioassay and protein misfolding cyclic amplification for variant creutzfeldt-jakob disease decontamination studies. mSphere 5:e00649

Benestad SL, Telling GC (2018) Chronic wasting disease: an evolving prion disease of cervids. Handb Clin Neurol 153:135–151

Benestad SL, Arsac JN, Goldmann W, Noremark M (2008) Atypical/Nor98 scrapie: properties of the agent, genetics, and epidemiology. Vet Res 39:19

Bessen RA, Kocisko DA, Raymond GJ, Nandan S, Lansbury PT Jr, Caughey B (1995) Nongenetic propagation of strain-specific phenotypes of scrapie prion protein. Nature 375:698–700

Bongianni M, Orrù CD, Groveman BR, Sacchetto L, Fiorini M, Tonoli G, Triva G, Capaldi S, Testi S, Ferrari S, Cagnin A, Ladogana A, Poleggi A, Colaizzo E, Tiple D, Vaianella L, Castriciano S, Marchioni D, Hughson AG, Imperiale D, Cattaruzza T, Fabrizi G, Pocchiari M, Monaco S, Caughey B, Zanusso G (2017) Diagnosis of human prion disease using real-time quaking-induced conversion testing of olfactory mucosa and cerebrospinal fluid samples. JAMA Neurol 74:1–8

Bossers A, De Vries R, Smits MA (2000) Susceptibility of sheep for scrapie as assessed by in vitro conversion of nine naturally occurring variants of PrP. J Virol 74:1407–1414

Brown P, Farrell M (2015) A practical approach to avoiding iatrogenic Creutzfeldt-Jakob disease (CJD) from invasive instruments. Infect Control Hosp Epidemiol 36:1–5

Brown P, Brandel JP, Sato T, Nakamura Y, Mackenzie J, Will RG, Ladogana A, Pocchiari M, Leschek EW, Schonberger LB (2012) Iatrogenic Creutzfeldt-Jakob disease, final assessment. Emerg Infect Dis 18:901–907

Cassmann ED, Greenlee JJ (2020) Pathogenesis, detection, and control of scrapie in sheep. Am J Vet Res 81:600–614

Castilla J, Saa P, Hetz C, Soto C (2005) In vitro generation of infectious scrapie prions. Cell 121:195–206

Caughey B, Sim VL (2024) Prions. In: Howley PM, Knipe DM (eds) Fields virology fundamentals, 7th edn. Wolters Kluwer, Philadelphia

Caughey B, Standke HG, Artikis E, Hoyt F, Kraus A (2022) Pathogenic prion structures at high resolution. PLoS Pathog 18:e1010594

Caughey B, Artikis E, Kraus A (2023) PrP prion structures. In: Zou W-Q, Gambetti P (eds) Prions and diseases, 2nd edn. Springer Nature

Chen ZY, Shi Q, Xiao K, Kong Y, Liang DL, Wang YH, Min R, Zhang J, Wang Z, Ye H, Gao R, Chu M, Nan HT, Jiang DM, Li JJ, Wang L, Zou WQ, Wu LY, Dong XP (2024) Multisite skin biopsies vs cerebrospinal fluid for prion seeding activity in the diagnosis of prion diseases. JAMA Neurol 81:1263–1273

Cheng YC, Hannaoui S, John TR, Dudas S, Czub S, Gilch S (2016) Early and non-invasive detection of chronic wasting disease prions in elk feces by real-time quaking induced conversion. PLoS One 11:e0166187

Christenson PR, Li M, Rowden G, Schwabenlander MD, Wolf TM, Oh SH, Larsen PA (2022) A field-deployable diagnostic assay for the visual detection of misfolded prions. Sci Rep 12:12246

Christenson PR, Li M, Rowden G, Larsen PA, Oh SH (2023) Nanoparticle-enhanced RT-QuIC (Nano-QuIC) diagnostic assay for misfolded proteins. Nano Lett 23:4074–4081

Christenson PR, Jeong H, Li M, Ahn H, Schmeichel AM, Misra P, Li D, Savica R, Low PA, Singer W, Larsen PA, Park HY, Oh SH (2024) Blood-based nanoparticle-enhanced quaking-induced conversion (Nano-QuIC): inhibitor-resistant detection of seeding activity in patients diagnosed with Parkinson's disease. Nano Lett 24:15016–15024

Colby DW, Zhang Q, Wang S, Groth D, Legname G, Riesner D, Prusiner SB (2007) Prion detection by an amyloid seeding assay. Proc Natl Acad Sci USA 104:20914–20919

Cooper SK, Hoover CE, Henderson DM, Haley NJ, Mathiason CK, Hoover EA (2019) Detection of CWD in cervids by RT-QuIC assay of third eyelids. PLoS One 14:e0221654

Cramm M, Schmitz M, Karch A, Mitrova E, Kuhn F, Schroeder B, Raeber A, Varges D, Kim YS, Satoh K, Collins S, Zerr I (2016) Stability and reproducibility underscore utility of RT-QuIC for diagnosis of Creutzfeldt-Jakob disease. Mol Neurobiol 53:1896

Cresta D, Warren DC, Quirouette C, Smith AP, Lane LC, Smith AM, Beauchemin CAA (2021) Time to revisit the endpoint dilution assay and to replace the TCID50 as a measure of a virus sample's infection concentration. PLoS Comput Biol 17:e1009480

Darish JR, Kaganer AW, Hanley BJ, Schuler KL, Schwabenlander MD, Wolf TM, Ahmed MS, Rowden GR, Larsen PA, Kobashigawa E, Tewari D, Lichtenberg S, Pedersen JA, Zhang S, Sreevatsan S (2024) Inter-laboratory comparison of real-time quaking-induced conversion (RT-QuIC) for the detection of chronic wasting disease prions in white-tailed deer retropharyngeal lymph nodes. J Vet Diagn Invest 10406387241285165:86

Deleault NR, Lucassen RW, Supattapone S (2003) RNA molecules stimulate prion protein conversion. Nature 425:717–720

Deleault NR, Harris BT, Rees JR, Supattapone S (2007) Formation of native prions from minimal components *in vitro*. Proc Natl Acad Sci USA 104:9741–9746

Deleault NR, Piro JR, Walsh DJ, Wang F, Ma J, Geoghegan JC, Supattapone S (2012) Isolation of phosphatidylethanolamine as a solitary cofactor for prion formation in the absence of nucleic acids. Proc Natl Acad Sci USA 109:8546–8551

Denkers ND, Henderson DM, Mathiason CK, Hoover EA (2016) Enhanced prion detection in biological samples by magnetic particle extraction and real-time quaking-induced conversion. J Gen Virol 97:2023–2029

Ding M, Teruya K, Zhang W, Lee HW, Yuan J, Oguma A, Foutz A, Camacho MV, Mitchell M, Greenlee JJ, Kong Q, Doh-Ura K, Cui L, Zou WQ (2021) Decrease in skin prion-seeding activity of prion-infected mice treated with a compound against human and animal prions: a first possible biomarker for prion therapeutics. Mol Neurobiol 58:4280–4292

Dougherty RM (1964) Animal virus titration techniques. In: Harris RJC (ed) Techniques in experimental virology. Academic, New York

Elder AM, Henderson DM, Nalls AV, Wilham JM, Caughey BW, Hoover EA, Kincaid AE, Bartz JC, Mathiason CK (2013) In vitro detection of prionemia in TSE-infected cervids and hamsters. PLoS One 8:e80203

Favole A, Mazza M, Vallino Costassa E, D'angelo A, Lombardi G, Marconi P, Crociara P, Berrone E, Gallo M, Palmitessa C, Orru CD, Caughey B, Acutis PL, Caramelli M, Casalone C, Corona C (2019) Early and pre-clinical detection of prion seeding activity in cerebrospinal fluid of goats using real-time quaking-induced conversion assay. Sci Rep 9:6173

Ferreira NC, Charco JM, Plagenz J, Orru CD, Denkers ND, Metrick MA 2nd, Hughson AG, Griffin KA, Race B, Hoover EA, Castilla J, Nichols TA, Miller MW, Caughey B (2021) Detection of chronic wasting disease in mule and white-tailed deer by RT-QuIC analysis of outer ear. Sci Rep 11:7702

Foliaki ST, Groveman BR, Yuan J, Walters R, Zhang S, Tesar P, Zou W, Haigh CL (2020) Pathogenic prion protein isoforms are not present in cerebral organoids generated from asymptomatic donors carrying the E200K mutation associated with familial prion disease. Pathogens 9:482

Foliaki ST, Smith A, Schwarz B, Bohrnsen E, Bosio CM, Williams K, Orru CD, Lachenauer H, Groveman BR, Haigh CL (2023) Altered energy metabolism in fatal familial insomnia cerebral organoids is associated with astrogliosis and neuronal dysfunction. PLoS Genet 19:e1010565

Foster JD, Goldmann W, Hunter N (2013) Evidence in sheep for pre-natal transmission of scrapie to lambs from infected mothers. PLoS One 8:e79433

Foutz A, Appleby BS, Hamlin C, Liu X, Yang S, Cohen Y, Chen W, Blevins J, Fausett C, Wang H, Gambetti P, Zhang S, Hughson A, Tatsuoka C, Schonberger LB, Cohen ML, Caughey B, Safar JG (2017) Diagnostic and prognostic value of human prion detection in cerebrospinal fluid. Ann Neurol 81:79–92

Green AJE (2019) RT-QuIC: a new test for sporadic CJD. Pract Neurol 19:49–55

Green AJE (2022) Editorial: utility of protein aggregation assays from laboratory to clinical application. Front Aging Neurosci 14:998136

Green AJE, Zanusso G (2018) Prion protein amplification techniques. Handb Clin Neurol 153:357–370

Groveman BR, Orru CD, Hughson AG, Bongianni M, Fiorini M, Imperiale D, Ladogana A, Pocchiari M, Zanusso G, Caughey B (2017a) Extended and direct evaluation of RT-QuIC assays for Creutzfeldt-Jakob disease diagnosis. Ann Clin Transl Neurol 4:139–144

Groveman BR, Raymond GJ, Campbell KJ, Race B, Raymond LD, Hughson AG, Orru CD, Kraus A, Phillips K, Caughey B (2017b) Role of the central lysine cluster and scrapie templating in the transmissibility of synthetic prion protein aggregates. PLoS Pathog 13:e1006623

Groveman BR, Foliaki ST, Orru CD, Zanusso G, Carroll JA, Race B, Haigh CL (2019) Sporadic Creutzfeldt-Jakob disease prion infection of human cerebral organoids. Acta Neuropathol Commun 7:90

Groveman BR, Ferreira NC, Foliaki ST, Walters RO, Winkler CW, Race B, Hughson AG, Zanusso G, Haigh CL (2021) Human cerebral organoids as a therapeutic drug screening model for Creutzfeldt-Jakob disease. Sci Rep 11:5165

Groveman BR, Race B, Foliaki ST, Williams K, Hughson AG, Baune C, Zanusso G, Haigh CL (2023) Sporadic Creutzfeldt-Jakob disease infected human cerebral organoids retain the original human brain subtype features following transmission to humanized transgenic mice. Acta Neuropathol Commun 11:28

Groveman BR, Williams K, Race B, Foliaki S, Thomas T, Hughson AG, Walters RO, Zou W, Haigh CL (2024) Lack of transmission of chronic wasting disease prions to human cerebral organoids. Emerg Infect Dis 30:1193–1202

Haley NJ, Siepker C, Hoon-Hanks LL, Mitchell G, Walter WD, Manca M, Monello RJ, Powers JG, Wild MA, Hoover EA, Caughey B, Richt JA (2016a) Seeded amplification of chronic wasting disease prions in nasal brushings and recto-anal mucosa associated lymphoid tissues from elk by real time quaking-induced conversion. J Clin Microbiol 54:1117

Haley NJ, Siepker C, Walter WD, Thomsen BV, Greenlee JJ, Lehmkuhl AD, Richt JA (2016b) Antemortem detection of chronic wasting disease prions in nasal brush collections and rectal biopsy specimens from white-tailed deer by real-time quaking-induced conversion. J Clin Microbiol 54:1108–1116

Haley NJ, Donner R, Henderson DM, Tennant J, Hoover EA, Manca M, Caughey B, Kondru N, Manne S, Kanthasamay A, Hannaoui S, Chang SC, Gilch S, Smiley S, Mitchell G, Lehmkuhl AD, Thomsen BV (2020) Cross-validation of the RT-QuIC assay for the antemortem detection of chronic wasting disease in elk. Prion 14:47–55

Hallinan GI, Ozcan KA, Hoq MR, Cracco L, Vago FS, Bharath SR, Li D, Jacobsen M, Doud EH, Mosley AL, Fernandez A, Garringer HJ, Jiang W, Ghetti B, Vidal R (2022) Cryo-EM structures of prion protein filaments from Gerstmann-Straussler-Scheinker disease. Acta Neuropathol 144:509–520

Henderson DM, Manca M, Haley NJ, Denkers ND, Nalls AV, Mathiason CK, Caughey B, Hoover EA (2013) Rapid antemortem detection of CWD prions in deer saliva. PLoS One 8:e74377

Henderson DM, Denkers ND, Hoover CE, Garbino N, Mathiason CK, Hoover EA (2015) Longitudinal detection of prion shedding in saliva and urine by chronic wasting disease-infected deer by real-time quaking-induced conversion. J Virol 89:9338–9347

Hermann P, Appleby B, Brandel JP, Caughey B, Collins S, Geschwind MD, Green A, Haik S, Kovacs GG, Ladogana A, Llorens F, Mead S, Nishida N, Pal S, Parchi P, Pocchiari M, Satoh K, Zanusso G, Zerr I (2021) Biomarkers and diagnostic guidelines for sporadic Creutzfeldt-Jakob disease. Lancet Neurol 20:235–246

Hoover CE, Davenport KA, Henderson DM, Pulscher LA, Mathiason CK, Zabel MD, Hoover EA (2016) Detection and quantification of CWD prions in fixed paraffin embedded tissues by real-time quaking-induced conversion. Sci Rep 6:25098

Houston F, Andreoletti O (2019) Animal prion diseases: the risks to human health. Brain Pathol 29:248–262

Hoyt F, Alam P, Artikis E, Schwartz CL, Hughson AG, Race B, Baune C, Raymond GJ, Baron GS, Kraus A, Caughey B (2022a) Cryo-EM of prion strains from the same genotype of host identifies conformational determinants. PLoS Pathog 18:e1010947

Hoyt F, Standke HG, Artikis E, Schwartz CL, Hansen B, Li K, Hughson AG, Manca M, Thomas OR, Raymond GJ, Race B, Baron GS, Caughey B, Kraus A (2022b) Cryo-EM structure of anchorless RML prion reveals variations in shared motifs between distinct strains. Nat Commun 13:4005

Hughson AG, Race B, Kraus A, Sangare LR, Robins L, Groveman BR, Saijo E, Phillips K, Contreras L, Dhaliwal V, Manca M, Zanusso G, Terry D, Williams JF, Caughey B (2016) Inactivation of prions and amyloid seeds with hypochlorous acid. PLoS Pathog 12:e1005914

Inzalaco HN, Bravo-Risi F, Morales R, Walsh DP, Storm DJ, Pedersen JA, Turner WC, Lichtenberg SS (2023) Ticks harbor and excrete chronic wasting disease prions. Sci Rep 13:7838

John TR, Schatzl HM, Gilch S (2013) Early detection of chronic wasting disease prions in urine of pre-symptomatic deer by real-time quaking-induced conversion assay. Prion 7:253–258

Kocisko DA, Come JH, Priola SA, Chesebro B, Raymond GJ, Lansbury PT, Caughey B (1994) Cell-free formation of protease-resistant prion protein. Nature 370:471–474

Kocisko DA, Priola SA, Raymond GJ, Chesebro B, Lansbury PT Jr, Caughey B (1995) Species specificity in the cell-free conversion of prion protein to protease-resistant forms: a model for the scrapie species barrier. Proc Natl Acad Sci USA 92:3923–3927

Konold T, Moore SJ, Bellworthy SJ, Terry LA, Thorne L, Ramsay A, Salguero FJ, Simmons MM, Simmons HA (2013) Evidence of effective scrapie transmission via colostrum and milk in sheep. BMC Vet Res 9:99

Kraus A, Hoyt F, Schwartz CL, Hansen B, Artikis E, Hughson AG, Raymond GJ, Race B, Baron GS, Caughey B (2021) High-resolution structure and strain comparison of infectious mammalian prions. Mol Cell 81:4540–4551

Kuznetsova A, Mckenzie D, Banser P, Siddique T, Aiken JM (2014) Potential role of soil properties in the spread of CWD in western Canada. Prion 8:92–99

Lee DJ, Christenson PR, Rowden G, Lindquist NC, Larsen PA, Oh S-H (2023) Microfluidic Quaking-Induced Conversion (Micro-QuIC) for rapid on-site amplification and detection of misfolded proteins. bioRxiv. https://doi.org/10.1101/2023.07.17.549283

Ligios C, Cancedda MG, Carta A, Santucciu C, Maestrale C, Demontis F, Saba M, Patta C, Demartini JC, Aguzzi A, Sigurdson CJ (2011) Sheep with scrapie and mastitis transmit infectious prions through the milk. J Virol 85:1136–1139

Luk CC, Mathiason CK, Orru CD, Jansen GH, Thiele A, Caughey B, Sim VL (2021) Creutzfeldt-Jakob disease in pregnancy: the use of modified RT-QuIC to determine infectivity in placental tissues. Prion 15:107–111

Mammana A, Baiardi S, Rossi M, Franceschini A, Donadio V, Capellari S, Caughey B, Parchi P (2020) Detection of prions in skin punch biopsies of Creutzfeldt-Jakob disease patients. Ann Clin Transl Neurol 7:559–564

Manka SW (2025) Breaking the mould: how the first structure of a deer prion suggests the framework for interspecies strain diversity and transmission barriers. J Neurochem 169:e70050

Manka SW, Zhang W, Wenborn A, Betts J, Joiner S, Saibil HR, Collinge J, Wadsworth JD (2022) 2.7 Å cryo-EM structure of ex vivo RML prion fibrils. Nat Commun 13:4004

Manka SW, Wenborn A, Betts J, Joiner S, Saibil HR, Collinge J, Wadsworth JDF (2023) A structural basis for prion strain diversity. Nat Chem Biol 19:607–613

Masujin K, Orru CD, Miyazawa K, Groveman BR, Raymond LD, Hughson AG, Caughey B (2016) Detection of atypical H-type bovine spongiform encephalopathy and discrimination of bovine prion strains by real-time quaking-induced conversion. J Clin Microbiol 54:676–686

McGuire LI, Peden AH, Orru CD, Wilham JM, Appleford NE, Mallinson G, Andrews M, Head MW, Caughey B, Will RG, Knight RSG, Green AJE (2012) RT-QuIC analysis of cerebrospinal fluid in sporadic Creutzfeldt-Jakob disease. Ann Neurol 72:278–285

McGuire LI, Poleggi A, Poggiolini I, Suardi S, Grznarova K, Shi S, De Vil B, Sarros S, Satoh K, Cheng K, Cramm M, Fairfoul G, Schmitz M, Zerr I, Cras P, Equestre M, Tagliavini F, Atarashi R, Knox D, Collins S, Haik S, Parchi P, Pocchiari M, Green A (2016) Cerebrospinal fluid real-time quaking-induced conversion is a robust and reliable test for sporadic creutzfeldt-jakob disease: an international study. Ann Neurol 80:160–165

McNulty E, Nalls AV, Mellentine S, Hughes E, Pulscher L, Hoover EA, Mathiason CK (2019) Comparison of conventional, amplification and bio-assay detection methods for a chronic wasting disease inoculum pool. PLoS One 14:e0216621

Mead S, Lloyd S, Collinge J (2019) Genetic factors in mammalian prion diseases. Annu Rev Genet 53:117–147

Metrick MA 2nd, Do Carmo Ferreira N, Saijo E, Hughson AG, Kraus A, Orru C, Miller MW, Zanusso G, Ghetti B, Vendruscolo M, Caughey B (2019) Million-fold sensitivity enhancement in proteopathic seed amplification assays for biospecimens by Hofmeister ion comparisons. Proc Natl Acad Sci USA 116(46):23029–23039

Meyer RK, Mckinley MP, Bowman KA, Braunfeld MB, Barry RA, Prusiner SB (1986) Separation and properties of cellular and scrapie prion protein. Proc Natl Acad Sci USA 83:2310–2314

Mok TH, Nihat A, Majbour N, Sequeira D, Holm-Mercer L, Coysh T, Darwent L, Batchelor M, Groveman BR, Orr CD, Hughson AG, Heslegrave A, Laban R, Veleva E, Paterson RW, Keshavan A, Schott JM, Swift IJ, Heller C, Rohrer JD, Gerhard A, Butler C, Rowe JB, Masellis M, Chapman M, Lunn MP, Bieschke J, Jackson GS, Zetterberg H, Caughey B, Rudge P, Collinge J, Mead S (2023) Seed amplification and neurodegeneration marker trajectories in individuals at risk of prion disease. Brain 146:2570–2583

Mori T, Atarashi R, Furukawa K, Takatsuki H, Satoh K, Sano K, Nakagaki T, Ishibashi D, Ichimiya K, Hamada M, Nakayama T, Nishida N (2016) A direct assessment of human prion adhered to steel wire using real-time quaking-induced conversion. Sci Rep 6:24993

Moudjou M, Castille J, Passet B, Herzog L, Reine F, Vilotte JL, Rezaei H, Beringue V, Igel-Egalon A (2020) Improving the predictive value of prion inactivation validation methods to minimize the risks of iatrogenic transmission with medical instruments. Front Bioeng Biotechnol 8:591024

Nalls AV, Mcnulty E, Powers J, Seelig DM, Hoover C, Haley NJ, Hayes-Klug J, Anderson K, Stewart P, Goldmann W, Hoover EA, Mathiason CK (2013) Mother to offspring transmission of chronic wasting disease in reeves' muntjac deer. PLoS One 8:e71844

Nalls AV, Mcnulty E, Hoover CE, Pulscher LA, Hoover EA, Mathiason CK (2017) Infectious prions in the pregnancy microenvironment of chronic wasting disease-infected reeves' muntjac deer. J Virol 91:e00501

Nalls AV, Mcnulty EE, Mayfield A, Crum JM, Keel MK, Hoover EA, Ruder MG, Mathiason CK (2021) Detection of chronic wasting disease prions in fetal tissues of free-ranging white-tailed deer. Viruses 13:2430

Orrú CD, Wilham JM, Hughson AG, Raymond LD, Mcnally KL, Bossers A, Ligios C, Caughey B (2009) Human variant Creutzfeldt-Jakob disease and sheep scrapie PrP(res) detection using seeded conversion of recombinant prion protein. Protein Eng Des Sel 22:515–521

Orru CD, Wilham JM, Raymond LD, Kuhn F, Schroeder B, Raeber AJ, Caughey B (2011) Prion disease blood test using immunoprecipitation and improved quaking-induced conversion. mBio 2:e00078-11

Orru CD, Hughson AG, Race B, Raymond GJ, Caughey B (2012) Time course of prion seeding activity in cerebrospinal fluid of scrapie-infected hamsters after intratongue and intracerebral inoculations. J. Clin. Microbiol. 50:1464–1466

Orru CD, Bongianni M, Tonoli G, Ferrari S, Hughson AG, Groveman BR, Fiorini M, Pocchiari M, Monaco S, Caughey B, Zanusso G (2014) A test for Creutzfeldt-Jakob disease using nasal brushings. N Engl J Med 371:519–529

Orru CD, Favole A, Corona C, Mazza M, Manca M, Groveman BR, Hughson AG, Acutis PL, Caramelli M, Zanusso G, Casalone C, Caughey B (2015a) Detection and discrimination of classical and atypical L-type bovine spongiform encephalopathy by real-time quaking-induced conversion. J Clin Microbiol 53:1115–1120

Orru CD, Groveman BR, Hughson AG, Zanusso G, Coulthart MB, Caughey B (2015b) Rapid and sensitive RT-QuIC detection of human Creutzfeldt-Jakob disease using cerebrospinal fluid. MBio 6:e02451-14

Orru CD, Groveman BR, Raymond LD, Hughson AG, Nonno R, Zou W, Ghetti B, Gambetti P, Caughey B (2015c) Bank vole prion protein as an apparently universal substrate for RT-QuIC-based detection and discrimination of prion strains. PLoS Pathog 11:e1004983

Orru CD, Hughson AG, Groveman BR, Campbell KJ, Anson KJ, Manca M, Kraus A, Caughey B (2016) Factors That Improve RT-QuIC Detection of Prion Seeding Activity. Viruses 8:140

Orru CD, Yuan J, Appleby BS, Li B, Li Y, Winner D, Wang Z, Zhan YA, Rodgers M, Rarick J, Wyza RE, Joshi T, Wang GX, Cohen ML, Zhang S, Groveman BR, Petersen RB, Ironside JW, Quinones-Mateu ME, Safar JG, Kong Q, Caughey B, Zou WQ (2017) Prion seeding activity and infectivity in skin samples from patients with sporadic Creutzfeldt-Jakob disease. Sci Transl Med 9:eaam7785

Orru CD, Soldau K, Cordano C, Llibre-Guerra J, Green AJ, Sanchez H, Groveman BR, Edland SD, Safar JG, Lin JH, Caughey B, Geschwind MD, Sigurdson CJ (2018) Prion seeds distribute throughout the eyes of sporadic Creutzfeldt-Jakob disease patients. Mbio 9:e02095

Orru CD, Isiofia O, Hughson AG, Caughey B (2023) Real-time quaking-induced conversion (QuIC) assays for the detection and diagnosis of human prion diseases. In: Zou W-Q, Gambetti P (eds) Prions and diseases, 2nd edn. Switzerland, Springer Nature

Orru CD, Groveman BR, Hughson AG, Barrio T, Isiofia K, Race B, Ferreira NC, Gambetti P, Schneider DA, Masujin K, Miyazawa K, Ghetti B, Zanusso G, Caughey B (2024) Sensitive detection of pathological seeds of alpha-synuclein, tau and prion protein on solid surfaces. PLoS Pathog 20:e1012175

Peden AH, Mcguire LI, Appleford NE, Mallinson G, Wilham JM, Orru CD, Caughey B, Ironside JW, Knight RS, Will RG, Green AJ, Head MW (2012) Sensitive and specific detection of sporadic Creutzfeldt-Jakob disease brain prion protein using real-time quaking induced conversion. J Gen Virol 93:438–449

Pritzkow S, Morales R, Lyon A, Concha-Marambio L, Urayama A, Soto C (2018) Efficient prion disease transmission through common environmental materials. J Biol Chem 293:3363–3373

Raymond GJ, Hope J, Kocisko DA, Priola SA, Raymond LD, Bossers A, Ironside J, Will RG, Chen SG, Petersen RB, Gambetti P, Rubenstein R, Smits MA, Lansbury PT Jr, Caughey B (1997) Molecular assessment of the transmissibilities of BSE and scrapie to humans. Nature 388:285–288

Raymond GJ, Race B, Orru CD, Raymond LD, Bongianni M, Fiorini M, Groveman BR, Ferrari S, Sacchetto L, Hughson AG, Monaco S, Pocchiari M, Zanusso G, Caughey B (2020) Transmission of CJD from nasal brushings but not spinal fluid or RT-QuIC product. Ann Clin Transl Neurol 7:932–944

Russo MJ, Orru CD, Concha-Marambio L, Giaisi S, Groveman BR, Farris CM, Holguin B, Hughson AG, Lafontant DE, Caspell-Garcia C, Coffey CS, Mollon J, Hutten SJ, Merchant K, Heym RG, Soto C, Caughey B, Kang UJ (2021) High diagnostic performance of independent alpha-synuclein seed amplification assays for detection of early Parkinson's disease. Acta Neuropathol Commun 9:179

Saa P, Castilla J, Soto C (2006a) Presymptomatic detection of prions in blood. Science 313:92–94

Saa P, Castilla J, Soto C (2006b) Ultra-efficient replication of infectious prions by automated protein misfolding cyclic amplification. J Biol Chem 281:35245–35252

Saborio GP, Permanne B, Soto C (2001) Sensitive detection of pathological prion protein by cyclic amplification of protein misfolding. Nature 411:810–813

Sakudo A (2020) Chronic Wasting Disease: Current Assessment of Transmissibility. Curr Issues Mol Biol 36:13–22

Satoh K, Fuse T, Nonaka T, Dong T, Takao M, Nakagaki T, Ishibashi D, Taguchi Y, Mihara B, Iwasaki Y, Yoshida M, Nishida N (2019) Postmortem quantitative analysis of prion seeding activity in the digestive system. Molecules 24:4601

Saunders SE, Shikiya RA, Langenfeld K, Bartelt-Hunt SL, Bartz JC (2011) Replication efficiency of soil-bound prions varies with soil type. J Virol 85:5476–5482

Scheres SH, Zhang W, Falcon B, Goedert M (2020) Cryo-EM structures of tau filaments. Curr Opin Struct Biol 64:17–25

Schmitz M, Silva Correia S, Hermann P, Maass F, Goebel S, Bunck T, Correia A, Lingor P, Fischer A, Zerr I (2023) Detection of Prion Protein Seeding Activity in Tear Fluids. N Engl J Med 388(19):1816–1817. https://doi.org/10.1056/NEJMc2214647

Sigurdson CJ, Bartz JC, Glatzel M (2019) Cellular and Molecular Mechanisms of Prion Disease. Annu Rev Pathol 14:497–516

Simmons SM, Payne VL, Hrdlicka JG, Taylor J, Larsen PA, Wolf TM, Schwabenlander MD, Yuan Q, Bartz JC (2024) Rapid and sensitive determination of residual prion infectivity from prion-decontaminated surfaces. mSphere 9:e0050424

Smith A, Groveman BR, Winkler C, Williams K, Walters R, Yuan J, Zou W, Peterson K, Foliaki ST, Haigh CL (2022) Stress and viral insults do not trigger E200K PrP conversion in human cerebral organoids. PLoS One 17:e0277051

Sola D, Betancor M, Marco Lorente PA, Perez Lazaro S, Barrio T, Sevilla E, Marin B, Moreno B, Monzon M, Acin C, Bolea R, Badiola JJ, Otero A (2023) Diagnosis in scrapie: conventional methods and new biomarkers. Pathogens 12:1399

Soto P, Bravo-Risi F, Benavente R, Lichtenberg S, Lockwood M, Reed JH, Morales R (2023) Identification of chronic wasting disease prions in decaying tongue tissues from exhumed white-tailed deer. mSphere 8:e0027223

Spiropoulos J, Hawkins SA, Simmons MM, Bellworthy SJ (2014) Evidence of in utero transmission of classical scrapie in sheep. J Virol 88:4591–4594

Srivastava A, Wang Q, Orru CD, Fernandez M, Compta Y, Ghetti B, Zanusso G, Zou WQ, Caughey B, Beauchemin CAA (2024) Enhanced quantitation of pathological alpha-synuclein in patient biospecimens by RT-QuIC seed amplification assays. PLoS Pathog 20:e1012554

Vascellari S, Orru CD, Hughson AG, King D, Barron R, Wilham JM, Baron GS, Race B, Pani A, Caughey B (2012) Prion seeding activities of mouse scrapie strains with divergent PrPSc protease sensitivities and amyloid plaque content using RT-QuIC and eQuIC. PLoS One 7:e48969

Vascellari S, Orru CD, Caughey B (2022) Real-time quaking- induced conversion assays for prion diseases, synucleinopathies, and tauopathies. Front Aging Neurosci 14:853050

Wang Z, Manca M, Foutz A, Camacho MV, Raymond GJ, Race B, Orru CD, Yuan J, Shen P, Li B, Lang Y, Dang J, Adornato A, Williams K, Maurer NR, Gambetti P, Xu B, Surewicz W, Petersen RB, Dong X, Appleby BS, Caughey B, Cui L, Kong Q, Zou WQ (2019) Early preclinical detection of prions in the skin of prion-infected animals. Nat Commun 10:247

Wilham JM, Orrú CD, Bessen RA, Atarashi R, Sano K, Race B, Meade-White KD, Taubner LM, Timmes A, Caughey B (2010) Rapid end-point quantitation of prion seeding activity with sensitivity comparable to bioassays. PLoS Pathogens 6:e1001217

Williams K, Foliaki ST, Race B, Smith A, Thomas T, Groveman BR, Haigh CL (2023) Neural cell engraftment therapy for sporadic Creutzfeldt-Jakob disease restores neuroelectrophysiological parameters in a cerebral organoid model. Stem Cell Res Ther 14:348

Wyckoff AC, Kane S, Lockwood K, Seligman J, Michel B, Hill D, Ortega A, Mangalea MR, Telling GC, Miller MW, Vercauteren K, Zabel MD (2016) Clay Components in soil dictate environmental stability and bioavailability of Cervid prions in mice. Front Microbiol 7:1885

Xu S, Reuter T, Gilroyed BH, Mitchell GB, Price LM, Dudas S, Braithwaite SL, Graham C, Czub S, Leonard JJ, Balachandran A, Neumann NF, Belosevic M, Mcallister TA (2014) Biodegradation of prions in compost. Environ Sci Technol 48:6909–6918

Yuan Q, Telling G, Bartelt-Hunt SL, Bartz JC (2018) Dehydration of prions on environmentally relevant surfaces protects them from inactivation by freezing and thawing. J Virol 92:e02191-17

Yuan Q, Rowden G, Wolf TM, Schwabenlander MD, Larsen PA, Bartelt-Hunt SL, Bartz JC (2022) Sensitive detection of chronic wasting disease prions recovered from environmentally relevant surfaces. Environ Int 166:107347

Zabel M, Ortega A (2017) The ecology of prions. Microbiol Mol Biol Rev 81:e00001-17

Zanusso G, Bongianni M, Caughey B (2014) A test for Creutzfeldt-Jakob disease using nasal brushings. N Engl J Med 371:1842–1843

Zerr I, Parchi P (2018) Sporadic Creutzfeldt-Jakob disease. Handb Clin Neurol 153:155–174

Zerr I, Ladogana A, Mead S, Hermann P, Forloni G, Appleby BS (2024) Creutzfeldt-Jakob disease and other prion diseases. Nat Rev Dis Primers 10:14

Zhang W, Orru CD, Foutz A, Ding M, Yuan J, Shah SZA, Zhang J, Kotobelli K, Gerasimenko M, Gilliland T, Chen W, Tang M, Cohen M, Safar J, Xu B, Hong DJ, Cui L, Hughson AG, Schonberger LB, Tatsuoka C, Chen SG, Greenlee JJ, Wang Z, Appleby BS, Caughey B, Zou WQ (2024) Large-scale validation of skin prion seeding activity as a biomarker for diagnosis of prion diseases. Acta Neuropathol 147:17

Chapter 8
Reactive Microglia and Astrocytes as Therapeutic Targets in Prion Diseases

Natallia Makarava, Rajesh Kushwaha, and Ilia V. Baskakov

Abstract Prion diseases, also known as transmissible spongiform encephalopathies, are a group of fatal, transmissible neurodegenerative disorders affecting both humans and animals, with no available therapeutic treatments. Recent research highlights the critical involvement of reactive microglia and astrocytes in the pathogenesis of these diseases. This chapter will review emerging evidence on the harmful roles of reactive microglia and astrocytes in prion disease progression and explore potential therapeutic strategies aimed at targeting their reactive states. A particular focus will be on the therapeutic potential of modulating microglial phagocytic activity and its associated pathways. We will address the challenges in designing effective therapies, including the phenotypic diversity of glial cells, regional differences in brain strain tropism, intercellular communication between microglia and astrocytes, dynamic changes in microglial role throughout disease progression, and the identification of optimal intervention windows. Finally, we will consider the possibility that different mechanisms underlie neurodegeneration across prion strains and Creutzfeldt-Jakob disease (CJD) subtypes, highlighting the importance of developing combination therapies targeting multiple pathways.

Keywords Prion diseases · Reactive microglia · Reactive astrocytes · Neuroinflammation · Sialylation · Phagocytosis

N. Makarava · R. Kushwaha · I. V. Baskakov (✉)
Department of Neurobiology, University of Maryland School of Medicine, Baltimore, MD, USA
e-mail: baskakov@som.umaryland.edu

G. Legname, F. Moda (eds.), *Biomarkers and Therapeutical Targets for Prion Diseases*, Subcellular Biochemistry 112, https://doi.org/10.1007/978-3-031-97055-9_8

Introduction

Prion diseases, also known as transmissible spongiform encephalopathies, comprise a group of transmissible neurodegenerative disorders affecting both humans and animals (Prusiner 1998). These diseases are lethal and currently lack any therapeutic treatment (Giles et al. 2017). Prion diseases are initiated by prions, or PrP^{Sc}, which represent the misfolded, aggregated form of a cellular sialoglycoprotein known as the prion protein, or PrP^{C} (Deleault et al. 2007; Legname et al. 2004; Makarava et al. 2010; Prusiner 1982; Wang et al. 2010). The pathogenic mechanism involves the replication and dissemination of prions throughout the central nervous system (CNS), achieved by recruiting and converting host-expressed PrP^{C} molecules into misfolded, β-sheet-rich PrP^{Sc} states (Cohen and Prusiner 1998).

Exploring the mechanisms responsible for PrP^{Sc} toxicity has been a primary focus of the field for the last three decades (Brandner et al. 1996; Collinge and Clarke 2007; Fang et al. 2016, 2018; Forloni et al. 1993; Harris and True 2006; Lakkaraju et al. 2022; Le et al. 2019; Mallucci et al. 2002, 2003; Novitskaya et al. 2006, 2007; Radford and Mallucci 2010; Solforosi et al. 2004). A wealth of data has provided solid experimental evidence that PrP^{Sc} is neurotoxic, resulting in neuronal dysfunction and loss. For the toxic effects of PrP^{Sc} to occur, PrP^{C} expression on neuronal surfaces is required (Brandner et al. 1996; Chesebro et al. 2005; Lakkaraju et al. 2022; Mallucci et al. 2003; Rambold et al. 2008; Solforosi et al. 2004). Reviews summarizing PrP^{C}-mediated toxicity of PrP^{Sc} along with downstream neurotoxic pathways can be found elsewhere (Le et al. 2019). Not surprisingly, the development of therapeutic strategies against prion diseases has mostly focused on inhibiting PrP^{Sc} replication or suppressing PrP^{C} expression. However, drugs developed against PrP^{Sc} have been shown to be effective against one or a few prion strains but ineffective against others (Berry et al. 2013). Moreover, targeting PrP^{Sc} replication or PrP^{Sc}-PrP^{C} interactions with antiprion compounds has resulted in the acquisition of resistance to antiprion drugs and the emergence of new prion strains (Berry et al. 2013; Bian et al. 2014; Burke et al. 2020; Ghaemmaghami et al. 2009). Indeed, it has been well established that prion strains are able to mutate and adapt to new replication environments (Burke et al. 2020; Ghaemmaghami et al. 2009; Gonzalez-Montalban et al. 2013; Katorcha et al. 2018; Mahal et al. 2010). Strain mutation and adaptation to new environments occur via a deformed templating and selective amplification of a state that fits better to a new environment (Baskakov 2014; Makarava and Baskakov 2013; Makarava et al. 2013), and/or through the selection of preexisting minor variants from a cloud of PrP^{Sc} structural variants (Li et al. 2010).

The mechanism of direct PrP^{Sc} toxicity does not adequately explain why prion strains that do not colocalize with neurons exhibit neurotoxicity. Moreover, transcriptome analysis revealed that neuronal transcripts do not change until the terminal stage of the disease, questioning whether the disease is driven by changes within neurons (Makarava et al. 2020c; Sorce et al. 2020). Several previous studies on PrP^{C}-mediated toxicity have utilized a single prion strain, often RML (Brandner et al. 1996; Lakkaraju et al. 2022; Mallucci et al. 2003), raising concerns about the

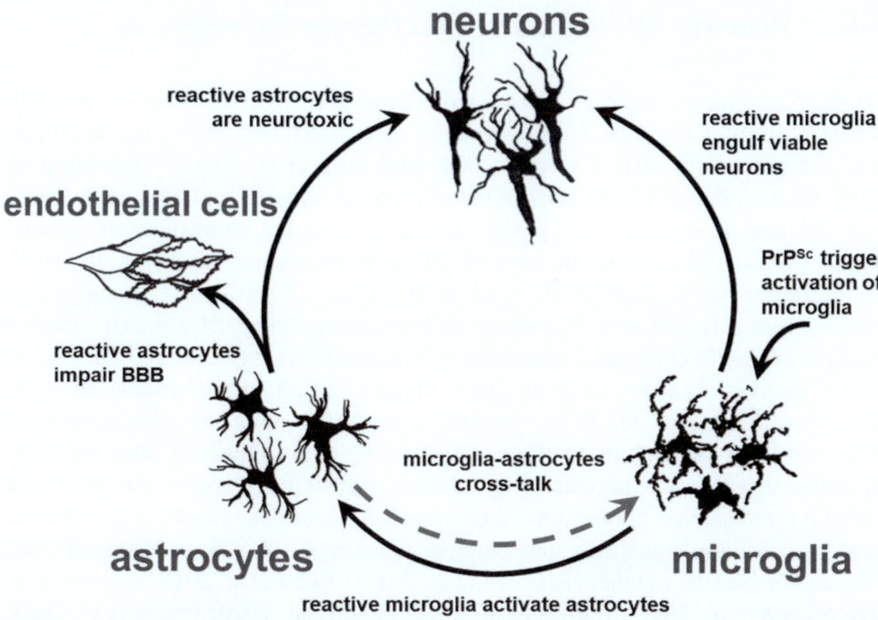

Fig. 8.1 Schematic diagram illustrating crosstalk between microglia, astrocytes, and neurons in prion diseases. PrP^{Sc} triggers a proinflammatory phenotype in microglia (Srivastava et al. 2018). Reactive microglia activate astrocytes (Kushwaha et al. 2021). In a reactive state, astrocytes lose homeostatic functions that support neurons (Makarava et al. 2021). Reactive astrocytes associated with prion disease are neurotoxic (Kushwaha et al. 2021) and have deleterious effects on endothelial cells of BBB (Kushwaha et al. 2023). (The figure was adapted from Makarava et al. 2024a)

uniformity of this mechanism across CJD subtypes or prion strains with diverse cell tropism. Additionally, the mechanism of PrP^{Sc} toxicity does not satisfactorily account for the varying vulnerability of individual brain regions to different prion strains (Karapetyan et al. 2009; Makarava et al. 2020b). For these reasons, it is worth considering alternative therapeutic targets for drug development.

Recent advances in the field suggest that reactive microglia and astrocytes not only respond to prion infection but may also play an active role in driving disease pathogenesis (Fig. 8.1) (Makarava et al. 2024a). In this chapter, we will survey emerging data on the deleterious effects of reactive microglia and astrocytes in prion diseases, as well as review data on the crosstalk between these cell types. We will discuss approaches to target the reactive states of microglia and astrocytes as potential therapeutic targets against prion diseases. We will raise issues attributed to the phenotypic diversity of astrocytes that are important for developing effective therapeutic approaches. Finally, we will review the possibility that different mechanisms may be responsible for neurodegeneration associated with different prion strains. Thus, we will discuss the importance of employing combination therapy targeting various targets.

Role of Reactive Microglia in Prion Disease Pathogenesis

The transformation of microglia into reactive states is recognized as one of the earliest events in the pathogenesis of neurodegenerative diseases, including prion diseases (Carroll et al. 2016; Lu et al. 2004; Makarava et al. 2020c; Vincenti et al. 2016). Remarkably, a global shift in the expression pattern of glial-specific genes predicted age in humans with greater precision than the expression of neuron-specific genes, underscoring the role of glia in normal aging (Soreq et al. 2017). Whole-transcriptome analysis of whole-brain tissues, as well as the analysis of selective gene sets, illustrated strong proinflammatory characteristics as a common signature of reactive microglia associated with prion diseases (Baker and Manuelidis 2003; Carroll et al. 2016; Lu et al. 2004; Majer et al. 2019, Sorensen et al. 2008; Tribouillard-Tanvier et al. 2009; Vincenti et al. 2016). With the advancement of single-cell RNA-sequencing (scRNAseq), transcriptome analysis of other neurodegenerative diseases and normal aging revealed that both microglia and astrocytes exhibit region-specific homeostatic transcriptional identities, which, under chronic neurodegeneration, transform into reactive phenotypes in a disease-specific and often region-specific manner (Boisvert et al. 2018; Clarke et al. 2018; Grabert et al. 2016; Habib et al. 2020; Mathys et al. 2017; Olah et al. 2020; Prater et al. 2023; Soreq et al. 2017; Wheeler et al. 2020; Zeisel et al. 2018). Transcriptome analysis of four brain regions affected by prions (thalamus, cortex, hypothalamus, and hippocampus) demonstrated that with disease progression, the region-specific homeostatic transcriptome signatures in microglia are replaced by a uniform, region-independent neuroinflammation signature (Makarava et al. 2020c). The neuroinflammation signature was not only region-independent but also uniform across prion strains with different cell tropisms (Makarava et al. 2020c). The neuroinflammation signature identified in prion-infected animals only partially overlapped with the microglia degenerative phenotype (MGnD) and the disease-associated microglia phenotype (DAM) reported previously in mouse models of other neurodegenerative diseases (Butovsky and Weiner 2018; Carroll et al. 2020; Keren-Shaul et al. 2017; Krasemann et al. 2017; Makarava et al. 2020c). Among the top differentially expressed genes upregulated in prion diseases were genes involved in phagocytic function and synapse pruning (*C1qa*, *C1qb*, *C1qc*, *C3ar1*, *C3*, *C4a*, *Dock2*, *Fgr*, *Fcgr1*, *Fcgr2b*, and *Fcgr3*), modifications of the extracellular matrix (*Serpina3n*), genes encoding proinflammatory chemokines that could contribute to neurotoxicity and apoptosis (*Il1α*, *Il1β*, *TNFα*, *Cxcl10*, *Cxcl9*, *Cxcl13*, *Ccl2*, *Ccl4*, *Ccl5*, *Ccl8*, *Ccl12*, *Il1rn*, *IL6*, *Il12β*), a proinflammatory transcription factor *Stat1*, and natural killer cell-mediated neurotoxicity *Fcgr2b* (Carroll et al. 2016; Makarava et al. 2020c; Sorce et al. 2020).

Several studies employing animals, postmortem human brains, or in vitro approaches have demonstrated that activation and proliferation of microglia occur in regions with PrP^{Sc} accumulation and in response to it, rather than as a consequence of neuronal death (Bate et al. 2002; Everbroeck et al. 2004; Giese et al. 1998; Greenlee et al. 2016; Kercher et al. 2007; Puoti et al. 2005; Sandberg et al.

2014; Vincenti et al. 2016; Williams et al. 1997). Notably, widespread activation and proliferation of microglia occur at much earlier stages than synaptic loss (Baker and Manuelidis 2003; Carroll et al. 2016; Gomez-Nicola et al. 2013; Lu et al. 2004; Sandberg et al. 2014), which is considered an early neuron-specific pathological sign (Hilton et al. 2013; Jeffrey et al. 2000). Using purified, brain-derived, or cultured cell-derived PrP^{Sc}, we demonstrated that PrP^{Sc} can directly trigger a proinflammatory response in primary microglia, with the chemical nature of the carbohydrate groups on the N-linked glycans of PrP^{Sc} being critical for defining the degree of microglial activation (Srivastava et al. 2018).

Over the years, a consensus has emerged that microglia constitute the primary host defense against prions that involves phagocytic clearance of PrP^{Sc} (Aguzzi and Zhu 2017; Carroll and Chesebro 2019; Mabbott et al. 2020). Indeed, several studies have shown that ablation of microglia either before prion infection or during the initial stages of the disease significantly accelerates disease progression (Bradford et al. 2021; Carroll et al. 2018, 2020; Zhu et al. 2016). Microglia depletion increases PrP^{Sc} deposition in organotypic cultured slices (Zhu et al. 2016). Knockout of MFGE8, a factor secreted by microglia that mediates the phagocytosis of apoptotic bodies, accelerates prion pathogenesis by 40 days, and elevates PrP^{Sc} levels (Kranich et al. 2010). TREM2 (triggering receptor expressed on myeloid cells-2), which is involved in the phagocytosis of apoptotic neurons, is also upregulated in prion-infected mice (Zhu et al. 2015).

However, contrary to the view that microglia are protective, partial inhibition of microglial proliferation and reactivity during the late subclinical stage (98–126 days postinfection with the ME7 strain) delays the onset of behavioral signs and extends survival by 26 days (Gomez-Nicola et al. 2013). Moreover, inhibition of microglial activation through the administration of an immunosuppressant just before or at the disease onset suppresses reactive gliosis and prolongs survival in humanized mice infected with human prions (Nakagaki et al. 2020). Vice versa, peripheral inflammation or intracerebral challenge of prion-infected mice with LPS exacerbates proinflammatory response in microglia, elevating production of the proinflammatory cytokines TNF-α, IL-1β, and IL-6, which leads to accelerated neurodegeneration and disease progression (Cunningham et al. 2005, 2009; Lunnon et al. 2011). Additionally, the knockout of galectin-3 (*Lgals3*), an opsonin that binds to the Mertk receptor on myeloid cells and mediates phagocytosis, prolongs survival in prion-infected mice (Mok et al. 2007). Finally, knockout of *Il1r1* (IL1 receptor type 1), a receptor of proinflammatory interleukin-1β expressed by microglia, attenuates astrocyte reactivity, delays the clinical onset of the disease, and prolongs survival of prion-infected mice (Schultz et al. 2004).

The phagocytic activity of microglia is implicated in the clearance of PrP^{Sc} and is assumed to play a positive role. However, a recent study links the overactivation of phagocytic pathways to synaptic loss and accelerated neurodegeneration. In response to prion infection, microglia upregulate IFN-1 (interferon type 1), which activates phagolysosomal pathways (Nazmi et al. 2019). Surprisingly, the knockout of the *IFNAR1* gene, which encodes a receptor for IFN-1, ameliorates clinical signs, improves neuronal survival and synaptic density, and prolongs survival in

ME7-infected mice (Nazmi et al. 2019). Contrary to the protective hypothesis, this work suggests that the activation of the phagocytic phenotype has a negative impact.

To reconcile apparently conflicting results on the protective versus deleterious roles of microglia, here we propose that the role of microglia changes from a net positive at early stages to a net negative at later stages of the disease. Indeed, microglia may have a net beneficial or deleterious impact depending on their capacity for PrP^{Sc} clearance versus the combined detrimental effects attributed to the loss of homeostatic functions, overproduction of proinflammatory cytokines toxic to neurons, triggering of reactive astrocytes with a neurotoxic phenotype, and direct phagocytic attack on synapses and/or viable neurons.

Microglial Phagocytic Pathways as Possible Therapeutic Targets

Consistent with the protective role of microglia in PrP^{Sc} clearance, flow cytometry analysis found PrP^{Sc} associated with microglia originating from prion-infected animals (Yamasaki et al. 2018). Moreover, using microglia isolated from prion-infected animals and live-cell imaging, we demonstrated considerable upregulation of phagocytic activity of reactive microglia (Sinha et al. 2021). Activated phagocytosis was observed regardless of the substrates tested—whether synaptosomes or myelin debris—indicating that the upregulated phagocytosis is determined by the reactive states rather than by specific substrates. Surprisingly, in the reactive state, microglia exhibited no discrimination between synaptosomes purified from prion-infected brains or normal synaptosomes (Sinha et al. 2021). Consistent with their protective role, microglia effectively phagocytosed PrP^{Sc} in mouse brains starting from the early preclinical stages (Makarava et al. 2024b). However, at the late preclinical stage, a critical shift occurs in microglial activity from phagocytosis of PrP^{Sc} to establishing extensive neuron-microglia contacts that resemble engulfment (Makarava et al. 2024b). This change transpires prior to the manifestation of clinical symptoms and is followed by a rapid accumulation of total PrP^{Sc}, suggesting a potential link to neuronal dysfunction and behavioral deficits. Interestingly, the engulfed neurons do not exhibit apoptotic markers, indicating that microglia are targeting viable neurons (Makarava et al. 2024b). These findings suggest that sustained upregulated phagocytic activity, initially a defensive response to prion infection, may eventually become detrimental due to the upregulation of phagocytic pathways that lead to an assault on viable neurons or synapses. The timing of the switch in phagocytic targets is crucial, as it presents opportunities for therapeutic intervention beginning at the onset of the disease.

Several phagocytic pathways are worth considering for exploring potential therapeutic targets. Many phagocytic pathways evolved evolutionarily to optimize neural circuitry during neurodevelopment or adult neurogenesis by eliminating excessive synapses and neurons (Alonso Bellido et al. 2023; Schafer et al. 2012; Stevens et al.

2007; Van Ryzin et al. 2019; Wakselman et al. 2008). However, these same pathways have been found to be upregulated in neurodegenerative diseases (Butler et al. 2021; Hong et al. 2016).

The Galectin-3—Mertk pathway. Galectin-3 is released by activated myeloid cells and acts as an opsonin by binding to galactose residues on the cell surface and to the Mertk receptor on phagocytes (Caberoy et al. 2012; Nomura et al. 2017). Galectin-3 was found to be upregulated in prion-infected mice. Moreover, global knockout of Galectin-3 prolonged the survival of prion-infected mice (Mok et al. 2007). The mechanism responsible for prolonged survival has not been investigated. Because Galectin-3 has pleiotropic effects, including its role in the repair and elimination of damaged lysosomes (Hoyer et al. 2022), this strategy that targets Galectin-3 should be considered with caution.

The C3b-CR3 pathway. The complement factor C3b tags neurons and synapses, driving phagocytosis through interaction with the complement receptor 3 (CR3), consisting of CD18 and CD11b (Hong et al. 2016; Schafer et al. 2012; Stevens et al. 2007). The CD11b-dependent pathway is responsible for the phagocytosis of neurons during development (Van Ryzin et al. 2019; Wakselman et al. 2008). A significant upregulation of *C3* and *Itgam* (the gene encoding CD11b) was observed in prion-infected mice and sCJD individuals (Makarava et al. 2020a; Ugalde et al. 2020).

The C1q—LRP1/CR3 pathway. C1q opsonizes synapses and drives phagocytosis via interaction with LRP1 or CR3 expressed by myeloid cells (Fouët et al. 2020; Linnartz et al. 2012; Ogden et al. 2001). Consistent with this mechanism, key components of the complement cascade, including C1qa, C1qb, C1qc, C3, and C3ar1, were found to be upregulated in prion-infected mice (Makarava et al. 2020a, c; Vincenti et al. 2016). Elimination of synapses, involving the tagging of synapses by C1q and their engulfment via an interaction with the C3 receptor, has been implicated in Alzheimer's disease, frontotemporal dementia, and normal aging (Hansen et al. 2018; Hong et al. 2016; Lui et al. 2016; Stephan et al. 2013). The inhibiting effects of C1q or C3 deficiency on prion transmission via peripheral routes have been well established (Klein et al. 2001; Mabbott et al. 2001); however, their role of glia-expressed C1q or C3 on prion pathogenesis in the CNS has not been tested.

Whether phagocytosis has net beneficial or detrimental impacts may change during the course of neurodegeneration. Simply blocking or boosting phagocytosis would not offer an effective therapeutic strategy. Instead, it is necessary to learn how to block the phagocytosis of specific targets (neurons and synapses) at specific stages of the disease, which presents significant challenges. Several important questions need to be addressed to define an effective therapeutic strategy targeting microglial phagocytosis. Does excessive microglial phagocytosis contribute to neurodegeneration? Do reactive microglia discriminate between viable neurons and apoptotic neurons damaged by PrP^{Sc}? What receptors are involved in the phagocytic uptake of neurons? Is phagocytic activity regulated in a selective fashion to target only a subset of phagocytic substrates? If so, can we selectively suppress the phagocytosis of viable neurons?

Microglia-Astrocytes Crosstalk

Recent years have unveiled intricate crosstalk between microglia and astrocytes (Linnerbauer et al. 2020; Matejuk and Ransohoff 2020). The reactive states of both astrocytes and microglia appear to be mutually dependent and regulated through multiple signaling pathways (Linnerbauer et al. 2020; Matejuk and Ransohoff 2020). Nevertheless, the question of how reactive microglia induce reactive phenotypes in astrocytes remains unsettled. According to Barres' hypothesis, microglia-derived factors TNF-α, IL-1α, and C1qa drive a neurotoxic, A1-reactive state in astrocytes (Liddelow et al. 2017). However, in prion-infected mice, the elimination of these factors had only modest effects in suppressing A1-specific markers (Hartmann et al. 2019). Moreover, contrary to expectations, triple $TNF^{-/-}/IL1a^{-/-}/C1qa^{-/-}$ knockout accelerated the progression of prion diseases (Hartmann et al. 2019). This study questioned the role of microglia-produced TNF-α, IL-1α, and C1qa as the main drivers of the neurotoxic phenotypes in astrocytes in prion diseases (Hartmann et al. 2019). Furthermore, contrary to the Barres hypothesis, the depletion of microglia resulted in an exacerbated reactive astrocyte phenotype and accelerated disease progression (Bradford et al. 2021; Carroll et al. 2018, 2020; Zhu et al. 2016). These results support the idea of crosstalk between reactive astrocytes and microglia and suggest that exuberant proinflammatory astrocyte reactivity might compensate for the lack of reactive microglia.

In animals infected with three mouse-adapted prion strains, 22L, RML, and ME7, the upregulation of microglia- and astrocyte-specific genes in specific brain regions follows consistent ranking orders, suggesting a tightly coupled activation between microglia and astrocytes (Makarava et al. 2020a, c, 2021). Upon exposure to cell media conditioned by reactive microglia isolated from prion-infected animals, primary astrocytes from normal animals acquire hypertrophic morphology and downregulate genes associated with synaptogenic functions such as *Thbs1*, *Thbs4*, and *Sparcl1* (Kushwaha et al. 2021). Thus, in addition to direct neurotoxicity, proinflammatory factors secreted by reactive microglia can induce reactive phenotypes in astrocytes, contributing to non-cell-autonomous neuronal dysfunction (Fig. 8.1).

Do reactive astrocytes influence the phenotype of microglia? Reactive astrocytes isolated from prion-infected animals secreted elevated levels of IL6, which is known to trigger microglial pathways linked to neurodegeneration (Rothaug et al. 2016). Conversely, inhibiting the activation of the transcription factor STAT3 selectively in astrocytes in mouse models of Alzheimer's and Huntington's disease suppressed astrocyte reactivity, leading to reduced neuroinflammation and microglial activation (Ben Haim et al. 2015b). In the reactive states, including those associated with prion diseases, astrocytes upregulate the expression of IL-33 and C3 (Kushwaha et al. 2021), known drivers of microglia-mediated phagocytosis and elimination of synapses (Schafer et al. 2012; Stevens et al. 2007; Vainchtein et al. 2018). The phagocytic functions of both microglia and astrocytes were shown to be influenced via alternative signaling pathways, illuminating multiple crosstalk mechanisms between microglia and astrocytes. In the absence of the tyrosine kinase receptor Mertk, which is expressed by both microglia and astrocytes, the engulfment of cell bodies by microglia was delayed, while astrocytes failed to polarize toward dying cells (Damisah et al. 2020).

Role of Reactive Astrocytes in Prion Disease Pathogenesis

In recent years, there has been a growing appreciation for the view that reactive astrocytes play an intimate role in chronic neurodegeneration (Acioglu et al. 2021; Ben Haim et al. 2015a; Habib et al. 2020; Oksanen et al. 2019). However, their precise function remains highly controversial, as discussed in previous reviews (Baskakov 2021; Ben Haim et al. 2015a; Escartin et al. 2021; Makarava et al. 2024a). In a healthy brain, astrocytes serve several crucial physiological functions (Dallérac et al. 2018; Santello et al. 2019; Sofroniew and Vinters 2010). They provide structural and trophic support to neurons, aiding in their growth and development. Additionally, astrocytes actively participate in regulating neurotransmitter levels, influencing synaptic communication. These glial cells also contribute to the establishment and stability of synapses between neurons. Furthermore, astrocytes play a role in maintaining cerebral blood flow, ensuring an adequate supply of nutrients and oxygen. They are involved in energy metabolism, providing nutrients to neighboring neurons. Importantly, astrocytes help maintain the integrity of the blood-brain barrier, protecting the brain from harmful substances. Not surprisingly, under normal conditions, astrocytes exhibit robust regional homeostatic identities, adapting to the specific needs of different brain regions (Makarava et al. 2021). In fact, seven developmentally predetermined subtypes of astrocytes that reside in different brain regions have been identified in mouse brains (Zeisel et al. 2018). In neurodegenerative diseases, including prion diseases, astrocytes acquire reactive phenotypes sustained throughout the disease progression (Ferrer 2017; Kaczmarczyk et al. 2022; Makarava et al. 2019; Phatnani and Maniatis 2015; Scheckel et al. 2020; Slota et al. 2022). Transcriptome analysis demonstrated that astrocytes responded to prion infection of the CNS much earlier and stronger than neurons (Kaczmarczyk et al. 2022; Makarava et al. 2020c; Scheckel et al. 2020; Slota et al. 2022).

A number of independent transcriptome studies have documented that in prion-infected mice, the activation of astrocytes does not follow the bidirectional A1-A2 model (Carroll et al. 2020; Hartmann et al. 2019; Makarava et al. 2020a, c, 2021; Scheckel et al. 2020). Our recent work has demonstrated that astrocytes respond to prion infection in a region-specific manner (Makarava et al. 2019, 2021, 2023). Pathway-specific heatmap analysis of prion-infected animals has revealed a global disturbance of genes across multiple astrocyte-specific functions, including but not limited to blood-brain barrier (BBB) regulation, transporters, myelination, energy metabolism, channels, extracellular matrix, growth factors/receptors/signaling, neuroprotection, and neurotoxicity (Makarava et al. 2021). Gene expression analysis suggests losses in neuronal support functions along with downregulation of genes involved in the formation and maintenance of synapses (*Nrxn1, Nlgn1, Cdh10, Gpc4, Gpc5*). While some neuroprotective pathways might be upregulated in response to prions, the net result of disturbances in neuroprotective/neurotoxic pathways, along with the global dysregulation across physiological functions, produces a neurotoxic phenotype. The neurotoxic reactive phenotype exhibits a universal gene signature regardless of a prion strain (Makarava et al. 2021). Remarkably, a very strong correlation between the gene sets reporting on the degree of astrocyte reactivity and the dysregulation in pathways associated with homeostatic functions suggests that the degree of astrocyte reactivity dictates the extent to which homeostatic functions are lost (Makarava et al. 2021). In agreement with this hypothesis, scoring of the differential gene expression across 17 animal groups inoculated with four prion strains via two routes reveals a very strong reverse correlation between the degree of astrocyte reactivity and the incubation time to the prion diseases (Makarava et al. 2021). Animal groups with the most severe astrocyte reactivity show the most rapid disease progression. In summary, the transcriptome analysis raises the possibility that phenotypic changes in reactive astrocytes contribute to the faster progression of diseases and perhaps even drive prion pathogenesis.

Infection of astrocytes with prions triggers neuronal dysfunction upon co-culturing of infected astrocytes with neurons (Cronier et al. 2012). However, it remained unclear whether the neurotoxic effects were due to astrocyte-produced PrP^{Sc} or factors released by astrocytes. In our recent work, reactive astrocytes isolated from prion-infected animals exerted adverse effects on primary neuronal cultures, resulting in a reduction in spine size and density, along with impairment of neuronal growth and synapse integrity (Kushwaha et al. 2021). The media conditioned by the reactive astrocytes also exhibited deleterious effects on primary neurons, including a reduction in the density and size of dendritic spines, disintegration of synapses, reduced expression of pre- and postsynaptic proteins, along with a decrease in neuronal viability (Kushwaha et al. 2021). Selective targeting of the unfolded protein response, which is exacerbated in reactive astrocytes, via inhibition of PERK signaling, was found to prolong the incubation time to terminal disease in mice (Smith et al. 2020).

Under normal conditions, astrocytes play an essential role in the development and maintenance of the BBB (Araya et al. 2008; Heithoff et al. 2021; Lee et al. 2003; Siddharthan et al. 2007). Increased BBB permeability was noticed nearly 40 years ago and was found to be common among animals infected with different prion strains (Wisniewski et al. 1983). Significant caspase immunoreactivity of blood vessels, indicative of endothelial cell death, was also reported in prion diseases (Haigh et al. 2014). In mice infected with prions, transcriptome analysis revealed significant perturbations in the expression of astrocyte-specific genes involved in BBB maintenance (Makarava et al. 2021). In support of changes seen from transcriptome analysis, the localization of aquaporin 4 (AQP4), the most prevalent water channel that normally localizes on astrocytic endfeet, changed dramatically upon the transition of astrocytes into a reactive state (Kushwaha et al. 2023; Makarava et al. 2021). The loss of BBB integrity and aberrant localization of AQP4, a sign of retraction of astrocytic endfeet from blood vessels, were observed prior to disease onset (Kushwaha et al. 2023). These changes suggest a loss of astrocyte polarization and possible dysregulation of astrocyte functions responsible for BBB maintenance. Indeed, a recent study demonstrated that reactive astrocytes or media conditioned by reactive astrocytes isolated from prion-infected mice induced a disease-associated phenotype in endothelial cells originating from noninfected adult mice (Kushwaha et al. 2023). IL-6, secreted by reactive astrocytes, was identified as a proinflammatory factor responsible for mediating their deleterious effect on the BBB (Kushwaha et al. 2023). Surprisingly, extracellular vesicles produced by normal astrocytes partially reversed the disease-associated phenotype of endothelial cells isolated from prion-infected mice (Kushwaha et al. 2023).

To summarize, astrocyte reactivity, originating as a physiological response to prion infection, gives rise to a disease-associated phenotype that interferes with astrocyte homeostatic functions. Indeed, a global dysregulation across multiple physiological functions of astrocytes, including loss of neuronal support, was observed in prion disease. The inverse correlation between the degree of astrocyte reactivity and the disease incubation time suggests that phenotypic changes in astrocytes contribute to faster disease progression. Reactive astrocytes isolated from prion-infected mice had deleterious effects on primary neurons and endothelial cells that constitute the BBB. Thus, in addition to the direct toxicity of PrP^{Sc}, neurodegeneration is also driven by a non-cell-autonomous astrocyte-dependent mechanism. This mechanism might act via downregulation of synaptogenic factors and/or upregulation of proinflammatory factors.

Regulators of Astrocyte Reactivity as Possible Therapeutic Targets

The question of whether the astrocyte reactive phenotype could be reversed remains a subject of debate. Studies involving the optic nerve subjected to mild injury, induced by brief ocular pressure, have indicated that astrocyte reactivity can be fully resolved if the insult is removed (Sun et al. 2013). However, in more severe insults such as spinal cord injury, which lead to the formation of glial scars consisting of reactive astrocytes, phenotypic changes have long been considered irreversible. Recent studies, however, have shown that reactive astrocytes isolated from injured spinal cords revert their phenotype upon transplantation into a naïve spinal cord, and vice versa (Hara et al. 2017). This suggests that the preservation of reactive phenotypes relies on persistent stimulus or the presence of environmental factors.

Is it possible to reverse reactive states in the presence of a persistent stimulus? The activation of the STAT3 transcription factor has been identified as a universal feature of astrocyte reactivity in neurodegenerative diseases, shared across different species, brain regions, and types of illnesses in Ben Haim et al. (2015a) and Yan et al. (2018). Selective inhibition of the STAT3 pathway in astrocytes has been found to suppress astrocyte activation or reverse their reactive phenotype, improving disease outcomes in animal models of neurodegenerative diseases, including Alzheimer's and Huntington's diseases (Ben Haim et al. 2015b; Ceyzériat et al. 2018; Reichenbach et al. 2019). Activation of STAT3 has also been observed in animals infected with prions (Na et al. 2007); however, its role in driving astrocyte reactive states associated with prion diseases has not yet been examined. Nevertheless, STAT3 represents one of the main targets for reversing astrocyte reactivity. Whether inhibiting STAT3 rescues important homeostatic functions of astrocytes and delays or ameliorates disease progression remains to be established.

A recent study demonstrated that in optic nerve injury, a balance between neurotoxic, C3-positive, and neuroprotective, C3-negative reactive astrocyte populations is regulated by distinct pools of compartmentalized cyclic adenosine monophosphate (cAMP) (Cameron et al. 2023). Raising nuclear or depleting cytoplasmic cAMP in reactive astrocytes promoted retinal ganglion cell survival. Targeting cAMP in a compartment-dependent manner might represent an alternative strategy for manipulating the reactive phenotype of astrocytes. However, it remains to be determined whether this strategy is suitable for a broad spectrum of neurodegenerative diseases or only applicable to optic nerve injury.

Consideration of Region-Specific Differences in Reactive Phenotype

Individual prion strains follow their own, strain-specific timelines of disease progression, exhibiting strain-specific affinity to different brain regions (Karapetyan et al. 2009; Makarava et al. 2020b, c). Several important issues must be considered for developing effective therapeutic strategies targeting reactive states of glia. Do glial cells respond to pathological insults in a strain-specific manner by adopting distinct, strain-specific states? Are glial reactive phenotypes region-specific or uniform across brain regions? What role does region-specific homeostatic identity play in dictating glial reactive phenotype? How do the differences in the timeline of disease progression between different brain regions affect the phenotypic diversity of glia across the whole brain? Only some of these questions have been addressed at this time.

As discussed above, transcriptome analyzes of microglia-specific genes revealed that in prion-infected mice, region-specific homeostatic signatures are replaced with a uniform neuroinflammation signature (Makarava et al. 2020c). The same gene sets are activated in animals infected with different prion strains, regardless of prion strain cell tropism (Carroll et al. 2016; Makarava et al. 2020c). However, the timing and degree of activation in different brain regions are dictated by individual strains.

In contrast to microglia, astrocytes exhibit significant regional specificity in their reactive states (Makarava et al. 2021, 2023). Region-specific phenotypic diversity of reactive astrocytes is found not only in mice infected with prions but also in animals subjected to other neurological insults. Our recent work examined the region-specific response of genes associated with astrocyte functions and reactivity to insults of diverse natures, including prion infection, mechanical injury (TBI), genetic mutations associated with familial Alzheimer's disease using the 5XFAD mouse model, ischemic insult, and normal aging (Makarava et al. 2023). Surprisingly, under pathological insults, the expression of genes associated with astrocytes preserved region-specific signatures, suggesting that astrocytes respond to insults in a region-specific manner (Fig. 8.2) (Makarava et al. 2023). In fact, principal component analysis (PCA), which considers not only the number of changed genes but also the extent of their changes, revealed a better separation into distinct clusters based on brain region rather than the nature of an insult (Makarava et al. 2023). For instance, even within the same insult, the reactive phenotypes of cortical and thalamic clusters were clearly distinctive (Fig. 8.2). Within the same region, the reactive phenotypes for individual insults showed considerable overlaps. While PCA did not dismiss the idea regarding the existence of insult-specific phenotypes, the insult-specific populations did not separate well from each other and instead partially overlapped, forming continuums of phenotypes. The continuums of phenotypes were region-specific, suggesting that in defining reactive phenotypes, the role of region-specific homeostatic identity is perhaps as important as the nature of an insult (Fig. 8.2). These results illustrate that region-specific homeostatic identities are critical in shaping the astrocyte response to pathological insults of diverse natures.

Are region-specific populations of astrocytes equally susceptible to insults in terms of their phenotypic changes? Judging by changes in the transcriptome, the rates of astrocyte aging under normal conditions varied across different brain regions (Boisvert et al. 2018). In prion diseases, the thalamus is considered the most susceptible brain region, exhibiting the most profound neuroinflammation. The vulnerability of thalamic astrocytes in prion diseases could be attributed to the intrinsic tropism of prions to this region, as the thalamus is impaired at early stages and most severely affected at advanced stages (Carroll et al. 2016; Makarava et al. 2020c; Sandberg et al. 2014). Surprisingly, the thalamus shows a profound astrocytic response even in experimental insults not targeting the thalamus. In 5XFAD mice, deposition of Aβ plaques and signs of reactive astrocytes occur first in the cortex and hippocampus at younger ages (Oakley et al. 2006). However, by 14 months of age, the thalamus shows the highest load of Aβ plaques in 5XFAD mice (Frost et al. 2020). Remarkably, thalamic astrocytes respond not only to insults targeting the thalamus but also to injuries in other brain regions. Indeed, in traumatic brain injury (TBI) in mice, the thalamus exhibits the highest GSA scores for astrocyte reactivity and function, despite the cortex being the primary site of injury. These results align with previous clinical findings in humans. Up to 17 years after severe TBI, individuals showed profound chronic neuroinflammation in the thalamus, attributed to damages in the thalamocortical tract (Ramlackhansingh et al. 2011; Scott et al. 2015). Moreover, neuroinflammation in the thalamus has been proposed as a marker of cortical injury and subsequent long-term cognitive deficits (Necula et al. 2021). Analysis of astrocyte-specific genes in aged mice revealed the highest GSA scores for functional gene sets in the thalamus. Given that the thalamus receives reciprocal

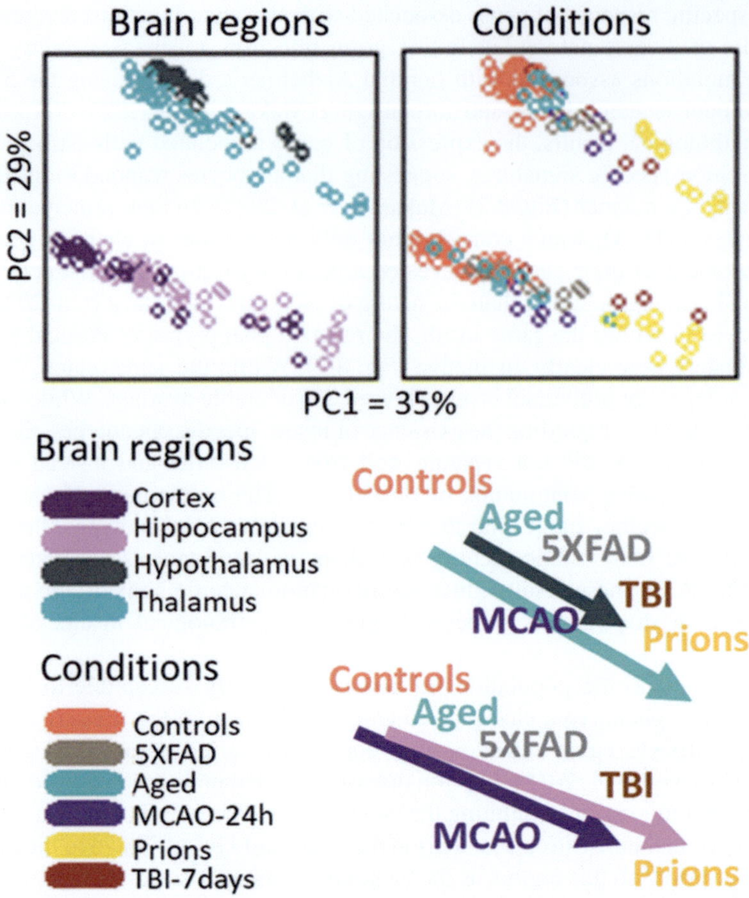

Fig. 8.2 Principal component analysis (PCA) of region-specific insult-elicited differences in gene expression across five animal groups: 10-month-old 5XFAD mouse model, aged 24-month-old mice, and mice subjected to ischemic insult (middle cerebral artery occlusion or MCAO) and analyzed 24 h post-insult, prion-infected mice, and mice subjected to TBI and analyzed 7 days post-injury. PCA was performed using the panel of astrocyte-specific genes for samples from five experimental animal groups and corresponding controls. Four brain regions were analyzed in each group: cortex, hippocampus, thalamus, and hypothalamus. The distribution of samples according to brain region and condition is shown on the top left and top right, respectively. A schematic illustrating the continuum of astrocytic phenotypes is shown on the bottom right. Each dot represents an individual animal, with different colors representing different brain regions or experimental conditions. (The figure was adapted from Makarava et al. 2023)

projections from the entire cerebral cortex, its vulnerability could significantly contribute to understanding pathology and predicting disease outcomes (Grossman and Inglese 2015; Scott et al. 2015). In summary, thalamic astrocytes are considered the most responsive to insults, reacting not only to direct insults within the thalamus but also to injuries in other brain regions.

Neuroinflammation Versus Direct PrP^{Sc} Toxicity as Drivers of Neurodegeneration

In mice, prions accumulate in different cell types depending on strain-specific cell tropism. The 22L strain is mainly associated with astrocytes, ME7 is primarily found in neurons, and SSLOW is predominantly colocalized with microglia, whereas the cell association of RML depends on the brain region, showing either neuron- or astrocyte-specific localization (Carroll et al. 2016; Makarava et al. 2020a). According to the hypothesis on direct neurotoxicity of PrP^{Sc}, the incubation time to disease is expected to be the shortest in ME7-inoculated mice. Contrary to this expectation, ME7 exhibits the longest incubation time among the four strains (SSLOW, RML, 22L, and ME7). In several previous studies that established the mechanism of PrP^{C}-mediated toxicity of PrP^{Sc}, single prion strains were employed, often RML (Brandner et al. 1996; Lakkaraju et al. 2022; Mallucci et al. 2003), raising questions about the generality of this mechanism.

Here, we propose that neurotoxicity attributed to neuroinflammation contributes to neurodegeneration in parallel with the mechanism of direct toxicity of PrP^{Sc} and that these two mechanisms are not mutually exclusive. Moreover, depending on strain-specific features, either neuroinflammation or direct toxicity of PrP^{Sc} could be the major driver of the disease (Fig. 8.3). We hypothesize that in strains with short incubation times, the disease is predominantly driven by neuroinflammation, whereas in strains with long incubation times, PrP^{C}-mediated PrP^{Sc} toxicity constitutes the major mechanism responsible for neurodegeneration (Fig. 8.2). Indeed, among the four mouse-adapted strains (SSLOW, 22L, RML, and ME7), SSLOW induces the most profound neuroinflammation and has the shortest incubation time for the disease, whereas ME7 displays the most attenuated neuroinflammation and the longest incubation time (Makarava et al. 2020a, 2021). One should not dismiss the possibility that the impact of the two mechanisms on neuronal loss or dysfunction could be brain region-specific. For instance, in RML-infected mice, the neuronal loss associated with PrP^{Sc} toxicity appears to be limited to the hippocampus.

Further support for this concept comes from a recent study employing a new mouse model where the expression of PrP^{C} was restricted exclusively to neurons

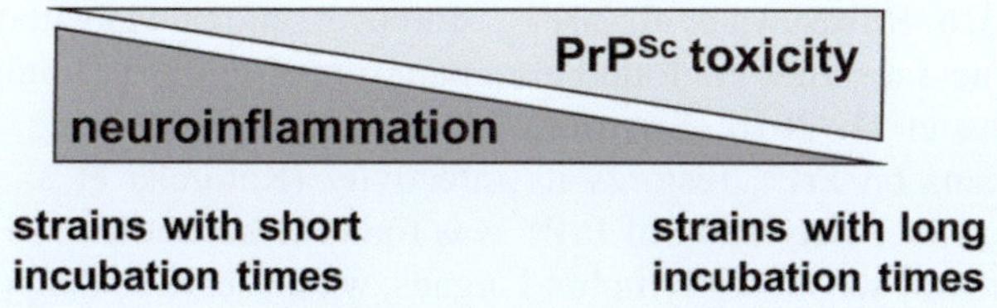

Fig. 8.3 Schematic diagram illustrating that the pathogenesis of prion diseases is driven by two mechanisms: direct toxicity of PrP^{Sc} to neurons and neuroinflammation. Neuroinflammation is proposed to be a predominant mechanism in strains with fast disease progression, whereas the mechanism that involves PrP^{Sc} toxicity plays a major role in strains with long incubation time to disease

(Lakkaraju et al. 2022). In response to prion infection, this mouse model did not exhibit any activation of microglia or astrocytes while showing a very long incubation time ranging between 400 and 500 days post inoculation (Lakkaraju et al. 2022). The disease seems to be driven entirely by PrP^{Sc} toxicity mediated via PrP^{C} expressed in neurons. The next chapter discusses PrP^{Sc} features important for triggering neuroinflammation.

The Sialylation Status of PrP^{Sc} Dictates the Degree of Neuroinflammation

PrP^{C} undergoes posttranslational modification with a GPI anchor and one or two sialylated N-linked glycans (Bolton et al. 1985; Stahl et al. 1993; Stahl et al. 1987; Turk et al. 1988). N-linked glycans exhibit extreme diversity concerning their structure and composition, resulting in the expression of several hundred PrP^{C} sialoglycoforms within a cell (Endo et al. 1989; Rudd et al. 1999; Stimson et al. 1999). Upon the conversion of PrP^{C} into PrP^{Sc}, the N-linked glycans are retained, leading to PrP^{Sc} decorated with N-glycans (Baskakov 2017; Baskakov et al. 2018; Rudd et al. 1999; Turk et al. 1988). Our studies demonstrate that individual prion strains selectively recruit PrP^{C} sialoglycoforms based on a strain structure and the sialylation status of individual PrP^{C} molecules (Baskakov and Katorcha 2016; Katorcha et al. 2015). This selective recruitment gives rise to strain-specific patterns of carbohydrate epitopes on PrP^{Sc} (Baskakov et al. 2018).

Sialic acid residues, located terminally on N-linked glycans, along with the underlying galactose residues, play a critical role in defining the response of the innate immune system. Sialylation of glycans serves as a component of a self-associated molecular pattern, aiding the innate immune system in distinguishing between "self," "altered self," or "non-self" (Brown and Neher 2014; Varki 2008). Removal of sialic acid residues exposes galactose residues, which then serve as "eat me" signals for both professional and nonprofessional macrophages, including microglia. Consistent with this mechanism, PrP^{Sc} produced via Protein Misfolding Cyclic Amplification using desialylated PrP^{C} as a substrate does not induce prion disease in animals after intracranial or intraperitoneal administrations (Katorcha et al. 2014, 2016; Srivastava et al. 2017). Moreover, animals inoculated with PrP^{Sc} lacking sialic acid residues are found to remain free of prions throughout their lifetime (Katorcha et al. 2016; Srivastava et al. 2017). Reinstating sialylation of N-linked glycans on PrP^{Sc} restores its infectivity (Katorcha et al. 2016). Upon peripheral exposure, sialylation of PrP^{Sc} was found to be critical for its trafficking and colonization of secondary lymphoid organs, which serve as the primary sites of prion replication (Srivastava et al. 2017). Following peripheral administration, PrP^{Sc} lacking sialylation is transported to the liver instead of secondary lymphoid organs (Srivastava et al. 2017). These studies suggest that sialylation protects PrP^{Sc} against clearance by macrophages and microglia and is critical in determining the outcome of prion infection (Makarava and Baskakov 2023).

Strain-specific differences in sialylation status of PrP^{Sc} due to the selective recruitment of PrP^{C} sialoglycoforms have been well established (Baskakov and Katorcha 2016; Katorcha et al. 2015; Srivastava et al. 2015). Multiple lines of evidence support the hypothesis that the sialylation status of PrP^{Sc} dictates the degree of neuroinflammation.

First, PrP^{Sc} purified from animal brains triggers a proinflammatory response in primary microglia (Srivastava et al. 2018). The strength of this response is determined by the level of exposed galactose in N-linked glycans of PrP^{Sc}. Partial desialylation of PrP^{Sc} enhances the proinflammatory response (Srivastava et al. 2018).

Second, recent analyzes of the sialylation status of PrP^{Sc} from different brain regions revealed a reverse correlation between the level of sialylation of PrP^{Sc} and the degree of neuroinflammation in those regions (Makarava et al. 2020b). Thalamic PrP^{Sc} exhibits lower sialylation compared to PrP^{Sc} in the hippocampus or cortex (Makarava et al. 2020b). The thalamus is the first to develop neuroinflammation and is the most severely affected in the terminal stage (Makarava et al. 2020b, c, 2021).

Third, among the four prion strains compared (SSLOW, RML, 22L, and ME7), SSLOW PrP^{Sc} exhibits the lowest level of sialylation and causes the most profound and widespread neuroinflammation (Makarava et al. 2020a). Conversely, ME7 PrP^{Sc} is sialylated at the highest level and is associated with mild neuroinflammation. Indeed, correlations exist between the level of sialylation, the degree of neuroinflammation, and the incubation time to disease (Makarava et al. 2020a).

Fourth, consistent with the hypothesis that carbohydrate groups of N-linked glycans are important determinants of glia activation, atypical PrP^{Sc} with low glycosylation status does not trigger neuroinflammation in animals nor does it cause clinical disease while replicating in the brain (Kovacs et al. 2013; Makarava et al. 2012, 2015, 2016).

It remains to be established whether targeting PrP^{C} sialylation could serve as a therapeutic approach. In mammals, sialylation of glycans occurs in the trans-Golgi and is catalyzed by 20 sialyltransferases (STs) (Audry et al. 2011). In PrP^{Sc}, sialic acid residues are linked via both α2-3 and α2-6 linkages, with α2-6 linkage being predominant (Endo et al. 1989; Katorcha and Baskakov 2017). Among the 20 STs, three enzymes (ST3Gal3, ST3Gal4, and ST3Gal6) sialylate N-linked glycans via an α2-3 linkage, whereas only two (ST6Gal1 and ST6Gal2) sialylate via an α2-6 linkage (Audry et al. 2011; Takashima 2008). Knockout of ST6Gal1 in mice did not alter the sialylation of PrP^{Sc}, nor did it affect the incubation time to disease or disease pathology (Makarava et al. 2022). These results highlight redundancy in sialylation and the challenges in targeting this pathway. It may be worth considering alternative targets, such as receptors and opsonins expressed by myeloid cells and astrocytes involved in the recognition of sialic acid residues and galactose.

Conclusion

Recent advances in the field suggest that reactive microglia and astrocytes are intimately involved in the pathogenesis of prion diseases. Therefore, multiple mechanisms behind neurodegeneration in prion diseases should be considered. Investigating pathways that target the reactive phenotypes of astrocytes and microglia is worthwhile for elucidating potential therapeutic strategies.

Several challenges must be taken into consideration when developing effective therapeutic strategies that target reactive astrocytes and microglia. First, it is likely that the reactive phenotypes, as well as the roles played by reactive astrocytes and microglia, change with disease progression. Defining the optimal time window for therapeutic intervention is important. Second, prion strains target different brain areas. Since astrocytes respond to prions in a region-specific manner, it might be challenging to manipulate their reactive phenotype uniformly across the whole brain. Testing potential therapeutic strategies using several prion strains is essential. Third, due to the crosstalk between microglia and astrocytes, suppressing or altering the reactive state of one cell type will likely change the reactive phenotype of another cell type. Furthermore, the nature of microglia-astrocyte interaction might change with disease progression, presenting additional challenges. Defining the effect of potential drug treatments on the reactive phenotypes of both astrocytes and microglia is worth the effort. Fourth, the disease progresses at different rates in different brain regions. Consequently, phenotypically diverse subpopulations of glia are expected to reside in different regions within the same disease stage. For testing the effects of potential drugs, it is critical to define glial responses in a region-specific manner.

References

Acioglu C, Li L, Elkabes S (2021) Contribution of astrocytes to neuropathology of neurodegenerative diseases. Brain Res 1758:147291

Aguzzi A, Zhu C (2017) Microglia in prion diseases. J Clin Invest 127(9):3230–3239

Alonso Bellido IM, Posada-Pérez M, Hernández-Rasco F, Vázquez-Reyes S, Cabanillas M, Herrera AJ, Bachiller S, Soldán-Hidalgo J, Espinosa-Oliva AM, Joseph B, de Pablos RM, Venero JL, Ruiz R (2023) Microglial caspase-3 is essential for modulating hippocampal neurogenesis. Brain Behav Immun 112:206–219

Araya R, Kudo M, Kawano M, Ishii K, Hashikawa T, Iwasato T, Itohara S, Terasaki T, Oohira A, Mishina Y, Yamada M (2008) Bmp signaling through bmpria in astrocytes is essential for proper cerebral angiogenesis and formation of the blood-brain-barrier. Mol Cell Neurosci 38(3):417–430

Audry M, Jeanneau C, Imberty A, Harduin-Lepers A, Delannoy P, Breton C (2011) Current trend in the structure-activity relationships of sialylatransferases. Glycobiology 21(6):716–726

Baker CA, Manuelidis L (2003) Unique inflammatory rna profiles of microglia in Creutzfeldt-Jakob disease. Proc Natl Acad Sci USA 100(2):675–679

Baskakov IV (2014) The many shades of prion strain adaptation. Prion 8(2):169–172

Baskakov IV (2017) Limited understanding of the functional diversity of n-linked glycans as a major gap of prion biology. Prion 11(2):82–88

Baskakov IV (2021) On the reactive states of astrocytes in prion diseases. Prion 15(1):87–93

Baskakov IV, Katorcha E (2016) Multifaceted role of sialylation in prion diseases. Front Neurosci 10(1):e358

Baskakov IV, Katorcha E, Makarava N (2018) Prion strain-specific structure and pathology: a view from the perspective of glycobiology. Viruses 10(12):723

Bate C, Boshuizen RS, Langeveld JPM, Williams A (2002) Temporal and spatial relationship between the death of PrP-damaged neurons and microglial activation. Neuroreport 13(13):1695–1700

Ben Haim L, Carrillo-de Sauvage M-A, Ceyzériat K, Escartin C (2015a) Elusive roles for reactive astrocytes in neurodegenerative diseases. Front Cell Neurosci 9:278

Ben Haim L, Ceyzériat K, Carrillo-de Sauvage MA, Aubry F, Auregan G, Guillermier M, Ruiz M, Petit F, Houitte D, Faivre E, Vandesquille M, Aron-Badin R, Dhenain M, Déglon N, Hantraye P, Brouillet E, Bonvento G, Escartin C (2015b) The jak/stat3 pathway is a common inducer of astrocyte reactivity in alzheimer's and Huntington's diseases. J Neurosci 35(6):2817–2829

Berry DB, Lu D, Geva M, Watts JC, Bhardwaj S, Oehler A, Renslo AR, De Armond SJ, Prusiner SB, Giles K (2013) Drug resistance confounding prion therapeutics. Proc Natl Acad Sci USA 110(44):E4160–E4169

Bian J, Kang HE, Telling GC (2014) Quinacrine promotes replication and conformational mutation of chronic wasting disease prions. Proc Natl Acad Sci USA 111(16):6028–6033

Boisvert MM, Erikson GA, Shokhirev MN, Allen NJ (2018) The aging astrocyte transcriptome from multiple regions of the mouse brain. Cell Rep 22:269–285

Bolton DC, Meyer RK, Prusiner SB (1985) Scrapie PrP 27-30 is a sialoglycoprotein. J Virol 53(2):596–606

Bradford BM, McGuire LI, Hume DA, Pridans C, Mabbott NA (2021) Complete microglia deficiency accelerates prion disease without enhancing CNS prion accumulation. bioRxiv:2021.2001.2005.425436

Brandner S, Isenmann S, Raeber A, Fischer M, Sailer A, Kobayashi Y, Marino S, Weissmann C, Aguzzi A (1996) Normal host prion protein necessary for scrapie-induced neurotoxicity. Nature 379:339–343

Brown GC, Neher JJ (2014) Microglial phagocytosis of live neurons. Nat Rev Neurosci 15(4):209–216

Burke CM, Mark KMK, Kun J, Beauchemin KS, Supattapone S (2020) Emergence of prions selectively resistant to combination drug therapy. PLoS Pathog 16(5):e1008581

Butler CA, Popescu AS, Kitchener EJA, Allendorf DH, Puigdellívol M, Brown GC (2021) Microglial phagocytosis of neurons in neurodegeneration, and its regulation. J Neurochem 158(3):621–639

Butovsky O, Weiner HL (2018) Microglial signatures and their role in health and disease. Nat Rev Neurosci 19(10):622–635

Caberoy NB, Alvarado G, Bigcas J-L, Li W (2012) Galectin-3 is a new mertk-specific eat-me signal. J Cell Physiol 227(2):401–407

Cameron EG, Nahmou M, Toth AB, Heo L, Tanasa B, Dalal R, Yan W, Nallagatla P, Xia X, Hay S, Knasel C, Stiles TL, Douglas C, Atkins M, Sun C, Ashouri M, Bian M, Chang KC, Russano K, Shah S, Woodworth MB, Galvao J, Nair RV, Kapiloff MS, Goldberg JL (2023) A molecular switch for neuroprotective astrocyte reactivity. Nature. https://doi.org/10.1038/s41586-023-06935-3

Carroll JA, Chesebro B (2019) Neuroinflammation, microglia, and cell-association during prion disease. Viruses 11(1):65

Carroll JA, Striebel JF, Rangel A, Woods T, Phillips K, Peterson KE, Race B, Chesebro B (2016) Prion strain differences in accumulation of PrPSc on neurons and glia are associated with similar expression profiles of neuroinflammatory genes: comparison of three prion strains. PLoS Pathog 12(4):e1005551

Carroll JA, Race B, Williams K, Striebel J, Chesebro B (2018) Microglia are critical in host defense against prion disease. J Virol 92(15):e00549–e00518

Carroll JA, Race B, Williams K, Striebel J, Chesebro B (2020) Rna-seq and network analysis reveal unique glial gene expression signatures during prion infection. Mol Brain 13(1):71

Ceyzériat K, Ben Haim L, Denizot A, Pommier D, Matos M, Guillemaud O, Palomares MA, Abjean L, Petit F, Gipchtein P, Gaillard MC, Guillermier M, Bernier S, Gaudin M, Aurégan G, Joséphine C, Déchamps N, Veran J, Langlais V, Cambon K, Bemelmans AP, Baijer J, Bonvento G, Dhenain M, Deleuze JF, Oliet SHR, Brouillet E, Hantraye P, Carrillo-de Sauvage MA, Olaso R, Panatier A, Escartin C (2018) Modulation of astrocyte reactivity improves functional deficits in mouse models of alzheimer's disease. Acta Neuropathol Commun 6(1):104

Chesebro B, Trifilo M, Race M, Meade-White K, Teng C, LaCasse R, Raymond L, Favara C, Baron G, Priola S, Caughey B, Masliah E, Oldstone M (2005) Anchorless prion protein result in infectious amyloid disease without clinical scrapie. Science 308:1435–1439

Clarke LE, Liddelow SA, Chakraborty C, Munich AE, Heiman M, Barres BA (2018) Normal aging induces a1-like astrocyte reactivity. Proc Natl Acad Sci USA 115(8):E1896–E1905

Cohen FE, Prusiner SB (1998) Pathologic conformations of prion proteins. Annu Rev Biochem 67:793–819

Collinge J, Clarke AR (2007) A general model of prion strains and their pathogenicity. Science 318(5852):930–936

Cronier S, Carimalo J, Schaeffer B, Jaumain E, Béringue V, Miquel MC, Laude H, Peyrin JM (2012) Endogenous prion protein conversion is required for prion-induced neuritic alterations and neuronal death. FASEB J 26(9):3854–3861

Cunningham C, Wilcockson DC, Campion S, Lunnon K, Perry VH (2005) Central and systemic endotoxin challenges exacerbate the local inflammatory response and increase neuronal death during chronic neurodegeneration. J Neurosci 25(40):9275–9284

Cunningham C, Campion S, Lunnon K, Murray CL, Woods JF, Deacon RM, Rawlins JN, Perry VH (2009) Systemic inflammation induces acute behavioral and cognitive changes and accelerates neurodegenerative disease. Biol Psychiatry 65(4):304–312

Dallérac G, Zapata J, Rouach N (2018) Versatile control of synaptic circuits by astrocytes: where, when and how? Nat Rev Neurosci 19(12):729–743

Damisah EC, Hill RA, Rai A, Chen F, Rothlin CV, Ghosh S, Grutzendler J (2020) Astrocytes and microglia play orchestrated roles and respect phagocytic territories during neuronal corpse removal in vivo. Sci Adv 6(26):eaba3239

Deleault NR, Harris BT, Rees JR, Supattapone S (2007) Formation of native prions from minimal components in vitro. Proc Acad Natl Sci USA 104(23):9741–9746

Endo T, Groth D, Prusiner SB, Kobata A (1989) Diversity of oligosaccharide structures linked to asparagines of the scrapie prion protein. Biochemistry 28(21):8380–8388

Escartin C, Galea E, Lakatos A, O'Callaghan JP, Petzold GC, Serrano-Pozo A, Steinhäuser C, Volterra A, Carmignoto G, Agarwal A, Allen NJ, Araque A, Barbeito L, Barzilai A, Bergles DE, Bonvento G, Butt AM, Chen WT, Cohen-Salmon M, Cunningham C, Deneen B, De Strooper B, Díaz-Castro B, Farina C, Freeman M, Gallo V, Goldman JE, Goldman SA, Götz M, Gutiérrez A, Haydon PG, Heiland DH, Hol EM, Holt MG, Iino M, Kastanenka KV, Kettenmann H, Khakh BS, Koizumi S, Lee CJ, Liddelow SA, MacVicar BA, Magistretti P, Messing A, Mishra A, Molofsky AV, Murai KK, Norris CM, Okada S, Oliet SHR, Oliveira JF, Panatier A, Parpura V, Pekna M, Pekny M, Pellerin L, Perea G, Pérez-Nievas BG, Pfrieger FW, Poskanzer KE, Quintana FJ, Ransohoff RM, Riquelme-Perez M, Robel S, Rose CR, Rothstein JD, Rouach N, Rowitch DH, Semyanov A, Sirko S, Sontheimer H, Swanson RA, Vitorica J, Wanner IB, Wood LB, Wu J, Zheng B, Zimmer ER, Zorec R, Sofroniew MV, Verkhratsky A (2021) Reactive astrocyte nomenclature, definitions, and future directions. Nat Neurosci 24(3):312–325

Everbroeck BV, Dobbeleir I, Waele MD, Leenheir ED, Lubke U, Martin J-J, Cras P (2004) Extracellular protein deposition correlates with glial activation and oxidative stress in Creutzfeldt-Jakob and Alzheimer's disease. Acta Neuropathol 108(3):194–200

Fang C, Imberdis T, Garza MC, Wille H, Harris DA (2016) A neuronal culture system to detect prion synaptotoxicity. PLoS Pathog 12(5):e1005623

Fang C, Wu B, Le NTT, Imberdis T, Mercer RCC, Harris DA (2018) Prions activate a p38 mapk synaptotoxic signaling pathway. PLoS Pathog 14(9):e1007283

Ferrer I (2017) Diversity of astroglial responses across human neurodegenerative disorders and brain aging. Brain Pathol 27(5):645–674

Forloni G, Angeretti N, Chiesa R, Monzani E, Salmona M, Bugiani O, Tagliavini F (1993) Neurotoxicity of a prion protein fragment. Nature 362:543–546

Fouët G, Gout E, Wicker-Planquart C, Bally I, De Nardis C, Dedieu S, Chouquet A, Gaboriaud C, Thielens NM, Kleman J-P, Rossi V (2020) Complement c1q interacts with LRP1 clusters II and IV through a site close but different from the binding site of its C1R and C1S-associated proteases. Front Immunol 11:583754

Frost GR, Longo V, Li T, Jonas LA, Judenhofer M, Cherry S, Koutcher J, Lekaye C, Zanzonico P, Li Y-M (2020) Hybrid pet/mri enables high-spatial resolution, quantitative imaging of amyloid plaques in an alzheimer's disease mouse model. Sci Rep 10(1):10379

Ghaemmaghami S, Ahn M, Lessard P, Giles K, Legname G, DeArmond SJ, Prusiner SB (2009) Continuous quinacrine treatment results in the formation of drug-resistant prions. PLoS Pathog 5(11):e1000673

Giese A, Brown DR, Groschup MH, Feldmann C, Haist I, Kretzschmar HA (1998) Role of microglia in neuronal cell death in prion disease. Brain Pathol 8(3):449–457

Giles K, Olson SH, Prusiner SB (2017) Developing therapeutics for PrP prion diseases. Cold Spring Harb Perspect Med 7(4):a023747

Gomez-Nicola D, Fransen NL, Suzzi S, Perry VH (2013) Regulation of microglial proliferation during chronic neurodegeneration. J Neurosci 33(6):2481–2493

Gonzalez-Montalban N, Lee YJ, Makarava N, Savtchenko R, Baskakov IV (2013) Changes in prion replication environemnt cause prion strain mutation. FASEB J 27(9):3702–3710

Grabert K, Michoel T, Karavolos MH, Clohisey S, Baillie JK, Stevens MP, Freeman TC, Summers KM, McColl BW (2016) Microglial brain region−dependent diversity and selective regional sensitivities to aging. Nat Neurosci 19:504

Greenlee MHW, Lind M, Kokemuller R, Mammadova N, Kondru N, Manne S, Smith J, Kanthasamy A, Greenlee J (2016) Temporal resolution of misfolded prion protein transport, accumulation, glial activation, and neuronal death in the retinas of mice inoculated with scrapie. Am J Pathol 186(9):2302–2309

Grossman EJ, Inglese M (2015) The role of thalamic damage in mild traumatic brain injury. J Neurotrauma 33(2):163–167

Habib N, McCabe C, Medina S, Varshavsky M, Kitsberg D, Dvir-Szternfeld R, Green G, Dionne D, Nguyen L, Marshall JL, Chen F, Zhang F, Kaplan T, Regev A, Schwartz M (2020) Disease-associated astrocytes in alzheimer's disease and aging. Nat Neurosci 23(6):701–706

Haigh CL, Lawson VA, Drew SC (2014) Blood vessel cell death during prion disease: implications for disease management and infection control. Exp Hematol 42(11):939–940

Hansen DV, Hanson JE, Sheng M (2018) Microglia in alzheimer's disease. J Cell Biol 217(2):459–472

Hara M, Kobayakawa K, Ohkawa Y, Kumamaru H, Yokota K, Saito T, Kijima K, Yoshizaki S, Harimaya K, Nakashima Y, Okada S (2017) Interaction of reactive astrocytes with type i collagen induces astrocytic scar formation through the integrin–n-cadherin pathway after spinal cord injury. Nat Med 23(7):818–828

Harris DA, True HL (2006) New insights into prion structure and toxicity. Neuron 50(3):353–357

Hartmann K, Sepulveda-Falla D, Rose IVL, Madore C, Muth C, Matschke J, Butovsky O, Liddelow S, Glatzel M, Krasemann S (2019) Complement 3+-astrocytes are highly abundant in prion diseases, but their abolishment led to an accelerated disease course and early dysregulation of microglia. Acta Neuropathol Commun 7(1):83

Heithoff BP, George KK, Phares AN, Zuidhoek IA, Munoz-Ballester C, Robel S (2021) Astrocytes are necessary for blood-brain barrier maintenance in the adult mouse brain. Glia 69(2):436–472

Hilton KJ, Cunningham C, Reynolds RA, Perry VH (2013) Early hippocampal synaptic loss precedes neuronal loss and associates with early behavioural deficits in three distinct strains of prion disease. PLoS One 8(6):e68062

Hong S, Beja-Glasser VF, Nfonoyim BM, Frouin A, Li S, Ramakrishnan S, Merry KM, Shi Q, Rosenthal A, Barres BA, Lemere CA, Selkoe DJ, Stevens B (2016) Complement and microglia mediate early synapse loss in alzheimer mouse models. Science 352(6286):712–716

Hoyer MJ, Swarup S, Harper JW (2022) Mechanisms controlling selective elimination of damaged lysosomes. Curr Opin Physio 29:100590

Jeffrey M, Halliday WG, Bell J, Johnston AR, MacLeod NK, Ingham C, Sayers AR, Brown DA, Fraser JR (2000) Synapse loss associated with abnormal PrP precedes neuronal degeneration in the scrapie-infected murine hippocampus. Neuropath Appl Neurobiol 26:41–54

Kaczmarczyk L, Schleif M, Dittrich L, Williams RH, Koderman M, Bansal V, Rajput A, Schulte T, Jonson M, Krost C, Testaquadra FJ, Bonn S, Jackson WS (2022) Distinct translatome changes in specific neural populations precede electroencephalographic changes in prion-infected mice. PLoS Pathog 18(8):e1010747

Karapetyan YE, Saa P, Mahal SP, Sferrazza GF, Sherman A, Sales N, Weissmann C, Lasmezas CI (2009) Prion strain discrimination based on rapid in vivo amplification and analysis by the cell panel assay. PLoS One 4(5):e5730

Katorcha E, Baskakov IV (2017) Analyses of n-linked glycans of PrPSc revealed predominantly 2,6-linked sialic acid residues. FEBS J 284(21):3727–3738

Katorcha E, Makarava N, Savtchenko R, D'Azzo A, Baskakov IV (2014) Sialylation of prion protein controls the rate of prion amplification, the cross-species barrier, the ratio of PrPSc glycoform and prion infectivity. PLoS Pathog 10(9):e1004366

Katorcha E, Makarava N, Savtchenko R, Baskakov IV (2015) Sialylation of the prion protein glycans controls prion replication rate and glycoform ratio. Sci Rep 5(1):16912

Katorcha E, Daus ML, Gonzalez-Montalban N, Makarava N, Lasch P, Beekes M, Baskakov IV (2016) Reversible off and on switching of prion infectivity via removing and reinstalling prion sialylation. Sci Rep 6:33119

Katorcha E, Gonzalez-Montalban N, Makarava N, Kovacs GG, Baskakov IV (2018) Prion replication environment defines the fate of prion strain adaptation. PLoS Pathog 14(6):e10007093

Kercher L, Favara C, Striebel JF, LaCasse R, Chesebro B (2007) Prion protein expression differences in microglia and astroglia influence scrapie-induced neurodegeneration in the retina and brain of transgenic mice. JVirol 81(19):10340–10351

Keren-Shaul H, Spinrad A, Weiner A, Matcovitch-Natan O, Dvir-Szternfeld R, Ulland TK, David E, Baruch K, Lara-Astaiso D, Toth B, Itzkovitz S, Colonna M, Schwartz M, Amit I (2017) A unique microglia type associated with restricting development of Alzheimer's disease. Cell 169(7):1276–1290.e1217

Klein MA, Kaeser PS, Schwarz P, Weyd H, Xenarios I, Zinkernagel RM, Carroll MC, Verbeek JS, Botto M, Walport MJ, Molina H, Kalinke U, Acha-Orbea H, Aguzzi A (2001) Complement facilitates early prion pathogenesis. Nat Med 7(4):488–492

Kovacs GG, Makarava N, Savtchenko R, Baskakov IV (2013) Atypical and classical forms of the disease-associated state of the prion protein exhibit distinct neuronal tropism, deposition patterns, and lesion profiles. Am J Pathol 183(5):1539–1547

Kranich J, Krautler NJ, Falsig J, Ballmer B, Li S, Hutter G, Schwarz P, Moos R, Julius C, Miele G, Aguzzi A (2010) Engulfment of cerebral apoptotic bodies controls the course of prion disease in a mouse strain-dependent manner. J Exp Med 207(10):2271–2281

Krasemann S, Madore C, Cialic R, Baufeld C, Calcagno N, El Fatimy R, Beckers L, O'Loughlin E, Xu Y, Fanek Z, Greco DJ, Smith ST, Tweet G, Humulock Z, Zrzavy T, Conde-Sanroman P, Gacias M, Weng Z, Chen H, Tjon E, Mazaheri F, Hartmann K, Madi A, Ulrich JD, Glatzel M, Worthmann A, Heeren J, Budnik B, Lemere C, Ikezu T, Heppner FL, Litvak V, Holtzman DM, Lassmann H, Weiner HL, Ochando J, Haass C, Butovsky O (2017) The trem2-apoe pathway drives the transcriptional phenotype of dysfunctional microglia in neurodegenerative diseases. Immunity 47(3):566–581.e569

Kushwaha R, Sinha A, Makarava N, Molesworth K, Baskakov IV (2021) Non-cell autonomous astrocyte-mediated neuronal toxicity in prion diseases. Acta Neuropathol Commun 9(1):22

Kushwaha R, Li Y, Makarava N, Pandit NP, Molesworth K, Birukov KG, Baskakov IV (2023) Reactive astrocytes associated with prion disease impair the blood brain barrier. Neurobiol Dis 185:106264

Lakkaraju AKK, Sorce S, Senatore A, Nuvolone M, Guo J, Schwarz P, Moos R, Pelczar P, Aguzzi A (2022) Glial activation in prion diseases is selectively triggered by neuronal PrP(Sc). Brain Pathol 32(5):e13056

Le NTT, Wu B, Harris DA (2019) Prion neurotoxicity. Brain Pathol 29(2):263–277

Lee SW, Kim WJ, Choi YK, Song HS, Son MJ, Gelman IH, Kim YJ, Kim KW (2003) Ssecks regulates angiogenesis and tight junction formation in blood-brain barrier. Nat Med 9(7):900–906

Legname G, Baskakov IV, Nguyen HOB, Riesner D, Cohen FE, DeArmond SJ, Prusiner SB (2004) Synthetic mammalian prions. Science 305(5684):673–676

Li J, Browning S, Mahal SP, Oelschlegel AM, Weissmann C (2010) Darwinian evolution of prions in cell culture. Science 327:869–872

Liddelow SA, Guttenplan KA, Clarke LE, Bennett FC, Bohlen CJ, Schirmer L, Bennett ML, Munch AE, Chung WS, Peterson TC, Wilton DK, Frouin A, Napier BA, Panicker N, Kumar M, Buckwalter MS, Rowitch DH, Dawson VL, Dawson TM, Stevens B, Barres BA (2017) Neurotoxic reactive astrocytes are induced by activated microglia. Nature 541:481–487

Linnartz B, Kopatz J, Tenner AJ, Neumann H (2012) Sialic acid on the neuronal glycocalyx prevents complement c1 binding and complement receptor-3-mediated removal by microglia. J Neurosci 32(3):946–952

Linnerbauer M, Wheeler MA, Quintana FJ (2020) Astrocyte crosstalk in CNS inflammation. Neuron 108(4):608–622

Lu ZY, Baker CA, Manuelidis L (2004) New molecular markers of early and progressive cjd brain infection. J Cell Biochem 93:644–652

Lui H, Zhang J, Makinson Stefanie R, Cahill Michelle K, Kelley Kevin W, Huang H-Y, Shang Y, Oldham Michael C, Martens Lauren H, Gao F, Coppola G, Sloan Steven A, Hsieh Christine L, Kim Charles C, Bigio Eileen H, Weintraub S, Mesulam M-M, Rademakers R, Mackenzie Ian R, Seeley William W, Karydas A, Miller Bruce L, Borroni B, Ghidoni R, Farese Robert V, Paz Jeanne T, Barres Ben A, Huang Eric J (2016) Progranulin deficiency promotes circuit-specific synaptic pruning by microglia via complement activation. Cell 165(4):921–935

Lunnon K, Teeling JL, Tutt AL, Cragg MS, Glennie MJ, Perry VH (2011) Systemic inflammation modulates fc receptor expression on microglia during chronic neurodegeneration. J Immunol 186(12):7215–7224

Mabbott NA, Bruce ME, Botto M, Walport MJ, Pepys MB (2001) Temporary depletion of complement component c3 or genetic deficiency of c1q significantly delays onset of scrapie. Nat Med 7:485–487

Mabbott NA, Bradford BM, Pal R, Young R, Donaldson DS (2020) The effects of immune system modulation on prion disease susceptibility and pathogenesis. Int J Mol Sci 21(19)

Mahal SP, Browning S, Li J, Suponitsky-Kroyter I, Weissmann C (2010) Transfer of a prion strain to different hosts leads to emergence of strain variants. Proc Acad Natl Sci USA 107(52):22653–22658

Majer A, Medina SJ, Sorensen D, Martin MJ, Frost KL, Phillipson C, Manguiat K, Booth SA (2019) The cell type resolved mouse transcriptome in neuron-enriched brain tissues from the hippocampus and cerebellum during prion disease. Sci Rep 9(1):1099

Makarava N, Baskakov IV (2013) The evolution of transmissible prions: the role of deformed templating. PLoS Pathog 9(12):e1003759

Makarava N, Baskakov IV (2023) Role of sialylation of n-linked glycans in prion pathogenesis. Cell Tissue Res 392(1):201–214

Makarava N, Kovacs GG, Bocharova OV, Savtchenko R, Alexeeva I, Budka H, Rohwer RG, Baskakov IV (2010) Recombinant prion protein induces a new transmissible prion disease in wild type animals. Acta Neuropathol 119(2):177–187

Makarava N, Kovacs GG, Savtchenko R, Alexeeva I, Ostapchenko VG, Budka H, Rohwer RG, Baskakov IV (2012) A new mechanism for transmissible prion diseases. J Neurosci 32(21):7345–7355
Makarava N, Savtchenko R, Baskakov IV (2013) Selective amplification of classical and atypical prions using modified protein misfolding cyclic amplification. J Biol Chem 288(1):33–41
Makarava N, Savtchenko R, Baskakov IV (2015) Two alternative pathways for generating transmissible prion disease de novo. Acta Neuropathol Commun 3(1):e69
Makarava N, Savtchenko R, Alexeeva I, Rohwer RG, Baskakov IV (2016) New molecular insight into mechanism of evolution of mammalian synthetic prions. Am J Pathol 186(4):1006–1014
Makarava N, Chang JC-Y, Kushwaha R, Baskakov IV (2019) Region-specific response of astrocytes to prion infection. Front Neurosci 13(1):e1048
Makarava N, Chang JC-Y, Molesworth K, Baskakov IV (2020a) Posttranslational modifications define course of prion strain adaptation and disease phenotype. J Clin Invest 130(8):4382–4395
Makarava N, Chang JC-Y, Baskakov IV (2020b) Region-specific sialylation pattern of prion strains provides novel insight into prion neurotropism. Int J Mol Sci 21(3):828
Makarava N, Chang JC-Y, Molesworth K, Baskakov IV (2020c) Region-specific glial homeostatic signature in prion diseases is replaced by a uniform neuroinflammation signature, common for brain regions and prion strains with different cell tropism. Neurobiol Dis 137(1):e104783
Makarava N, Mychko O, Chang JC-Y, Molesworth K, Baskakov IV (2021) The degree of astrocyte activation is predictive of the incubation time to prion disease. Acta Neuropathol Commun 9(1):87
Makarava N, Katorcha E, Chang JC, Lau JTY, Baskakov IV (2022) Deficiency in ST6GAL1, one of the two α2,6-sialyltransferases, has only a minor effect on the pathogenesis of prion disease. Front Mol Biosci 9:1058602
Makarava N, Mychko O, Molesworth K, Chang JC, Henry RJ, Tsymbalyuk N, Gerzanich V, Simard JM, Loane DJ, Baskakov IV (2023) Region-specific homeostatic identity of astrocytes is essential for defining their response to pathological insults. Cells 12(17)
Makarava N, Kushwaha R, Baskakov IV (2024a) Reactive astrocytes in prion diseases: friend or foe? PLoS Pathog 20(6):e1012286
Makarava N, Safadi T, Bocharova O, Mychko O, Pandit NP, Molesworth K, Baiardi S, Zhang L, Parchi P, Baskakov IV (2024b) Reactive microglia partially envelop viable neurons in prion diseases. J Clin Invest 134(23)
Mallucci GR, Ratté S, Asante EA, Linehan J, Gowland I, Jefferys JG, Collinge J (2002) Post-natal knockout of prion protein alters hippocampal CA1 properties, but does not result in neurodegeneration. EMBO J 21(3):202–210
Mallucci G, Dickinson A, Linehan J, Klohn PC, Brandner S, Collinge J (2003) Depleting neuronal PrP in prion infection prevents disease and reverses spongiosis. Science 302:871–874
Matejuk A, Ransohoff RM (2020) Crosstalk between astrocytes and microglia: an overview. Front Immunol 11:1416
Mathys H, Adaikkan C, Gao F, Young JZ, Manet E, Hemberg M, De Jager PL, Ransohoff RM, Regev A, Tsai L-H (2017) Temporal tracking of microglia activation in neurodegeneration at single-cell resolution. Cell Rep 21(2):366–380
Mok SWF, Riemer C, Madela K, Hsu DK, Liu F-T, Gultner S, Heise I, Baier M (2007) Role of galectin-3 in prion infections of the CNS. Biochem Biophys Res Commun 359:672–678
Na Y-J, Jin J-K, Kim J-I, Choi E-K, Carp RI, Kim Y-S (2007) JAK-stat signaling pathway mediates astrogliosis in brains of scrapie-infected mice. J Neurochem 103(2):637–649
Nakagaki T, Ishibashi D, Mori T, Miyazaki Y, Takatsuki H, Tange H, Taguchi Y, Satoh K, Atarashi R, Nishida N (2020) Administration of fk506 from late stage of disease prolongs survival of human prion-inoculated mice. Neurotherapeutics 17(4):1850–1860
Nazmi A, Field RH, Griffin EW, Haugh O, Hennessy E, Cox D, Reis R, Tortorelli L, Murray CL, Lopez-Rodriguez AB, Jin L, Lavelle EC, Dunne A, Cunningham C (2019) Chronic neurodegeneration induces type I interferon synthesis via STING, shaping microglial phenotype and accelerating disease progression. Glia 67(7):1254–1276

Necula D, Cho FS, He A, Paz JT (2021) Secondary thalamic neuroinflammation after focal cortical stroke and traumatic injury mirrors corticothalamic functional connectivity. J Comp Neurol. https://doi.org/10.1002/cne.25259

Nomura K, Vilalta A, Allendorf DH, Hornik TC, Brown GC (2017) Activated microglia desialylate and phagocytose cells via neuraminidase, galectin-3, and mer tyrosine kinase. J Immunol 198(12):4792–4801

Novitskaya V, Bocharova OV, Bronstein I, Baskakov IV (2006) Amyloid fibrils of mammalian prion protein are highly toxic to cultured cells and primary neurons. J Biol Chem 2006:13828–13836

Novitskaya V, Makarava N, Sylvester I, Bronstein IB, Baskakov IV (2007) Amyloid fibrils of mammalian prion protein induce axonal degeneration in ntera2-derived terminally differentiated neurons. J Neurochem 102:398–407

Oakley H, Cole SL, Logan S, Maus E, Shao P, Craft J, Guillozet-Bongaarts A, Ohno M, Disterhoft J, Van Eldik L, Berry R, Vassar R (2006) Intraneuronal β-amyloid aggregates, neurodegeneration, and neuron loss in transgenic mice with five familial alzheimer's disease mutations: potential factors in amyloid plaque formation. J Neurosci 26(40):10129–10140

Ogden CA, de Cathelineau A, Hoffmann PR, Bratton D, Ghebrehiwet B, Fadok VA, Henson PM (2001) C1q and mannose binding lectin engagement of cell surface calreticulin and CD91 initiates macropinocytosis and uptake of apoptotic cells. J Exp Med 194(6):781–795

Oksanen M, Lehtonen S, Jaronen M, Goldsteins G, Hämäläinen RH, Koistinaho J (2019) Astrocyte alterations in neurodegenerative pathologies and their modeling in human induced pluripotent stem cell platforms. Cell Mol Life Sci 76(14):2739–2760

Olah M, Menon V, Habib N, Taga MF, Ma Y, Yung CJ, Cimpean M, Khairallah A, Coronas-Samano G, Sankowski R, Grün D, Kroshilina AA, Dionne D, Sarkis RA, Cosgrove GR, Helgager J, Golden JA, Pennell PB, Prinz M, Vonsattel JPG, Teich AF, Schneider JA, Bennett DA, Regev A, Elyaman W, Bradshaw EM, De Jager PL (2020) Single cell rna sequencing of human microglia uncovers a subset associated with alzheimer's disease. Nat Commun 11(1):6129

Phatnani H, Maniatis T (2015) Astrocytes in neurodegenerative disease. Cold Spring Harb Perspect Biol 7(6)

Prater KE, Green KJ, Mamde S, Sun W, Cochoit A, Smith CL, Chiou KL, Heath L, Rose SE, Wiley J, Keene CD, Kwon RY, Snyder-Mackler N, Blue EE, Logsdon B, Young JE, Shojaie A, Garden GA, Jayadev S (2023) Human microglia show unique transcriptional changes in Alzheimer's disease. Nat Aging 3(7):894–907

Prusiner SB (1982) Novel proteinaceous infectious particles cause scrapie. Science 216(4542):136–144

Prusiner SB (1998) Prions. Proc Natl Acad Sci USA 95:13363–13383

Puoti G, Giaccone G, Mangieri M, Limido L, Fociani P, Zerbi P, Suardi S, Rossi G, Iussich D, Capobianco R, Fede GD, Marcon G, Cotrufo R, Filippini G, Bugiani O, Tagliavini F (2005) Sporadic Creutzfeldt-Jakob disease: the extent of microglia activation is dependent on the biochemical type of PrPSc. J Neuropathol Exp Neurol 64(10):902–909

Radford HE, Mallucci GR (2010) The role of GPI-anchored PrP C in mediating the neurotoxic effect of scrapie prions in neurons. Curr Issues Mol Biol 12(2):119–127

Rambold AS, Müller V, Ron U, Ben-Tal N, Winklhofer KF, Tatzelt J (2008) Stress-protective signalling of prion protein is corrupted by scrapie prions. EMBO J 27(14):1974–1984

Ramlackhansingh AF, Brooks DJ, Greenwood RJ, Bose SK, Turkheimer FE, Kinnunen KM, Gentleman S, Heckemann RA, Gunanayagam K, Gelosa G, Sharp DJ (2011) Inflammation after trauma: microglial activation and traumatic brain injury. Ann Neurol 70(3):374–383

Reichenbach N, Delekate A, Plescher M, Schmitt F, Krauss S, Blank N, Halle A, Petzold GC (2019) Inhibition of Stat3-mediated astrogliosis ameliorates pathology in an Alzheimer's disease model. EMBO Mol Med 11(2):e9665

Rothaug M, Becker-Pauly C, Rose-John S (2016) The role of interleukin-6 signaling in nervous tissue. Biochim Biophys Acta (BBA) – Mol Cell Res 1863(6, Part A):1218–1227

Rudd PM, Endo T, Colominas C, Groth D, Wheeler SF, Harvey DJ, Wormald MR, Serban H, Prusiner SB, Kobata A, Dwek RA (1999) Glycosylation differences between the normal and pathogenic prion protein isoforms. Proc Natl Acad Sci USA 96:13044–13049

Sandberg MK, Al-Doujaily H, Sharps B, De Oliveira MW, Schmidt C, Richard-Londt A, Lyall S, Linehan JM, Brandner S, Wadsworth JD, Clarke AR, Collinge J (2014) Prion neuropathology follows the accumulation of alternate prion protein isoforms after infective titre has peaked. Nat Commun 5:e4347

Santello M, Toni N, Volterra A (2019) Astrocyte function from information processing to cognition and cognitive impairment. Nat Neurosci 22(2):154–166

Schafer Dorothy P, Lehrman Emily K, Kautzman Amanda G, Koyama R, Mardinly Alan R, Yamasaki R, Ransohoff Richard M, Greenberg Michael E, Barres Ben A, Stevens B (2012) Microglia sculpt postnatal neural circuits in an activity and complement-dependent manner. Neuron 74(4):691–705

Scheckel C, Imeri M, Schwarz P, Aguzzi A (2020) Ribosomal profiling during prion disease uncovers progressive translational derangement in glia but not in neurons. elife 9:e62911

Schultz J, Schwarz A, Neidhold S, Burwinkel M, Riemer C, Simon D, Kopf M, Otto M, Baier M (2004) Role of interleukin-1 in prion disease-associated astrocyte activation. Am J Pathol 165:671–678

Scott G, Hellyer PJ, Ramlackhansingh AF, Brooks DJ, Matthews PM, Sharp DJ (2015) Thalamic inflammation after brain trauma is associated with thalamo-cortical white matter damage. J Neuroinflammation 12(1):224

Siddharthan V, Kim YV, Liu S, Kim KS (2007) Human astrocytes/astrocyte-conditioned medium and shear stress enhance the barrier properties of human brain microvascular endothelial cells. Brain Res 1147:39–50

Sinha A, Kushwaha R, Molesworth K, Mychko O, Makarava N, Baskakov IV (2021) Phagocytic activities of reactive microglia and astrocytes associated with prion diseases are dysregulated in opposite directions. Cells 10(7):1728

Slota JA, Sajesh BV, Frost KF, Medina SJ, Booth SA (2022) Dysregulation of neuroprotective astrocytes, a spectrum of microglial activation states, and altered hippocampal neurogenesis are revealed by single-cell RNA sequencing in prion disease. Acta Neuropathol Commun 10(1):161

Smith HL, Freeman OJ, Butcher AJ, Holmqvist S, Humoud I, Schätzl T, Hughes DT, Verity NC, Swinden DP, Hayes J, de Weerd L, Rowitch DH, Franklin RJM, Mallucci GR (2020) Astrocyte unfolded protein response induces a specific reactivity state that causes non-cell-autonomous neuronal degeneration. Neuron 105(5):855–866.e855

Sofroniew MV, Vinters HV (2010) Astrocytes: biology and pathology. Acta Neuropathol 119(1):7–35

Solforosi L, Criado JR, McGavern DB, Wirz S, Sanchez-Alavez M, Sugama S, DeGiorgio LA, Volpe BT, Wiseman E, Abalos G, Masliah E, Gilden D, Oldstone MB, Conti B, Williamson RA (2004) Cross-linking cellular prion protein triggers neuronal apoptosis in vivo. Science 303:1514–1516

Sorce S, Nuvolone M, Russo G, Chincisan A, Heinzer D, Avar M, Pfammatter M, Schwarz P, Delic M, Müller M, Hornemann S, Sanoudou D, Scheckel C, Aguzzi A (2020) Genome-wide transcriptomics identifies an early preclinical signature of prion infection. PLoS Pathog 16(6):e1008653

Sorensen G, Medina S, Parchaliuk D, Phillipson C, Robertson C, Booth SA (2008) Comprehensive transcriptional profiling of prion infection in mouse models reveals networks of responsive genes. BMC Genomics 9(1):114

Soreq L, Consortium UBE, Consortium NABE, Rose J, Soreq E, Hardy J, Trabzuni D, Cookson MR, Smith C, Ryten M, Patani R, Ule J (2017) Major shifts in glial regional identity are a transcriptional hallmark of human brain aging. Cell Rep 18(2):557–570

Srivastava S, Makarava N, Katorcha E, Savtchenko R, Brossmer R, Baskakov IV (2015) Post-conversion sialylation of prions in lymphoid tissues. Proc Acad Natl Sci USA 112(48):E6654–E6662

Srivastava S, Katorcha E, Daus ML, Lasch P, Beekes M, Baskakov IV (2017) Sialylation controls prion fate in vivo. J Biol Chem 292(6):2359–2368

Srivastava S, Katorcha E, Makarava N, Barrett JP, Loane DJ, Baskakov IV (2018) Inflammatory response of microglia to prions is controlled by sialylation of PrPSc. Sci Rep 8(1):e11326

Stahl N, Borchelt DR, Hsiao K, Prusiner SB (1987) Scrapie prion protein contains a phosphatidylinositol glycolipid. Cell 51(2):229–240

Stahl N, Baldwin MA, Teplow DB, Hood L, Gibson BW, Burlingame AL, Prusiner SB (1993) Structural studies of the scrapie prion protein using mass spectrometry and amino acid sequencing. Biochemistry 32:1991–2002

Stephan AH, Madison DV, Mateos JM, Fraser DA, Lovelett EA, Coutellier L, Kim L, Tsai H-H, Huang EJ, Rowitch DH, Berns DS, Tenner AJ, Shamloo M, Barres BA (2013) A dramatic increase of c1q protein in the CNS during normal aging. J Neurosci 33(33):13460–13474

Stevens B, Allen NJ, Zazquez LE, Howell GR, Christopherson KS, Nouri N, Micheva KD, Mehalow AK, Huberman AD, Stafford B, Sher A, Litke AM, Lambris JD, Smith SJ, John SW, Barres BA (2007) The classical complement cascade mediates CNS synapse elimination. Cell 131(6):1164–1178

Stimson E, Hope J, Chong A, Burlingame AL (1999) Site-specific characterization of the n-linked glycans of murine prion protein by high-performance liquid chromatography/electrospray mass spectrometry and exoglycosidase digestions. Biochemistry 38:4885–4895

Sun D, Qu J, Jakobs TC (2013) Reversible reactivity by optic nerve astrocytes. Glia 61(8):1218–1235

Takashima S (2008) Characterization of mouse sialylatransferase gene: their evolution and diversity. Biosci Biotechnol Biochem 72:1155–1167

Tribouillard-Tanvier D, Striebel JF, Peterson KE, Chesebro B (2009) Analysis of protein levels of 24 cytokines in scrapie agent-infected brain and glial cell cultures from mice differing in prion protein expression levels. J Virol 83(21):11244–11253

Turk E, Teplow DB, Hood LE, Prusiner SB (1988) Purification and properties of the cellular and scrapie hamster prion proteins. Eur J Biochem 176(1):21–30

Ugalde CL, Lewis V, Stehmann C, McLean CA, Lawson VA, Collins SJ, Hill AF (2020) Markers of a1 astrocytes stratify to molecular sub-types in sporadic Creutzfeldt-Jakob disease brain. Brain Commun 2(2):fcaa029

Vainchtein ID, Chin G, Cho FS, Kelley KW, Miller JG, Chien EC, Liddelow SA, Nguyen PT, Nakao-Inoue H, Dorman LC, Akil O, Joshita S, Barres BA, Paz JT, Molofsky AB, Molofsky AV (2018) Astrocyte-derived interleukin-33 promotes microglial synapse engulfment and neural circuit development. Science 359(6381):1269–1273

Van Ryzin JW, Marquardt AE, Argue KJ, Vecchiarelli HA, Ashton SE, Arambula SE, Hill MN, McCarthy MM (2019) Microglial phagocytosis of newborn cells is induced by endocannabinoids and sculpts sex differences in juvenile rat social play. Neuron 102(2):435–449.e436

Varki A (2008) Sialic acids in human health and disease. Trends Mol Med 14(8):351–360

Vincenti JE, Murphy L, Grabert K, McColl BW, Cancellotti E, Freeman TC, Manson JC (2016) Defining the microglia response during the time course of chronic neurodegeneration. J Virol 90(6):3003–3017

Wakselman S, Béchade C, Roumier A, Bernard D, Triller A, Bessis A (2008) Developmental neuronal death in hippocampus requires the microglial cd11b integrin and dap12 immunoreceptor. J Neurosci 28(32):8138

Wang F, Wang X, Yuan CG, Ma J (2010) Generating a prion bacterially expressed recombinant prion protein. Science 327(5969):1132–1135

Wheeler MA, Clark IC, Tjon EC, Li Z, Zandee SEJ, Couturier CP, Watson BR, Scalisi G, Alkwai S, Rothhammer V, Rotem A, Heyman JA, Thaploo S, Sanmarco LM, Ragoussis J, Weitz DA, Petrecca K, Moffitt JR, Becher B, Antel JP, Prat A, Quintana FJ (2020) MAFG-driven astrocytes promote CNS inflammation. Nature 578(7796):593–599

Williams A, Lucassen PJ, Ritchie D, Bruce M (1997) PrP deposition, microglial activation, and neuronal apoptosis in murine scrapie. Exp Neurol 144:433–438

Wisniewski HM, Lossinsky AS, Moretz RC, Vorbrodt AW, Lassmann H, Carp RI (1983) Increased blood-brain barrier permeability in scrapie-infected mice. J Neuropathol Exp Neurol 42(6):615–626

Yamasaki T, Suzuki A, Hasebe R, Horiuchi M (2018) Flow cytometric detection of PrP(Sc) in neurons and glial cells from prion-infected mouse brains. J Virol 92(1):e01457-17

Yan Z, Gibson SA, Buckley JA, Qin H, Benveniste EN (2018) Role of the JAK/STAT signaling pathway in regulation of innate immunity in neuroinflammatory diseases. Clin Immunol 189:4–13

Zeisel A, Hochgerner H, Lonnerberg P, Johnsson A, Memic F, van der Zwan J, Harig M, Braun E, Borm LE, La Manno G, Codeluppi S, Furlan A, Lee K, Skene N, Harris KD, Hjerling-Leffler J, Arenas E, Ernfors P, Marklund U, Linnarsson S (2018) Molecular architecture of the mouse nervous system. Cell 174(4):999–1014

Zhu C, Herrmann US, Li B, Abakumova I, Moos R, Schwarz P, Rushing EJ, Colonna M, Aguzzi A (2015) Triggering receptor expressed on myeloid cells-2 is involved in prion-induced microglial activation but does not contribute to prion pathogenesis in mouse brains. Neurobiol Aging 36(5):1994–2003

Zhu C, Herrmann US, Falsig J, Abakumova I, Nuvolone M, Schwarz P, Frauenknecht K, Rushing EJ, Aguzzi A (2016) A neuroprotective role for microglia in prion diseases. J Exp Med 213(6):1047–1059

Chapter 9
Biological Fluid Biomarkers in Human Prion Diseases with a Note on Biosafety

Isidro Ferrer

Abstract The definitive diagnosis of human prion diseases can only be obtained postmortem by combining clinical symptoms, neuropathology and PrP immunohistochemistry, Western blotting of PrP types, zygosity of codon 129, and genetic study of *PRNP*. Premortem diagnosis is strongly sustained by one positive prion-specific assay, commonly protein misfolded cyclic amplification (PMCA) or real-time quaking-induced conversion (RT-QuIC), principally in CSF samples. Surrogate biomarkers 14-3-3, t-tau, P-tau, βA4, and total-PrP levels in the CSF help discriminate other neurodegenerative diseases, but their sensitivity and specificity are variable depending on the prion disease. Other altered proteins in the CSF, such as neurofilament light chain (NfL), calcium-binding protein S100β, neuron-specific enolase, α-synuclein and β-synuclein, neurogranin and SNAP-25, triggering receptor expressed on myeloid cells 2 (TREM2), cytokines, astroglial markers, and microRNAs, need further validation. Total-tau and NfL levels in the blood may serve to monitor disease progression, whereas the value of total-PrP, synuclein, S100β, TREM2, and peripheral inflammatory markers in the blood is limited. Since the products of positive PMCA and PrP^{Sc} are present in several tissues in CJD, special care and biosafety conditions must be applied in managing and processing human biological samples of suspected prion disease. Regarding RT-QuIC products, further experimental studies are needed to elucidate their seeding capacity.

Keywords Prion diseases · Biomarkers · CSF · Blood · RT-QuIC · PMCA · Biosafety

I. Ferrer (✉)
Department of Pathology and Experimental Therapeutics, University of Barcelona, CIBERNED, 08907 Hospitalet de Llobregat, Barcelona, Spain

G. Legname, F. Moda (eds.), *Biomarkers and Therapeutical Targets for Prion Diseases*, Subcellular Biochemistry 112, https://doi.org/10.1007/978-3-031-97055-9_9

Introduction

Prion diseases are spongiform encephalopathies linked to the transformation of the normal prion proteins (PrP^C), encoded by the *PRNP* gene, into abnormally folded protease-resistant isoforms (currently named PrP^{Sc}), which accumulate in the brain (and other tissues and fluids), causing neuronal death (Prusiner 1994, 1998).

Human prion diseases are sporadic Creutzfeldt–Jakob disease (sCJD) (80–90%), genetic prion diseases (10–15%), and acquired CJD (less than 1%). Animal prion diseases are scrapie in sheep, goats, and mufflons; bovine spongiform encephalopathy (BSE) in cattle; feline spongiform encephalopathy in felines, including domestic cats; transmissible mink encephalopathy (TME); and chronic wasting disease (CWD) in deer and elk (Sy et al. 2002; McKintosh et al. 2003; Johnson 2005; Ironside et al. 2017; Whitechurch et al. 2017; Baiardi et al. 2019, 2021; Orge et al. 2021). Prion diseases are transmissible to vulnerable donors when prions are inoculated (or administered with other procedures). The resulting pathological brain damage is specific for different prion strains, which depend, in part, on the conformational structure of the prion (Aguzzi et al. 2007; Carta and Aguzzi 2022; Block and Bartz 2023).

Initially, the transmission between human and animal prion diseases was considered exceptional, if present, due to species-specific prion barriers. However, experimental transmission assays, and more dramatically, the accidentally produced BSE and its transmission to humans, have turned the creed of interspecies prion transmission barrier unoperational (Aguzzi and Falsig 2012; Houston and Andréoletti 2019).

Sporadic Creutzfeldt–Jakob Disease

sCJD is categorized into several subtypes depending on the characteristics of the prion (type 1 and type 2) and the genotype of codon 129 of *PRNP* (homozygous or heterozygous for valine and methionine: MM, MV, VV). PrP type 1 has a primary cleavage site at residue 82 and a molecular weight of about 21 kDa and type 2 has a molecular mass of 19 kDa and a primary cleavage site at residue 97. Additional characteristics are the glycosylation state and the protease resistance of prions. PrP type 2 in sCJD is currently type 2A. Molecular and genetic variations are manifested with particular clinical symptoms and characteristic neuropathological traits that include regional vulnerability, variable spongiosis, and individual PrP deposits in the brain. The main six molecular and genotypic subtypes are MM1, MV1, VV1, MM2, MV2, and VV2. However, MM1 and MV1 are phenotypically indistinguishable, and they are usually referred to as MM(V)1 subtype; the MM2 group includes the histopathological cortical (MM2C) and thalamic (MM2T or sporadic fatal insomnia: sFI) subtypes; and the MV2 group comprises a subtype with kuru-like plaques in the cerebellum (MV2K) and a cortical subtype (MV2C) (Baiardi et al.

2019, 2021; Cali et al. 2006; Fiorini et al. 2017; Parchi et al. 1996, 1999, 2000, 2009a, 2012). About 35% of sCJD cases show a combination of subtypes 1 and 2 in different brain regions (Baiardi et al. 2019; Parchi et al. 2009b; Cali et al. 2020). New subtypes are reported (Kobayashi et al. 2008; Gelpi et al. 2022). Other atypical cases have divergent clinical, neuropathological, molecular, and strain-specific profiles. A rare, independent subtype, denominated variably protease-sensitive prionopathy (VPSPr), is characterized by variable protease-resistance prion deposition and unique clinical, molecular, and neuropathological features (Gambetti et al. 2008; Zou et al. 2010; Baiardi et al. 2022).

Transmission studies reveal that the six main sCJD subtypes are due to five different strains named M1, M2C, M2T, V2, and V1 based on the codon 129 genotype and the specific pathological patterns elicited following transmission (Baiardi et al. 2019; Bishop et al. 2010; Parchi et al. 2010; Sigurdson et al. 2019). New strains are identified (Galeno et al. 2017).

Genetic Prion Diseases

Genetic prion diseases are genetic Creutzfeldt–Jakob disease (gCJD), fatal familial insomnia (FFI), Gerstmann–Sträussler–Scheinker (GSS) disease, PrP-cerebral amyloid angiopathy (PrP-CAA) and PrP-systemic amyloidosis (PrP-SA) (Baiardi et al. 2019, 2021; Collins et al. 2001; Schmitz et al. 2017).

gCJD is linked to point mutations or octapeptide repeat insertions in *PRNP*. Clinical, neuropathological, biochemical, and transmission characteristics of gCJD are similar to those seen in sCJD (Baiardi et al. 2021). However, the neuropathological features of PrP^{Sc} deposits and differences in the amount of monoglycosylated and diglycosylated isoforms, together with the presence of doublets in the nonglycosylated prion, vary depending on the mutation (Baiardi et al. 2021; Parchi et al. 2000; Schmitz et al. 2017; Grasbon-Frodl et al. 2004; Hill et al. 2006; Mead et al. 2007). Recently, a new classification has been proposed, including subtypes M1-mutation, V2 (K) mutation with or without kuru plaques, V1-mutation, M2C-mutation (cortical variant), M2T-D178N-mutation (thalamic variant) or FFI, and M-intermediate-E200K (Baiardi et al. 2021). Mixed and atypical phenotypes are seldom reported (Baiardi et al. 2021; Shintaku et al. 2021).

FFI is a particular form of genetic prion disease linked to the *PRNP* D178N mutation and 129MM homozygosity; cases bearing *PRNP* D178 mutation and 129VV homozygosity are similar to gCJD, whereas the same mutation in patients with 129MV is manifested as a combined form and long duration. Neuropathological changes in FFI are consistently found in anteroventral and dorsomedial thalamic nuclei and inferior olives. PrP^{Sc} accumulation is low and partially sensitive to proteinase. Western blots of the PrP^{Sc} show CJD prion type 2, but the reticular/synaptic deposition pattern of the prion aggregates resembles that found in sCJD type 1 (Jürgens-Wemheuer et al. 2021). Moreover, FFI may resemble a prion type with its conformation-sharing properties, partly with type 1 and type 2 prions

(Jürgens-Wemheuer et al. 2021). Strong monoglycosylated and diglycosylated bands are typical in FFI (Parchi et al. 1998). Comparative studies of PrP conformers in FFI and sFI (sCJD-MM2T) suggest novel strain properties distinguishing sFI and FFI (Cracco et al. 2017; Takeuchi et al. 2019).

GSS is linked to mutations at distinct codons, stop mutations, and insertional mutations of octapeptide repeats. The most common form is associated with the P102L mutation. The clinical course is reminiscent of a neurodegenerative disease with ataxia and dementia. Large plaque-like PrP^{Sc} deposits in the cerebrum and cerebellum and discrete spongiform change characterize the neuropathology. Neurofibrillary tangles identical to those seen in Alzheimer's disease (AD) occur in cases with determined point mutations (Baiardi et al. 2019; Bugiani et al. 2000; Ghetti et al. 2003). PrP^{Sc} in Western blots shows a ladder-like band pattern and a proteinase K-resistant fragment of 7–10 kDa (Schmitz et al. 2017; Ghetti et al. 2003). The limited number of cases and the diversity of *PRNP* mutations causing GSS have impeded the categorization of prion strains in GSS (Rossi et al. 2019).

Infectious Human Prion Diseases

Acquired prion diseases are the extinct Kuru, variant Creutzfeldt–Jakob disease (vCJD), and iatrogenic CJD (iCJD). Kuru was due to the ingestion of prion-contaminated corpses in the context of ritual cannibalism in certain tribes of Papua New Guinea. vCJD is linked to the human ingestion of infected meal products from cattle with BSE (Ironside et al. 2017), and it is associated with PrP type 2B (Parchi et al. 2009a). iCJD is produced after contaminated dura mater allografts, corneal transplants, treatment with growth hormone and gonadotropin administration obtained from postmortem contaminated hypophysis, and accidents (punctures) with contaminated material usually linked to medical practice and prion research (Brown et al. 2012; Haïk and Brandel 2014; Ritchie et al. 2017). Blood transfusions from cases incubating vCJD have been a rare cause of iCJD. Kuru and vCJD have typical clinical, neuropathological, and biochemical features. iCJD may resemble sCJD MM(V)1; the MM2C, MM2T, and VV1 subtypes have never been reported. Yet, a "plaque-like" subtype characterized by florid and kuru-like plaques and frequent stellate cells is almost exclusive to a subgroup of cases with iCJD; the prion in these cases has a molecular weight intermediate of type 1 and type 2 (MMiK subtype) (Orge et al. 2021; Kretzschmar et al. 2003; Kobayashi et al. 2007; Cali et al. 2015; Kobayashi et al. 2016). All acquired prion diseases have been linked to M1, V2, or BSE-derived strains (Orge et al. 2021).

Diagnosis of Human Prion Diseases

The diagnosis of prion diseases may be suspected during life with variable levels of accuracy. Still, the definitive diagnosis is proved following a comprehensive neuropathological, immunohistochemical, and molecular study at postmortem. The neuropathological and PrP immunohistochemical study, usually performed on formalin-fixed, paraffin-embedded, and dewaxed tissue sections, permits the identification of the different sCJD and gCJD subtypes, specific forms of iCJD, and vCJD based on the localization and distribution of neuron loss, type and distribution of spongiform change (small vacuoles, large confluent vacuoles), and particular patterns of PrP^{Sc} deposition in the brain (synaptic, plaque-like, perineuronal, peripheral to confluent vacuoles, kuru-like plaques, florid plaques, radial deposits, and others). Homogenates of frozen samples processed for Western blotting allow the identification of prion types. DNA extraction is used to assess codon 129 zygosity and mutations in *PRNP* (Brandel and Knight 2018; Gambetti et al. 2003; Ghetti et al. 1996; Ironside et al. 2017; Kobayashi et al. 2016; Orge et al. 2021; Parchi et al. 1999, 2009b, 2012).

Postmortem studies are also necessary to categorize GSS, PrP-CAA, and PrP-SA (Bugiani et al. 2000; Ghetti et al. 1996, 2003; Piccardo et al. 1998; Cracco et al. 2019).

Several complementary probes are helpful in the clinical diagnosis of human prion diseases during life, including neuroimaging, positron emission tomography, electroencephalography, genetics, and analysis of selected biomarkers in biological fluids. Other techniques have restricted applications: cerebral biopsies are currently very rare due to the invasive procedure, together with the limited information in many cases. However, PrP immunohistochemistry in tonsillar biopsies is helpful in the diagnosis of vCJD (Ramasamy et al. 2003).

Magnetic resonance imaging (MRI) may show, often asymmetrical, focal restricted diffusion at least in two cortical areas and the caudate nucleus, thalamus, and putamen in MM(V)1 subtype and more selective in the thalamus in VV2, MV2, and MM2T subtypes. A high FLAIR signal in the posterior thalamus (pulvinar nucleus) and lower intensity in the anterior putamen are characteristic alterations in vCJD. Sensitivity varies from 80% to 98%, and specificity from 78% to 98%, depending on the centers (Hermann et al. 2021). Neuroimaging in other subtypes of sCJD is less precise. [18F] Fluoro-2-deoxy-D-glucose positron emission tomography (FDG-PET) may show nonspecific changes except early-reduced thalamic glucose metabolism in the MM2T subtype.

[(11)C]PiB PET detects prion protein in the brain and is helpful in the detection of prion accumulation in selected brain nuclei years before the appearance of clinical symptoms in GSS (Kepe et al. 2010; Deters et al. 2016). [18F]Flortaucipir PET visualizes brain tau deposition in certain forms of GSS (Risacher et al. 2018).

Abnormal electroencephalographic (EEG) signals reminiscent of nonconvulsive status epilepticus are frequent in advanced sCJD. However, the presence of periodic sharp-wave complexes (PEWCs) is considered typical of sCJD in MM(V)1 and less frequent in MV2, VV2, and MM2 cases (Steinhoff et al. 2004).

Genetic studies of codon 129 in *PRNP* are clues to categorize human prion disease subtypes. Looking at mutations (point mutations, stop mutations, and octapeptide repeats) in *PRNP* is essential in diagnosing genetic prion diseases.

Biochemical changes found in the CSF and other biological fluids in CJD and other human prion diseases are the subject of many reviews (Hermann et al. 2021; Zanusso et al. 2016; Zerr et al. 2017; Lattanzio et al. 2017; Thompson and Mead 2019; Abu-Rumeileh et al. 2020; Vallabh et al. 2020; Mok and Mead 2020; Eraña et al. 2020; Cazzaniga et al. 2020; Figgie and Appleby 2021; Satoh 2022; Zerr 2022; Atarashi 2023).

The present review first deals with CSF and other fluid (mainly blood and plasma) surrogate biomarkers or complementary indirect probes that can help the clinical diagnosis of probable prion disease but do not demonstrate the presence of PrP^{Sc}. Western blotting and ELISA are currently employed to study proteins in the CSF. The SIMOA (single molecule array) analyzer, using a digital version of the ELISA technique, procures an accurate and precise detection of small amounts of proteins in plasma.

Second, we will discuss biomarkers that may identify the presence of PrP^{Sc} in biological fluids and tissues during life. The central PrP^{Sc} detection assays are Protein Misfolding Cyclic Amplification (PMCA) and real-time quaking-induced conversion (RT-QuIC).

Finally, biosafety measures are considered regarding the potential infectivity of biological samples learned following the research use and the clinical implementation of protein amplification and conversion assays in the study of prions.

CSF Surrogate Biomarkers

Proteomic differences exist between ventricular and lumbar CSF (Rostgaard et al. 2023). S100β, total-tau (t-tau), and phosphorylated tau (P-tau) are higher, whereas neurofilament light chain (NfL) and β-amyloid ($A\beta_{40}$, $A\beta_{42}$) levels are lower in the ventricular compartment (Rostgaard et al. 2023). Data presented in this review correspond to studies with CSF obtained from lumbar punctures used for diagnostic purposes.

CSF proteins selected as surrogate prion disease biomarkers are mainly 14-3-3, t-tau, P-tau, total-PrP, NfL, S100β, neuron-specific enolase (NSE), α-synuclein, β-synuclein, neurogranin, SNAP-25, and thymosin β4 (Figgie and Appleby 2021). In addition, β-amyloid is advantageous in the differential diagnosis of Alzheimer's disease (AD).

Other complementary assays are centered on glial cell markers, neuroinflammation, mitochondria, microRNAs, and miscellaneous molecules.

sCJD

14-3-3

Increased 14-3-3 levels in the CSF in sCJD were first detected in 1996 by immunoassay (Hsich et al. 1996) and Western blotting (Zerr et al. 1996). Increased 14-3-3ε and 14-3-3γ CSF levels are the most sensitive isoforms to sCJD (Takahashi et al. 1999; Satoh et al. 2010). One of the two methods (ELISA and Western blotting) is currently used in most basic laboratories to screen sCJD.

High CSF 14-3-3 protein levels have sensitivity between 80% and 100% in most sCJD cases (Orge et al. 2021; Lattanzio et al. 2017; Castellani et al. 2004; Sanchez-Juan et al. 2006; Collins et al. 2006; Stoeck et al. 2012; Schmitz et al. 2016b). However, their sensitivity is reduced to 60% in sCJD MM2 and VV2 (Lattanzio et al. 2017; Castellani et al. 2004; Sanchez-Juan et al. 2006; Collins et al. 2006). Sensitivity is highest in cases bearing VV2 (100%), relatively high in MM(V)1 (92.8%), low in MV2K (50%), very low in MM2C (30%), and inconsistent in MM2T (0%) (Lattanzio et al. 2017).

The specificity varies from 40% to 96% (Lattanzio et al. 2017; Sanchez-Juan et al. 2006; Stoeck et al. 2012; Fiorini et al. 2020). The variability of published results is related, in part, to the method employed; higher levels of positivity are obtained with ELISA compared with Western blotting; substantial variations among laboratories are additional factors of variability. Moreover, 14-3-3 protein levels can be lower at early stages of the disease and may also decrease in cases with long disease duration (Gmitterová et al. 2009). However, another study has shown 14-3-3 increased sensitivity from onset to the advanced stage in the MV129 heterozygous group (Sanchez-Juan et al. 2007).

CSF 14-3-3 protein levels may also increase in many diseases accompanied by neuronal cell death, such as hypoxic brain damage, herpes simplex encephalitis, atypical encephalitis, intracerebral metastases, metabolic encephalopathies, vascular dementia, AD, dementia with Lewy bodies, other rapidly progressive dementias, and following traumatic lumbar punctures (Sanchez-Juan et al. 2006; Stoeck et al. 2012; Zerr et al. 1998; Satoh et al. 1999; Chapman et al. 2000; Hamlin et al. 2012; Kong et al. 2023a).

Total Tau and Phosphorylated Tau (P-Tau)

The CSF's total-tau (t-tau) levels are increased in sCJD (Otto et al. 2002). Further studies mainly based on ELISA kits from different sources show variable sensitivity from 84% to 100% and specificity from 54% to 96% (Lattanzio et al. 2017; Sanchez-Juan et al. 2006; Abu-Rumeileh et al. 2019a; Fiorini et al. 2020; Hamlin et al. 2012; Riemenschneider et al. 2003; Karch et al. 2015; Franceschini et al. 2017; Li et al. 2018; Blennow et al. 2019; Llorens et al. 2020a; Leitão et al. 2016). Cutoff values are variable depending on the laboratories and the kit employed, usually ranging from ˃1030 pg/mL to ˃1300 pg/mL in the case of Innogenetics/Fujirebio Europe kits. As for 14-3-3, t-tau levels are lower in MM2 subtypes and at the early stages of the disease (Sanchez-Juan et al. 2006; Karch et al. 2015; Cohen et al. 2016). More precisely, the sensitivity was 100% in VV2, 92.8% in MM(V1), 76.9% in MV2K, 40% in MM2C, and 33% in MM2T in another study (Lattanzio et al. 2017). Increased CSF levels of t-tau mainly involve isoforms containing exons 2 and 10 (Chen et al. 2016).

T-tau in the CSF also increases in other neurological diseases such as malignancies, encephalitis, vascular dementia, and AD (Lattanzio et al. 2017). Yet, CSF t-tau and nonphosphorylated tau values are currently higher in sCJD than in AD (Hamlin et al. 2012; Blennow et al. 2019; Coulthart et al. 2011; Ermann et al. 2018; Lehmann et al. 2019). NT1-tau is also increased in the CSF (Mengel et al. 2021). Other studies propose CSF t-tau levels as predictors of survival (Staffaroni et al. 2019; Rübsamen et al. 2020).

Phosphorylated tau (P-tau) is also increased in the CSF in sCJD (67). Higher P-tau levels are found in VV2 and MV2K than in MM(V)1 and MM2C subtypes (Lattanzio et al. 2017). Interestingly, increased CSF P-tau correlates with tiny P-tau-positive neurites in VV2 and MV2K subtypes rather than with the presence of associated AD and aging-related tau astrogliopathy (ARTAG) pathology, which was common in aged cases with sCJD (Lattanzio et al. 2017). The ratio between t-tau and P-tau is higher in sCJD than in AD, with a specific range from 94% to 97% and a sensitivity ranging from 75% to 94% (Riemenschneider et al. 2003; Skinningsrud et al. 2008; Baldeiras et al. 2009; Skillbäck et al. 2014; Dorey et al. 2015; Llorens et al. 2016; Bruzova et al. 2021). CSF t-tau is a reliable biomarker for sCJD, but false positive results may occur, especially in rapid AD and acute encephalopathies (Hermann et al. 2022b).

Aβ42

Aβ42 concentration in the CSF is slightly decreased in sCJD cases (Lattanzio et al. 2017; Abu-Rumeileh et al. 2017); higher amyloid deposition scores in the brain correlate with lower Aβ42 concentration in the CSF, thus indicating that CSF Aβ42 levels in sCJD (and gCJD) are linked to advanced stages of β-amyloid plaques and higher Thal phases (Lattanzio et al. 2017). Highly increased CSF tau protein and decreased Aβ42 help to discriminate sCJD from AD (Kapaki et al. 2001).

Total Prion Protein (T-PrP)

t-PrP in the CSF is reduced in the CSF in sCJD compared with other rapid dementias (Dorey et al. 2015; Meyne et al. 2009; Torres et al. 2012; Villar-Piqué et al. 2019b; Minikel et al. 2019; Vallabh et al. 2019). Combined analysis of CSF t-PrP, t-tau, P-tau, and Aβ42 is clinically helpful in the differential diagnosis between sCJD and AD (Abu-Rumeileh et al. 2017). All six human PrP peptides, spanning the N- and C-terminal domains of PrP, were reduced (Minikel et al. 2019). In serial lumbar punctures obtained at different disease stages of sCJD patients, t-PrP concentrations inversely correlate with disease progression (Villar-Piqué et al. 2019b). Differences in disease progression have prompted CSF t-PrP quantification as a tool for prion disease drug development (Vallabh et al. 2020).

Neurofilament Light Chain (NfL)

CSF NfL levels are increased in sCJD (Kovacs et al. 2017; Zerr et al. 2018; Abu-Rumeileh et al. 2018a; Kanata et al. 2019). The sensitivity of NfL in the CSF is 90–96%, and the specificity is 80–85% (Abu-Rumeileh et al. 2018a; Steinacker et al. 2016; Abu-Rumeileh and Parchi 2021). In contrast to 14-3-3 and t-tau, NfL levels are higher in MV2 and VV2 subtypes, yet tau levels diverge from NfL levels depending on the subtype, degree of subcortical involvement, and disease duration; MM1 cases show a significantly lower concentration of CSF NfL than those with sCJD MV2, despite the much higher t-tau levels and the more rapid clinical course (Abu-Rumeileh et al. 2018a).

Increased NfL values are also seen in other neurodegenerative dementias, thus reducing the sole weight of NfL in the CSF as a diagnostic biomarker of sCJD (Abu-Rumeileh et al. 2019a; Abu-Rumeileh and Parchi 2021; Khalil et al. 2018).

Calcium-Binding Protein S100β

CSF S100 protein levels are increased in sCJD with a sensitivity of 78–94% and a specificity of 81–87% (Beaudry et al. 1999).

Neuron-Specific Enolase (NSE)

CSF NSE levels are increased in sCJD; the sensitivity ranges from 50% to 80%, and the specificity from 83% to 98% (Zerr et al. 1995; Aksamit et al. 2001; Kohira et al. 2000). High levels of tau protein, NSE, and S100β are associated with shorter survival times (Sanchez-Juan et al. 2007).

α-Synuclein and β-Synuclein

CSF *α-synuclein* levels, using commercial ELISA kits, are elevated in sCJD with high sensitivity (94–98%) and specificity (96–97%) (Llorens et al. 2017a, 2018a; Kruse et al. 2018). Phosphorylated α-synuclein is also elevated in sCJD compared with other neurodegenerative diseases, including dementia with Lewy bodies. The area under the curve (AUC) value for α-synuclein is higher for the discrimination of sCJD from dementias associated with Lewy bodies than phosphorylated α-synuclein (Schmitz et al. 2019). A meta-analysis and recent studies support the utility of α-synuclein study in the CSF in CJD (Kong et al. 2022).

CSF *β-synuclein* levels are also increased in sCJD. A single study shows that β-synuclein performs better than 14-3-3 (AUC 0.95 vs. 0.89) and, to a lesser extent, than total tau (AUC 0.92) (Abu-Rumeileh et al. 2023).

Neurogranin and SNAP-25

CSF neurogranin levels are increased in sCJD and significantly higher than in AD. sCJD MM1/MV1 subtypes show higher neurogranin levels than VV2 cases. Neurogranin is increased at early sCJD clinical disease stages and is a good prognostic marker of survival time in CJD (Blennow et al. 2019).

Increased neurogranin levels in sCJD were also supported in another study, in parallel with increased levels of SNAP-25 compared with other neurodegenerative diseases (Bentivenga et al. 2023). Moreover, SNAP-25 is a better survival marker than neurogranin in sCJD (Bentivenga et al. 2023).

Other Prion Diseases: vCJD, iCJD, gCJD, GSS, FFI

vCJD

CSF protein 14-3-3 is not as useful a marker for vCJD as it is for sCJD. Increased concentration of CSF tau is increased in vCJD, but they do not discriminate with other forms of dementia. S100β and NSE in the CSF are not useful markers in diagnosing vCJD (Green et al. 2001).

iCJD

14-3-3, t-tau, P-tau/t-tau ratio, NfL, and α-synuclein protein levels in the CSF in patients with iCJD did not differ from those described for sCJD (Llorens et al. 2020b).

gCJD, GSS, FFI

14-3-3, tau protein, S100β, α-synuclein, and NSE protein levels are increased in the CSF in most gCJD, particularly those associated with the E200K, V210I, T188K, and E196A mutations, but the sensitivity of those biomarkers is low in most cases of FFI and GSS disease (Kovacs et al. 2005; Ladogana et al. 2009; Chen et al. 2019; Schmitz et al. 2022b). 14-3-3 and t-tau sensitivity is higher in patients carrying the V210I-129 haplotype than those carrying E200K-129 M (Lattanzio et al. 2017).

Increases of CSF tau isoforms with exon-2 and exon-10 segments are found in gCJD bearing E200K and T188K mutations but not in the cases of FFI (Chen et al. 2019). P-tau/t-tau is also reduced in certain genetic prion diseases, including some cases of GSS, but not in gCJD linked to D178MV and FFI linked to D178MM (Llorens et al. 2016). In genetic prion diseases related to octapeptide repeat insertions, the biomarker sensitivity correlates with the number of repeats (Schmitz et al. 2022b). t-PrP levels in the CSF also decrease in gCJD linked to E200K, V210I mutations, and FFI linked to D178N-129 M but are not modified in GSS linked to P102L mutation (Villar-Piqué et al. 2019b).

Other Changes in the CSF in Human Prion Diseases

Cytokines

IL-8, IL-4, and IL-10 levels are elevated, and TGFβ2 decreased in the CSF in sCJD (Stoeck et al. 2005, 2006).

Triggering Receptor Expressed on Myeloid Cells 2 (TREM2)

Elevated CSF soluble TREM2, considered a microglial cell marker, is elevated in the CSF from patients with sCJD, gCJD with mutations E200K and V210I, and iatrogenic CJD; soluble TREM2 levels are not modified in the CSF in FFI (Diaz-Lucena et al. 2021).

Astroglial Markers

Increased CSF YKL-40 levels, parallel with increased numbers of YKL-40-positive inflammatory astrocytes in the brain, are found in AD and sCJD (Llorens et al. 2017b). In the same line, increased levels of astroglial markers glial acidic protein (GFAP), chitotriosidase 1 (CHIT1), and chitinase-3-like protein 1 (YKL-40) are found in the CSF in sCJD, mainly in the VV2 subtype (Abu-Rumeileh et al. 2019b). These modifications have limited diagnostic value but may serve to monitor putative changes during disease progression and to monitor astroglial inflammatory response to putative treatments (Llorens et al. 2017b; Abu-Rumeileh et al. 2019b).

Increased YKL-40 levels are found in FFI and patients carrying *PRNP*-E200K mutations (Llorens et al. 2017b).

Mitochondria

Malate Dehydrogenase 1 (MDH1): Elevated CSF levels of MDH1 in sCJD suggest impaired malate oxidation and altered metabolism involving the citric acid cycle linked to neuron death in prion diseases (Schmitz et al. 2016c).

Miscellaneous

TDP-43 protein levels are significantly lower in sCJD (Bruzova et al. 2021). Increased CSF *calmodulin* levels are found in a percentage of sCJD cases, but calmodulin positivity also occurs to a lesser extent in non-PrP diseases (Chen et al. 2021).

Multiple *microRNA*s (miRNAs) show altered expression in the CSF in prion diseases (Llorens et al. 2018b; Kanata et al. 2018). However, no correlation exists between brain and CSF miRNA profiles in sCJD, indicating that CSF miRNA profiles do not faithfully mirror miRNA alterations detected in brain tissue of human prion diseases (Llorens et al. 2018b). Therefore, the interpretation, diagnostic value, and correlation of CSF miRNA alterations with brain function in sCJD are still in the preliminary stage.

The analysis *of extracellular vesicles* in the CSF is a candidate tool for diagnosing prion diseases (Khadka et al. 2023). Increased PrP has been detected in CSF *exosomes* from ovine CSF fluid (Vella et al. 2008) and prion (PrP^{Sc}), revealed by PMCA, in most CSF-derived exosomes from scrapie-infected sheep (López-Pérez et al. 2021). Using exosomes as a diagnostic tool in human prion diseases has not been tested.

Plasma Surrogate Biomarkers

sCJD

Total-Tau

T-tau increases in plasma or serum in sCJD (Kovacs et al. 2017; Steinacker et al. 2016; Thompson et al. 2021). The sensitivity of serum tau is 84.6%, and the specificity is 96.2% (Steinacker et al. 2016). The AUC values for plasma t-tau are 0.94 for sCJD, 0.82 for AD, and 0.83 for other neurological diseases (Kovacs et al. 2017). These values are similar to those obtained in another study (AUC values 0.93 for sCJD) (Zerr et al. 2021). Plasma t-tau depends on the *PRNP* codon 129 genotype and has a moderate prediction survival (Zerr et al. 2021).

NT1-tau is increased in the plasma in sCJD compared with AD and controls; moreover, NT-1 plasma levels correlate with the stage and rate of disease progression (Mengel et al. 2021).

Neurofilament Light Chain (NfL)

NfL levels in plasma are increased in sCJD (Abu-Rumeileh and Parchi 2021; Gu et al. 2023; Kovacs et al. 2017; Schmitz et al. 2022a; Staffaroni et al. 2019; Steinacker et al. 2016; Thompson et al. 2021; Zerr et al. 2021). NfL sensitivity for serum NF-L is 100%, and the specificity is 85.5% (133). The AUC is 0.99 for sCJD, 0.99 for AD, and 0.96 for plasma NfL (Kovacs et al. 2017). AUC values are similar in another study comparing sCJD versus non-CJD dementias (Zerr et al. 2021). Nf-L discriminates better than t-tau between CJD and rapid AD but has no prediction capacity for survival (Zerr et al. 2021).

Total-PrP

Plasma t-PrP concentrations are elevated in sCJD and other neurodegenerative dementias (Llorens et al. 2020c). Curiously, t-PrP in plasma correlates with CSF markers of neurodegeneration but not with t-PrP levels in sCJD (Llorens et al. 2020c). Plasma t-PrP has a limited value in disease discrimination.

β-Synuclein

A single study has shown increased β-synuclein in plasma in sCJD (AUC 0.91), performing better than plasma t-tau (AUC 0.79) and NfL (AUC 0.65) (Abu-Rumeileh et al. 2023). Based on these findings, plasma β-synuclein has been proposed to diagnose sCJD (Abu-Rumeileh et al. 2023). Yet further data from other laboratories are needed to confirm this assumption.

Other Changes in Plasma in Human Prion Diseases

Serum *S100B* shows high sensitivity (84.2%) but lower specificity (63%) in sCJD (Steinacker et al. 2016).

YKL-40 serum levels are significantly elevated in sCJD compared with other dementia-related diseases (169). YKL-40 levels augment with disease progression (Villar-Piqué et al. 2019a).

Soluble *TREM2* increases in sCJD in plasma compared with controls, showing positive correlations with plasma t-tau, NfL, and YKL-40 (Diaz-Lucena et al. 2021).

Peripheral Inflammatory Biomarkers peripheral neutrophil to lymphocyte ratio (NLR), monocyte to HDL ratio (MHR), and neutrophil to HDL ratio (NHR) are significantly associated with disease severity in sCJD. Higher NHR and lower high-density lipoprotein (HDL) are associated with shorter survival times (Kong et al. 2023b).

Genetic Prionopathies

t-tau and *NfL* plasma levels are increased in *genetic prionopathies* (Kovacs et al. 2017; Thompson et al. 2021). The AUC values were estimated at 0.932 for plasma t-tau and 1.00 for plasma NfL (Kovacs et al. 2017).

In one study, t-tau and NfL levels did not show significant differences between *PRNP* mutation carriers and controls during a follow-up of 1 year (Vallabh et al. 2020). However, another study reports a progressive NfL rise about two years before the onset of clinical symptoms in patients with *PRNP* mutations associated with a slowly progressive clinical course (Thompson et al. 2021).

No modifications in t-tau, GFAP, YKL-40, and calcium-binding protein B levels in plasma were found in *FFI* (Hermann et al. 2022a). However, FFI cases showed increased NfL levels in plasma compared with controls; moreover, higher levels were associated with the stage and duration of the disease (Hermann et al. 2022a).

CSF vs Plasma Surrogate Biomarkers in Prion Diseases

14-3-3, t-tau, and NSE are used as markers of cell death; NfL as a marker of axon integrity; neurogranin, SNAP-25, α-synuclein, and β-synuclein are linked to synapses; GFAP and S100β are astrocyte markers; and cytokines, TREM2, and YKL-40 are associated with neuroinflammation. In addition, the t-PrP assay is helpful as it inversely correlates with the presence of prion protein in the brain. In contrast, Aβ is advantageous in the differential diagnosis of AD. In clinical practice, 14-3-3, t-tau, P-tau, NfL, and Aβ levels in the CSF are the most widely used surrogate biomarkers for diagnosis. CSF t-tau, 14-3-3, and NfL markers are significantly associated with survival. However, the sensitivity and specificity are variable from one biomarker to another, mainly depending on the type of prion disease. NfL and t-tau in plasma help monitor disease progression and response to possible treatments in prion diseases (Hermann et al. 2021; Thompson and Mead 2019; Abu-Rumeileh et al. 2020; Figgie and Appleby 2021).

Prion-Specific Assays: PrP^{Sc}-Seeded Aggregation Assays

Novel prion strains were created with recombinant mouse prion protein produced in *E. coli* polymerized into amyloid fibrils; intracerebral inoculation into transgenic mice expressing MoPrP(89-231) produced a neurologic dysfunction between 380 and 660 days and expression of protease-resistant in brain extracts following inoculation (Legname et al. 2004). Brain extracts transmitted disease to wild-type FVB mice and Tg mice over-expressing PrP, with 150- and 90-day incubation times, respectively (Legname et al. 2004). Protease-sensitive synthetic prions have also been produced (Colby et al. 2010). The ability of synthetic prions to induce pathology in animals under particular conditions is vital to understanding the process of prion protein conversion into prions (Legname and Moda 2017). Moreover, constructing new synthetic-based prion strains opens its application to studying new biomarkers. In parallel with these helpful applications, additional measures must be taken before this new source of prions in the laboratories (Castilla et al. 2008).

Protein Misfolded Cyclic Amplification (PMCA)

PMCA is based on the capacity of PrP^{Sc} to induce a conformational change of PrP^{C}, used as a substrate, leading to the amplification of minute fragments of PrP^{Sc} of the assessed sample to quantifiable amounts of prion. The method applies a cyclic process of alternative steps of incubation and sonication of the mixture in microtubes; this step is usually followed by the detection of prion aggregates by Western blotting (Saborio et al. 2001; Moda 2017; Soto and Pritzkow 2018). The test detects PrP^{Sc} in peripheral tissues (Giaccone and Moda 2020). It has excellent sensitivity for detecting PrP^{Sc} in the CSF of scrapie-infected mice and CSF, plasma, and urine of patients with vCJD (Kanata et al. 2018; Atarashi et al. 2007; Moda et al. 2014; Bougard et al. 2018; Concha-Marambio et al. 2016; Oshita et al. 2016; Barria et al. 2018; Cali et al. 2019a). However, certain modifications of the PMCA method, for example, employing partially deglycosylated PrP^{C}, may produce unexpected results (Makarava et al. 2013). PMCA has low sensitivity in sCJD (Hermann et al. 2021). PrP^{Sc} was detected in CSF, but not in urine or blood, in sCJD patients using PMCA, followed by a sensitive immunoassay, SOFIA (Rubenstein and Chang 2013). However, more recent PMCA studies have detected prions in the urine of 29 of 81 patients with sCJD (mainly MM1 and VV2) (Pritzkow et al. 2023) and vCJD (Cali et al. 2019b).

Real-Time Quaking-Induced Conversion (RT-QuIC)

Crucial predecessors of RT-QuIC were amyloid assay seeding (ASA) (Colby et al. 2007) and Quaking-Induced Conversion (QuIC) (Atarashi et al. 2008).

RT-QuiC uses different recombinant PrPs (rPrP) as substrates to amplify small amounts of PrP^{Sc} in the sample under study. The mixture is subjected to rapid intermittent shaking or agitation to aggregate and form fibrils. The product, coupled with a fluorescent dye, usually thioflavin, is visualized as amyloid, and the process can be monitored in real-time using a fluorescence plate reader (Atarashi 2023; Atarashi et al. 2011; Candelise et al. 2017; Cramm et al. 2016; Da Silva Correia et al. 2023; Fiorini et al. 2020; Makarava et al. 2013; McGuire et al. 2012; Orrú et al. 2015a, b; Peden et al. 2012; Poleggi et al. 2022; Rhoads et al. 2020; Satoh 2022; Satoh et al. 2017; Schmitz et al. 2016a; Wilham et al. 2010; Zerr et al. 2020).

The interpretation of the RT-QuIC in the CSF is affected by the presence of red and white cells and elevated total protein concentrations (Cramm et al. 2016). CSF samples are recommended to be transparent and colorless, with a white cell count of <10 ×10^6/L and a total protein concentration of <1 g/L (Green 2019). An RT-QuIC variant (eQuiC) incorporates an immunoprecipitation step with the specific antibodies added after incubating with recombinant PrP (Orrú et al. 2011)). However, eQuIC is positive in plasma in vCJD but not in sCJD (Orrú et al. 2011). On the other hand, RT-QuIC has low efficiency in vCJD compared with PMCA.

It has also been stressed that the seeded conversion efficiency and the diagnostic accuracy of the RT-QuIC assay strongly depend on the kind of recombinant PrP substrate (Da Silva Correia et al. 2023; Bélondrade et al. 2021).

Lower sensitivity is found at the early disease stage and in cases with prolonged survival; inflammation can also be a source of false positivity (Hermann et al. 2023; Foutz et al. 2017).

RT-QuIC in the CSF has a high sensitivity and specificity in sCJD with slight variations depending on the substrate and methodology used. The sensitivity also varies depending on the CJD subtype, being higher in MM1/MV1, MV2, VV1, and VV2 subtypes (between 78% and 100%) and lower in MM2 (between 42% and 78% depending on the substrate) (Lattanzio et al. 2017; Green 2019). The CSF RT-QuIC differentiates 94% of cases of sCJD MM1 from the sCJD MM2 phenotype and 80% of sCJD VV2 from sCJD VV1 (Bélondrade et al. 2021). Interestingly, RT-QuIC is positive in 97% of MV2K sCJD cases, while 14-3-3 protein and total-tau positive tests show 52.6 and 75.9% positivity, respectively (Baiardi et al. 2023).

Other studies report a sensitivity of 92% and a specificity of 100% of RT-QuIC applied to CSF samples of sCJD (Fiorini et al. 2020; Watson et al. 2022; Bsoul et al. 2023). RT-QuIC proved near 100% sensitivity for sCJD but also for iatrogenic and most familial CJD phenotypes (Llorens et al. 2020b; Mok et al. 2023).

Despite various caveats and conditions that must be considered to improve the accuracy of RT-QuIC in the CSF of patients with suspected prion diseases, concordant RT-QuIC values in the CSF are obtained across the European CJD network (McKenzie et al. 2022).

RT-QuIC is also a helpful method in diagnosing variably protease-sensitive prionopathy (VPSPr), in which surrogate biomarkers in the CSF are unworthy; RT-QuIC has a sensitivity of 66% in VPSPr (Rhoads et al. 2020). However, RT-QuIC has much lower sensitivity than PMCA in vCJD (Franceschini et al. 2017).

RT-QuIC in sporadic fatal insomnia (sFI) has a sensitivity of 60% (Abu-Rumeileh et al. 2018b).

The detection sensitivities of RT-QuIC in genetic prion diseases are variable (Hermann et al. 2021; Abu-Rumeileh et al. 2019a; Llorens et al. 2016; Schmitz et al. 2016a; Rhoads et al. 2020; Foutz et al. 2017). RT-QuIC positivity accounts for 78% in GSS, 100% in FFI, 87% in gCJD E200K, and 100% in gCJD V203I (Sano et al. 2013). High sensitivity (100%) is also corroborated in E200K-129 M and E200K-129 V and slightly lower (95.2%) in V210I-129 M. Yet cases bearing other *PRNP* mutations show negative results (Lattanzio et al. 2017). RT-QuIC is also helpful in detecting prion seeding in the CSF in pre-symptomatic E200K carriers

(Mok et al. 2023). However, patients bearing other *PRNP* mutations are not detected with RT-QuIC (Mok et al. 2023).

RT-QuIC has also been applied in the olfactory mucosa (Fiorini et al. 2020; Orrú et al. 2014). Interestingly, RT-QuIC assays using nasal brushing offer a sensitivity of 97% (Orrú et al. 2014). Nasal brushing displays a more robust and faster RT-QuIC response than CSF samples (Orrú et al. 2014; Duan et al. 2023). A variant of RT-QuIC may detect the capacity of seeding of infected blood in the context of vCJD (Thomas et al. 2023). A modified PMCA method has also identified prions in the olfactory mucosa in sCJD (Cazzaniga et al. 2022).

Infectious Properties of PrP^{Sc}-Amplified Products

Although the information on this matter is still beginning, prion infectivity occurs after incubation of misfolded PrP subjected to serial annealing or following PMCA in scrapie-infected mice or hamsters under particular situations (Moudjou et al. 2013; Weber et al. 2007; Shikiya and Bartz 2011; Makarava et al. 2012; Moda et al. 2015). Additional studies support that PMCA products may be infective under appropriate conditions (Castilla et al. 2005; Chianini et al. 2012; Gao et al. 2017; Bistaffa et al. 2017, 2021).

Unfortunately, studies centered on the infectivity of RT-QuIC products obtained with different protocols using distinct substrates (including full-length recombinant versus truncated recombinant PrP), pretreatments, cycles, and temperatures are limited and contradictory. One study reported no infectivity of RT-QuIC products obtained by seeding hamster recombinant PrP with hamster-scrapie brain homogenates following inoculation in hamsters (Schmitz et al. 2016a, referred to as JM Wilham and B Caughey, unpublished data). However, another study reported that RT-QuIC products obtained from two different mouse prion strains were able to trigger the aggregation of rPrP with the formation of amyloid rPrP fibrils; when injected in wild-type mice, these amyloids give rise to a mutant strain able to induce prion-like pathology (Bistaffa et al. 2017; Sano et al. 2014). On the other hand, efficient propagation of vCJD has been elicited with samples containing plasma and heparin using the cell–protein misfolding cyclic amplification technique (Oshita et al. 2016).

The application of PrP^{Sc}-seeded aggregation assays to other biological samples in sCJD has shown the detection of PrP in the urine (Luk et al. 2016) and detection and infectivity in skin samples (Orrú et al. 2017; Mammana et al. 2020; Xiao et al. 2021). RT-QuIC analyses have shown PrP amplification, although lower than in fresh specimens, in formalin-fixed, formic-acid-treated sCJD brain tissue (Dong et al. 2021). Spontaneous generation of infectious prion disease in transgenic mice under particular experimental conditions further alerts us about additional hazards with prions (Torres et al. 2013). Moreover, fine-tuning the PMCA method consistently induces spontaneous misfolding of recombinant PrP into bona fide prions within hours (Torres et al. 2013).

Based on these new revelations, special precautions must be taken when handling prion-contaminated material, biological samples, and synthetically generated PrP products (Bistaffa et al. 2017; Eraña et al. 2023; Pritzkow et al. 2018; Pritzkow et al. 2021). The manipulation of biological fluids must be conducted by trained technicians with appropriate equipment and performed in biosecure facilities; safety measures must be further considered using PRP^{Sc}-seeded aggregation assays (Bistaffa et al. 2017). Biosafety level 3 (BSL-3) facilities are recommended as a precautionary measure.

Conclusion and Future Perspectives

The definite diagnosis of prion diseases needs, in the majority of cases, the information provided by the postmortem neuropathological study, which includes morphological and immunohistochemical data of prion deposition in the brain and other tissues, codon 129 genotyping, prion typing, and *PRNP* gene analysis. The combination of these data, the clinical manifestations during life, and the information about antecedents linked to risk factors are also necessary to evaluate a subgroup of iatrogenic cases. Magnetic resonance imaging, including MRI diffusion-weighted imaging (DWI) (Kishida et al. 2023), tau-PET, electroencephalography, and selected surrogate biomarkers in biological fluids help approximate prion disease during life and putatively monitor the progression of the disease in clinical trials testing future treatments. The most precise tools for the detection of prions are prion-specific amplification assays, principally protein misfolded cyclic amplification (PMCA) and real-time quaking-induced conversion (RT-QuIC) in biological fluids and tissues, and the detection of prions using prion-PET. The availability of such advanced techniques is currently restricted to specialized reference laboratories. The sensibility and specificity of PrP^{Sc} seeding methods are very high, but they do not discriminate against prion disease subtypes. Combined methods are commonly necessary to advance the subtype of prionopathy in any particular patient. Therefore, additional procedures are needed to identify precise subtypes of prionopathies during life. Another critical point is the categorization of prion strains that mark the characteristics of prion seeding and spreading of different PrP^{Sc} species. Cryoelectron microscopy and computational methods for 3D reconstruction of amyloids permit the high-resolution definition of different types of prion fibrils (Manka et al. 2023a, b). Yet, further efforts must be applied to recognize new PrP^{Sc} strains in human prion diseases.

PrP^{Sc} amplification methods applied to different regions trace the progression of prions in the brain and other tissues. Moreover, prions have been detected in tissues and materials in contact with prions whose presence was not suspected before the application of these methods. Since the products of positive PMCA assays are infectious, and PrP^{Sc} is present in several tissues in CJD, special care and biosafety conditions must be applied in managing and processing human biological samples of suspected prion disease. Regarding RT-QuIC products, further experimental studies are needed to elucidate their seeding capacity. Particular attention must be paid to protecting the personnel working with the novel seeding methods that generate new prions.

Acknowledgments Personal observations were funded, in part, by ERDF under the program Interreg Poctefa: RedPrion 148/16.

Compliance with Ethical Standards The authors have no conflicts of interest to declare that are relevant to the content of this chapter.

References

Abu-Rumeileh S, Parchi P (2021) Cerebrospinal fluid and blood neurofilament light chain protein in prion disease and other rapidly progressive dementias: current state of the art. Front Neurosci 15:648743

Abu-Rumeileh S, Lattanzio F, Stanzani Maserati M et al (2017) Diagnostic accuracy of a combined analysis of cerebrospinal fluid t-PrP, t-tau, p-tau, and Aβ42 in the differential diagnosis of Creutzfeldt-Jakob disease from Alzheimer's disease with emphasis on atypical disease variants. J Alzheimers Dis 55:1471–1480

Abu-Rumeileh S, Capellari S, Stanzani-Maserati M et al (2018a) The CSF neurofilament light signature in rapidly progressive neurodegenerative dementias. Alzheimers Res Ther 10:3

Abu-Rumeileh S, Redaelli V, Baiardi S et al (2018b) Sporadic Fatal insomnia in Europe: phenotypic features and diagnostic challenges. Ann Neurol 84:347–360

Abu-Rumeileh S, Baiardi S, Polischi B et al (2019a) Diagnostic value of surrogate CSF biomarkers for Creutzfeldt-Jakob disease in the era of RT-QuIC. J Neurol 266:3136–3143

Abu-Rumeileh S, Steinacker P, Polischi B et al (2019b) CSF biomarkers of neuroinflammation in distinct forms and subtypes of neurodegenerative dementia. Alzheimers Res Ther 12:2

Abu-Rumeileh S, Baiardi S, Ladogana A et al (2020) Comparison between plasma and cerebrospinal fluid biomarkers for the early diagnosis and association with survival in prion disease. J Neurol Neurosurg Psychiatry 91:1181–1188

Abu-Rumeileh S, Halbgebauer S, Bentivenga GM et al (2023) High diagnostic performance of plasma and cerebrospinal fluid β-synuclein for sporadic Creutzfeldt-Jakob disease. Ann Clin Transl Neurol 10:1904–1909

Aguzzi A, Falsig J (2012) Prion propagation, toxicity and degradation. Nat Neurosci 15:936–939

Aguzzi A, Heikenwalder M, Polymenidou M (2007) Insights into prion strains and neurotoxicity. Nat Rev Mol Cell Biol 8:552–561

Aksamit AJ, Preissner CM, Homburger HA (2001) Quantitation of 14-3-3 and neuron-specific enolase proteins in CSF in Creutzfeldt-Jakob disease. Neurology 57:728–730

Atarashi R (2023) RT-QuiC as ultrasensitive method for prion detection. Cell Tissue Res 392:295–300

Atarashi R, Moore RA, Sim VL et al (2007) Ultrasensitive detection of scrapie prion protein using seeded conversion of recombinant prion protein. Nat Methods 4:645–650

Atarashi R, Wilham JM, Christensen L et al (2008) Simplified ultrasensitive prion detection by recombinant PrP conversion with shaking. Nat Methods 5:211–212

Atarashi R, Satoh K, Sano K et al (2011) Ultrasensitive human prion detection in CSF by real-time quaking-induced conversion. Nat Med 17:175–178

Baiardi S, Rossi M, Capellari S, a. (2019) Recent advances in the histo-molecular pathology of human prion disease. Brain Pathol 29:278–300

Baiardi S, Rossi M, Mammana A et al (2021) Phenotypic diversity of genetic Creutzfeldt-Jakob disease: a histo-molecular-based classification. Acta Neuropathol 142:707–728

Baiardi S, Mammana A, Rossi M et al (2022) Phenotypic heterogeneity of variably protease-sensitive prionopathy: a report of three cases carrying different genotypes at PRNP codon 129. Viruses 14:367

Baiardi S, Mammana A, Dellavalle S et al (2023) Defining the phenotypic spectrum of sporadic Creutzfeldt-Jakob disease MV2K: the kuru plaque type. Brain 146:3289–3300

Baldeiras IE, Ribeiro MH, Pacheco P et al (2009) Diagnostic value of CSF protein profile in a Portuguese population of sCJD patients. J Neurol 256:1540–1550

Barria MA, Lee A, Green AJ et al (2018) Rapid amplification of prions from variant Creutzfeldt-Jakob disease cerebrospinal fluid. J Pathol Clin Res 4:86–92

Beaudry P, Cohen P, Brandel JP et al (1999) 14-3-3 protein, neuron-specific enolase, and S-100 protein in cerebrospinal fluid of patients with Creutzfeldt-Jakob disease. Dement Geriatr Cogn Disord 10:40–46

Bélondrade M, Nicot S, Mayran C et al (2021) Sensitive protein misfolding cyclic amplification of sporadic Creutzfeldt-Jakob disease prions is strongly seed and substrate dependent. Sci Rep 11:4058

Bentivenga GM, Baiardi S, Mastrangelo A et al (2023) Diagnostic and prognostic value of cerebrospinal fluid SNAP-25 and neurogranin in Creutzfeldt-Jakob disease in a clinical setting cohort of rapidly progressive dementias. Alzheimers Res Ther 15:150

Bishop MT, Will RG, Manson JC (2010) Defining sporadic Creutzfeldt-Jakob disease strains and their transmission properties. Proc Natl Acad Sci USA 107:12005–12010

Bistaffa E, Rossi M, De Luca CMG et al (2017) Biosafety of Prions. Prog Mol Biol Transl Sci 150:455–485

Bistaffa E, Marín-Moreno A, Juan Carlos Espinosa JC et al (2021) PMCA-generated prions from the olfactory mucosa of patients with fatal familial insomnia cause prion disease in mice. elife 10:e65311

Blennow K, Diaz-Lucena D, Zetterberg H et al (2019) Neurogranin as a neuronal damage marker in CJD: a comparative study with AD. J Neurol Neurosurg Psychiatry 90:846–853

Block AJ, Bartz JC (2023) Prion strains: shining new light on old concepts. Cell Tissue Res 392:113–133

Bougard D, Belondrade M, Mayran C et al (2018) Diagnosis of methionine/valine variant Creutzfeldt–Jakob disease by protein misfolding cyclic amplification. Emerg Infect Dis 24:1364–1366

Brandel JP, Knight R (2018) Variant Creutzfeldt-Jakob disease. Handb Clin Neurol 153:191–205

Brown P, Brandel J-P, Sato T et al (2012) Iatrogenic Creutzfeldt-Jakob disease, final assessment. Emerging Infect Dis 18:901–907

Bruzova M, Rusina R, Stejskalova Z et al (2021) Autopsy-diagnosed neurodegenerative dementia cases support the use of cerebrospinal fluid protein biomarkers in the diagnostic work-up. Sci Rep 11:10837

Bsoul R, Lund EL, Burns K et al (2023) Improved real-time quaking induced conversion for early diagnostics of Creutzfeldt-Jakob disease in Denmark. Int J Mol Sci 24:6098

Bugiani O, Giaccone G, Piccardo P et al (2000) Neuropathology of Gerstmann-Sträussler-Scheinker disease. Microsc Res Tech 50:10–15

Cali I, Castellani R, Yuan J et al (2006) Classification of sporadic Creutzfeldt-Jakob disease revisited. Brain 129:2266–2277

Cali I, Miller CJ, Parisi JE et al (2015) Distinct pathological phenotypes of Creutzfeldt-Jakob disease in recipients of prion contaminated growth hormone. Acta Neuropathol Commun 3:37

Cali I, Lavrich J, Moda F et al (2019a) PMCA-replicated PrP(D) in urine of vCJD patients maintains infectivity and strain characteristics of brain PrP(D): transmission study. Sci Rep 9:5191

Cali I, Lavrich J, Moda F, et al. et al (2019b) PMCA-replicated PrPD in urine of vCJD patients maintains infectivity and strain characteristics of brain PrPD: Transmission study. Sci Rep 26(9):5191

Cali I, Puoti G, Smucny J et al (2020) Co-existence of PrPD types 1 and 2 in sporadic Creutzfeldt-Jakob disease of the VV subgroup: phenotypic and prion protein characteristics. Sci Rep 10:1503
Candelise N, Schmitz M, Da Silva Correia SM et al (2017) Applications of the real-time quaking-induced conversion assay in diagnosis, prion strain-typing, drug pre-screening and other amyloidopathies. Expert Rev Mol Diagn 17:897–904
Carta M, Aguzzi A (2022) Molecular foundations of prion strain diversity. Curr Opin Neurobiol 72:22–31
Castellani RJ, Colucci M, Xie Z et al (2004) Sensitivity of 14-3-3 protein test varies in subtypes of sporadic Creutzfeldt-Jakob disease. Neurology 63:436–442
Castilla J, Saa P, Hetz C et al (2005) In vitro generation of infectious scrapie prions. Cell 121:195–206
Castilla J, Morales R, Saá P et al (2008) Cell-free propagation of prion strains. EMBO J 27:2557–2566
Cazzaniga FA, De Luca CMG, Bistaffa E et al (2020) Cell-free amplification of prions: where do we stand? Prog Mol Biol Transl Sci 175:325–358
Cazzaniga FA, Bistaffa E, De Luca CMG et al (2022) PMCA-based detection of prions in the olfactory mucosa of patients with sporadic Creutzfeldt-Jakob disease. Front Aging Neurosci 14:848991
Chapman T, McKeel DW, Morris JC (2000) Misleading results with the 14-3-3 assay for the diagnosis of Creutzfeldt-Jakob disease. Neurology 55:1396–1397
Chen C, Zhou W, Lv Y et al (2016) The levels of tau isoforms containing exon-2 and exon-10 segments increased in the cerebrospinal fluids of the patients with sporadic Creutzfeldt-Jakob disease. Mol Neurobiol 53:3999–4009
Chen C, Hu C, Shi Q et al (2019) Profiles of 14-3-3 and totasl tau in CSF samples of Chinese patients of different prion diseases. Front Neurosci 13:934
Chen C, Hu C, Zhou W et al (2021) Calmodulin level is significantly increased in the cerebrospinal fluid of patients with sporadic Creutzfeldt-Jakob disease. Eur J Neurol 28:1134–1141
Chianini F, Fernandez-Borges N, Vidal E et al (2012) Rabbits are not resistant to prion infection. Proc Natl Acad Sci USA 109:5080–5085
Cohen OS, Chapman J, Korczyn AD et al (2016) CSF tau correlates with CJD disease severity and cognitive decline. Acta Neurol Scand 133:119–123
Colby DW, Zhang Q, Wang S et al (2007) Prion detection by an amyloid seeding assay. Proc Natl Acad Sci USA 104:20914–20919
Colby DW, Wain R, Baskakov IV et al (2010) Protease-sensitive synthetic prions. PLoS Pathog 6:e1000736
Collins S, McLean CA, Masters CL (2001) Gerstmann-Sträussler-Scheinker syndrome, fatal familial insomnia, and kuru: a review of these less common human transmissible spongiform encephalopathies. J Clin Neurosci 8:387–397
Collins SJ, Sanchez-Juan P, Masters CL et al (2006) Determinants of diagnostic investigation sensitivities across the clinical spectrum of sporadic Creutzfeldt-Jakob disease. Brain 129:2278–2287
Concha-Marambio L, Pritzkow S, Moda F et al (2016) Detection of prions in blood from patients with variant Creutzfeldt-Jakob disease. Sci Transl Med 8:370ra183
Coulthart MB, Jansen GH, Olsen E et al (2011) Diagnostic accuracy of cerebrospinal fluid protein markers for sporadic Creutzfeldt-Jakob disease in Canada: a 6-year prospective study. BMC Neurol 11:133
Cracco L, Notari S, Cali I et al (2017) Novel strain properties distinguishing sporadic prion diseases sharing prion protein genotype and prion type. Sci Rep 7:38280
Cracco L, Xiao X, Nemani SK et al (2019) Gerstmann-Sträussler-Scheinker disease revisited: accumulation of covalently-linked multimers of internal prion protein fragments. Acta Neuropathol Commun 7:85
Cramm M, Schmitz M, Karch A et al (2016) Stability and reproducibility underscore utility of RT-QuIC for diagnosis of Creutzfeldt-Jakob disease. Mol Neurobiol 53:1896–1904

Da Silva Correia SM, Schmitz M, Fischer A et al (2023) Role of different recombinant PrP substrates in the diagnostic accuracy of the CSF RT-QuIC assay in Creutzfeldt-Jakob disease. Cell Tissue Res 392:301–306

Deters KD, Risacher SL, Yoder KK et al (2016) [(11)C]PiB PET in Gerstmann-Sträussler-Scheinker disease. Am J Nucl Med Mol Imaging 6:84–93

Diaz-Lucena D, Kruse N, Thüne K et al (2021) TREM2 expression in the brain and biollogical fluids in prion diseases. Acta Neuropathol 141:841–859

Dong TT, Akagi A, Nonaka T et al (2021) Formalin RT-QuIC assay detects prion-seeding activity in formalin-fixed brain samples from sporadic Creutzfeldt-Jakob disease patients. Neurobiol Dis 159:105504

Dorey A, Tholance Y, Vighetto A et al (2015) Association of cerebrospinal fluid prion protein levels and the distinction between Alzheimer disease and Creutzfeldt-Jakob disease. JAMA Neurol 72:267–275

Duan S, Yang J, Cui Z et al (2023) Seed amplification assay of nasal swab extracts for accurate and non-invasive molecular diagnosis of neurodegenerative diseases. Transl Neurodegener 12:13

Eraña H, Charco JM, González-Miranda E et al (2020) Detection of pathognomonic biomarker PrPSc and the contribution of cell free-amplification techniques to the diagnosis of prion diseases. Biomol Ther 10:469

Eraña H, Díaz-Domínguez CM, Charco JM et al (2023) Understanding the key features of the spontaneous formation of bona fide prions through a novel methodology that enables their swift and consistent generation. Acta Neuropathol Commun 11:145

Ermann N, Lewczuk P, Schmitz M et al (2018) CSF nonphosphorylated tau as a biomarker for the discrimination of AD from CJD. Ann Clin Transl Neurol 5:883–887

Figgie MP, Appleby BS (2021) Clinical use of improved diagnostic testing for detection of prion disease. Viruses 13:789

Fiorini M, Bongianni M, Monaco S et al (2017) Chemical characterization of prions. Prog Mol Biol Transl Sci 150:389–407

Fiorini M, Iselle G, Perra D et al (2020) High diagnostic accuracy of RT-QuIC assay in a prospective study of patients with suspected sCJD. Int J Mol Sci 21:880

Foutz A, Appleby BS, Hamlin C et al (2017) Diagnostic and prognostic value of human prion detection in cerebrospinal fluid. Ann Neurol 81:79–92

Franceschini A, Baiardi S, Hughson AG et al (2017) High diagnostic value of second generation CSF RT-QuIC across the wide spectrum of CJD prions. Sci Rep 7:10655

Galeno R, Di Bari MA, Nonno R et al (2017) Prion strain characterization of a novel subtype of Creutzfeldt-Jakob disease. J Virol 91:e02390

Gambetti P, Kong Q, Zou W et al (2003) Sporadic and familial CJD: classification and characterisation. Br Med Bull 66:213–239

Gambetti P, Dong Z, Yuan J et al (2008) A novel human disease with abnormal prion protein sensitive to protease. Ann Neurol 63:697–708

Gao C, Han J, Zhang J et al (2017) Protein Misfolding cyclic amplification cross-species products of mouse-adapted scrapie strain 139A and hamster-adapted scrapie strain 263K with brain and muscle tissues of opposite animals generate infectious prions. Mol Neurobiol 54:3771–3782

Gelpi E, Baiardi S, Nos C et al (2022) Sporadic Creutzfeldt-Jakob disease VM1: phenotypic and molecular characterization of a novel subtype of human prion disease. Acta Neuropathol Commun 10:114

Ghetti B, Piccardo P, Frangione B et al (1996) Prion protein amyloidosis. Brain Pathol 6:127–145

Ghetti B, Tagliavini F, Takao M et al (2003) Hereditary prion protein amyloidoses. Clin Lab Med 23:65–85

Giaccone G, Moda F (2020) PMCA applications for prion detection in peripheral tissues of patients with variant Creutzfeldt-Jakob disease. Biomol Ther 10:405

Gmitterová K, Heinemann U, Bodemer M et al (2009) 14-3-3 CSF levels in sporadic Creutzfeldt-Jakob disease differ across molecular subtypes. Neurobiol Aging 30:1842–1850

Grasbon-Frodl E, Lorenz H, Mann U et al (2004) Loss of glycosylation associated with the T183A mutation in human prion disease. Acta Neuropathol 108:476–484
Green AJE (2019) RT-QuIC: a new test for sporadic CJD. Pract Neurol 19:49–55
Green AJ, Thompson EJ, Stewart GE et al (2001) Use of 14-3-3 and other brain-specific proteins in CSF in the diagnosis of variant Creutzfeldt-Jakob disease. J Neurol Neurosurg Psychiatry 70:744–748
Gu L, Shu H, Wang Y et al (2023) Blood neurofilament light chain in different types of dementia. Curr Alzheimer Res 20:149–160
Haïk S, Brandel JP (2014) Infectious prion diseases in humans: cannibalism, iatrogenicity and zoonoses. Infect Genet Evol 26:303–312
Hamlin C, Puoti G, Berri S et al (2012) A comparison of tau and 14-3-3 protein in the diagnosis of Creutzfeldt-Jakob disease. Neurology 79:547–552
Hermann P, Appleby B, Brandel JP et al (2021) Biomarkers and diagnostic guidelines for sporadic Creutzfeldt-Jakob disease. Lancet Neurol 20:235–246
Hermann P, Canaslan S, Villar-Piqué A et al (2022a) Plasma neurofilament light chain as a biomarker for fatal familial insomnia. Eur J Neurol 29:1841–1846
Hermann P, Haller P, Goebel S et al (2022b) Total and phosphorylated cerebrospinal fluid tau in the differential diagnosis of sporadic Creutzfeldt-Jakob disease and rapidly progressive Alzheimer's disease. Viruses 14:276
Hermann P, Schmitz M, Cramm M et al (2023) Application of real-time quaking-induced conversion in Creutzfeldt-Jakob disease surveillance. J Neurol 270:2149–2161
Hill AF, Joiner S, Beck JA et al (2006) Distinct glycoform ratios of protease resistant prion protein associated with PRNP point mutations. Brain 129:676–685
Houston F, Andréoletti O (2019) Animal prion diseases: the risks to human health. Brain Pathol 29:248–262
Hsich G, Kenney K, Gibbs CJ et al (1996) The 14-3-3 brain protein in cerebrospinal fluid as a marker for transmissible spongiform encephalopathies. N Engl J Med 335:924–930
Ironside JW, Ritchie DL, Head MW (2017) Prion diseases. Handb Clin Neurol 145:393–403
Johnson RT (2005) Prion diseases. Lancet Neurol 4:635–642
Jürgens-Wemheuer W, Wrede A, Schulz-Schaeffer W (2021) Defining the prion type of fatal familial insomnia. Pathogens 10:1293
Kanata E, Thüne K, Xanthopoulos K et al (2018) MicroRNA alterations in the brain and body fluids of humans and animal prion disease models: current status and perspectives. Front Aging Neurosci 10:220
Kanata E, Golanska E, Villar-Piqué A et al (2019) Cerebrospinal fluid neurofilament light in suspected sporadic Creutzfeldt-Jakob disease. J Clin Neurosci 60:124–127
Kapaki E, Kilidireas K, Paraskevas GP et al (2001) Highly increased CSF tau protein and decreased beta-amyloid (1-42) in sporadic CJD: a discrimination from Alzheimer's disease? J Neurol Neurosurg Psychiatry 71:401–403
Karch A, Hermann P, Ponto C et al (2015) Cerebrospinal fluid tau levels are a marker for molecular subtype in sporadic Creutzfeldt-Jakob disease. Neurobiol Aging 36:1964–1968
Kepe V, Ghetti B, Farlow MR et al (2010) PET of brain prion protein amyloid in Gerstmann-Sträussler-Scheinker disease. Brain Pathol 20:419–430
Khadka A, Spiers JG, Cheng L et al (2023) Extracellular vesicles with diagnostic and therapeutic potential for prion diseases. Cell Tissue Res 392:247–267
Khalil M, Teunissen CE, Otto M et al (2018) Neurofilaments as biomarkers in neurological disorders. Nat Rev Neurol 14:577–589
Kishida H, Ueda N, Tanaka F (2023) The advances in the early and accurate diagnosis of Creutzfeldt-Jakob disease and other prion diseases: where are we today? Expert Rev Neurother 23:803–817
Kobayashi A, Asano M, Mohri S et al (2007) Cross-sequence transmission of sporadic Creutzfeldt-Jakob disease creates a new prion strain. J Biol Chem 282:30022–30028

Kobayashi A, Arima K, Ogawa M et al (2008) Plaque-type deposition of prion protein in the damaged white matter of sporadic Creutzfeldt-Jakob disease MM1 patients. Acta Neuropathol 116:561–566

Kobayashi A, Parchi P, Yamada M et al (2016) Neuropathological and biochemical criteria to identify acquired Creutzfeldt-Jakob disease among presumed sporadic cases. Neuropathology 36:305–310

Kohira I, Tsuji T, Ishizu H et al (2000) Elevation of neuron-specific enolase in serum and cerebrospinal fluid of early stage Creutzfeldt–Jakob disease. Acta Neurol Scand 102:385–387

Kong Y, Chen Z, Wang X et al (2022) Diagnostic utility of cerebrospinal fluid α-synuclein in Creutzfeldt-Jakob disease: a systematic review and meta-analysis. J Alzheimers Dis 89:493–503

Kong Y, Chen Z, Shi Q et al (2023a) Clinical correlates of cerebrospinal fluid 14-3-3 protein in non-prion rapid progressive dementia. J Alzheimers Dis 91:263–272

Kong Y, Chen Z, Zhang J, Wu L (2023b) Neutrophil to High-density Lipoprotein ratio (NHR) as a potential predictor of disease severity and survival time in Creutzfeldt-Jakob disease. BMC Neurol 23:34

Kovacs GG, Puopolo M, Ladogana A et al (2005) Genetic prion disease: the EUROCJD experience. Hum Genet 118:166–174

Kovacs GG, Andreasson U, Liman V et al (2017) Plasma and cerebrospinal fluid Tau and neurofilament concentrations in rapidly progressive neurological syndromes: a neuropathology-based cohort. Eur J Neurol 24:1326–1e77

Kretzschmar HA, Sethi S, Földvári Z et al (2003) Iatrogenic Creutzfeldt-Jakob disease with florid plaques. Brain Pathol 13:245–249

Kruse N, Heslegrave A, Gupta V et al (2018) Interlaboratory validation of cerebrospinal fluid α-synuclein quantification in the diagnosis of sporadic Creutzfeldt-Jakob disease. Alzheimers Dement 10:461–470

Ladogana A, Sanchez-Juan P, Mitrová E et al (2009) Cerebrospinal fluid biomarkers in human genetic transmissible spongiform encephalopathies. J Neurol 256:1620–1628

Lattanzio F, Abu-Rumeileh S, Franceschini A et al (2017) Prion-specific and surrogate CSF biomarkers in Creutzfeldt-Jakob disease: diagnostic accuracy in relation to molecular subtypes and analysis of neuropathological correlates of p-tau and Aβ42 levels. Acta Neuropathol 133:559–578

Legname G, Moda F (2017) The prion concept and synthetic prions. Prog Mol Biol Transl Sci 150:147–156

Legname G, Baskakov IV, Nguyen HO et al (2004) Synthetic mammalian prions. Science 305:673–676

Lehmann S, Paquet C, Malaplate-Armand C et al (2019) Diagnosis associated with tau higher than 1200 pg/mL: insights from the clinical and laboratory practice. Clin Chim Acta 495:451–456

Leitão MJ, Baldeiras I, Almeida MR et al (2016) CSF Tau proteins reduce misdiagnosis of sporadic Creutzfeldt-Jakob with inconclussive 14-3-3 result. J Neurol 263:1847–1861

Li QX, Varghese S, Sarros S et al (2018) CSF Tau supplements 14-3-3 protein detection for sporadic Creutzfeldt-Jakob disease diagnosis while transitioning to next generation diagnostics. J Clin Neurosci 50:292–293

Llorens F, Schmitz M, Karch A et al (2016) Comparative analysis of cerebrospinal fluid biomarkers in the differential diagnosis of neurodegenerative dementia. Alzheimers Dement 12:577–589

Llorens F, Kruse N, Schmitz M et al (2017a) Evaluation of α-synuclein as a novel cerebrospinal fluid biomarker in different forms of prion diseases. Alzheimers Dement 13:710–719

Llorens F, Thüne K, Tahir W et al (2017b) YKL-40 in the brain and cerebrospinal fluid of neurodegenerative dementias. Mol Neurodegener 12:83

Llorens F, Kruse N, Karch A et al (2018a) Validation of α-synuclein as a CSF biomarker for sporadic Creutzfeldt-Jakob disease. Mol Neurobiol 55:2249–2257

Llorens F, Thune K, Marti E et al (2018b) Regional and subtype-dependent miRNA signatures in sporadic CreutzfeldtJakob disease are accompanied by alterations in miRNA silencing machinery and biogenesis. PLoS Pathog 14:e1006802

Llorens F, Villar-Piqué A, Hermann P et al (2020a) Cerebrospinal fluid non-phosphorylated tau in the differential diagnosis of Creutzfeldt-Jakob disease: a comparative prospective study with 14-3-3. J Neurol 267:543–550

Llorens F, Villar-Piqué A, Hermann P et al (2020b) Diagnostic accuracy of prion disease biomarkers in iatrogenic Creutzfeldt-Jakob disease. Biomol Ther 10:290

Llorens F, Villar-Piqué A, Schmitz M et al (2020c) Plasma total prion protein as a potential biomarker for neurodegenerative dementia: diagnostic accuracy in the spectrum of prion diseases. Neuropathol Appl Neurobiol 46:240–254

López-Pérez Ó, Sanz-Rubio D, Hernaiz A et al (2021) Cerebrospinal fluid and plasma small extracellular vesicles and miRNAs as biomarkers for prion diseases. Int J Mol Sci 22:6822

Luk C, Jones S, Thomas C et al (2016) Diagnosing sporadic Creutzfeldt-Jakob disease by the detection of abnormal prion protein in patient urine. JAMA Neurol 73:1454–1460

Makarava N, Kovacs GG, Savtchenko R et al (2012) A new mechanism for transmissible prion diseases. J Neurosci 32:7345–7355

Makarava N, Savtchenko R, Baskakov IV (2013) Selective amplification of classical and atypical prions using modified protein misfolding cyclic amplification. J Biol Chem 288:33–41

Mammana A, Baiardi S, Rossi M et al (2020) Detection of prions in skin punch biopsies of Creutzfeldt-Jakob disease patients. Ann Clin Transl Neurol 7:559–564

Manka SW, Wenborn A, Betts J et al (2023a) A structural basis for prion strain diversity. Nat Chem Biol 19:607–613

Manka SW, Wenborn A, Collinge J et al (2023b) Prion strains viewed through the lens of cryo-EM. Cell Tissue Res 392:167–178

McGuire LI, Peden AH, Orrú CD et al (2012) Real time quaking-induced conversion analysis of cerebrospinal fluid in sporadic Creutzfeldt-Jakob disease. Ann Neurol 72:278–285

McKenzie N, Piconi G, Culeux A et al (2022) Concordance of cerebrospinal fluid real-time quaking-induced conversion across the European Creutzfedt-Jakob disease surveillance network. Eur J Neurol 29:2431–2438

McKintosh E, Tabrizi SJ, Collinge J (2003) Prion diseases. J Neurovirol 9:183–193

Mead S, Webb TE, Campbell TA et al (2007) Inherited prion disease with 5-OPRI: phenotype modification by repeat length and codon 129. Neurology 69:730–738

Mengel D, Mok TH, Nihat A et al (2021) NT1-tau is increased in CSF and plasma of CJD patients, and correlates with disease progression. Cells 10:3514

Meyne F, Gloeckner SF, Ciesielczyk B et al (2009) Total prion protein levels in the cerebrospinal fluid are reduced in patients with various neurological disorders. J Alzheimers Dis 17:863–873

Minikel EV, Kuhn E, Cocco A et al (2019) Domain-specific quantification of prion protein in cerebrospinal fluid by targeted mass spectrometry. Mol Cell Proteomics 18:2388–2400

Moda F (2017) Protein misfolding cyclic amplification of infectious prions. Prog Mol Biol Transl Sci 150:361–374

Moda F, Gambetti P, Notari S et al (2014) Prions in the urine of patients with variant Creutzfeldt-Jakob disease. N Engl J Med 371:530–539

Moda F, Le TN, Aulic S et al (2015) Synthetic prions with novel strain-specified properties. PLoS Pathog 11:e1005354

Mok TH, Mead S (2020) Preclinical biomarkers of prion infection and neurodegeneration. Curr Opin Neurobiol 61:82–88

Mok TH, Nihat A, Majbour N et al (2023) Seed amplification and neurodegeneration marker trajectories in individuals at risk of prion disease. Brain 146:2570–2583

Moudjou M, Sibille P, Fichet G et al (2013) Highly infectious prions generated by a single round of microplate-based protein misfolding cyclic amplification. MBio 5:e00829–e00813

Orge L, Lima C, Machado C et al (2021) Neuropathology of animal prion diseases. Biomolecules 11:466

Orrú CD, Wilham JM, Raymond LD et al (2011) Prion disease blood test using immunoprecipitation and improved quaking-induced conversion. MBio 2:e00078–e00011

Orrú CD, Bongianni M, Tonoli G et al (2014) A test for Creutzfeldt-Jakob disease using nasal brushings. N Engl J Med 371:519–529
Orrú CD, Groveman BR, Hughson AG et al (2015a) Rapid and sensitive RT-QuIC detection of human Creutzfeldt-Jakob disease using cerebrospinal fluid. MBio 6:e02451–e02414
Orrú CD, Groveman BR, Raymond LD et al (2015b) Bank vole prion protein as an apparently universal substrate for RT-QuIC-based detection and discrimination of prion strains. PLoS Pathog 11:e1004983
Orrú CD, Yuan J, Appleby BS et al (2017) Prion seeding activity and infectivity in skin samples from patients with sporadic Creutzfeldt-Jakob disease. Sci Transl Med 9:eaam7785
Oshita M, Yokoyama T, Takei Y et al (2016) Efficient propagation of variant Creutzfeldt-Jakob disease prion protein using the cell-protein misfolding cyclic amplification technique with samples containing plasma and heparin. Transfusion 56:223–230
Otto M, Wiltfang J, Cepek L et al (2002) Tau protein and 14-3-3 protein in the differential diagnosis of Creutzfeldt–Jakob disease. Neurology 58:192–197
Parchi P, Castellani R, Capellari S et al (1996) Molecular basis of phenotypic variability in sporadic Creutzfeldt-Jakob disease. Ann Neurol 39:767–778
Parchi P, Petersen RB, Chen SG et al (1998) Molecular pathology of fatal familial insomnia. Brain Pathol 8:539–548
Parchi P, Giese A, Capellari S et al (1999) Classification of sporadic Creutzfeldt-Jakob disease based on molecular and phenotypic analysis of 300 subjects. Ann Neurol 46:224–233
Parchi P, Zou W, Wang W et al (2000) Genetic influence on the structural variations of the abnormal prion protein. Proc Natl Acad Sci USA 97:10168–10172
Parchi P, Notari S, Weber P, a. (2009a) Inter-laboratory assessment of PrPSc typing in Creutzfeldt-Jakob disease: a Western blot study within the NeuroPrion consortium. Brain Pathol 19:384–391
Parchi P, Strammiello R, Notari S et al (2009b) Incidence and spectrum of sporadic Creutzfeldt-Jakob disease variants with mixed phenotype and co-occurrence of PrPSc types: an updated classification. Acta Neuropathol 118:659–671
Parchi P, Cescatti M, Notari S et al (2010) Agent strain variation in human prion disease: insights from a molecular and pathological review of the National Institutes of Health series of experimentally transmitted disease. Brain 133:3030–3042
Parchi P, de Boni L, Saverioni D, a. (2012) Consensus classification of human prion disease histotypes allows reliable identification of molecular subtypes: an inter-rater study among surveillance centres in Europe and USA. Acta Neuropathol 124:517–529
Peden AH, McGuire LI, Appleford NEJ et al (2012) Sensitive and specific detection of sporadic Creutzfeldt-Jakob disease brain prion protein using real-time quaking-induced conversion. J Gen Virol 93:438–449
Piccardo P, Dlouhy SR, Lievens PM et al (1998) Phenotypic variability of Gerstmann-Sträussler-Scheinker disease is associated with prion protein heterogeneity. J Neuropathol Exp Neurol 57:979–988
Poleggi A, Baiardi S, Ladogana A et al (2022) The use of real-time quaking-induced conversion for the diagnosis of human prion diseases. Front Aging Neurosci 14:874734
Pritzkow S, Morales R, Lyon A et al (2018) Efficient prion disease transmission through common environmental materials. J Biol Chem 293:3363–3373
Pritzkow S, Gorski D, Ramirez F et al (2021) Prion dissemination through the environment and medical practices: facts and risks for human health. Clin Microbiol Rev 34:e0005919
Pritzkow S, Ramirez F, Lyon A et al (2023) Detection of prions in the urine of patients affected by sporadic Creutzfeldt-Jakob disease. Ann Clin Transl Neurol 10:2316–2323
Prusiner SB (1994) Biology and genetics of prion diseases. Ann Rev Microbiol 48:655–686
Prusiner SB (1998) Prions. Proc Natl Acad Sci USA 95:13363–13383
Ramasamy I, Law M, Collins S et al (2003) Organ distribution of prion proteins in variant Creutzfeldt-Jakob disease. Lancet Infect Dis 3:214–222
Rhoads DD, Wrona A, Foutz A et al (2020) Diagnosis of prion diseases by RT-QuIC results in improved surveillance. Neurology 95:e1017–e1026

Riemenschneider M, Wagenpfeil S, Vanderstichele H et al (2003) Phospho-tau/total tau ratio in cerebrospinal fluid discriminates Creutzfeldt–Jakob disease from other dementias. Mol Psychiatry 8:343–347

Risacher SL, Farlow MR, Bateman DR et al (2018) Detection of tau in Gerstmann-Sträussler-Scheinker disease (PRNP F198S) by [18F]Flortaucipir PET. Acta Neuropathol Commun 6:114

Ritchie DL, Barria MA, Peden AH et al (2017) Iatrogenic Creutzfeldt-Jakob disease: investigating human prion transmission across genotypic barriers using human tissue-based and molecular approaches. Acta Neuropathol 133:579–595

Rossi M, Baiardi S, Parchi P (2019) Understanding prion strains: evidence from studies of the disease forms affecting humans. Viruses 11:309

Rostgaard N, Olsen MH, Ottenheijm M et al (2023) Differential proteomic profile of lumbar and ventricular cerebrospinal fluid. Fluids Barriers CNS 20:6

Rubenstein R, Chang B (2013) Re-assessment of PrP(Sc) distribution in sporadic and variant CJD. PLoS One 8:e66352

Rübsamen N, Llorens F, Hermann P et al (2020) A prognostic model for overall survival in sporadic Creutzfeldt-Jakob disease. Alzheimers Dement 16:1438–1447

Saborio GP, Permanne B, Soto C (2001) Sensitive detection of pathological prion protein by cyclic amplification of protein misfolding. Nature 411:810–813

Sanchez-Juan P, Green A, Ladogana A et al (2006) CSF tests in the differential diagnosis of Creutzfeldt-Jakob disease. Neurology 67:637–643

Sanchez-Juan P, Sánchez-Valle R, Green A et al (2007) Influence of timing on CSF tests value for Creutzfeldt-Jakob disease diagnosis. J Neurol 254:901–906

Sano K, Satoh K, Atarashi R et al (2013) Early detection of abnormal prion protein in genetic human prion diseases now possible using real-time QuIC assay. PLoS One 8:e54915

Sano K, Atarashi R, Ishibashi D et al (2014) Conformational properties of prion strains can be transmitted to recombinant prion protein fibrils in real-time quaking-induced conversion. J Virol 88:11791–11801

Satoh K (2022) CSF biomarkers for prion diseases. Neurochem Int 155:105306

Satoh J, Kurohara K, Yukitake M et al (1999) 14-3-3 protein detectable in the cerebrospinal fluid of patients with prion-unrelated neurological diseases is expressed constitutively in neurons and glial cells in culture. Eur Neurol 41:216–225

Satoh K, Tobiume M, Matsui Y et al (2010) Establishment of a standard 14-3-3 protein assay of cerebrospinal fluid as a diagnostic tool for Creutzfeldt–Jakob disease. Lab Investig 90:1637–1644

Satoh K, Atarashi R, Nishida N (2017) Real-time quaking-induced conversion for diagnosis of prion disease. Methods Mol Biol 1658:305–310

Schmitz M, Cramm M, Llorens F et al (2016a) The real-time quaking-induced conversion assay for detection of human prion disease and study of other protein misfolding diseases. Nat Protoc 11:2233–2242

Schmitz M, Ebert E, Stoeck K et al (2016b) Validation of 14-3-3 protein as a marker in sporadic Creutzfeldt–Jakob disease diagnostic. Mol Neurobiol 53:2189–2199

Schmitz M, Llorens F, Pracht A et al (2016c) Regulation of human cerebrospinal fluid malate dehydrogenase 1 in sporadic Creutzfeldt-Jakob disease patients. Aging 8:2927–2935

Schmitz M, Dittmar K, Llorens F et al (2017) Hereditary human prion diseases: an update. Mol Neurobiol 54:4138–4149

Schmitz M, Villar-Piqué A, Llorens F et al (2019) Cerebrospinal fluid total and phosphorylated α-synuclein in patients with Creutzfeldt–Jakob disease and synucleinopathy. Mol Neurobiol 56:3476–3483

Schmitz M, Canaslan S, Espinosa JC et al (2022a) Validation of plasma and CSF neurofilament light chain as an early marker for sporadic Creutzfeldt-Jakob disease. Mol Neurobiol 59:1–9

Schmitz M, Villar-Piqué A, Hermann P et al (2022b) Diagnostic accuracy of cerebrospinal fluid biomarkers in genetic prion diseases. Brain 145:700–712

Shikiya RA, Bartz JC (2011) Vitro generation of high-titer prions. J Virol 85:13439–13442

Shintaku M, Nakamura T, Kaneda D et al (2021) Genetic Creutzfeldt-Jakob disease-M232R with the cooccurrence of multiple prion strains, M1 +M2C +M2T: report of an autopsy case. Neuropathology 41:206–213

Sigurdson CJ, Bartz JC, Glatzel M (2019) Cellular and molecular mechanisms of prion disease. Annu Rev Pathol 14:497–516

Skillbäck T, Rosén C, Asztely F et al (2014) Diagnostic performance of cerebrospinal fluid total tau and phosphorylated tau in Creutzfeldt-Jakob disease: results from the Swedish mortality registry. JAMA Neurol 71:476–483

Skinningsrud A, Stenset V, Gundersen AS et al (2008) Cerebrospinal fluid markers in Creutzfeldt-Jakob disease. Cereb Fluid Res 5:14

Soto C, Pritzkow S (2018) Protein misfolding, aggregation, and conformational strains in neurodegenerative diseases. Nat Neurosci 21:1332–1340

Staffaroni AM, Kramer AO, Casey M et al (2019) Association of blood and cerebrospinal fluid tau level and other biomarkers with survival time in sporadic Creutzfeldt-Jakob disease. JAMA Neurol 76:969–977

Steinacker P, Blennow K, Halbgebauer S et al (2016) Neurofilaments in blood and CSF for diagnosis and prediction of onset in Creutzfeldt–Jakob disease. Sci Rep 6:38737

Steinhoff BJ, Zerr I, Glatting M et al (2004) Diagnostic value of periodic complexes in Creutzfeldt-Jakob disease. Ann Neurol 56:702–708

Stoeck K, Bodemer M, Ciesielczyk B et al (2005) Interleukin 4 and interleukin 10 levels are elevated in the cerebrospinal fluid of patients with Creutzfeldt-Jakob disease. Arch Neurol 62:1591–1594

Stoeck K, Bodemer M, Zerr I (2006) Pro- and anti-inflammatory cytokines in the CSF of patients with Creutzfeldt-Jakob disease. J Neuroimmunol 172:175–181

Stoeck K, Sanchez-Juan P, Gawinecka J et al (2012) Cerebrospinal fluid biomarker supported diagnosis of Creutzfeldt-Jakob disease and rapid dementias: a longitudinal multicentre study over 10 years. Brain 135:3051–3061

Sy MS, Gambetti P, Wong BS (2002) Human prion diseases. Med Clin North Am 86:551–571

Takahashi H, Iwata T, Kitagawa Y et al (1999) Increased levels of epsilon and gamma isoforms of 14-3-3 proteins in cerebrospinal fluid in patients with Creutzfeldt-Jakob disease. Clin Diagostic Lab Immunol 6:983–9855

Takeuchi A, Mohri S, Kai H et al (2019) Two distinct prions in fatal familial insomnia and its sporadic form. Brain Commun 1:fcz045

Thomas CM, Salamat MKF, de Wolf C et al (2023) Development of a sensitive real-time quaking-induced conversion (RT-QuIC) assay for application in prion-infected blood. PLoS One 18:e0293845

Thompson AGB, Mead SH (2019) Review: fluid biomarkers in the human prion diseases. Mol Cell Neurosci 97:81–92

Thompson AGB, Anastasiadis P, Druyeh R et al (2021) Evaluation of plasma tau and neurofilament light chain biomarkers in a 12-year clinical cohort of human prion diseases. Mol Psychiatry 26:5955–5966

Torres M, Cartier L, Matamala JM et al (2012) Altered prion protein expression pattern in CSFas a biomarker for Creutzfeldt-Jakob disease. PLoS One 7:e36159

Torres JM, Castilla J, Pintado B et al (2013) Spontaneous generation of infectious prion disease in transgenic mice. Emerg Infect Dis 19:1938–1947

Vallabh SM, Nobuhara CK, Llorens F et al (2019) Prion protein quantification in human cerebrospinal fluid as a tool for prion disease drug development. Proc Natl Acad Sci USA 116:7793–7798

Vallabh SM, Minikel EV, Williams VJ et al (2020) Cerebrospinal fluid and plasma biomarkers in individuals at risk for genetic prion diseases. BMC Med 18:140

Vella LJ, Greenwood DL, Cappai R et al (2008) Enrichment of prion protein in exosomes derived from ovine cerebral spinal fluid. Vet Immunol Immunopathol 124:385–393

Villar-Piqué A, Schmitz M, Hermann P et al (2019a) Plasma YKL-40 in the spectrum of neurodegenerative dementia. J Neuroinflammation 16:145

Villar-Piqué A, Schmitz M, Lachmann I et al (2019b) Cerebrospinal fluid total prion protein in the spectrum of Prion diseases. Mol Neurobiol 56:2811–2811

Watson N, Hermann P, Ladogana A et al (2022) Validation of revised international Creutzfeldt-Jakob disease surveillance network diagnostic criteria for sporadic Creutzfeldt-Jakob disease. JAMA Netw Open 5:e2146319

Weber P, Giese A, Piening N et al (2007) Generation of genuine prion infectivity by serial PMCA. Vet Microbiol 123:346–357

Whitechurch BC, Welton JM, Collins SJ, Lawson VA (2017) Prion diseases. Adv Neurobiol 15:335–364

Wilham JM, Orrú CD, Bessen RA et al (2010) Rapid end-point quantitation of prion seeding activity with sensitivity comparable to bioassays. PLoS Pathog 6:e1001217

Xiao K, Yang X, Zhou W et al (2021) Validation and application of skin RT-QuIC to patients in China with probable CJD. Pathogens 10:1642

Zanusso G, Monaco S, Pocchiari M et al (2016) Advanced tests for early and accurate diagnosis of Creutzfeldt-Jakob disease. Nat Rev Neurol 12:325–333

Zerr I (2022) Laboratory diagnosis of Creutzfeldt–Jakob disease. N Engl J Med 386:1345–1350

Zerr I, Bodemer M, Räcker S et al (1995) Cerebrospinal fluid concentration of neuron-specific enolase in diagnosis of Creutzfeldt-Jakob disease. Lancet 345:1609–1610

Zerr I, Bodemer M, Otto M et al (1996) Diagnosis of Creutzfeldt–Jakob disease by two-dimensional gel electrophoresis of cerebrospinal fluid. Lancet 348:846–849

Zerr I, Bodemer M, Gefeller O et al (1998) Detection of 14-3-3 protein in the cerebrospinal fluid supports the diagnosis of Creutzfeldt-Jakob disease. Ann Neurol 43:32–40

Zerr I, Zafar S, Schmitz M et al (2017) Cerebrospinal fluid in Creutzfeldt-Jakob disease. Handb Clin Neurol 146:115–124

Zerr I, Schmitz M, Karch A et al (2018) Cerebrospinal fluid neurofilament light levels in neurodegenerative dementia: evaluation of diagnostic accuracy in the differential diagnosis of prion diseases. Alzheimers Dement 14:751–763

Zerr I, Cramm M, da Silva Correia SM et al (2020) Optimization of the real-time quaking-induced conversion assay for prion disease diagnosis. Front Bioeng Biotechnol 8:586890

Zerr I, Villar-Piqué A, Hermann P et al (2021) Diagnostic and prognostic value of plasma neurofilament light and total-tau in sporadic Creutzfeldt-Jakob disease. Alzheimers Res Ther 13:86

Zou WQ, Puoti G, Xiao X et al (2010) Variably protease-sensitive prionopathy: a new sporadic disease of the prion protein. Ann Neurol 68:162–172

Chapter 10
Human Genetic Evidence for New Targets in Prion Diseases: Opportunities and Challenges

Elizabeth Hill and Simon Mead

Abstract Human genetics offers a powerful, causally grounded approach to therapeutic target identification in prion diseases. In this chapter, we examine strategies used to discover and evaluate gene candidates, emphasizing the complementary roles of literature-based hypotheses, expression profiling, cellular screening, and in vivo models. Although candidate selection using cellular and transcriptomic systems has yielded limited translational success due to their inability to fully model the human disease, they remain valuable tools for mechanistic validation. Animal models retain key pathological features of prion disease and provide insights into cell-type-specific and non-cell-autonomous mechanisms, particularly involving glial contributions.

We focus on the translational implications of human genetic discoveries, notably the validation of *PRNP* as a therapeutic target supported by both Mendelian and GWAS data and the identification of novel risk loci, including *STX6* and *GAL3ST1*. Functional evaluation of these genes across multiple experimental platforms has highlighted the complexity of moving from genetic association to therapeutic intervention. Distinctions in prion disease initiation versus propagation versus neurotoxicity stages, and the need for cell- and stage-specific models, are key considerations.

The chapter concludes by outlining a framework for assessing the relevance for target development, integrating genetic evidence, functional models, and safety data. While *PRNP-targeting* therapies are now entering clinical trials, further research is needed to delineate mechanisms for other candidates. With advances in multiomic data integration and experimental modeling, genetically guided approaches hold great promise in expanding the therapeutic landscape for these currently untreatable neurodegenerative disorders.

The original version of the chapter has been revised. A correction to this chapter can be found at https://doi.org/10.1007/978-3-031-97055-9_14

E. Hill · S. Mead (✉)
University College London (UCL) Institute of Prion Diseases, Medical Research Council Prion Unit at UCL, London, UK
e-mail: s.mead@prion.ucl.ac.uk

G. Legname, F. Moda (eds.), *Biomarkers and Therapeutical Targets for Prion Diseases*, Subcellular Biochemistry 112, https://doi.org/10.1007/978-3-031-97055-9_10

Keywords Prion disease · CJD · Target identification · Therapeutic targets · Genetics

State of the Art for Target Identification

Challenges in Identifying New Targets in Prion Disease

There is a huge unmet clinical need in prion diseases, as the disease is universally fatal with no established disease-modifying treatments. This is partly driven by the lack of private investment into a disease that is rare, with the most common human prion disease, sporadic Creutzfeldt–Jakob disease (sCJD), having an annual incidence of 1–2/million/year, equating to a lifetime risk of ~1:5000 (Genevieve et al. 2013). This is compounded by the more general challenge in neurodegenerative diseases of the failure of clinical trials, despite in vitro, cell-based, and in vivo systems providing evidence for target engagement (Hay et al. 2014). A contributing factor could be that the target or model was irrelevant to the human disease process. For prion diseases, what is the best approach to identify a target and model?

Candidate Selection Based on Literature Searching

Historically, the basic method for candidate selection was literature searching. This led to biological studies of individual candidate genes that were hypothesized to have a role in disease pathogenesis based on a related function or interacting with a disease-associated protein. One example of this in the prion field is glycosaminoglycans (GAGs), which are polysaccharide side chains of proteoglycans, such as heparan, heparan sulfate, and chondroitin sulfate. Literature dating back to the 1980s demonstrated heparan sulfate proteoglycans are associated with both cellular PrP (Gabizon et al. 1993; Caughey et al. 1994) and PrP^{Sc} deposits in the disease context (Snow et al. 1989; McBride et al. 1998). Building on these reports, semisynthetic analogs of endogenous GAGs, otherwise known as polyanionic compounds, have been trialed in cell-based models and in vivo, with beneficial effects in reducing PrP^{Sc} levels and improving clinical outcomes, presumably from a competitive inhibition mechanism with endogenous GAGs. However, often this candidate-based approach has proven unsuccessful in translation, as was the case with polyanionic compounds, and is limited by its biased nature.

Candidate Selection Based on Expression Studies

Identifying differentially expressed genes in prion disease may provide mechanistic insight into the molecular players and biological pathways that play a role in pathogenesis, which may translate to potential biomarkers of disease and/or therapeutic targets.

Cellular Studies

Multiple studies have been conducted comparing gene expression profiles in uninfected and prion-infected cells dating back to the 1990s (Doh-ura et al. 1995). These studies aimed to elucidate molecular changes associated with prion replication in simplified, controlled in vitro environments. Early insights included gene regulatory networks linked to extracellular matrix components that modulate prion propagation (Marbiah et al. 2014), signaling-related genes (Greenwood et al. 2005), and pathways associated with cholesterol biosynthesis (Bach et al. 2009). However, these findings were often dependent on specific cell types and experimental conditions and failed to translate into meaningful targets for therapeutic development.

A critical limitation of cellular models in this context is their limited biological relevance to the human disease process. While some studies have identified gene expression changes suggestive of prion-related mechanisms, the overlap of these differentially expressed genes with those found in prion-infected mouse or human brain tissue is minor. This "minor overlap" refers to the presence of a few shared expression patterns or genes across systems, but not in a robust or reproducible manner that would support these models as reliable predictors of human disease targets.

Compounding this issue, one notable study reported a complete lack of universal transcriptional changes induced by prion infection across three different neuronal cell models (Julius et al. 2008). This suggests high variability in cellular responses to infection, influenced by cell line-specific factors, and points to the absence of a core prion-related transcriptional signature in vitro. Such inconsistencies highlight a central problem: gene expression changes observed in cellular systems often do not recapitulate the complex, multicellular, and non-cell-autonomous environment of the brain, where prion disease pathology unfolds.

As a result, cellular studies often fail to provide consistent or translatable targets for prion disease. The artificial nature of these models, including the use of non-physiological culture conditions, the absence of supporting glial and immune cells, and differences in prion strains or the host genotype, all contribute to this limitation. Although cellular studies remain valuable for screening or mechanistic follow-up, they should be interpreted cautiously and ideally integrated with in vivo and human data to increase confidence in target validity.

Animal Studies

Animal studies offer a more relevant system to interrogate gene expression changes in the complex brain microenvironment, taking into consideration the multiple cell types. Unlike cellular models, animal models maintain the intricate cell-cell interactions, regional brain architecture, and systemic immune responses that are fundamental to prion disease pathogenesis. This complexity allows researchers to observe disease-relevant processes such as neuroinflammation, glial activation, synaptic dysfunction, and neuronal degeneration within a biologically relevant context. Also, animal studies allow investigation of the temporal progression of disease in a way that more closely reflects human prion disease, including early preclinical changes and late-stage pathology. These features contribute to their higher translational value for identifying therapeutic targets and understanding disease mechanisms.

There have been many brain gene expression studies performed in prion-infected mice elucidating genes involved in proteolysis, protease inhibition, cell growth and maintenance, the immune response, signal transduction, cell adhesion, and molecular metabolism (Xiang et al. 2004); hippocampal neurogenesis (Slota et al. 2022a); immediate early gene response (Ojeda-Juárez et al. 2022); upregulation of cilium-related genes (Kim and Jeong 2022); and genes related to synaptic dysfunction (Slota et al. 2022b), among others. However, there is not always concordance in the differentially expressed genes identified across studies. This could be driven in part by the methodologies used, but studies also differ based on the prion strain used for infection, as well as the genetic background of the mouse model employed. In a comprehensive, system-based study involving the infection of six different genetic backgrounds of mice with two prion strains, with gene expression changes being assessed at 8–10 time points, Hwang and colleagues identified gene modules that appeared to be key to prion pathogenesis (Hwang et al. 2009). These included pathways such as complement activation, GAG metabolism, cholesterol homeostasis, sphingolipid metabolism, androgen metabolism, and lipid acceptors.

One central theme that has come from in vivo transcriptomic analysis has been the central role of glial pathophysiology as a driver of disease (Sorce et al. 2020). Distinctive microglia- and astrocyte-associated expression signatures have been identified during prion infection (Carroll et al. 2020) supporting moving away from "neuron-centric" targeting approaches. However, transcriptomic analyses of the brain are complex, with gene expression profiles differing based on brain region in prion-infected mice (Slota et al. 2022b). Furthermore, bulk brain transcriptomics loses the resolution of cell-type-specific gene expression differences. With the advent of single-cell RNA sequencing, the central role of glia has been solidified, with the gene expression of glial cell types being altered in the early stages of the disease (Dimitriadis et al. 2022; Majer et al. 2019). Specifically, there is dysregulation of neuroprotective astrocytes as well as a spectrum of microglial activation states in prion-infected mice (Slota et al. 2022a). As such, microglial- and astrocyte-enriched genes were shown to contribute a strong inflammatory profile consisting of inflammatory cytokines, genes encoding for extracellular matrix proteins as well as those related to phagocytosis and proteolysis (Majer et al. 2019). This may interplay with neuronal dysregulation with there being an early, net upregulation of transcription factors and stress-induced genes at preclinical stages of disease, as well as a general downregulation of transcripts related to neuronal communication. This includes glutamate receptors, phosphatase subunits, and numerous synapse-related markers at later stages of the disease (Majer et al. 2019).

Studies assessing the translatome have also highlighted a key role for glia. One cell-type-specific, genome-wide ribosomal profiling study identified striking translational alterations in glia in terminal, prion-infected mice but surprisingly limited changes in neurons (Scheckel et al. 2020). However, in an independent, related study, the results were different with a prominent altered translatome profile being detected in neurons in prion-infected mice even at early stages of the disease (Kaczmarczyk et al. 2022). The differences between these studies may have been driven by the different fold change cutoffs as well as technical differences.

Human Studies

The degree to which these mouse transcriptomics studies have relevance to the human population remains an open question. A degree of overlap in gene expression profiles in the studies provides some evidence (Ojeda-Juárez et al. 2022; Dimitriadis et al. 2022; Kanata et al. 2019). One study also tried to narrow the translational gap by using tg340-PRNP129MM mice infected with postmortem sCJD brain tissue, an sCJD model that faithfully recapitulates the molecular and pathological alterations of the human disease, uncovering epitranscriptomic alterations (Kanata et al. 2019). However, the majority of gene expression studies conducted in human patient brains have used a gene candidate approach (Vanni et al. 2017) as opposed to a global, genome-wide transcriptomic approach. The complex safety implications of working with fresh, frozen human brain may limit the translation of animal work to human gene expression studies.

Cellular Screens

The recent advent of technologies such as RNA interference (RNAi) and CRISPR has allowed genome-wide, unbiased, functional studies of genes with potential roles in various prion-related phenotypes. For example, a genome-wide RNAi functional screen conducted to detect cellular host factors that modulate prion propagation identified targets such as *Hnrnpk*, which translated to more complex models such as organotypic slice cultures and Drosophila (Avar et al. 2022). There was also concordance of these hits with a previously identified gene regulatory network associated with prion propagation (Marbiah et al. 2014), providing confidence that this screening approach was highlighting biologically meaningful targets.

Identifying factors that control PrP^C levels is also relevant for pathogenesis as PrP^C is the substrate for prion replication. Indeed, genome-wide interference screens have also identified modulators of prion protein biogenesis, demonstrating an intricate regulatory network controlling PrP^C expression. In a high-throughput arrayed whole-transcriptome RNA interference screen, 743 candidate regulators of PrP^C levels were identified, followed by recursive candidate attrition through multiple secondary screens prioritizing six gene candidates (Heinzer et al. 2021). In a complementary study investigating the involvement of the microRNAome in modulating PrP^C levels, PrP^C protein levels in cells were assayed following subjection to a genome-wide library encompassing 2019 miRNA mimics across three cell lines (Pease et al. 2019). The three screens yielded 17 overlapping high-confidence miRNA mimic hits, either directly or indirectly modulating PrP^C levels. In an independent screen aimed at investigating the effect of transcription factors on PrP^C

expression, 24 upregulators and 12 downregulators of PrP^C expression were identified (Jiang-An et al. 2022). These studies in combination have provided additional targets that could have therapeutic promise. Screening strategies have also been employed to identify regulators of PrP intracellular trafficking. One RNA interference study showed that depletion of AP2M1, RAB5A, VPS35, and M6PR prevented PrP^C internalization, whereas downregulation of GIT2 and VPS28 increased PrP^C internalization (Ballmer et al. 2017). PrP^C cell-surface expression was reduced by downregulation of RAB5A, VPS28, and VPS35 and enhanced by silencing EHD1.

Conclusions

Although the aforementioned approaches have been valuable to the field and facilitated our general understanding of disease mechanisms, they also come with some limitations (Table 10.1). Perhaps the two most striking barriers of these approaches are the relevance to potential targets of human disease (literature searching, cellular screens, expression studies in non-human model systems) or the question of whether they are causal or secondary to the disease.

Table 10.1 Strengths and limitations of state-of-the-art approaches to identify new targets for prion diseases

Strengths	Limitations
Literature searching	
Widely accessible	Biased approach
No cost restraints	Potential of bad reporting, bias to positive data, or unsound experimental design
Available published data is extensive	Conflicting studies
Expression studies	
Potential to act as biomarkers and be targeted therapeutically	Parsing driver genes from passenger genes is a challenge
Human relevance if conducted in human postmortem brain	Cellular and animal studies may not translate to human, and human brain studies are only possible at end stage
Single-cell technologies increase the cellular resolution of RNA-seq	Complex—brain region and cell-type-specific expression changes
Cellular screens	
Unbiased, inclusive approach with precedent for identifying biologically plausible targets that drive disease-relevant phenotypes	Lack of a human cell line propagating human prions necessitates prion infection studies being done in rodent cell lines or human cell lines propagating nonhuman prions
High throughput with recent technological advances (CRISPR, RNAi)	Potential of false negatives/positives with multiplicity of testing

Alternative Methods to Identify New Targets with Human Genetic Evidence

Human genetics evidence is a powerful approach due to the implicit causality of genetics in the target selection process and its direct relevance to human patients. Indeed, human genetic evidence has been shown to increase the likelihood of successful drug development programs (Ochoa et al. 2022; Nelson et al. 2015; Plenge et al. 2013).

Candidate Gene Studies

To date, around 70 genetic variants have been described in the *PRNP* gene, encoding PrP^C (reviewed in reference (Mead et al. 2019)). Some of these are known to be disease-causing Mendelian variants with mutation of *PRNP* being first highlighted in 1989 by linkage analysis in two independent reports studying families segregating autosomal dominant forms of prion disease (Hsiao et al. 1989; Owen et al. 1989). Since this discovery, multiple other genetic changes in *PRNP* have come to light by candidate gene studies, including a multitude of other missense mutations, premature truncations, alterations in octapeptide repeat number, and also frameshift mutations. Collectively, these account for ~10–15% of the annual incidence of prion disease, with each of these being associated with variable degrees of penetrance and often diverse clinical presentations and progression rates. Non-disease-causing single nucleotide polymorphisms (SNPs) in the *PRNP* gene have also been identified as modulating the risk and progression of prion disease. The most well recognized are variations at codon 129, which have been firmly established to have a modifying effect across all of the different subtypes of prion disease as well as all of the different disease stages (Collinge et al. 1991; Collinge et al. 2006; Minikel et al. 2019; Mead et al. 2006; Palmer et al. 1991; Webb et al. 2008, 2009).

Although linkage analysis has a genome-wide element, subsequent candidate gene screening biases discovery either to the *PRNP* locus or to loci that proved not to be replicable. However, in reality, there is likely a diverse array of other genetic loci that influence disease risk and outcomes. The advancement of sequencing technologies has facilitated unbiased, large-scale, genome-wide studies leading to the identification of other genetic factors, which are implicitly causal.

Case-Control GWAS Studies in sCJD

Case-control genome-wide association studies (GWAS) in sCJD patients can provide insight into the genetic factors affecting the spontaneous formation of prions in the absence of pathogenic mutations in *PRNP* or an acquired challenge.

Polymorphisms at codons 127, 129, and 219 within the *PRNP* gene, which alter the amino acid sequence, have been robustly associated with prion disease risk and progression due to their large effect size (reviewed in reference (Mead et al. 2019)). However, prior to 2020, there was an absence of non-*PRNP* risk factors for sCJD despite two GWAS studies being conducted, most likely as a result of lack of power due to limited sample availability (Mead et al. 2009, 2012; Sanchez-Juan et al. 2012, 2014).

However, in 2020, an international, collaborative effort enabled an adequately powered GWAS study (5208 sCJD patients) to allow the detection of genetic variants with more modest effects on disease risk (Jones et al. 2020). This work reproducibly identified novel SNPs in and near *STX6* (rs3747957; $P = 9.7 \times 10^{-9}$) and *GAL3ST1* (rs2267161; $P = 8.6 \times 10^{-10}$) as conferring risk for the disease at genome-wide significance, as well as independently validating and defining the risk at the *PRNP* locus (rs1799990; $P = 2.7 \times 10^{-15}$). This discovery furthered our knowledge of the etiology of sCJD, implicating intracellular trafficking (*STX6*) and sphingolipid metabolism (*GAL3ST1*) as potential novel causal disease mechanisms. Furthermore, although *PRNP* has been the focus of disease therapeutics for many years, this work unveiled untapped therapeutic potential.

Case-Only GWAS Study of Clinical Phenotypes in sCJD

Phenotype GWAS studies in sCJD patients have the potential to provide insight into the genetic factors affecting clinical parameters, such as clinical duration and age of onset, which are strikingly variable in the patient population (Thompson et al. 2013; Nihat et al. 2022). In 2023, a GWAS was conducted to explore variability in clinical duration (3773 sCJD patients) and age of onset (3767 sCJD patients) (Hummerich et al. 2023). SNPs at the *PRNP* locus were robustly associated with clinical duration (top SNP rs1799990, $P = 3.45 \times 10^{-36}$), with subsequent analysis suggesting that codon 129 was the underlying driver. No genome-wide significant genetic determinants of age at onset were identified, although the *HS6ST3* gene was significant ($P = 1.93 \times 10^{-6}$) in a gene-based test.

Apart from *PRNP*, it is noteworthy that there was no evidence of genome-wide genetic correlation with the case-only GWAS (determinants of clinical phenotypes) with the case-control GWAS (disease risk factors). Neither did novel risk alleles in or near the *STX6* and *GAL3ST1* genes affect the disease duration of sCJD or age of onset. As such, one could speculate that the effects at these risk loci may be involved in the initial generation of a prion seed or the likelihood of a single or small number of infected cells further spreading and establishing infection. This is in contrast to being involved in the subsequent generalized prion propagation in the brain and downstream neurotoxicity. Indeed, a recent study in Alzheimer's disease (AD) has suggested that the genetic determinants that affect disease susceptibility do not necessarily overlap with those that influence subsequent pathological load (Leonenko et al. 2019). Taken together, human genetic studies have offered insight into prion

Table 10.2 Strengths and limitations of GWAS to identify new targets for prion diseases underpinned by human genetic evidence

Strengths	Limitations
Unbiased and genome-wide, leading to the discovery of novel candidate loci (in or near to *STX6, GAL3ST1)* as well as validating *PRNP* codon 129 as the driver of disease risk	Reliance on natural variation and low effect sizes: *PRNP* odds ratio (OR) 1.23 (95% CI 1.2–1.3)*; STX6* OR 1.16 (1.1–1.2); *GAL3ST1* OR 1.18 (1.12–1.25)
Nominated intracellular trafficking and sphingolipid metabolism as novel causal disease mechanisms	Limitations on sample size due to the rarity of sCJD—limited power to detect more variants
Implicitly causal role in prion disease—higher likelihood of therapeutic translation	Limited to European ancestry at present. Many countries do not have active surveillance methods, making it hard to obtain more diverse samples
Common variants: high relevance to the wider population	Challenge validating *STX6* and G*AL3ST1* genes in disease models and moving towards disease mechanisms

CI confidence interval

disease pathophysiology, having noteworthy strengths compared to other approaches but also some limitations (Table 10.2).

Strengths and Limitations of Strategies Employed to Evaluate New Targets with Human Genetic Evidence

Although the field has certainly moved forward in our understanding of the genetic architecture underlying the risk of human prion diseases, the challenge then shifts to understanding the mechanistic underpinnings of the susceptibility genes identified (*PRNP, STX6, GAL3ST1).* This will not only further our understanding of the molecular mechanisms underlying prion disease pathogenesis but also facilitate potential translation to therapeutics.

Moving from a Risk Locus to a Disease Hypothesis Using Bioinformatic Approaches and Harmonization with Other Association Studies

Bioinformatic approaches can be employed to prioritize causal risk genes and variants at the susceptibility locus and further decipher the genetic mechanisms through which they may be acting. Furthermore, integrating these findings with gene expression data and other gene-phenotype association studies (transcriptome-WAS, proteome-WAS, lipidomics GWAS) can be highly informative in delineating the direction of effects and establishing how risk is conferred, allowing causal hypotheses to be developed.

Galactose-3-O-Sulfotransferase 1 (GAL3ST1)
GAL3ST1 is a predominantly oligodendrocyte-expressed gene, which encodes an enzyme involved in the synthesis of sulfatide, a major lipid constituent of the myelin sheath (Takahashi and Suzuki 2012). Association at the *GAL3ST1* locus comprised two causal genetic variants in strong linkage disequilibrium (LD), including a valine to methionine missense variant at codon 29 (V29M), which is associated with increased sulfatide levels in blood (total and five distinct classes of sulfatides, $P = 2.5 \times 10^{-15}$ to 2.7×10^{-37}) (Cadby et al. 2022). This coding change provides a strong genetic mechanism for the association signal at this locus. However, the challenge that remains is deciphering the impact of this coding change on the enzyme, which may have implications for enzyme activity, substrate preference, and/or differences in flux through the sulfatide synthesis pathway. Experimental validation is therefore needed to better understand the association of the coding change with prion disease pathogenesis.

Syntaxin-6 (STX6)
STX6 encodes a SNARE protein principally acting in retrograde trafficking from early endosomes to the *trans*-Golgi network (Bock et al. 1996, 1997). At the *STX6* locus, the causal set of 23 highly correlated, non-coding genetic variants makes it challenging to identify a specific causal variant (Jones et al. 2020). By combining variant-to-gene algorithms with colocalization and expression data, the *STX6* risk SNP haplotype was shown to have strong eQTL effects across multiple brain regions, with the risk haplotype increasing brain expression of *STX6 (*Jones et al. 2020*;* Lonsdale et al. 2013*)*. Focusing on the putamen and caudate, which showed the strongest colocalization of GWAS data with eQTLs, the risk allele was associated with 15% and 13% increases in *STX6* expression, respectively (Lonsdale et al. 2013). Genetic upregulation of both gene and protein expression of syntaxin-6 has also more recently been supported by transcriptome and proteome-wide association studies (Küçükali et al. 2025). Strengthening this hypothesis, studies in tauopathies, where the same set of risk variants at the *STX6* locus drive disease risk, have correlated genetic risk loci with transcriptomic (Chen et al. 2018) and epigenomic data (Fodder et al. 2023) in progressive supranuclear palsy (PSP) and proteomic data in AD (Wingo et al. 2021). The pleiotropic risk effects that syntaxin-6 exerts in multiple protein misfolding diseases and the multiple, independent lines of molecular evidence that risk effects are driven through increased syntaxin-6 levels increase confidence in this being the prevailing genetic mechanism.

Prion disease is the result of many complex cellular interactions acting through non-cell autonomous mechanisms with highly cell-type-specific effects (Kushwaha et al. 2021; Smith et al. 2020; Tahir et al. 2022). Extending on the eQTL analysis using data derived from heterogeneous bulk brain tissues lacking single-cell resolution, Bryois and colleagues recently performed an eQTL analysis using single nuclei RNA-seq from 196 individuals in eight brain cell types, which allowed interrogation of the brain cell-type syntaxin-6 may be exerting its risk (Bryois et al.

2021). This revealed striking cell-type-specific effects in the genetic control of *STX6* gene expression by the lead *STX6* risk SNP identified in the sCJD GWAS study (Jones et al. 2020) (rs3747957), with the eQTL effect predominating in oligodendrocytes ($P = 1.49 \times 10^{-26}$) and to a lower degree in excitatory neurons ($P = 0.003$). This risk-associated gene upregulation specific to oligodendrocytes (Küçükali et al. 2025) is certainly intriguing given that this cell type has been proposed to be intrinsically resistant to prion infection (Prinz et al. 2004). It also provides convergence with the other novel, GWAS-identified candidate gene, *GAL3ST1,* which is a predominantly oligodendrocyte-expressed gene.

The value of the association of *STX6* risk variants with increased syntaxin-6 expression is two-fold: not only does it provide a clear direction of effect based on human genetics information but it also lessens the need to establish the exact, single variant associated with disease risk due to the knowledge of the genetic mechanism through which that variant is working. Furthermore, in terms of experimental follow-up, a functional genetic approach can be used whereby manipulated syntaxin-6 expression can be directly correlated to disease phenotypes in animal models and cells.

Strengths and Limitations of Cell-Free In Vitro Methodologies to Assess Gene Candidates

In vitro studies offer a powerful, cell-free system to explore the effect of different candidate genes in prion seeding and replication. Infectivity can be generated in cell-free conditions using the protein misfolding cyclic amplification assay (PMCA), whereby the conversion of PrP^{C} into PK-resistant PrP^{Sc} is enhanced by using cyclic bursts of sonication (Saborio et al. 2001). This approach can be modified to use a brain homogenate substrate derived from a mouse model with knockout of a candidate target gene to assess how this modifies prion conversion and replication.

One example of a study using such an approach was focused on Hsp70, whereby PMCA reactions were seeded with RML prions in the substrate from either *Hsp70*$^{-/-}$ or *Hsp70*$^{+/+}$ mice (Mays et al. 2019). Interestingly, higher levels of PK-resistant PrP^{Sc} were produced with the *Hsp70*$^{+/+}$ substrate compared to the *Hsp70*$^{-/-}$ substrate, suggesting that the seeding kinetics were altered by Hsp70. The CJD genetic risk factor, *STX6*, has also been explored using this approach, although no differences in prion replication were seen in *Stx6*$^{+/+}$ and *Stx6*$^{-/-}$ substrates seeded with RML prions (Sangar et al. 2024). This argues against an indirect effect of syntaxin-6 expression acting through the abundance of a cofactor for prion replication. The result also needs to be interpreted with caution considering PMCA mimics no aspects of cellular trafficking, the functional role of syntaxin-6. Furthermore, there may be some functional redundancy at the protein level of other related proteins, which is especially likely with the syntaxin family of proteins.

Additionally, in standard PMCA reactions, a prion seed is supplied, making it a model of established propagation as opposed to providing a system to study initial prion conversion. The latter may be required for gene candidates identified by GWAS, hypothesized to be involved in disease susceptibility. In the literature, PMCA has been shown to be able to reproduce the spontaneous generation of prions (Deleault et al. 2007; Barria et al. 2009). However, translating this approach to assess initial conversion events in brain homogenate derived from wild-type mice compared to mice with altered expression of a gene candidate would be challenging. First and foremost, due to spontaneous conversion being a rare event, it would require a large number of independent replicates to power such a genetic study. Second, there is always the doubt that the conversion event could be initiated by contamination of the laboratory environment. Recently, a PMCA-based method for the spontaneous generation of *bona fide* prions was reported (Eraña et al. 2024), which may be able to overcome some of these barriers. However, this method requires a very specific set of experimental conditions that may or may not be generalizable to disease risk.

Other in vitro methodologies, such as real-time quaking-induced conversion (RT-QuIC) and PrP aggregation assays, are also established, which allow the assessment of relative seed levels or the aggregation propensity of PrP, respectively. However, these systems are artificial and do not generate infectious prions so results should be interpreted with caution.

Taken together, in vitro techniques, such as PMCA, offer a powerful platform to explore the role of gene candidates on seeded prion replication, although more work is needed to extend this to assess spontaneous misfolding (Table 10.3).

Table 10.3 Strengths and limitations of PMCA to evaluate new targets with human genetics evidence

Strengths	Limitations
Replicates specific infectivity in a simplified, high-throughput setup	Model of established propagation, not initiation so the value of approach depends on the mechanism of the candidate
Allows faithful replication of prion strains across many species of prions	Not suitable for all candidate genes—dependent on localization and/or function
Replicates species barriers	More promiscuous than in vivo—species barrier can be overcome
Rapid results and cost-effective	Cell-free, reductionist system that requires knockout brain homogenate
Modular system where minor changes can address different hypotheses	Potential for compensatory changes in knockout brain homogenate.
Can reproduce the spontaneous generation of prions	Confounded if PrP^C levels differ between substrates

Strengths and Limitations of Cellular Models to Assess Gene Candidates

While cellular models have significant limitations in the initial discovery of candidate genes, particularly due to their inability to replicate the full complexity of prion disease pathology, they regain value when used to evaluate and functionally test targets identified through other approaches such as GWAS or in vivo transcriptomics. In these contexts, cell-based systems offer a controlled, reductionist environment that enables detailed mechanistic interrogation of gene function, perturbation-response profiling, and hypothesis-driven follow-up. Recognizing this distinction is crucial: although cellular models are insufficient as standalone discovery tools, they are instrumental in validating and de-risking candidate targets when deployed in combination with more physiologically relevant systems.

Prion-susceptible mouse lines have provided valuable insights into the molecular and cellular events controlling prion propagation with susceptible cell lines stably promoting replication of PrP^{Sc} upon subpassaging, allowing high levels of infectivity to be generated (Nishida et al. 2000; Schätzl et al. 1997). In 2003, the landmark development of the scrapie cell assay (SCA) revolutionized the study of prion infectivity in permissive cell lines, allowing robust, quantitative determination of prion propagation with high reproducibility (Klöhn et al. 2003). In combination with gene manipulation studies, this methodology provides an elegant approach to assess the role of potential disease modifier genes on prion replication, with the key proof of principle being demonstrated with knockdown of *Prnp* in prion-susceptible cell lines inhibiting prion propagation, resulting in a reduced spot count (Goold et al. 2011; Bhamra et al. 2022). Indeed, in 2015 this system was used to perform an in vitro screen of 14 candidate genes thought to be associated with prion disease susceptibility at the time (Brown et al. 2014). Half of these genes (*Zbtb38, Sorcs1, Stmn2, Hspa13, Fkbp9, Actr10,* and *Plg*) showed highly significant changes in prion infectivity when overexpressed or knocked down, suggesting their involvement in prion propagation and/or clearance but also implying a sensitivity of this system to perturbation.

As discussed in section "Case-control GWAS studies in sCJD", two novel genetic risk factors were discovered in sCJD by GWAS in 2020: *STX6* and *GAL3ST1* (Jones et al. 2020). Considering *GAL3ST1* is predominantly an oligodendrocyte-expressed gene, the SCA is not necessarily the ideal system to assess its role in prion propagation as the compatible cell lines are all neuronal in origin. *STX6* has broad expression across the brain cell types, making this a feasible approach to assess its role in prion propagation. However, when syntaxin-6 was knocked down in neuroblastoma cells, there was no consistent effect on prion propagation (Jones et al. 2020).

However, most susceptible cell lines only support prion propagation of strains that have been experimentally adapted to rodents to overcome the species barrier. Adapted prion strains have been developed in the laboratory setting, whereby scrapie, BSE, or CJD prions have been serially passaged in experimental animals of specific genetic backgrounds (reviewed in (Igel-Egalon et al. 2018)). Therefore, a limitation to cellular work concerns the inability of cells to stably propagate human prions. Furthermore, the GWAS-identified risk factors were associated with the risk of sCJD, yet these cellular systems rely on acquired infection, which may be driven by distinct mechanisms to spontaneous prion conversion.

Although the SCA offers a gold-standard cell-based assay for prion propagation, the development of cell-based assays of prion-induced neurotoxicity has proven more challenging. Indeed, neurotoxicity in prion disease is complex, with there not necessarily being a single mechanism (reviewed in (Le et al. 2019; Mercer and Harris 2023)). As such there is a bewildering number of ways neurotoxicity can be explored in cell culture making it difficult to establish connections between candidate disease genes and neurotoxic phenotypes. Although multiple valuable models have been developed (Benilova et al. 2020; Fang et al. 2016; Foliaki et al. 2020), there is often a lack of concordance of phenotypes between models, and the majority can only be considered to measure one aspect of neurotoxicity. Taken together, although a reductionist system with notable limitations (Table 10.4), cellular models provide a valuable platform to independently investigate the specific role of gene candidates in prion propagation and certain aspects of prion-induced neurotoxicity in a controlled setting.

Strengths and Limitations of In Vivo Models to Assess Gene Candidates

Mouse models offer the unique advantage that both prion propagation and neurotoxicity can be assessed in the same system. It is important to reiterate here that, as with cellular models, the utility of in vivo systems also varies depending on the stage of investigation. While cellular and animal models may have limitations in unbiased candidate discovery, they gain increasing importance in the downstream

Table 10.4 Strengths and limitations of cellular models to evaluate new targets with human genetics evidence

Strengths	Limitations
SCA provides robust, endpoint quantitative measurements of prion propagation	Compatible cell lines with the SCA are neuronal in origin, precluding the analysis of glial cell types
Assess strain-specific effects of candidate genes on propagation by using cell lines susceptible to >1 strain	Inability of susceptible cell lines to stably propagate human prions
Cost-effective and easy to maintain	Acquired models of infection
Simplified system to test direct, mechanistic hypotheses in a well-controlled setting	Reductionist system not incorporating potential non-cell autonomous mechanisms/recapitulating the in vivo situation
Cell-based assays of specific aspects of neurotoxicity exist	No gold-standard model of prion-induced toxicity

evaluation and functional validation of genes identified through human genetics, transcriptomics, or other discovery-driven approaches. In this context, mouse models are indispensable for assessing gene function within the full complexity of the brain's multicellular environment.

Many disease characteristics relevant to assessing gene candidates can only be accurately studied in the context of a multicellular organism. Such characteristics include susceptibility to disease, incubation time, as well as strain-specific neuropathological and clinical phenotypes. Furthermore, by incorporating the complex contributions of all the brain cell types and capturing non-cell autonomous mechanisms, mouse models are particularly adept at exploring the role of *STX6* and *GAL3ST1* in prion disease pathogenesis, with both hypothesized to be exerting their risk effects in oligodendrocytes. To date, there are no published prion-related in vitro or cellular models of oligodendrocytes, necessitating the use of animal models to capture the effect of this cell type on prion pathobiology.

Acquired Mouse Models of Prion Disease

Mice challenged with prions are an invaluable model for studying prion disease as wild-type mice are not only naturally susceptible but also develop *bona fide* prion disease, remarkably recapitulating all of the key aspects of the human disease at both the biochemical and neuropathological levels. Despite the etiology of the disease being acquired, given the overlap of the downstream disease hallmark features, this remains a strong model system to explore the role of candidate genes in established disease, with the value of this paradigm being demonstrated by the multitude of different studies already conducted using this approach. Furthermore, animals intracerebrally infected with the typical dose of 1% (w/v) prion-infected brain homogenate are generally diagnosed with scrapie sickness within a tight time window, which is advantageous when considering powering in vivo studies assessing modifying "factors." With the advancement of genomic engineering technologies, it is now relatively straightforward to implement a functional genetics approach in vivo by manipulating the expression of a gene of interest to explore the effect on prion disease pathogenesis in prion-infected mice.

The main clinical endpoint from these studies is the incubation period, defined as the time from prion inoculation to scrapie sickness diagnosis. However, a key consideration is what is a biologically robust and meaningful change. Often studies report and emphasize minor changes in the incubation period, typically ranging between 10% and 20%, which can be associated with highly significant p-values. However, although there is a statistical difference, this does not necessarily translate to a biologically meaningful difference. Indeed, it has been suggested that these magnitudes of differences could be within the natural variability that might be expected with inoculations being done at different time points (Tamgüney et al. 2008). As such, it is important to be cautious when interpreting statistically significant differences in modestly changed incubation periods, as the statistical tests could be considered to be anticonservative. Despite blinding being implemented, it is challenging to completely eliminate unconscious observer bias, which may artificially inflate the p-values. For example, once one animal is diagnosed with scrapie sickness, there may be a degree of confirmation bias, resulting in hypervigilance of staff to cage mates for welfare reasons, leading to animals getting diagnosed in a tight time frame. Taking this together, large effect sizes in incubation period differences are needed to firmly establish a clear role of a gene candidate in prion disease pathogenesis, which is ideally supported by other pathological outcome measures and validated in other experimental models. *Prnp* provides a good example, with *Prnp* knockout in mice rendering them resistant to prion disease, preventing the development of the hallmark features of prion disease at the clinical, biochemical, or neuropathological level (Büeler et al. 1993; Sailer et al. 1994).

Seeing sizeable differences (>20%) in incubation periods may be challenging with some gene candidates, with *STX6* and *GAL3ST1* modifier genes, for example, only having modest effects in human prion disease (*STX6* odds ratio (95% CI) = 1.15 (1.05–1.26); *GAL3ST1* odds ratio = 1.11 (1.00–1.23) (Jones et al. 2020)). In this context, infection with the standard 1% (w/v) prion-infected brain homogenate may be too strong a dose of prions, overwhelming any modifying effect present. Although more potent manipulation of gene expression would be hypothesized to translate to a stronger modulation of the disease outcomes, this is not always the case with factors such as compensation further complicating matters. One way to potentially overcome this could be performing a limiting dilution titration of prions into mice with manipulated expression of the gene candidate, which better mimics the prion load at the incipient phases of the disease. This experimental setup better addresses whether gene candidates are working at the conception phase of prion infection, better capturing phenomena such as spread; however, this remains an acquired paradigm. If a gene candidate was involved in initial seed formation, which may be expected for GWAS-identified susceptibility factors, a mouse model that develops spontaneous disease would need to be employed to definitely explore this initial disease stage.

Prion infection in mice is proposed to proceed in two distinct mechanistic phases (Sandberg et al. 2011, 2014). There is an initial exponential increase in infectivity, which rapidly reaches a plateau marking the end of what is referred to as the "clinically silent" phase 1. This is followed by the second "neurotoxic" phase where neuropathology becomes established (Sanders et al. 2014). Therefore, the acquired mouse infection paradigm offers a valuable model to assess the roles of gene candidates in prion propagation (prion titers, PrP^{Sc} levels) and prion-induced neurotoxicity (biomarkers of neurodegeneration, neuroinflammation, spongiosis, synapse loss). In this acquired paradigm, disease is initiated by infection with an exogenous source of prions.

Spontaneous Mouse Models of Prion Disease

There exist many mouse models that are reported to develop spontaneous prion disease, normally driven by *Prnp* overexpression and often coupled with disease-causing mutations in *Prnp*. These genetically programmed mouse models with spontaneous PrP misfolding provide a valuable system to explore the impact of different disease-causing mutations in *PRNP* on pathophysiology as well as providing a platform to study the critical molecular determinants of prion initiation by manipulating the levels of gene candidates.

If relevant, conducting studies in a spontaneous model is important as there are likely important susceptibility differences across different prion subtypes when assessing different candidate genes. This has been demonstrated through studies using small molecules with Anle138b, for example, having different effects on survival in an acquired RML-infected mouse model (Wagner et al. 2013) relative to a spontaneous paradigm (Qin et al. 2019). The discordance in effects noted between the infection and spontaneous paradigm across multiple studies underscores the pathophysiological differences between them and thus the importance of cross-examination and validation of genetic candidates. However, spontaneous models that exist have limitations as they either have highly variable readouts of low penetrance (Vallabh et al. 2023), making it difficult to power a study to detect subtle differences with manipulation of a risk gene, or they do not model *bona fide* prion disease (reviewed in (Watts and Prusiner 2014)).

In terms of spontaneous paradigms, it is important to first make the distinction between human and nonhuman PrP mouse models. Practically all attempts to produce transgenic mouse models for human genetic prion disease using the human PrP sequence had been unsuccessful with the exception of one recent GSS mouse model with the A117V mutation (Asante et al. 2020). The latter was deemed the first "authentic" transgenic model of an inherited human prion disease as it develops spontaneous disease and exhibits disease-relevant, transmissible PrP assemblies without being inoculated with any infectious brain tissue. However, whether this would be an appropriate model to explore the role of gene candidates in pathophysiology is questionable with the <100% attack rate and the variable disease endpoints.

The rest of the successful transgenic mouse models for prion diseases have been made using the mouse or bank vole PrP sequences or in chimeric mouse/human PrP molecules, which seem to be more prone to spontaneous misfolding. However, although these models develop spontaneous disease, they are not a complete model for human prion disease due to their lack of transmissibility or absence of pathological features characteristic of the disease (reviewed in (Watts and Prusiner 2014)). Furthermore, the physiological and biological relevance of all of these transgenic lines is questionable given the gross overexpression of PrP as well as the use of heterologous promoters, which result in PrP expression patterns distinct from its normal distribution. To combat this, CRISPR-Cas9 technology has been harnessed to generate PrP knock-in models, although these lack robust endpoints. Indeed, one therapeutic study using such a model concluded they were unable to rigorously evaluate the efficacy of the small molecule because none of the selected endpoints provided a clear, quantitative measure of disease (Vallabh et al. 2023).

There are some robust models of spontaneous prion disease, such as transgenic mice expressing wild-type and mutant bank vole prion protein (BVPrP), which have been reported to spontaneously generate transmissible prions and have robust disease endpoints that occur in a relatively tight time window (Watts et al. 2012, 2016). Although these mice express an artificial sequence of PrP at levels much higher than physiological levels, and as such are not the best models of human disease, they may be sufficient to address whether candidate genes modify the spontaneous generation of prions. Notably, transgenic mice expressing BVPrP containing the E200K mutation (Watts et al. 2016) may be optimal as patients with the E200K mutation are very similar to sCJD in their clinical presentation, with both types of prion diseases being late-onset and fast-acting.

Taken together, spontaneous mouse models provide a valuable model to evaluate the impact of manipulation of genetic candidates on prion initiation as well as prion propagation and neurotoxicity. However, despite their strengths conceptually, the current spontaneous mouse models available have their limitations, which have implications for the design of genetic studies exploring the role of risk genes (Table 10.5). There is thus an unmet need to generate more relevant spontaneous mouse models of prion disease.

Table 10.5 Strengths and limitations of prion disease mouse models to evaluate new targets with human genetics evidence

Strengths	Limitations
Acquired	
Recapitulate all of the neuropathological features of the human disease	Typical infecting dose of 1% prion-infected brain homogenate may overwhelm the effects of candidate genes
Robust clinical endpoints with little variability	Observer/confirmation bias in the clinical endpoints—anticonservative statistics
Same core disease process—initiator irrelevant in most therapeutic cases	Acquired model of prion disease, thus a different disease initiator than the more common forms of prion disease
Biochemical signature of human prion disease	Inoculation does not reflect a normal disease initiation and is prone to the experimental artifact
Inoculation with a limiting dose of prions allows probing of initial events	Effect of gene candidates may not be translatable across paradigms/strains
Inoculation of humanized models with human strains increases translational reach	Biosafety concerns when using humanized PrP mouse models
Spontaneous	
Allows study of prion initiation in addition to prion replication/toxicity—relevance for sporadic disease	Less robust, quantitative endpoints and measures of disease tend to be more variable—harder to power studies
Directly relevant for inherited prion diseases if model has a genetic mutation	Models tend to overexpress PrP or have disease-associated mutations—physiological relevance?
Preferred model system for the study of genetic susceptibility factors	Questions surrounding the authenticity of the spontaneous mouse models available
Emergence of "authentic" spontaneous mouse models which produce transmissible prions	Generalisability of mutant PrP molecules to non-genetic forms of the disease
	Multiple spontaneous conversion events as opposed to the prevailing hypothesis of there being one in sporadic disease

Conclusions

In conclusion, there are a multitude of different models that can be used to evaluate new targets in prion disease underpinned by human genetics evidence. The success of the approach taken largely relies on the disease stage in which the gene candidate is acting in prion disease. This is not necessarily known *a priori* but can be inferred using a functional genetics approach, whereby the levels of the gene candidate are manipulated and the functional effects examined. However, using *STX6* as an example, such gross modulation of gene expression typically used in these sorts of studies far exceeds the magnitude seen in human patients with the risk allele only driving modest, subtle changes in gene expression. This approach could also be questioned for *PRNP* and *GAL3ST1,* which are not hypothesized to be working through expression changes.

A famous quote from Howard Skipper, an American oncologist, stated, "a model is a lie that helps you see the truth." Indeed, every model system discussed, whether it is reductionist or complex, has strengths and limitations. Therefore, using a multipronged strategy of investigation could be considered the gold-standard approach if time and resources allow.

Moving from Causal Hypotheses to Therapeutic Strategies

Following the evaluation of candidate genes with human genetic evidence in different experimental paradigms, how does one weigh up collectively the evidence and decide whether it is of therapeutic relevance? There are multiple criteria that should, or should preferably, be fulfilled (Table 10.6).

PRNP serves as the perfect illustrative example of a candidate gene, underpinned by both Mendelian and GWAS genetic evidence, which fulfills all of these criteria and has since been explored in a therapeutic setting. Here, there is a defined mechanism of action with reduced *PRNP* levels lowering the levels of the substrate for conversion as well as countering its other proposed roles central to disease pathophysiology, such as mediating neurotoxicity. There is evidence for a protective effect across multiple experimental paradigms *in vitro* (Concha-Marambio et al. 2023), in cell-based systems (Goold et al. 2011; Bhamra et al. 2022) as well as in prion-infected mice using both genetic (Büeler et al. 1993, 1994; Mallucci et al. 2003; Prusiner et al. 1993; Sailer et al. 1994; Sakaguchi et al. 1995) and pharmacological strategies (Minikel et al. 2020a; Raymond et al. 2019) to lower its levels. Studies in prion-infected mice treated with *Prnp*-targeting ASOs have provided evidence for a protective effect across prion disease stages and strains with a striking magnitude of benefit (Minikel et al. 2020a). Importantly, it seems that lowering the levels of PrP is safe with *Prnp*$^{-/-}$ mice being largely healthy (Büeler et al. 1992) except for minor and unreproducible phenotypes (Bremer et al. 2010). Considering any therapeutic strategy will only result in a partial reduction, this would not be expected to cause a problem with heterozygote knockout animals not exhibiting a

Table 10.6 Criteria for success in evaluating gene candidates with human genetics evidence for translational value in prion disease

Criteria	Considerations
Evidence across multiple, diverse experimental paradigms	In vitro data does not always translate to the cellular context; similarly, reductionist cellular models do not always recapitulate the complex, multicellular environment of an in vivo model incorporating non-cell autonomous mechanisms
Magnitude of benefit	To be therapeutically useful, the approach needs to have a sizeable magnitude of effect at a disease stage which is meaningful in a clinical context
Evidence across prion strains	Builds confidence in the therapeutic value as well as providing suggestive evidence modifying the gene candidate will have universality across subtypes of prion disease, increasing its translational reach
Timing of benefit	This will inform the relevant target population (at-risk mutation carriers vs. symptomatic patients), which will heavily depend on the stage of the disease process the gene candidate is implicated
Candidate gene "de-risking": Safety	Tolerability of targeting the candidate gene in vivo is necessary for therapeutic translation, which can be guided by phenotypes seen in knockout animals or examination of human loss-of-function genetic variation
Mechanism	Although not essential, grasping a mechanism of action of the gene candidate in prion disease pathogenesis will facilitate therapeutic translation, guiding decisions such as when treatment would be useful

phenotype. Furthermore, inactivating mutations in the *PRNP* gene in humans in the heterozygote state seem to be tolerated (Minikel et al. 2016; Minikel et al. 2020b), further supporting the notion that in the human context, reducing PrP levels will be tolerated in a therapeutic setting. As such *PRNP* lowering or blocking is an appealing therapeutic hypothesis, resulting in it being pursued in a translational setting using different therapeutic modalities such as *PRNP*-targeting ASOs (Phase 1/2a trial employing *ION717, NCT06153966)* or by employing PrP-targeting antibodies (Mead et al. 2022).

What about the other GWAS-identified genetic candidates? *STX6* has been the most followed-up. However, prion infection of *Stx6*$^{-/-}$ mice with 1% (w/v) RML prions had no, or minimal, effect on disease incubation times suggesting modulating its expression has no effect on established disease (Jones et al. 2024). It is therefore unlikely that syntaxin-6 reduction would be useful in symptomatic patients. However, if it is found to be involved in prion initiation, which would support its identification as a susceptibility factor in GWAS, it cannot be discounted that lowering its levels at a sufficiently early disease stage could delay the emergence of prion disease, which would be of relevance to the at-risk patient population. However, the current model systems that exist make this a difficult hypothesis to pursue (see section "Strengths and limitations of strategies employed to evaluate new targets with human genetic evidence"), and, in any case, it does not have the concrete evidence base as demonstrated with *PRNP,* which has been extensively de-risked dating back to the 1990s (Büeler et al. 1992). Although there is suggestive evidence reduction of syntaxin-6 would be tolerated, with *Stx6*$^{-/-}$ mice being overtly healthy and with evidence from human loss-of-function variation, the susceptibility mechanism remains to be established making therapeutic translation unlikely in the near future.

A wide range of experimental models have been employed to support both candidate selection and target evaluation in prion disease research. Literature-based hypotheses, transcriptomic profiling, and genetic screening in cellular systems have guided early discovery efforts, although these models often fall short of faithfully recapitulating the complexity of human disease. More sophisticated animal models, both acquired and spontaneous, have enabled the evaluation of candidate genes in the context of prion propagation and neurotoxicity, particularly capturing non-cell autonomous mechanisms and cell-type specific contributions, such as those involving glial cells. In vitro systems like PMCA and cell-based propagation assays have further provided rapid, reductionist platforms for mechanistic testing, though with limitations in modeling initiation and human relevance.

Taken together, the genetically inspired targets have been hugely influential in the prion disease field. Among them, *PRNP* stands out as a genetically validated drug target supported by both Mendelian and GWAS evidence and is now being pursued in clinical trials. Meanwhile, the identification of additional GWAS risk loci, such as *STX6* and *GAL3ST1*, has advanced our understanding of novel pathogenic pathways in human prion disease. The next critical step is to develop a more granular, mechanistic understanding of how these susceptibility loci confer risk, and to perform higher-powered genome sequencing studies across diverse prion disease subtypes. These efforts will be key to unlocking new translational opportunities and expanding the therapeutic landscape for these currently untreatable conditions.

Acknowledgements This work was supported by funding from the UK Medical Research Council (core funding for the MRC Prion Unit at UCL).

References

Asante EA et al (2020) Spontaneous generation of prions and transmissible PrP amyloid in a humanised transgenic mouse model of A117V GSS. PLoS Biol 18:e3000725. https://doi.org/10.1371/journal.pbio.3000725

Avar M et al (2022) An arrayed genome-wide perturbation screen identifies the ribonucleoprotein hnRNP K as rate-limiting for prion propagation. bioRxiv:2022.2003.2003.482765. https://doi.org/10.1101/2022.03.03.482765

Bach C et al (2009) Prion-induced activation of cholesterogenic gene expression by Srebp2 in neuronal cells. J Biol Chem 284:31260–31269. https://doi.org/10.1074/jbc.M109.004382

Ballmer BA et al (2017) Modifiers of prion protein biogenesis and recycling identified by a highly parallel endocytosis kinetics assay. J Biol Chem 292:8356–8368. https://doi.org/10.1074/jbc.M116.773283

Barria MA, Mukherjee A, Gonzalez-Romero D, Morales R, De Soto C (2009) Novo generation of infectious prions in vitro produces a new disease phenotype. PLoS Pathog 5:e1000421. https://doi.org/10.1371/journal.ppat.1000421

Benilova I et al (2020) Highly infectious prions are not directly neurotoxic. Proc Natl Acad Sci 117:23815. https://doi.org/10.1073/pnas.2007406117

Bhamra S et al (2022) Prion propagation is dependent on key amino acids in Charge cluster 2 within the prion protein. bioRxiv:2022.2008.2008.503133. https://doi.org/10.1101/2022.08.08.503133

Bock JB, Lin RC, Scheller RH (1996) A new syntaxin family member implicated in targeting of intracellular transport vesicles. J Biol Chem 271:17961–17965. https://doi.org/10.1074/jbc.271.30.17961

Bock JB, Klumperman J, Davanger S, Scheller RH (1997) Syntaxin 6 functions in trans-Golgi network vesicle trafficking. Mol Biol Cell 8:1261–1271. https://doi.org/10.1091/mbc.8.7.1261

Bremer J et al (2010) Axonal prion protein is required for peripheral myelin maintenance. Nat Neurosci 13:310–318. https://doi.org/10.1038/nn.2483

Brown CA et al (2014) In vitro screen of prion disease susceptibility genes using the scrapie cell assay. Hum Mol Genet 23:5102–5108. https://doi.org/10.1093/hmg/ddu233

Bryois J et al (2021) Cell-type specific cis-eQTLs in eight brain cell-types identifies novel risk genes for human brain disorders. medRxiv:2021.2010.2009.21264604. https://doi.org/10.1101/2021.10.09.21264604

Büeler H et al (1992) Normal development and behaviour of mice lacking the neuronal cell-surface PrP protein. Nature 356:577–582. https://doi.org/10.1038/356577a0

Büeler H et al (1993) Mice devoid of PrP are resistant to scrapie. Cell 73:1339–1347. https://doi.org/10.1016/0092-8674(93)90360-3

Büeler H et al (1994) High prion and PrPSc levels but delayed onset of disease in scrapie-inoculated mice heterozygous for a disrupted PrP gene. Mol Med 1:19–30. https://doi.org/10.1007/BF03403528

Cadby G et al (2022) Comprehensive genetic analysis of the human lipidome identifies loci associated with lipid homeostasis with links to coronary artery disease. Nat Commun 13:3124. https://doi.org/10.1038/s41467-022-30875-7

Carroll JA, Race B, Williams K, Striebel J, Chesebro B (2020) RNA-seq and network analysis reveal unique glial gene expression signatures during prion infection. Mol Brain 13:71. https://doi.org/10.1186/s13041-020-00610-8

Caughey B, Brown K, Raymond GJ, Katzenstein GE, Thresher W (1994) Binding of the protease-sensitive form of PrP (prion protein) to sulfated glycosaminoglycan and Congo red [corrected]. J Virol 68:2135–2141. https://doi.org/10.1128/jvi.68.4.2135-2141.1994

Chen JA et al (2018) Joint genome-wide association study of progressive supranuclear palsy identifies novel susceptibility loci and genetic correlation to neurodegenerative diseases. Mol Neurodegener 13:41. https://doi.org/10.1186/s13024-018-0270-8

Collinge J, Palmer MS, Dryden AJ (1991) Genetic predisposition to iatrogenic Creutzfeldt-Jakob disease. Lancet 337:1441–1442. https://doi.org/10.1016/0140-6736(91)93128-V

Collinge J et al (2006) Kuru in the 21st century--an acquired human prion disease with very long incubation periods. Lancet 367:2068–2074. https://doi.org/10.1016/s0140-6736(06)68930-7

Concha-Marambio L et al (2023) Development of a methodology for large-scale production of prions for biological and structural studies. Front Mol Biosci 10. https://doi.org/10.3389/fmolb.2023.1184029

Deleault NR, Harris BT, Rees JR, Supattapone S (2007) Formation of native prions from minimal components in vitro. Proc Natl Acad Sci USA 104:9741–9746. https://doi.org/10.1073/pnas.0702662104

Dimitriadis A et al (2022) Single-nuclei transcriptomics of mammalian prion diseases identifies dynamic gene signatures shared between species. bioRxiv:2022.2009.2013.507650. https://doi.org/10.1101/2022.09.13.507650

Doh-ura K, Perryman S, Race R, Chesebro B (1995) Identification of differentially expressed genes in scrapie-infected mouse neuroblastoma cells. Microb Pathog 18:1–9

Eraña H et al (2024) A Protein Misfolding Shaking Amplification-based method for the spontaneous generation of hundreds of bona fide prions. Nat Commun 15:2112. https://doi.org/10.1038/s41467-024-46360-2

Fang C, Imberdis T, Garza MC, Wille H, Harris DA (2016) A neuronal culture system to detect prion Synaptotoxicity. PLoS Pathog 12:e1005623. https://doi.org/10.1371/journal.ppat.1005623

Fodder K et al (2023) Brain DNA methylomic analysis of frontotemporal lobar degeneration reveals OTUD4 in shared dysregulated signatures across pathological subtypes. Acta Neuropathol 146:77–95. https://doi.org/10.1007/s00401-023-02583-z

Foliaki ST et al (2020) Pathogenic Prion Protein Isoforms Are Not Present in Cerebral Organoids Generated from Asymptomatic Donors Carrying the E200K Mutation Associated with Familial Prion Disease. Pathogens (Basel, Switzerland) 9. https://doi.org/10.3390/pathogens9060482

Gabizon R, Meiner Z, Halimi M, Ben-Sasson SA (1993) Heparin-like molecules bind differentially to prion-proteins and change their intracellular metabolic fate. J Cell Physiol 157:319–325. https://doi.org/10.1002/jcp.1041570215

Genevieve MJAK et al (2013) Intensity of human prion disease surveillance predicts observed disease incidence. J Neurol Neurosurg Psychiatry 84:1372. https://doi.org/10.1136/jnnp-2012-304820

Goold R et al (2011) Rapid cell-surface prion protein conversion revealed using a novel cell system. Nat Commun 2:281. https://doi.org/10.1038/ncomms1282

Greenwood AD et al (2005) Cell line dependent RNA expression profiles of prion-infected mouse neuronal cells. J Mol Biol 349:487–500. https://doi.org/10.1016/j.jmb.2005.03.076

Hay M, Thomas DW, Craighead JL, Economides C, Rosenthal J (2014) Clinical development success rates for investigational drugs. Nat Biotechnol 32:40–51. https://doi.org/10.1038/nbt.2786

Heinzer D et al (2021) Novel regulators of PrPC biosynthesis revealed by genome-wide RNA interference. PLoS Pathog 17:e1010013. https://doi.org/10.1371/journal.ppat.1010013

Hummerich H et al (2023) Genome wide association study of clinical duration and age at onset of sporadic CJD. medRxiv:2023.2010.2017.23297050. https://doi.org/10.1101/2023.10.17.23297050

Hsiao K et al (1989) Linkage of a prion protein missense variant to Gerstmann-Sträussler syndrome. Nature 338:342–345. https://doi.org/10.1038/338342a0

Hwang D et al (2009) A systems approach to prion disease. Mol Syst Biol 5:252–252. https://doi.org/10.1038/msb.2009.10

Igel-Egalon A, Béringue V, Rezaei H, Sibille P (2018) Prion strains and transmission barrier phenomena. Pathogens (Basel, Switzerland) 7:5. https://doi.org/10.3390/pathogens7010005

Jiang-An Y et al (2022) Robust and versatile arrayed libraries for human genome-wide CRISPR activation, deletion and silencing. bioRxiv:2022.2005.2025.493370. https://doi.org/10.1101/2022.05.25.493370

Jones E et al (2020) Identification of novel risk loci and causal insights for sporadic Creutzfeldt-Jakob disease: a genome-wide association study. Lancet Neurol 19:840–848. https://doi.org/10.1016/s1474-4422(20)30273-8

Jones E et al (2024) Characterisation and prion transmission study in mice with genetic reduction of sporadic Creutzfeldt-Jakob disease risk gene Stx6. Neurobiol Dis 190:106363. https://doi.org/10.1016/j.nbd.2023.106363

Julius C et al (2008) Transcriptional stability of cultured cells upon prion infection. J Mol Biol 375:1222–1233. https://doi.org/10.1016/j.jmb.2007.11.003

Kaczmarczyk L et al (2022) Distinct translatome changes in specific neural populations precede electroencephalographic changes in prion-infected mice. PLoS Pathog 18:e1010747. https://doi.org/10.1371/journal.ppat.1010747

Kanata E et al (2019) RNA editing alterations define manifestation of prion diseases. Proc Natl Acad Sci 116:19727–19735. https://doi.org/10.1073/pnas.1803521116

Kim YC, Jeong BH (2022) Transcriptomic analysis identifies novel potential biomarkers and highlights cilium-related biological processes in the early stages of prion disease in mice. Prion 16:84–90. https://doi.org/10.1080/19336896.2022.2095186

Klöhn PC, Stoltze L, Flechsig E, Enari M, Weissmann C (2003) A quantitative, highly sensitive cell-based infectivity assay for mouse scrapie prions. Proc Natl Acad Sci USA 100:11666–11671. https://doi.org/10.1073/pnas.1834432100

Küçükali F et al (2025) Multiomic analyses direct hypotheses for Creutzfeldt-Jakob disease risk genes. Brain:awaf032. https://doi.org/10.1093/brain/awaf032

Kushwaha R, Sinha A, Makarava N, Molesworth K, Baskakov IV (2021) Non-cell autonomous astrocyte-mediated neuronal toxicity in prion diseases. Acta Neuropathol Commun 9:22. https://doi.org/10.1186/s40478-021-01123-8

Le NTT, Wu B, Harris DA (2019) Prion neurotoxicity. Brain Pathol (Zurich, Switzerland) 29:263–277. https://doi.org/10.1111/bpa.12694

Leonenko G et al (2019) Genetic risk for alzheimer disease is distinct from genetic risk for amyloid deposition. Ann Neurol 86:427–435. https://doi.org/10.1002/ana.25530

Lonsdale J et al (2013) The genotype-tissue expression (GTEx) project. Nat Genet 45:580–585. https://doi.org/10.1038/ng.2653

Majer A et al (2019) The cell type resolved mouse transcriptome in neuron-enriched brain tissues from the hippocampus and cerebellum during prion disease. Sci Rep 9:1099. https://doi.org/10.1038/s41598-018-37715-z

Mallucci G et al (2003) Depleting neuronal PrP in prion infection prevents disease and reverses spongiosis. Science (New York, NY) 302:871–874. https://doi.org/10.1126/science.1090187

Marbiah MM et al (2014) Identification of a gene regulatory network associated with prion replication. EMBO J 33:1527–1547. https://doi.org/10.15252/embj.201387150

Mays CE et al (2019) Prion disease is accelerated in mice lacking stress-induced heat shock protein 70 (HSP70). J Biol Chem 294:13619–13628. https://doi.org/10.1074/jbc.RA118.006186

McBride PA, Wilson MI, Eikelenboom P, Tunstall A, Bruce ME (1998) Heparan sulfate proteoglycan is associated with amyloid plaques and neuroanatomically targeted PrP pathology throughout the incubation period of scrapie-infected mice. Exp Neurol 149:447–454. https://doi.org/10.1006/exnr.1997.6740

Mead S et al (2006) Inherited prion disease with six octapeptide repeat insertional mutation–molecular analysis of phenotypic heterogeneity. Brain 129:2297–2317. https://doi.org/10.1093/brain/awl226

Mead S et al (2009) Genetic risk factors for variant Creutzfeldt-Jakob disease: a genome-wide association study. Lancet Neurol 8:57–66. https://doi.org/10.1016/s1474-4422(08)70265-5

Mead S et al (2012) Genome-wide association study in multiple human prion diseases suggests genetic risk factors additional to PRNP. Hum Mol Genet 21:1897–1906. https://doi.org/10.1093/hmg/ddr607

Mead S, Lloyd S, Collinge J (2019) Genetic factors in mammalian prion diseases. Annu Rev Genet 53:117–147. https://doi.org/10.1146/annurev-genet-120213-092352

Mead S et al (2022) Prion protein monoclonal antibody (PRN100) therapy for Creutzfeldt-Jakob disease: evaluation of a first-in-human treatment programme. Lancet Neurol 21:342–354. https://doi.org/10.1016/s1474-4422(22)00082-5

Mercer RCC, Harris DA (2023) Mechanisms of prion-induced toxicity. Cell Tissue Res 392:81–96. https://doi.org/10.1007/s00441-022-03683-0

Minikel EV et al (2016) Quantifying prion disease penetrance using large population control cohorts. Sci Transl Med 8:322ra329. https://doi.org/10.1126/scitranslmed.aad5169

Minikel EV et al (2019) Age at onset in genetic prion disease and the design of preventive clinical trials. Neurology 93:e125–e134. https://doi.org/10.1212/wnl.0000000000007745

Minikel EV et al (2020a) Prion protein lowering is a disease-modifying therapy across prion disease stages, strains and endpoints. Nucleic Acids Res 48:10615–10631. https://doi.org/10.1093/nar/gkaa616

Minikel EV et al (2020b) Evaluating drug targets through human loss-of-function genetic variation. Nature 581:459–464. https://doi.org/10.1038/s41586-020-2267-z

Nelson MR et al (2015) The support of human genetic evidence for approved drug indications. Nat Genet 47:856–860. https://doi.org/10.1038/ng.3314

Nihat A et al (2022) Development of novel clinical examination scales for the measurement of disease severity in Creutzfeldt-Jakob disease. J Neurol Neurosurg Psychiatry 93:404. https://doi.org/10.1136/jnnp-2021-327722

Nishida N et al (2000) Successful transmission of three mouse-adapted scrapie strains to murine neuroblastoma cell lines overexpressing wild-type mouse prion protein. J Virol 74:320–325. https://doi.org/10.1128/jvi.74.1.320-325.2000

Ochoa D et al (2022) Human genetics evidence supports two-thirds of the 2021 FDA-approved drugs. Nat Rev Drug Discov 21:551. https://doi.org/10.1038/d41573-022-00120-3

Ojeda-Juárez D et al (2022) Prions induce an early arc response and a subsequent reduction in mGluR5 in the hippocampus. Neurobiol Dis 172:105834. https://doi.org/10.1016/j.nbd.2022.105834

Owen F et al (1989) Insertion in prion protein gene in familial Creutzfeldt-Jakob disease. Lancet 1:51–52. https://doi.org/10.1016/s0140-6736(89)91713-3

Palmer MS, Dryden AJ, Hughes JT, Collinge J (1991) Homozygous prion protein genotype predisposes to sporadic Creutzfeldt–Jakob disease. Nature 352:340–342. https://doi.org/10.1038/352340a0

Pease D et al (2019) Genome-wide identification of microRNAs regulating the human prion protein. Brain Pathol (Zurich, Switzerland) 29:232–244. https://doi.org/10.1111/bpa.12679

Plenge RM, Scolnick EM, Altshuler D (2013) Validating therapeutic targets through human genetics. Nat Rev Drug Discov 12:581–594. https://doi.org/10.1038/nrd4051

Prinz M et al (2004) Intrinsic resistance of oligodendrocytes to prion infection. J Neurosci Off J Soc Neurosci 24:5974–5981. https://doi.org/10.1523/jneurosci.0122-04.2004

Prusiner SB et al (1993) Ablation of the prion protein (PrP) gene in mice prevents scrapie and facilitates production of anti-PrP antibodies. Proc Natl Acad Sci USA 90:10608–10612. https://doi.org/10.1073/pnas.90.22.10608

Qin K, Zhao L, Solanki A, Busch C, Mastrianni J (2019) Anle138b prevents PrP plaque accumulation in Tg(PrP-A116V) mice but does not mitigate clinical disease. J Gen Virol 100:1027–1037. https://doi.org/10.1099/jgv.0.001262

Raymond GJ et al (2019) Antisense oligonucleotides extend survival of prion-infected mice. JCI Insight 5:e131175. https://doi.org/10.1172/jci.insight.131175

Saborio GP, Permanne B, Soto C (2001) Sensitive detection of pathological prion protein by cyclic amplification of protein misfolding. Nature 411:810–813. https://doi.org/10.1038/35081095

Sailer A, Büeler H, Fischer M, Aguzzi A, Weissmann C (1994) No propagation of prions in mice devoid of PrP. Cell 77:967–968. https://doi.org/10.1016/0092-8674(94)90436-7

Sakaguchi S et al (1995) Accumulation of proteinase K-resistant prion protein (PrP) is restricted by the expression level of normal PrP in mice inoculated with a mouse-adapted strain of the Creutzfeldt-Jakob disease agent. J Virol 69:7586–7592. https://doi.org/10.1128/jvi.69.12.7586-7592.1995

Sanchez-Juan P et al (2012) Genome-wide study links MTMR7 gene to variant Creutzfeldt-Jakob risk. Neurobiol Aging 33(1487):e1421–e1488. https://doi.org/10.1016/j.neurobiolaging.2011.10.011

Sanchez-Juan P et al (2014) A genome wide association study links glutamate receptor pathway to sporadic Creutzfeldt-Jakob disease risk. PLoS One 10:e0123654. https://doi.org/10.1371/journal.pone.0123654

Sandberg MK, Al-Doujaily H, Sharps B, Clarke AR, Collinge J (2011) Prion propagation and toxicity in vivo occur in two distinct mechanistic phases. Nature 470:540–542. https://doi.org/10.1038/nature09768

Sandberg MK et al (2014) Prion neuropathology follows the accumulation of alternate prion protein isoforms after infective titre has peaked. Nat Commun 5:4347. https://doi.org/10.1038/ncomms5347

Sanders DW et al (2014) Distinct tau prion strains propagate in cells and mice and define different tauopathies. Neuron 82:1271–1288. https://doi.org/10.1016/j.neuron.2014.04.047

Sangar D et al (2024) Syntaxin-6 delays prion protein fibril formation and prolongs presence of toxic aggregation intermediates. eLife 13:e83320. https://doi.org/10.7554/eLife.83320

Schätzl HM et al (1997) A hypothalamic neuronal cell line persistently infected with scrapie prions exhibits apoptosis. J Virol 71:8821–8831. https://doi.org/10.1128/jvi.71.11.8821-8831.1997

Scheckel C, Imeri M, Schwarz P, Aguzzi A (2020) Ribosomal profiling during prion disease uncovers progressive translational derangement in glia but not in neurons. eLife 9:e62911. https://doi.org/10.7554/eLife.62911

Slota JA, Sajesh BV, Frost KF, Medina SJ, Booth SA (2022a) Dysregulation of neuroprotective astrocytes, a spectrum of microglial activation states, and altered hippocampal neurogenesis are revealed by single-cell RNA sequencing in prion disease. Acta Neuropathol Commun 10:161. https://doi.org/10.1186/s40478-022-01450-4

Slota JA, Medina SJ, Frost KL, Booth SA (2022b) Neurons and astrocytes elicit brain region specific transcriptional responses to prion disease in the murine CA1 and thalamus. Front Neurosci 16:918811. https://doi.org/10.3389/fnins.2022.918811

Smith HL et al (2020) Astrocyte unfolded protein response induces a specific reactivity state that causes non-cell-autonomous neuronal degeneration. Neuron 105:855–866.e855. https://doi.org/10.1016/j.neuron.2019.12.014

Snow AD, Kisilevsky R, Willmer J, Prusiner SB, DeArmond SJ (1989) Sulfated glycosaminoglycans in amyloid plaques of prion diseases. Acta Neuropathol 77:337–342. https://doi.org/10.1007/bf00687367
Sorce S et al (2020) Genome-wide transcriptomics identifies an early preclinical signature of prion infection. PLoS Pathog 16:e1008653–e1008653. https://doi.org/10.1371/journal.ppat.1008653
Tahir W, Thapa S, Schatzl HM (2022) Astrocyte in prion disease: a double-edged sword. Neural Regen Res 17:1659–1665. https://doi.org/10.4103/1673-5374.332202
Takahashi T, Suzuki T (2012) Role of sulfatide in normal and pathological cells and tissues. J Lipid Res 53:1437–1450. https://doi.org/10.1194/jlr.R026682
Tamgüney G et al (2008) Genes contributing to prion pathogenesis. J Gen Virol 89:1777–1788. https://doi.org/10.1099/vir.0.2008/001255-0
Thompson AG et al (2013) The Medical Research Council prion disease rating scale: a new outcome measure for prion disease therapeutic trials developed and validated using systematic observational studies. Brain 136:1116–1127. https://doi.org/10.1093/brain/awt048
Vallabh SM et al (2023) Therapeutic trial of anle138b in mouse models of genetic prion disease. J Virol 97:e0167222. https://doi.org/10.1128/jvi.01672-22
Vanni S et al (2017) Differential overexpression of SERPINA3 in human prion diseases. Sci Rep 7:15637. https://doi.org/10.1038/s41598-017-15778-8
Wagner J et al (2013) Anle138b: a novel oligomer modulator for disease-modifying therapy of neurodegenerative diseases such as prion and Parkinson's disease. Acta Neuropathol 125:795–813. https://doi.org/10.1007/s00401-013-1114-9
Watts JC, Prusiner SB (2014) Mouse models for studying the formation and propagation of prions. J Biol Chem 289:19841–19849. https://doi.org/10.1074/jbc.R114.550707
Watts JC et al (2012) Spontaneous generation of rapidly transmissible prions in transgenic mice expressing wild-type bank vole prion protein. Proc Natl Acad Sci USA 109:3498–3503. https://doi.org/10.1073/pnas.1121556109
Watts JC et al (2016) Towards authentic transgenic mouse models of heritable PrP prion diseases. Acta Neuropathol 132:593–610. https://doi.org/10.1007/s00401-016-1585-6
Webb TE et al (2008) Phenotypic heterogeneity and genetic modification of P102L inherited prion disease in an international series. Brain 131:2632–2646. https://doi.org/10.1093/brain/awn202
Webb TE, Whittaker J, Collinge J, Mead S (2009) Age of onset and death in inherited prion disease are heritable. Am J Med Genet B Neuropsychiatr Genet 150b:496–501. https://doi.org/10.1002/ajmg.b.30844
Wingo AP et al (2021) Integrating human brain proteomes with genome-wide association data implicates new proteins in Alzheimer's disease pathogenesis. Nat Genet 53:143–146. https://doi.org/10.1038/s41588-020-00773-z
Xiang W et al (2004) Identification of differentially expressed genes in scrapie-infected mouse brains by using global gene expression technology. J Virol 78:11051–11060. https://doi.org/10.1128/jvi.78.20.11051-11060.2004

Chapter 11
Advancing Prion Diagnostics: RT-QuIC Applications in Peripheral Tissues

Merve Begüm Bacınoğlu, Giuseppe Bufano, Federico Angelo Cazzaniga, Gianluigi Zanusso, Giuseppe Legname, and Fabio Moda

Abstract Prion diseases (PrDs) are fatal neurodegenerative disorders characterized by the accumulation of misfolded prion protein (PrP^{Sc}) in the central nervous system (CNS). This pathological isoform of the cellular prion protein drives disease pathogenesis through its unique ability to propagate itself via a template-directed misfolding mechanism. The definite diagnosis of PrDs relies on the detection of PrP^{Sc} in the CNS by invasive procedures or *postmortem* examination, limiting early detection and *antemortem* diagnostic investigations. Real-time quaking-induced conversion (RT-QuIC) has emerged as a revolutionary diagnostic tool, allowing ultrasensitive detection of PrP^{Sc} in cerebrospinal fluid (CSF) and other easily accessible tissues, including the olfactory mucosa, skin, and, more recently, tears. This assay exploits the autocatalytic amplification of misfolded prions, providing high sensitivity and specificity in the detection of peripheral PrP^{Sc}. This chapter explores the advancements and applications of RT-QuIC in diagnosing human PrDs.

M. B. Bacınoğlu · G. Bufano · F. A. Cazzaniga
Unit of Laboratory Medicine, Laboratory of Clinical Pathology, Fondazione IRCCS Istituto Neurologico Carlo Besta, Milan, Italy
e-mail: merve.bacinoglu@istituto-besta.it; giuseppe.bufano@istituto-besta.it; federico.cazzaniga@istituto-besta.it

G. Zanusso
Department of Neuroscience, Biomedicine and Movement Sciences, Università degli Studi di Verona, Verona, Italy
e-mail: gianluigi.zanusso@univr.it

G. Legname
Department of Neuroscience, Scuola Internazionale Superiore di Studi Avanzati, Trieste, Italy
e-mail: legname@sissa.it

F. Moda (✉)
Unit of Laboratory Medicine, Laboratory of Clinical Pathology, Fondazione IRCCS Istituto Neurologico Carlo Besta, Milan, Italy

Department of Medical Biotechnology and Translational Medicine, Università degli Studi di Milano, Milan, Italy
e-mail: fabio.moda@unimi.it

G. Legname, F. Moda (eds.), *Biomarkers and Therapeutical Targets for Prion Diseases*, Subcellular Biochemistry 112, https://doi.org/10.1007/978-3-031-97055-9_11

Keywords Human prion diseases · Prion · Real-time quaking-induced conversion · Peripheral biomarkers

Human Prion Diseases

Prion diseases (PrDs), also known as transmissible spongiform encephalopathies (TSEs), are a group of rare, rapidly progressive, fatal, and transmissible neurodegenerative disorders that affect both humans and animals (Aguzzi and Heppner 2000). Despite their highly heterogeneous nature, both clinically and neuropathologically, PrDs share common characteristics such as a long incubation period, very short clinical course, spongiform changes, and most importantly, the deposition of a disease-associated transmissible agent in the CNS (Prusiner 1998; Knight and Collins 2001). This agent, known as prion (PrP^{Sc}), is the abnormally folded form of the physiological cellular prion protein (PrP^{C}). The pathological and infectious features of the disease originate from the conformational conversion (misfolding) of PrP^{C} into PrP^{Sc} (Stahl and Prusiner 1991). PrP^{C} is a glycophosphatidylinositol (GPI)-anchored glycoprotein highly expressed in the CNS, whose function is still yet to be elucidated (Stahl et al. 1987; Castle and Gill 2017). It is encoded by the *PRNP* gene, and once synthesized, undergoes a series of crucial post-translational modifications. These include the addition of a GPI anchor to its C-terminus, the formation of a disulfide bridge between two C-terminal cysteine residues (Cys179-Cys214), and the N-linked glycosylation of asparagine residues (Asn181-Asn197) (Turk et al. 1988; Dear et al. 2007). The different degrees of PrP^{C} glycosylation result in the formation of three distinct isoforms of the protein: diglycosylated, monoglycosylated, and unglycosylated (Oesch et al. 1985; Haraguchi et al. 1989; Hill et al. 2006). PrP^{C} is characterized by an abundance of α-helices structures, a short antiparallel β-strand in the C-terminal, solubility in detergents, and sensitivity to proteolytic digestion with proteinase K (PK) (Caughey et al. 1989; Rudd et al. 1999; Wüthrich and Riek 2001; Bate et al. 2016). In contrast, PrP^{Sc} has higher amounts of β-sheet structures, which confer reduced solubility in mild detergents and partial resistance to proteolytic digestion with PK (Stahl and Prusiner 1991). This resistance leads to the generation of a 141 amino acids long core, PrP^{res}, showing truncated fragments of the di-, mono- and unglycosylated PrP^{Sc} that can be observed via biochemical analyses, particularly via Western blot (Wb) (Oesch et al. 1985; Haraguchi et al. 1989; Hill et al. 2006).

The initial mechanism of PrP^{Sc} formation is still unknown, however, the spreading of prions is dependent on the presence of their native form (Morales et al. 2007). Studies have demonstrated that infecting PrP^{C} knock-out mice with prions does not induce PrP^{Sc} spreading and disease progression, whereas, in the presence of the native protein, PrP^{Sc} can act as a template for conversion and spreading (Büeler et al. 1992). This spreading is sustained by an autocatalytic mechanism, as PrP^{Sc} oligomers incorporate and convert other PrP^{C} monomers, leading to the growth in size and eventually generating fibrils (Scheckel and Aguzzi 2018). These fibrils fragment, creating new sites for further recruitment and conversion of PrP^C monomers, amplify the reaction and lead to their accumulation (Prusiner 1991; Jarrett and Lansbury 1992; Kupfer et al. 2009). Interestingly, PrP^{Sc} can also acquire different conformations, referred to as "strains" (Caughey 2003). Strains can significantly impact the characteristics of PrDs, shaping clinical symptoms and neuropathological changes, particularly the patterns of PrP^{Sc} deposition in the brain (Bruce 2003). Particularly, PrDs can be classified according to the affected species, means of transmission (e.g., sporadic, genetic, acquired), clinical symptoms, the *PRNP* gene polymorphisms, and PrP^{Sc} isoforms (Parchi et al. 1997, 2000). A brief overview of the genetic, sporadic, and acquired forms of PrDs is provided in the paragraphs below (Collinge 2001).

Sporadic Human PrDs

Sporadic PrDs (sPrDs) include Creutzfeldt–Jakob disease (sCJD), fatal insomnia (sFI), and variably protease-sensitive prionopathy (VPSPr). sCJD is the most common human PrDs, accounting for 85% of all cases (Watson et al. 2022). Currently, sCJD is further classified into six major clinicopathological phenotypes that correlate with the polymorphism methionine (M) or valine (V) at codon 129 of the *PRNP* gene and the biochemical properties of PrP^{Sc}, resulting in two "PrP^{Sc} types" following PK digestion. Based on (1) the presence of either M or V at codon 129 and (2) PrP^{Sc} type corresponding to type 1 (unglycosylated PrP band migrating at 21 kDa) and type 2 (unglycosylated PrP band migrating at 19 kDa) PrP^{Sc}, sCJD can be classified into MM1, MM2, MV1, MV2, VV1, or VV2 (Parchi et al. 1999). Additionally, the MM2 subtype is even further divided following its histopathological features: MM2-cortical, mainly affecting the cerebral cortex, or MM2-thalamic, mainly affecting the thalamus, also known as the sFI (Moda et al. 2012; Chen et al. 2023). The most recently recognized sporadic PrDs is VPSPr, a disease that shows peculiar properties of the aggregates, revealing, after Wb analysis, a distinctive ladder-like profile of at least 5 PK-resistant fragments in the 7–26 kDa range, which indicates a minimal or complete lack of the di-glycosylated form (Zou et al. 2010).

Genetic Human PrDs

Genetic PrDs (gPrDs) are caused by pathogenic variations in the *PRNP* gene, inherited in an autosomal dominant pattern. To date, more than 60 different mutations have been described, including missense mutations, insertions or deletions of an octapeptide region of the N-terminal domain, and nonsense mutations (Appleby et al. 2022). The gPrDs account for approximately 10–15% of PrDs, and the most common mutations include the E200K, D178N, V210I, and P102L with variable penetrance (Kovács et al. 2005; Minikel et al. 2016). The incidence differs across countries, and the frequency of these mutations can vary between different populations. The three main phenotypes are genetic CJD (gCJD), fatal familial insomnia (FFI), and Gerstmann-Sträussler-Scheinker syndrome (GSS). gCJD clinicopathological features closely resemble those of the sCJD, with the E200K mutation being the most common genetic cause. Interestingly, the disease-causing mutations correlate with distinct symptoms, age of onset, and disease duration and can be influenced by the M/V polymorphism at codon 129 (Mead 2006). The influence of polymorphism at codon 129 on disease phenotype is particularly important in the case of D178N mutation; depending on whether codon 129 encodes V or M, subjects with D178N mutation are associated with either gCJD or FFI (Goldfarb et al. 1992; Baiardi et al. 2021). FFI represents the genetic counterpart of sFI, with the differential diagnosis primarily based on genetic testing. The GSS is the first gPrDs linked to a *PRNP* mutation and is most commonly associated with the P102L mutation (Hsiao et al. 1989).

Acquired Human PrDs

Acquired PrDs differ from sporadic and genetic forms through their external mode of transmission, including iatrogenic CJD (iCJD), variant CJD (vCJD), and Kuru. The iCJD stems from human-to-human transmission through medical operations, and the first recognized case was linked to corneal transplants (Duffy et al. 1974; Bonda et al. 2016). Since then, two main sources of iCJD have been acknowledged with over 500 cases worldwide: human growth hormone treatment and human duramater graft from undiagnosed CJD cadavers, causing an epidemic peak in 1995 and 1997, respectively (Brown et al. 2012). Although iCJD cases are almost extinguished, individuals are still diagnosed with iCJD due to the abnormally extended incubation time, and factors like patients' age at the time of the operation, route of infection, and genetic background all contribute to this variability in disease onset and progression. On the other hand, vCJD is caused by the consumption of contaminated food derived from animals infected with bovine spongiform encephalopathy (BSE) or after receiving blood transfusions from vCJD-affected individuals, occupational exposure to PrDs biological samples, and susceptible subjects were MM at codon 129 (Trevitt and Singh 2003; Liberski et al. 2019; Brandel et al. 2020). However, active surveillance of animals has prevented humans from BSE exposure. Since 2017, no vCJD cases have been reported (except those from occupational exposure), albeit a single patient of MV at codon 129 was reported as not clinically distinguishable from sCJD (Mok et al. 2017). Finally, Kuru, the first PrD to be diagnosed, is associated with the ritualistic practice of cannibalism among the Fore linguistic group and neighboring communities in Papua New Guinea (Gajdusek and Zigas 1959). Noteworthy, the biochemical analyses reveal that iCJD, Kuru, and sCJD are all associated with a predominant monoglycosylated PrP^{res} glycoform ratio, whereas vCJD and FFI are associated with a predominant diglycosylated band, underscoring the molecular diversity of PrDs (Parchi et al. 1999; Head et al.

2004). Given the rapid clinical progression and the transmissible nature, the acquired PrDs continue to generate widespread public interest and an urgency for early and accurate diagnosis.

Diagnosis of PrDs

A definite diagnosis of PrDs still requires *postmortem* biochemical and neuropathological analysis to detect PrP^{Sc} aggregates in the CNS (Eraña et al. 2020). Clinical diagnosis of PrDs is often challenging as PrDs show high clinical heterogeneity at disease onset and have variable disease progressions, survival times, and overlapping symptoms with other neurodegenerative diseases (ONDs). Currently, the diagnostic criteria for CJD are based on the clinical characteristics combined with in vivo diagnostic tools, including electroencephalographic findings, brain magnetic resonance imaging (MRI), and cerebrospinal fluid (CSF) or blood biomarker analyses (National CJD Research & Surveillance Unit, Diagnostic Criteria for Creutzfeldt–Jakob Disease, published January 2017, https://www.cjd.ed.ac.uk/sites/default/files/criteria.pdf; Hermann et al. 2018, 2021). From a laboratory perspective, the surrogate protein biomarkers 14-3-3 are used to support the clinical diagnosis of PrDs as well as total tau, phosphorylated tau, and neurofilament light chain (Abu-Rumeileh et al. 2018). However, these proteins lack the specificity for PrDs and are frequently used for other neurodegenerative conditions (Lattanzio et al. 2017). For gPrDs, the genetic analysis of the *PRNP* gene allows the diagnosis and categorization; nevertheless, the symptom onset is still unpredictable (Baiardi et al. 2023). Currently, PrP^{Sc} represents the only disease-specific biomarker for PrDs. The detection of PrP^{Sc} in peripheral tissues has been the focus of prion research for decades; however, the conventional techniques struggled to detect PrP^{Sc} *antemortem* due to its minute concentrations. In 2001, Soto and colleagues developed the protein misfolding cyclic amplification (PMCA) technique, which mimics in vitro the phenomenon of prion conversion that occurs in vivo. PMCA allows the detection of minute amounts of PrP^{Sc} (Saborio et al. 2001). This technique involves incubating a PrP^{C}-enriched substrate, typically brain homogenates (BHs), with samples that may contain even trace amounts of PrP^{Sc}. During incubation, PrP^{Sc} induces the conversion of PrP^{C} into PrP^{Sc}, leading to its aggregation. A subsequent step of sonication disrupts aggregates into smaller fragments, promoting further conversion of PrP^{C} to PrP^{Sc}. These two steps alternate cyclically, and at the end of the reaction, the amplified material can be diluted in a new substrate and subjected to further amplification cycles, referred to as a round of PMCA (Saborio et al. 2001; Cazzaniga et al. 2020; Giaccone and Moda 2020). After each round, samples are collected and analyzed by Wb after PK digestion (Castilla et al. 2004). However, a few years later, a test like PMCA, but easier to handle and not requiring the same biosafety procedures as PMCA, has been developed. This assay is called real-time quaking-induced conversion (RT-QuIC) and is described below. The operating principle is similar to that of PMCA but methodologically different.

The RT-QuIC Assay

In 2011, Atarashi and colleagues developed the RT-QuIC as an ultrasensitive assay for prion detection (Atarashi et al. 2011a, b). The technique has proven its remarkable utility in clinical applications, and in 2017, RT-QuIC analysis of CSF or other tissues officially entered the guidelines for the diagnosis of sCJD (National CJD Research & Surveillance Unit, Diagnostic Criteria for Creutzfeldt-Jakob Disease, published January 2017, https://www.cjd.ed.ac.uk/sites/default/files/criteria.pdf; Hermann et al. 2018, 2021). In detail, this in vitro technique uses bacterially produced recombinant PrP^C (recPrP) as a reaction substrate. For the analysis of human samples, the first-generation RT-QuIC (PQ-RT-QuIC) employed the full-length recPrP with the amino acid sequence of human ($recHu_{(23-231)}$) or hamster ($recSHa_{(23-231)}$) (Atarashi et al. 2011b; McGuire et al. 2012). However, it was later replaced by the second-generation RT-QuIC (IQ-RT-QuIC), where an N-terminally truncated $recSHa_{(90-231)}$ was used and significantly improved the sensitivity and reliability of the test in detecting PrP^{Sc} in a variety of human tissue samples (Atarashi et al. 2011b; McGuire et al. 2012; Orrú et al. 2015; Poleggi et al. 2022). Generally, the recPrP is diluted in a reaction mix containing salts, a fluorescent dye called Thioflavin T (ThT), pH stabilizers, and occasionally detergents to maintain protein stability and enhance the assay sensitivity (Orrú et al. 2020; Da Silva Correia et al. 2023). When a biological sample containing even minute amounts of PrP^{Sc} is added to the reaction mix, it promotes the aggregation of recPrP into amyloid-like structures (Saborio et al. 2001). This aggregation process, driven by the misfolding of recPrP, is monitored in real-time using ThT. ThT binds into amyloid aggregates, emitting a fluorescent signal that correlates with the formation of fibrils. The reaction occurs in cyclic incubation steps with intermittent shaking to favor the interaction between the PrP^{Sc} and recPrP (Schmitz et al. 2016). Samples are tested in 96-well plates, at least in quadruplicate, alongside positive and negative controls, at a controlled temperature. The assay is performed in a dedicated microplate reader. Fluorescence values are recorded in real-time at regular intervals (typically every 15 min) and then plotted on a graph, with time represented on the x-axis and fluorescence values on the y-axis. The graph displays the mean fluorescence values of the four replicates, resulting in a sigmoidal-shaped curve that shows three distinct phases: (1) a lag phase where the PrP^{Sc} starts to interact with recPrP, initiating the misfolding process (no/very low fluorescence emission); (2) an exponential growth phase where misfolded recPrP aggregates into oligomers and small fibrils (rapid fluorescence emission); and (3) a plateau phase where most of recPrP is aggregated into fibrils (maximum ThT fluorescence) (McGuire et al. 2012; Rossi et al. 2020). The technology is exceptionally sensitive, capable of detecting PrP^{Sc} at concentrations as low as attograms within a short timeframe (Eraña et al. 2020). Over the years, various forms of recPrP, including those with human, bank vole, and chimeric sheep-hamster amino acid sequences, have been tested. However, compared to the hamster protein, all showed a tendency toward self-aggregation, making the interpretation of the results more challenging (Cramm et al. 2016; Mok et al. 2021). Due to its versatility, reliability, and robustness, RT-QuIC is now widely used in specialized laboratories worldwide for research, diagnostics, and surveillance. Its applications extend beyond human prion diseases to include animal prion diseases as well (Atarashi et al. 2007; Dassanayake et al. 2016; Bistaffa et al. 2019; Haley et al. 2020; Harpaz et al. 2024). As summarized in Table 11.1 and detailed below, studies involving large cohorts of human control samples have consistently reported specificities close to 100%. Although false-positive results are extremely rare, some of the control samples that resulted in positive were not neuropathologically confirmed cases (Green 2019). In case of false-negative results, a retesting of diluted CSF samples may be useful to raise the sensitivity of the assay (Fiorini et al. 2020). The updated diagnostic criteria, which include RT-QuIC analysis of biological samples, have significantly improved the sensitivity for detecting sCJD while maintaining the same level of specificity (Watson et al. 2022). Definite diagnosis of PrDs still requires *postmortem* examination of the brain. It is important to note that technical specifications and details regarding the production methods for each recPrP substrate are not always provided (Hermann et al. 2018, 2021). International studies have demonstrated a high degree of inter-laboratory consistency, indicating that variations in recPrP production protocols have a minimal impact on the diagnostic performance of the assay (Cramm et al. 2016; McGuire et al. 2016).

RT-QuIC Applications for Peripheral Prion Detection

CSF has long been the primary sample analyzed using RT-QuIC due to its invariable implication in CNS disorders, and it remains the ideal fluid for excluding treatable conditions or for confirming the diagnosis in patients with a clinical suspicion of PrDs (Zanusso et al. 2003). Recent advancements in RT-QuIC have expanded its application to a variety of peripheral tissues. This shift toward peripheral tissue samples offers the potential for less invasive diagnostic options and the possibility of earlier disease detection. Several studies have investigated the use of olfactory mucosa (OM) and skin biopsies for RT-QuIC analysis, motivated by the observed accumulation of PrP^{Sc} in the olfactory epithelium and skin tissue of sCJD patients (Tabaton et al. 2004). More recently, the group led by Inga Zerr showed the ability to detect PrP^{Sc} in the tears of patients with PrDs (Schmitz et al. 2023). The following sections will provide an overview of the current RT-QuIC applications, their limitations, and the challenges that remain, organized by tissue type.

Cerebrospinal Fluid

In 2011, PQ-RT-QuIC was optimized to reliably detect PrP^{Sc} in CSF (PQ-CSF) from various PrDs patients, using the $recHu_{(23-231)}$ as the reaction substrates (Atarashi et al. 2011a; Hermann et al. 2021). In the pilot study, CSF samples from two

Table 11.1 Summary of the CSF RT-QuIC assay diagnostic performances

Reference	Year	Generation of the assay	Substrate of reaction	PrDs (n. cases)	Controls (n. cases)	Sensitivity (%)	Specificity (%)
Atarashi et al. (2011b)	2011	PQ-CSF	$recHu_{(23–231)}$	CJD (18)	CTRL (35)	83.3	100.0
				CJD (16)	OND (144)	87.5	100.0
McGuire et al. (2012)	2012	PQ-CSF	$recSHa_{(23–231)}$	sCJD (123)	Non-CJD (103)	89.0	99.0
Sano et al. (2013)	2013	PQ-CSF	$recHu_{(23–231)}$	GSS (20), FFI (12), gCJD-E200K (22), gCJD-V203I (2)	CTRL (1)	78.0–100.0	100.0
Orrú et al. (2014)	2014	PQ-CSF	$recSHa_{(23–231)}$	sCJD (28), gCJD-E200K (2)	Non-CJD (46)	77.0	100.0
Orrú et al. (2015)	2015	IQ-CSF	$recSHa_{(90–231)}$	sCJD (48)	Non-CJD (39)	95.8	100.0
Cramm et al. (2016)	2016	PQ-CSF	recHaSh	sCJD (64), gCJD-E200K (33), gCJD-V210I (6), FFI (7)	CTRL (400)	85.0	99.0
Groveman et al. (2017)	2016	PQ-CSF	$recSHa_{(23–231)}$	CJD (113)	Non-PrDs (64)	73.0	100.0
		IQ-CSF	$recSHa_{(90–231)}$			94.0	100.0
Park et al. (2016)	2016	PQ-CSF	$recHu_{(24–234)}$	sCJD (81)	Non-CJD (100)	76.5	100.0
Lattanzio et al. (2017)	2017	PQ-CSF	$recSHa_{(23–231)}$	sCJD (305), VPSPr (1), gCJD (46)	Non-CJD (348)	82.1	99.4
Franceschini et al. (2017)	2017	PQ-CSF	$recSHa_{(23–231)}$	CJD (199), gCJD (41)	non-CJD (100)	–	–
		IQ-CSF	$recSHa_{(90–231)}$			97.2	100.0
Bongianni et al. (2017)	2017	PQ-CSF	$recSHa_{(23–231)}$	sCJD (61), gCJD (8)	Non-CJD (17)	71.4	100.0
		IQ-CSF	$recSHa_{(90–231)}$			82.6	100.0
Foutz et al. (2017)	2017	IQ-CSF	$recSHa_{(90–231)}$	CJD (191)	RPD (81)	92.0–95.0	99.0–100.0
Hermann et al. (2018)	2018	PQ-CSF	recHaSh	sCJD (65)	Non-CJD (118)	89.0	100.0
Abu-Rumeileh et al. (2019)	2019	PQ-CSF	$recSHa_{(23–231)}$	sCJD (80), VPSPr (1), gCJD (22)	Non-CJD (109)	82.5	100.0
		IQ-CSF	$recSHa_{(90–231)}$	sCJD (62), VPSPr (1), gCJD (15)	Non-CJD (53)	97.4	100.0
Fiorini et al. (2020)	2020	IQ-CSF	$recSHa_{(90–231)}$	sCJD (102)	Non-CJD (80)	96.0	100.0
Llorens et al. (2020)	2020	PQ-CSF	recHaSh	iCJD (23)	–	85.7	–

Rhoads et al. (2020)	2020	IQ-CSF	recSHa$_{(90–231)}$	sCJD (439), sFI (5), VPSPr (3), gCJD (31), FFI (4), GSS (3)	CTRL (69)	90.3	98.5
Mammana et al. (2020)	2020	PQ-CSF	recSHa$_{(23–231)}$	sCJD (24), gCJD (3)	Non-CJD (12)	89.0	100.0
Orrú et al. (2020)	2021	IQ-CSF	recSHa$_{(90–231)}$	sCJD (55)	CTRL (45)	Concordance: 96–100%	
Mok et al. (2021)	2021	PQ-IQ-CSF	recBv$_{(23–231)}$	sCJD (79), vCJD (2), gPrDs (20)	Non-PrD (57)	88.6	91.2
Hermann et al. (2023)	2023	PQ-CSF	recHaSh	sCJD (888)	Non-PrD (371)	90.2	98.7
Baranová et al. (2024)	2024	IQ-CSF	recSHa$_{(90–231)}$	sCJD (33), VPSPr (1), gCJD-E200K (2), GSS (2)	CTRL (30)	100.0	100.0

PrDs prion diseases, *PQ-CSF* first generation of RT-QuIC assay using CSF samples, *recHu$_{(23–231)}$* recombinant full-length human prion protein [23–231], *CJD* Creutzfeldt-Jakob disease, *CTRL* control group, *OND* other neurodegenerative disorders, *recSHa$_{(23–231)}$* recombinant full-length Syrian hamster prion protein [23–231], *sCJD* sporadic Creutzfeldt–Jakob disease, *Non-CJD* individual/patient not affected by Creutzfeldt–Jakob disease, *GSS* Gerstmann-Sträussler-Scheinker disease, *FFI* fatal familial insomnia, *gCJD-E200K E200K* genetic Creutzfeldt–Jakob disease, *gCJD-V203I* V203I genetic Creutzfeldt–Jakob disease, *IQ-CSF* second generation of RT-QuIC assay using CSF samples, *recSHa$_{(90–231)}$* recombinant truncated Syrian hamster prion protein [90–231], *recHaSh*recombinant hamster-sheep chimeric prion protein, *gCJD-V210I* V210I genetic Creutzfeldt–Jakob disease, *Non-PrDs* individual/patient not affected by prion diseases, *recHu$_{(24–234)}$* recombinant full-length human prion protein [24–234], *VPSPr* variably protease-sensitive prionopathy, *gCJD* genetic Creutzfeldt–Jakob disease, *RPD* rapidly progressive dementia, *iCJD* iatrogenic Creutzfeldt–Jakob disease, *sFI* sporadic fatal insomnia, *recBv$_{(23–231)}$*:recombinant full-length bank vole prion protein [23–231] international cohorts including 18 confirmed cases of CJD and 35 controls with ONDs were tested, achieving a remarkable sensitivity of 83.3% and a specificity of 100% (Atarashi et al. 2011a). To validate the reliability of the assay, a blinded study was performed using CSF samples from 16 confirmed cases, 25 probable cases of CJD, and 144 cases of ONDs from Japan and Australia. None of the samples from OND patients tested positive, further demonstrating the robustness of the assay, with a sensitivity once again exceeding 80% (Atarashi et al. 2011b). A year later, McGuire and colleagues performed RT-QuIC assays using the recSHa$_{(23–231)}$. As in the earlier study, an exploratory and a confirmatory cohort were analyzed, comprising a total of 123 neuropathologically confirmed sCJD cases and 103 controls. The modified assay detected PrP^{Sc} in 109 of the 123 CSF samples, achieving an improved sensitivity of 89% while obtaining a specificity of 99% (McGuire et al. 2012). Intriguingly, when PQ-CSF was applied to a cohort of gPrDs (n = 56) for the first time, Sano and colleagues reported high sensitivities for GSS (78%), FFI (100%), gCJD-E200K (87%), and gCJD-V203I (100%), including a control sample, yielding a specificity of 100% (Sano et al. 2013). Subsequent studies further confirmed the feasibility of using both full-length recHu$_{(23–231)}$ and recSHa$_{(23–231)}$ in RT-QuIC assays. Between 2014 and 2017, multiple research groups reported sensitivities ranging from 73% to 77% and specificities of 100% across different cohorts of patients with sCJD (n = 222 in total) and non-sCJD controls (n = 210 in total), as indicated in Table 11.1 (Orrú et al. 2014; Park et al. 2016; Groveman et al. 2017). Notably, within these cohorts, two samples initially classified as positive were later identified as cases of gCJD caused by E200K mutation in the *PRNP* gene (gCJD-E200K). Additionally, one patient with gCJD-V180I and another with GSS syndrome consistently yielded negative results. In the same period, Lattanzio and colleagues analyzed 700 CSF samples, including definite, probable, and possible cases of CJD, as well as 46 cases of gCJD. At the chosen threshold, the sCJD results indicated a promising diagnostic utility, obtaining similar sensitivities for definite sCJD (82.7%), probable sCJD (79.4%), and possible sCJD (75.9%). Additionally, the results for definite gCJD resulted in 91.3% positivity, and notably, for the first time, one definite case of VPSPr was included, which tested negative. This study also showed that the sensitivity of PQ-CSF is unaffected by the *PRNP* codon 129 genotype alone but decreases in MM2 and VV2 subtypes when combined with the PrP^{res} glycotype (Lattanzio et al. 2017). Reaction kinetics, including lag phase and fluorescence intensity, were indicated to be consistent across sCJD subtypes, supporting earlier findings that the assay kinetics do not vary significantly among subtypes, except for faster reactions in MM1 compared to MV1 and VV1 (Cramm et al. 2016). Although the PQ-CSF protocol represented a major advancement, its sensitivity and reliability were not without limitations, triggering the development of the mentioned "second generation," also referred to as IQ-CSF assays, which demonstrated faster results and an improved sensitivity (94%) for CJD compared to PQ-CSF (73%), while maintaining the high specificity (Groveman et al. 2017).

Further studies confirmed these findings, showing that the optimized protocol was able to identify 81% of definite CJD cases that had previously tested negative by PQ-CSF. Interestingly, the IQ-CSF also allowed the detection of PrP^{Sc} in rare PrDs phenotypes, including VPSPr (100% sensitivity) (Franceschini et al. 2017). Similarly, Foutz and colleagues obtained high sensitivity and specificity in a large retrospective (92% and 98.5%, respectively) and prospective (95% and 100%, respectively) analysis of sCJD samples (Foutz et al. 2017). This study also revealed that the sensitivity varied according to the sCJD subtype, with the MM1, MV1, and VV2 subtypes having sensitivities of 96%, 100%, and 100%, respectively, while the VV1 and MM2-cortical subtypes exhibited lower sensitivities of 75% and 71%. Additionally, the study achieved a 95% probability of distinguishing aggressive MM1 forms from slower-progressing MM2 and an 80% probability of differentiating aggressive VV2 from slower-progressing VV1 by integrating IQ-CSF analyses findings with *PRNP* gene sequencing (Foutz et al. 2017). Two subsequent studies confirmed the different efficiency of the IQ-CSF assay in detecting prions across different sCJD subtypes. Notably, a ring trial specifically validated its reliability in identifying cases of MM1, MV1, and VV2 subtypes. Compared to PQ-CSF, the IQ-CSF protocol resulted in low sensitivity in detecting certain PrDs, including GSS (P102L) and FFI. Instead, the CSF of gCJD subjects associated with either E200K-129 M or V210I-129 M presented a sensitivity ranging from 81.8% to 100% (Foutz et al. 2017; Franceschini et al. 2017; Lattanzio et al. 2017). Recently, Vallabh and colleagues studied 22 subjects with the E200K mutation, and the detection of PrP^{Sc} was possible in the CSF (1–3 years before onset) of three individuals who subsequently converted to symptomatic and died of PrDs. This suggested, for the first time, that prion seeding activity in the CSF might be one of the earliest findings of disease onset in E200K carriers (Vallabh et al. 2024). Since 2016, four studies have explored two distinctive reaction substrates: (1) the recombinant Syrian hamster (residues 14–128)—sheep (residues 141–234 of the R_{154} Q_{171} polymorph) chimeric PrP and (2) recombinant full-length bank vole PrP (recHaSh and $recBv_{(23–231)}$, respectively). This also allowed the application of RT-QuIC to a variety of other PrDs beyond the sCJD and gCJD. In particular, the recHaSh demonstrated optimal sensitivity for detecting prion in the CSF of patients with acquired PrDs (i.e., iCJD 87.5%) (Llorens et al. 2020), while the $recBv_{(23–231)}$ was highly susceptible to a wide range of prion strains, including GSS and FFI. In the case of acquired CJD, the group led by A. Green described that approximately 67% of patients with iCJD following a cadaveric growth hormone administration were positive for CSF RT-QuIC (Green 2019). Although the sensitivity of the assay was not as high as that for sCJD, it still provided valuable diagnostic information for iCJD. For vCJD, PQ-CSF failed to amplify prions from CSF samples, suggesting that alternative substrates may be necessary for accurate diagnosis (Orrú et al. 2015). Indeed, a recent work using $recBv_{(23–231)}$ has demonstrated the ability to detect prions in vCJD brain tissue, raising the possibility that further modifications to the RT-QuIC assay may improve its utility in diagnosing vCJD and other PrDs using CSF samples (Mok et al. 2021). These findings demonstrate the significant advancement of sCJD diagnosis across different PrDs subtypes through RT-QuIC (Watson et al. 2022).

In recent years, several leading research groups in the RT-QuIC field have identified various pre-analytical factors that can affect assay performance. Specifically, the concentration of certain components in CSF has been highlighted: (i) elevated white cell counts (>10 × 10^6/L) can lead to false-positive results, and (ii) high red cell counts (>1250 cells/μL) may cause potentially false-negative outcomes. Additionally, studies using brain homogenate samples have suggested that higher concentrations of polar lipids can inhibit amyloid formation in the RT-QuIC reaction (Hoover et al. 2017). To address these challenges and minimize the impact of unknown inhibitory factors, a recent study has proposed a 10X dilution of CSF samples as a potential solution (Baranová et al. 2024).

Olfactory Mucosa

PrDs are known to affect specific brain regions, including the olfactory pathways, due to the neural connection between the olfactory bulb and other brain areas. Early in disease progression, prion aggregates can accumulate in the OM neuroepithelium of sCJD patients, making it a suitable site to investigate the presence of PrP^{Sc} (Zanusso et al. 2003). Driven by these considerations, in 2014, Orrú and colleagues introduced an optimized version of RT-QuIC for CJD diagnosis, utilizing nasal brushings to collect samples from the OM neuroepithelium of patients. The OM is located in the upper nasal cavity and can be accessed using noninvasive methods like nasal brushings or swabs. In the work of Orrú, a total of 31 OM samples collected from patients with a probable ($n = 14$) or definite ($n = 15$) diagnosis of sCJD, gCJD ($n = 2$), and ONDs ($n = 43$) used as negative controls were tested. The assay was performed using $recSHa_{23–231}$ as the reaction substrate, and for the first time, this test detected the presence of PrP^{Sc} in 30 out of 31 patients, with a remarkable sensitivity of 97% and 100% specificity, outperforming the traditional CSF testing, as summarized in Tables 11.1 and 11.2 (Orrú et al. 2014). The diagnostic accuracy of the OM assay was assessed in 2020 by analyzing in parallel CSF ($n = 182$) and OM ($n = 42$) from suspected sCJD patients and from those presenting rapidly progressive dementia (RPD). Among these, 102 patients were then diagnosed with probable or definite sCJD. Notably, by combining RT-QuIC testing of both CSF and OM samples, they achieved 100% sensitivity and specificity, highlighting the assay's superior diagnostic accuracy in identifying sCJD among patients with RPD (Fiorini et al. 2020). A similar study was carried out using $recSHa_{90–231}$ as a substrate, involving 86 patients with clinical diagnoses of probable, possible, or suspected sCJD. Again, both CSF and OM samples from patients with sCJD ($n = 61$), gCJD ($n = 6$), GSS ($n = 2$), and ONDs ($n = 50$) were analyzed to compare the assay efficiency (Bongianni et al. 2017). Moreover, they introduced a soft flocked swab for OM sampling, replacing the previously used brush. This improved the practicality of OM collection and achieved a diagnostic sensitivity of 97% and specificity of 100% for sCJD cases. In combination with the results of CSF analyses, the sensitivity and specificity reached 100%, suggesting once again that combining the RT-QuIC tests of the two biological samples offers a higher and more accurate approach for the diagnosis of sCJD (Bongianni et al. 2017). In the same year, Moda and colleagues tested OM samples collected from two patients carrying the *D178N* mutation in the *PRNP* gene, showing FFI clinical signs from 4 and 10 months at the moment of OM collection, and 26 OM samples from either patients with ONDs or healthy controls. For the first time, through RT-QuIC, the detection of PrP^{Sc} in the OM of FFI patients was possible, using $recBv_{90–231}$ as substrate. The results were also confirmed by analyzing the same OM samples with the PMCA technique (Redaelli et al. 2017). Despite the different substrates used over the years, the second-generation RT-QuIC has been proven to be the best tool in clinical use for OM analyses. A 2020 multi-center study involving different laboratories evaluated the reproducibility of second-generation RT-QuIC assays for diagnosing sCJD. Researchers analyzed CSF with OM brushings from 9 sCJD cases and 19 controls that were initially tested blind by a coordinating laboratory and then sent to the others. Blinded testing across the six laboratories revealed a 98–100% concordance rate in detecting the presence of prions with RT-QuIC in both samples. These findings indicate that the second-generation RT-QuIC assay on CSF samples and on OM samples is highly transferable, reproducible, and robust for sCJD diagnosis in clinical practice (Orrú et al. 2020).

Skin

Starting from the pioneering research of Notari and colleagues that describes the presence of PrP^{Sc} in skin biopsy from a vCJD patient, the potential of skin biopsies as biological samples able to induce prion seeding activity with the use of RT-QuIC was investigated (Notari et al. 2010). Researchers collected *postmortem* specimens of skin from sCJD ($n = 21$) and vCJD ($n = 1$) patients and non-CJD individuals ($n = 15$) as controls. Employing the first-generation RT-QuIC assay, using either the $recSHa_{23–231}$ or the $recBV_{23–230}$ as reaction substrate, they detected prion seeding activity in skin samples with a sensitivity and specificity of 100% (Table 11.2), at levels approximately 1,000–100,000 times lower than in corresponding brain tissues. Furthermore, when skin homogenates from sCJD patients were inoculated into transgenic mice expressing human PrP^{C}, all mice developed prion disease within 564 days, demonstrating the infectivity of the skin-derived prions (Orrú et al. 2017). These findings combined indicated that sCJD skin contains both prion seeding activity and infectivity, raising concerns about the potential iatrogenic transmission through surgical procedures (Orrú et al. 2017). Later in 2020, Mammana and colleagues employed the same RT-QuIC assay with the same substrates to analyze 71 skin punch biopsy samples collected from sCJD patients ($n = 35$), five assessed *ante-* and the others *postmortem*, and non-CJD patients ($n = 37$) as controls. The

Table 11.2 Summary of the sensitivities and specificities of RT-QuIC assays using peripheral tissues and fluids for PrDs' diagnosis

	Reference	Year	Substrate of reaction	PrDs (n. cases)	Controls (n. cases)	Sensitivity (%)	Specificity (%)
Olfactory mucosa	Orrú et al. (2014)	2014	recSHa$_{(23–231)}$	sCJD (29), gCJD-E200K (2)	Non-CJD (43)	97.0	100.0
	Bongianni et al. (2017)	2017	recSHa$_{(90–231)}$	sCJD (61), gCJD (6), GSS (2)	CTRL (50)	≈92.0	100.0
	Redaelli et al. (2017)	2017	recBv$_{(90–231)}$	FFI (2)	OND (16), HC (10)	100.0	100.0
	Fiorini et al. (2020)	2020	recSHa$_{(23–231)}$	sCJD (35)	CTRL (7)	91.5	100.0
	Orrú et al. (2020)	2020	recSHa$_{(90–231)}$	sCJD (9)	CTRL (19)	89.0–100.0	95.0–100.0
Skin	Orrú et al. (2017)	2017	recBv$_{(23–230)}$	sCJD (21), vCJD (1)	Non-CJD (15)	100.0	100.0
	Mammana et al. (2020)	2020	recSHa$_{(23–231)}$	sCJD (32), gCJD (3)	Non-CJD (37)	68.6	100.0
			recBv$_{(23–230)}$			88.6	100.0
	Zhang et al. (2024)	2024	recSHa$_{(90–231)}$	sCJD (225), gCJD (11), GSS (3), iCJD (1), sFI (1), FFI (1), VPSPr (3)	Non-CJD (94)	≈90.3	≈98.0
			recBv$_{(90–231)}$			87.3	94.7
	Baranová et al. (2024)	2024	recSHa$_{(90–231)}$	sCJD (33), VPSPr (1), gCJD-E200K (2), GSS (2)	CTRL (30)	89.5	100.0
	Chen et al. (2024)	2024	recSHa$_{(90–231)}$	sCJD (86), gCJD (13), FFI (2)	Non-PrDs (23)	79.1	95.7
Eye	Orrù et al. (2018)	2018	recSHa$_{(90–230)}$	sCJD (11)	Non-sCJD (6)	100.0	100.0
Tears	Schmitz et al. (2023)	2023	recE200K	sCJD (9), gCJD (4), mutation carriers (5)	CTRL (26)	84.0	100.0

(continued)

Table 11.2 (continued)

	Reference	Year	Substrate of reaction	PrDs (n. cases)	Controls (n. cases)	Sensitivity (%)	Specificity (%)
DS	Satoh et al. (2019)	2019	–	sCJD (4), gCJD-E200K (1), GSS (1)	–	–	–
PN	Baiardi et al. (2019)	2019	recSHa$_{(90–231)}$	sCJD (12)	Non-CJD (2)	100.0	100.0

PrDs prion diseases, *recSHa$_{(23–231)}$* recombinant full-length Syrian hamster prion protein [23–231], *sCJD* sporadic Creutzfeldt–Jakob disease, *gCJD-E200K E200K* genetic Creutzfeldt–Jakob disease, *Non-CJD* individual/patient not affected by Creutzfeldt–Jakob disease, *recSHa$_{(90–231)}$* recombinant truncated Syrian hamster prion protein [90–231], *sCJD* sporadic Creutzfeldt–Jakob disease, *gCJD* genetic Creutzfeldt-Jakob disease, *GSS* Gerstmann-Sträussler-Scheinker disease, *CTRL* control group, *recBv$_{(90–231)}$:* recombinant truncated bank vole prion protein [90–231], *FFI* fatal familial insomnia, *OND* other neurodegenerative disorders, *HC* health control, *recBv$_{(23–231)}$:* recombinant full-length bank vole prion protein [23–231], *recBv$_{(90–231)}$:* recombinant truncated bank vole prion protein [90–231], *vCJD* variant Creutzfeldt-Jakob disease, *iCJD* iatrogenic Creutzfeldt-Jakob disease, *sFI* sporadic fatal insomnia, *VPSPr* variably protease-sensitive prionopathy, *Non-PrDs* individual/patient not affected by prion diseases, *Non-sCJD* individual/patient not affected by sporadic Creutzfeldt-Jakob disease, *recE200K* recombinant human prion protein with E200K mutation, *mutation carriers* patient with a mutation in the *PRNP* gene

study reported a sensitivity of 89% and specificity of 100% for the RT-QuIC assay in diagnosing sCJD, pointing out that skin-based RT-QuIC testing could serve as a new, reliable, minimally invasive diagnostic tool (Mammana et al. 2020). Zhang et al., in 2024, performed a large-scale study analyzing a total of 875 skin samples collected from 339 neuropathologically confirmed PrDs cases across two cohorts, sampling 2–3 body areas per individual. The RT-QuIC analyses of these skin samples confirmed that their seeding activity is a viable biomarker for PrDs detection, with a sensitivity for the first cohort ranging between 87.3% and 91.3% and specificity between 94.7% and 100%, while in the second cohort a sensitivity and specificity of 89.4% and 95.5%, respectively. Notably, the sensitivities were subtype-dependent, the highest being sCJD VV1–2 subtype, followed by VV2, MV1/2, MV1, MV2, MM1, MM1/2, MM2, and VV1. The effectiveness of skin seeding activity has therefore proven to be influenced by factors such as the biochemical characteristics of the prion strain, the polymorphism at codon 129 of the *PRNP* gene, the specific body area sampled, and the duration of the disease (Zhang et al. 2024). Similarly, the group of Baranová et al. tried to detect a seeding activity in matched *postmortem* CSF and skin samples from PrDs ($n = 38$) patients and controls ($n = 30$). Skin sample analysis yielded a sensitivity of 89.5% and a 100% specificity. Interestingly, the median seeding dose in the skin was an order of magnitude higher than in CSF, despite the overall lower fluorescence signals. These results overall reinforced the idea of using skin biopsy samples in *antemortem* diagnostics (Baranová et al. 2024). Ultimately, another study conducted by Chen et al. aimed at analyzing samples from PrDs (n = 101) and non-PrDs (n = 23) patients and indicated that single-site skin biopsies had a sensitivity comparable to the one obtained by CSF analysis (79.1%). Combining the analysis of two or more skin biopsy sites, instead, the diagnostic sensitivity was significantly improved up to 95%, higher than the one obtained by CSF while maintaining the same specificity. Therefore, a multisite skin biopsy approach may offer a superior, minimally invasive, and alternative approach, especially in those cases where CSF analysis is inconclusive or lumbar puncture is not feasible (Chen et al. 2024).

Other Peripheral Tissues

The possibility of detecting prions in peripheral tissues extended the assay's applicability to all tissues hypothesized to be important in and/or affected by PrDs pathogenesis (Table 11.2). In a 2018 study, Orrú and colleagues explored prion accumulation in the eye using the second-generation RT-QuIC assay. This included a comprehensive range of ocular tissues, including the cornea, lens, ocular fluid, retina, choroid, sclera, optic nerve, and extraocular muscle. Particularly, the iCJD related to ocular transplantation and the visual or oculomotor symptoms of sCJD patients underscored a possible preventive and diagnostic utility. Indeed, the RT-QuIC analyses of *postmortem* sCJD samples (n = 11) resulted in a remarkable 100% sensitivity and 100% specificity. The prion aggregates were highly visible in the posterior retina and potentially could be exploited for an early diagnosis (Orrù et al. 2018). In fact, another study in 2023 demonstrated the presence of prions in tear samples from patients with sCJD (n = 9), gCJD (n = 4), and asymptomatic *PRNP* mutation carriers (n = 5). Even though the cohort included asymptomatic individuals, the assay resulted in 84% sensitivity for both sporadic and genetic forms. Additionally, a recombinant human PrP protein with E200K mutation (recE200K) has been used as the reaction substrate while sustaining a 100% specificity. As tears are noninvasive and easily collectible fluids, they offer unique advantages for routine diagnostic applications. They require no specialized handling protocols, can be collected routinely, and can be easily incorporated into large-scale sample collections (Schmitz et al. 2023). As another possible application of RT-QuIC, the analyses of *postmortem* collected digestive system tissues from PrDs patients revealed significant prion accumulation in non-neuronal organs, including the liver and kidney. Importantly, the analyses of esophagus samples demonstrated diagnostic potential, underlining the need to incorporate prion-specific safety measures into endoscopic procedures (Satoh et al. 2019). In the same year, *postmortem* samples from the peripheral nervous system of various PrDs patients provided insights into the phenotypic diversity of sCJD and its relationship with prion strains. Findings suggest that peripheral neuropathy in sCJD is influenced by prion strain characteristics, with a markedly lower prevalence in the typical MM(V)1 subtype compared to VV2 and MV2K subtypes (Baiardi et al. 2019). These observations highlight the complexity of sCJD pathogenesis and its strain-dependent variability, offering new avenues for understanding and diagnosing PrDs. Last, blood is tested as a crucial peripheral sample for diagnostic testing due to its noninvasive nature, its suitability for large-scale screening within healthcare systems, and its potential role in preventing secondary transmission of diseases due to blood transfusions. However, despite its promise, RT-QuIC analyses of blood samples from PrDs patients remain a significant challenge. In 2011, the RT-QuIC assay was integrated with immunoprecipitation, creating the enhanced QuIC (eQuIC) assay to facilitate the capture of prions from blood samples of sCJD patients. While eQuIC demonstrated remarkable sensitivity in detecting vCJD brain-derived spikes in human plasma (Orrú et al. 2011), it faced limitations in identifying prions in blood samples from patients with sPrDs (Nonaka et al. 2024). Meanwhile, in 2023, Thomas and colleagues indicated a sensitive detection of prion accumulation in vCJD-spiked sheep blood up to two years before clinical onset, leveraging iron oxide bead capture for enhanced sensitivity (Thomas et al. 2023). These findings underscore the promise of RT-QuIC to sensitively detect prions in blood, even if further optimizations and testing are needed.

Conclusions

The application of RT-QuIC to peripheral tissues has significantly broadened the diagnostic landscape for PrDs, providing a highly sensitive and specific tool for the detection of prion seeding activity outside the CNS. It is important to emphasize that the inclusion of RT-QuIC in the diagnostic criteria for prion diseases in 2017, along with the analysis of CSF and OM samples, has significantly improved the clinical diagnosis of PrDs. This also has profound implications for early diagnosis, particularly in minimally and noninvasive and *antemortem* settings, allowing a deeper understanding of prion accumulations in peripheral tissues and their relevance to disease progression. Despite these achievements, challenges remain in the standardization of protocols between laboratories and the improvement of sensitivity for less common prion strains and atypical disease presentations. The research and diagnostic validation of RT-QuIC for PrDs continues to expand possibilities in the study of prion-like protein disorders. Notably, this assay has been adapted to analyze biological samples from patients with common neurodegenerative diseases, including alpha-synucleinopathies, Alzheimer's disease, frontotemporal dementia, and amyotrophic lateral sclerosis, which share similar pathophysiological mechanisms with PrDs (Salvadores et al. 2014; Saijo et al. 2017, 2019, 2020; Bongianni et al. 2019, 2022; De Luca et al. 2019; Metrick et al. 2020; Scialò et al. 2020; Perra et al. 2021; Bargar et al. 2021; Coysh and Mead 2022; Bellomo et al. 2022; D'Andrea et al. 2023; Concha-Marambio et al. 2023; Kuzkina et al. 2023; Okuzumi et al. 2023; Vascellari et al. 2023; Brockmann et al. 2024; Fontana et al. 2024; Dellarole et al. 2024; Ma et al. 2024). Further optimization of the RT-QuIC technique holds the potential to enhance diagnostic accuracy while deepening our understanding of the pathophysiology of both PrDs and prion-like proteinopathies. These advancements will not only improve patient care and support the development of therapeutic interventions, but will also be crucial for disease monitoring, particularly in the context of pharmacological treatments. This progress is especially important for emerging pharmacological trials, including the PrProfile trial by IONIS (https://cjdisa.com/clinical-trial-update-for-prion-diseases/), where precise patient stratification plays a critical role in selecting suitable participants and maximizing the effectiveness of these studies.

Compliance with Ethical Standards The authors declare no competing financial or nonfinancial interests.

References

Abu-Rumeileh S, Capellari S, Stanzani-Maserati M et al (2018) The CSF neurofilament light signature in rapidly progressive neurodegenerative dementias. Alzheimers Res Ther 10:3. https://doi.org/10.1186/s13195-017-0331-1

Abu-Rumeileh S, Baiardi S, Polischi B et al (2019) Diagnostic value of surrogate CSF biomarkers for Creutzfeldt–Jakob disease in the era of RT-QuIC. J Neurol 266:3136–3143. https://doi.org/10.1007/s00415-019-09537-0

Aguzzi A, Heppner FL (2000) Pathogenesis of prion diseases: a progress report, vol 7. Cell Death Differ, p 889

Appleby BS, Shetty S, Elkasaby M (2022) Genetic aspects of human prion diseases. Front Neurol 13. https://doi.org/10.3389/fneur.2022.1003056

Atarashi R, Moore RA, Sim VL et al (2007) Ultrasensitive detection of scrapie prion protein using seeded conversion of recombinant prion protein. Nat Methods 4:645–650. https://doi.org/10.1038/nmeth1066

Atarashi R, Sano K, Satoh K, Nishida N (2011a) Real-time quaking-induced conversion. Prion 5:150–153. https://doi.org/10.4161/pri.5.3.16893

Atarashi R, Satoh K, Sano K et al (2011b) Ultrasensitive human prion detection in cerebrospinal fluid by real-time quaking-induced conversion. Nat Med 17:175. https://doi.org/10.1038/nm.2294

Baiardi S, Redaelli V, Ripellino P et al (2019) Prion-related peripheral neuropathy in sporadic Creutzfeldt-Jakob disease. J Neurol Neurosurg Psychiatry 90:424–427. https://doi.org/10.1136/jnnp-2018-319221

Baiardi S, Rossi M, Mammana A et al (2021) Phenotypic diversity of genetic Creutzfeldt–Jakob disease: a histo-molecular-based classification. Acta Neuropathol 142:707. https://doi.org/10.1007/s00401-021-02350-y

Baiardi S, Mammana A, Capellari S, Parchi P (2023) Human prion disease: molecular pathogenesis, and possible therapeutic targets and strategies. Expert Opin Ther Targets 27:1271

Baranová S, Moško T, Brůžová M et al (2024) Detection of prions in matching post-mortem skin and cerebrospinal fluid samples using second-generation real-time quaking-induced conversion assay. Sci Rep 14:6294. https://doi.org/10.1038/s41598-024-56789-6

Bargar C, De Luca CMG, Devigili G et al (2021) Discrimination of MSA-P and MSA-C by RT-QuIC analysis of olfactory mucosa: the first assessment of assay reproducibility between two specialized laboratories. Mol Neurodegener 16:82. https://doi.org/10.1186/s13024-021-00491-y

Bate C, Nolan W, McHale-Owen H, Williams A (2016) Sialic acid within the glycosylphosphatidylinositol anchor targets the cellular prion protein to synapses. J Biol Chem 291:17093–17101. https://doi.org/10.1074/jbc.M116.731117

Bellomo G, Giulia De Luca CM, Paoletti FP et al (2022) Alpha synuclein seed amplification assays for diagnosing synucleinopathies: the way forward. Neurology 99:195. https://doi.org/10.1212/WNL.0000000000200878

Bistaffa E, Vuong TT, Cazzaniga FA et al (2019) Use of different RT-QuIC substrates for detecting CWD prions in the brain of Norwegian cervids. Sci Rep 9:18595. https://doi.org/10.1038/s41598-019-55078-x

Bonda DJ, Manjila S, Mehndiratta P et al (2016) Human prion diseases: surgical lessons learned from iatrogenic prion transmission. Neurosurg Focus 41:E10. https://doi.org/10.3171/2016.5.FOCUS15126

Bongianni M, Orrù C, Groveman BR et al (2017) Diagnosis of human prion disease using real-time quaking-induced conversion testing of olfactory mucosa and cerebrospinal fluid samples. JAMA Neurol 74:155. https://doi.org/10.1001/jamaneurol.2016.4614

Bongianni M, Ladogana A, Capaldi S et al (2019) α-Synuclein RT-QuIC assay in cerebrospinal fluid of patients with dementia with Lewy bodies. Ann Clin Transl Neurol 6:2120. https://doi.org/10.1002/acn3.50897

Bongianni M, Catalan M, Perra D et al (2022) Olfactory swab sampling optimization for α-synuclein aggregate detection in patients with Parkinson's disease. Transl Neurodegener 11:37. https://doi.org/10.1186/s40035-022-00311-3

Brandel J-P, Vlaicu MB, Culeux A et al (2020) Variant Creutzfeldt–Jakob disease diagnosed 7.5 years after occupational exposure. N Engl J Med. https://doi.org/10.1056/nejmc2000687

Brockmann K, Lerche S, Baiardi S et al (2024) CSF α-synuclein seed amplification kinetic profiles are associated with cognitive decline in Parkinson's disease. NPJ Park Dis 10:24. https://doi.org/10.1038/s41531-023-00627-5

Brown P, Brandel J-P, Sato T et al (2012) Iatrogenic Creutzfeldt-Jakob disease, final assessment. Emerg Infect Dis 18:901–907. https://doi.org/10.3201/eid1806.120116

Bruce ME (2003) TSE strain variation. Br Med Bull 66:99

Büeler H, Fischer M, Lang Y et al (1992) Normal development and behaviour of mice lacking the neuronal cell-surface PrP protein. Nature 356:577–582. https://doi.org/10.1038/356577a0

Castilla J, Saá P, Soto C (2004) Cyclic amplification of prion protein misfolding. In: Techniques in prion research. Birkhäuser, Basel

Castle AR, Gill AC (2017) Physiological functions of the cellular prion protein. Front Mol Biosci 4

Caughey B (2003) Prion protein conversions: insight into mechanisms, TSE transmission barriers and strains. Br Med Bull 66:109

Caughey B, Race RE, Ernst D et al (1989) Prion protein biosynthesis in scrapie-infected and uninfected neuroblastoma cells. J Virol 63:175–181. https://doi.org/10.1128/jvi.63.1.175-181.1989

Cazzaniga FA, De Luca CMG, Bistaffa E et al (2020) Cell-free amplification of prions: where do we stand? Prog Mol Biol Transl Sci 175:325–358. https://doi.org/10.1016/bs.pmbts.2020.08.005

Chen Z, Kong Y, Zhang J et al (2023) Toward an early clinical diagnosis of MM2-type sporadic Creutzfeldt–Jakob disease. Ann Clin Transl Neurol 10:1209. https://doi.org/10.1002/acn3.51802

Chen Z, Shi Q, Xiao K et al (2024) Multisite skin biopsies vs cerebrospinal fluid for prion seeding activity in the diagnosis of prion diseases. JAMA Neurol 81:1263. https://doi.org/10.1001/jamaneurol.2024.3458

Collinge J (2001) Prion diseases of humans and animals: their causes and molecular basis. Annu Rev Neurosci 24:519–550. https://doi.org/10.1146/annurev.neuro.24.1.519

Concha-Marambio L, Pritzkow S, Shahnawaz M et al (2023) Seed amplification assay for the detection of pathologic alpha-synuclein aggregates in cerebrospinal fluid. Nat Protoc 18:1179–1196. https://doi.org/10.1038/s41596-022-00787-3

Coysh T, Mead S (2022) The future of seed amplification assays and clinical trials. Front Aging Neurosci 14

Cramm M, Schmitz M, Karch A et al (2016) Stability and reproducibility underscore utility of RT-QuIC for diagnosis of Creutzfeldt-Jakob disease. Mol Neurobiol 53:1896. https://doi.org/10.1007/s12035-015-9133-2

D'Andrea C, Cazzaniga FA, Bistaffa E et al (2023) Impact of seed amplification assay and surface-enhanced Raman spectroscopy combined approach on the clinical diagnosis of Alzheimer's disease. Transl Neurodegener 12:35. https://doi.org/10.1186/s40035-023-00367-9

Da Silva Correia SM, Schmitz M, Fischer A et al (2023) Role of different recombinant PrP substrates in the diagnostic accuracy of the CSF RT-QuIC assay in Creutzfeldt-Jakob disease. Cell Tissue Res 392:301

Dassanayake RP, Orrú CD, Hughson AG et al (2016) Sensitive and specific detection of classical scrapie prions in the brains of goats by real-time quaking-induced conversion. J Gen Virol 97:803–812. https://doi.org/10.1099/jgv.0.000367

De Luca CMG, Elia AE, Portaleone SM et al (2019) Efficient RT-QuIC seeding activity for α-synuclein in olfactory mucosa samples of patients with Parkinson's disease and multiple system atrophy. Transl Neurodegener 8:24. https://doi.org/10.1186/s40035-019-0164-x

Dear DV, Young DS, Kazlauskaite J et al (2007) Effects of post-translational modifications on prion protein aggregation and the propagation of scrapie-like characteristics in vitro. Biochim Biophys Acta Proteins Proteomics 1774:792–802. https://doi.org/10.1016/j.bbapap.2007.05.005

Dellarole IL, Vacchi E, Ruiz-Barrio I et al (2024) Tau seeding activity in skin biopsy differentiates tauopathies from synucleinopathies. NPJ Park Dis 10:116. https://doi.org/10.1038/s41531-024-00728-9

Duffy P, Wolf J, Collins G et al (1974) Letter: possible person-to-person transmission of Creutzfeldt-Jakob disease. N Engl J Med 290:692–693

Eraña H, Charco JM, González-Miranda E et al (2020) Detection of pathognomonic biomarker PrPSc and the contribution of cell free-amplification techniques to the diagnosis of prion diseases. Biomol Ther 10:469. https://doi.org/10.3390/biom10030469

Fiorini M, Iselle G, Perra D et al (2020) High diagnostic accuracy of RT-QuIC assay in a prospective study of patients with suspected sCJD. Int J Mol Sci 21:880. https://doi.org/10.3390/ijms21030880

Fontana E, Bongianni M, Benussi A et al (2024) Detection of TDP-43 seeding activity in the olfactory mucosa from patients with frontotemporal dementia. Alzheimers Dement 20:1156–1165. https://doi.org/10.1002/alz.13541

Foutz A, Appleby BS, Hamlin C et al (2017) Diagnostic and prognostic value of human prion detection in cerebrospinal fluid. Ann Neurol 81:79–92. https://doi.org/10.1002/ana.24833

Franceschini A, Baiardi S, Hughson AG et al (2017) High diagnostic value of second generation CSF RT-QuIC across the wide spectrum of CJD prions. Sci Rep 7:10655. https://doi.org/10.1038/s41598-017-10922-w

Gajdusek DC, Zigas V (1959) Kuru. Clinical, pathological and epidemiological study of an acute progressive degenerative disease of the central nervous system among natives of the eastern highlands of New Guinea. Am J Med 26:442

Giaccone G, Moda F (2020) PMCA applications for prion detection in peripheral tissues of patients with variant Creutzfeldt-Jakob disease. Biomol Ther 10:405. https://doi.org/10.3390/biom10030405

Goldfarb LG, Petersen RB, Tabaton M et al (1992) Fatal familial insomnia and familial Creutzfeldt-Jakob disease: disease phenotype determined by a DNA polymorphism. Science (80-) 258:806–808. https://doi.org/10.1126/science.1439789

Green AJE (2019) RT-QuIC: a new test for sporadic CJD. Pract Neurol 19:49. https://doi.org/10.1136/practneurol-2018-001935

Groveman BR, Orrú CD, Hughson AG et al (2017) Extended and direct evaluation of RT-QuIC assays for Creutzfeldt-Jakob disease diagnosis. Ann Clin Transl Neurol 4:139. https://doi.org/10.1002/acn3.378

Haley NJ, Donner R, Henderson DM et al (2020) Cross-validation of the RT-QuIC assay for the antemortem detection of chronic wasting disease in elk. Prion 14:47–55. https://doi.org/10.1080/19336896.2020.1716657

Haraguchi T, Fisher S, Olofsson S et al (1989) Asparagine-linked glycosylation of the scrapie and cellular prion proteins. Arch Biochem Biophys 274:1–13. https://doi.org/10.1016/0003-9861(89)90409-8

Harpaz E, Cazzaniga FA, Tran L et al (2024) Transmission of Norwegian reindeer CWD to sheep by intracerebral inoculation results in an unusual phenotype and prion distribution. Vet Res 55:94. https://doi.org/10.1186/s13567-024-01350-6

Head MW, Bunn TJR, Bishop MT et al (2004) Prion protein heterogeneity in sporadic but not variant Creutzfeldt–Jakob disease: U.K. cases 1991–2002. Ann Neurol 55:851–859. https://doi.org/10.1002/ana.20127

Hermann P, Laux M, Glatzel M et al (2018) Validation and utilization of amended diagnostic criteria in Creutzfeldt-Jakob disease surveillance. Neurology 91. https://doi.org/10.1212/WNL.0000000000005860

Hermann P, Appleby B, Brandel J-P et al (2021) Biomarkers and diagnostic guidelines for sporadic Creutzfeldt-Jakob disease. Lancet Neurol 20:235–246. https://doi.org/10.1016/S1474-4422(20)30477-4

Hermann P, Schmitz M, Cramm M et al (2023) Application of real-time quaking-induced conversion in Creutzfeldt–Jakob disease surveillance. J Neurol 270:2149–2161. https://doi.org/10.1007/s00415-022-11549-2

Hill AF, Joiner S, Beck JA et al (2006) Distinct glycoform ratios of protease resistant prion protein associated with PRNP point mutations. Brain 129:676–685. https://doi.org/10.1093/brain/awl013

Hoover CE, Davenport KA, Henderson DM et al (2017) Endogenous brain lipids inhibit prion amyloid formation in vitro. J Virol 91. https://doi.org/10.1128/JVI.02162-16

Hsiao K, Baker HF, Crow TJ et al (1989) Linkage of a prion protein missense variant to Gerstmann–Sträussler syndrome. Nature 338:342–345. https://doi.org/10.1038/338342a0

Jarrett JT, Lansbury PT (1992) Amyloid fibril formation requires a chemically discriminating nucleation event: studies of an amyloidogenic sequence from the bacterial protein OsmB. Biochemistry 31:12345–12352. https://doi.org/10.1021/bi00164a008

Knight R, Collins S (2001) Human prion diseases: cause, clinical and diagnostic aspects. Contrib, Microbiol

Kovács GG, László L, Bakos A et al (2005) Increased incidence of genetic human prion disease in Hungary. Neurology 65:1666–1669. https://doi.org/10.1212/01.wnl.0000184513.95290.80

Kupfer L, Hinrichs W, Groschup M (2009) Prion protein Misfolding. Curr Mol Med 9:826–835. https://doi.org/10.2174/156652409789105543

Kuzkina A, Rößle J, Seger A et al (2023) Combining skin and olfactory α-synuclein seed amplification assays (SAA)—towards biomarker-driven phenotyping in synucleinopathies. NPJ Park Dis 9:79. https://doi.org/10.1038/s41531-023-00519-8

Lattanzio F, Abu-Rumeileh S, Franceschini A et al (2017) Prion-specific and surrogate CSF biomarkers in Creutzfeldt-Jakob disease: diagnostic accuracy in relation to molecular subtypes and analysis of neuropathological correlates of p-tau and Aβ42 levels. Acta Neuropathol 133:559. https://doi.org/10.1007/s00401-017-1683-0

Liberski PP, Gajos A, Sikorska B, Lindenbaum S (2019) Kuru, the first human prion disease. Viruses 11:232. https://doi.org/10.3390/v11030232

Llorens F, Villar-piqué A, Hermann P et al (2020) Diagnostic accuracy of prion disease biomarkers in iatrogenic creutzfeldt-jakob disease. Biomol Ther 10. https://doi.org/10.3390/biom10020290

Ma Y, Farris CM, Weber S et al (2024) Sensitivity and specificity of a seed amplification assay for diagnosis of multiple system atrophy: a multicentre cohort study. Lancet Neurol 23:1225–1237. https://doi.org/10.1016/S1474-4422(24)00395-8

Mammana A, Baiardi S, Rossi M et al (2020) Detection of prions in skin punch biopsies of Creutzfeldt-Jakob disease patients. Ann Clin Transl Neurol 7:559–564. https://doi.org/10.1002/acn3.51000

McGuire LI, Peden AH, Orrú CD et al (2012) Real time quaking-induced conversion analysis of cerebrospinal fluid in sporadic Creutzfeldt-Jakob disease. Ann Neurol 72:278. https://doi.org/10.1002/ana.23589

McGuire LI, Poleggi A, Poggiolini I et al (2016) Cerebrospinal fluid real-time quaking-induced conversion is a robust and reliable test for sporadic creutzfeldt–jakob disease: an international study. Ann Neurol 80:160–165. https://doi.org/10.1002/ana.24679

Mead S (2006) Prion disease genetics. Eur J Hum Genet 14:273

Metrick MA, Ferreira NDC, Saijo E et al (2020) A single ultrasensitive assay for detection and discrimination of tau aggregates of Alzheimer and pick diseases. Acta Neuropathol Commun 8:22. https://doi.org/10.1186/s40478-020-0887-z

Minikel EV, Vallabh SM, Lek M et al (2016) Quantifying prion disease penetrance using large population control cohorts. Sci Transl Med 8:322ra9. https://doi.org/10.1126/scitranslmed.aad5169

Moda F, Suardi S, Di Fede G et al (2012) MM2-thalamic Creutzfeldt-Jakob disease: neuropathological, biochemical and transmission studies identify a distinctive prion strain. Brain Pathol 22:662. https://doi.org/10.1111/j.1750-3639.2012.00572.x

Mok T, Jaunmuktane Z, Joiner S et al (2017) Variant Creutzfeldt–Jakob Disease in a Patient with Heterozygosity at PRNP Codon 129. N Engl J Med 376:292–294. https://doi.org/10.1056/NEJMc1610003

Mok TH, Nihat A, Luk C et al (2021) Bank vole prion protein extends the use of RT-QuIC assays to detect prions in a range of inherited prion diseases. Sci Rep 11:5231. https://doi.org/10.1038/s41598-021-84527-9

Morales R, Abid K, Soto C (2007) The prion strain phenomenon: molecular basis and unprecedented features. Biochim Biophys Acta Mol Basis Dis 1772:681

Nonaka T, Iwasaki Y, Horiuchi H, Satoh K (2024) Detection limitations of prion seeding activities in blood samples from patients with sporadic prion disease. BMC Neurol 24:92. https://doi.org/10.1186/s12883-024-03590-7

Notari S, Moleres FJ, Hunter SB et al (2010) Multiorgan detection and characterization of protease-resistant prion protein in a case of variant CJD examined in the United States. PLoS One 5:e8765. https://doi.org/10.1371/journal.pone.0008765

Oesch B, Westaway D, Wälchli M et al (1985) A cellular gene encodes scrapie PrP 27-30 protein. Cell 40:735–746. https://doi.org/10.1016/0092-8674(85)90333-2

Okuzumi A, Hatano T, Matsumoto G et al (2023) Propagative α-synuclein seeds as serum biomarkers for synucleinopathies. Nat Med 29:1448–1455. https://doi.org/10.1038/s41591-023-02358-9

Orrú CD, Wilham JM, Raymond LD et al (2011) Prion disease blood test using immunoprecipitation and improved quaking-induced conversion. MBio 2:e00078. https://doi.org/10.1128/mBio.00078-11

Orrú CD, Bongianni M, Tonoli G et al (2014) A test for Creutzfeldt-Jakob disease using nasal brushings. N Engl J Med 371:519–529. https://doi.org/10.1056/NEJMoa1315200

Orrú CD, Groveman BR, Hughson AG et al (2015) Rapid and sensitive RT-QuIC detection of human creutzfeldt-jakob disease using cerebrospinal fluid. MBio 6. https://doi.org/10.1128/mBio.02451-14

Orrú CD, Yuan J, Appleby BS et al (2017) Prion seeding activity and infectivity in skin samples from patients with sporadic Creutzfeldt-Jakob disease. Sci Transl Med 9:eaam7785. https://doi.org/10.1126/scitranslmed.aam7785

Orrù CD, Soldau K, Cordano C et al (2018) Prion seeds distribute throughout the eyes of sporadic Creutzfeldt-Jakob disease patients. MBio 9. https://doi.org/10.1128/mBio.02095-18

Orrú CD, Groveman BR, Foutz A et al (2020) Ring trial of 2nd generation RT-QuIC diagnostic tests for sporadic CJD. Ann Clin Transl Neurol 7:2262–2271. https://doi.org/10.1002/acn3.51219

Parchi P, Capellari S, Chen SG et al (1997) Typing prion isoforms [5]. Nature 386:232

Parchi P, Giese A, Capellari S et al (1999) Classification of sporadic Creutzfeldt-Jakob disease based on molecular and phenotypic analysis of 300 subjects. Ann Neurol 46:224–233. https://doi.org/10.1002/1531-8249(199908)46:2<224::AID-ANA12>3.0.CO;2-W

Parchi P, Zou W, Wang W et al (2000) Genetic influence on the structural variations of the abnormal prion protein. Proc Natl Acad Sci 97:10168–10172. https://doi.org/10.1073/pnas.97.18.10168

Park J-H, Choi Y-G, Lee Y-J et al (2016) Real-time quaking-induced conversion analysis for the diagnosis of sporadic Creutzfeldt-Jakob disease in Korea. J Clin Neurol 12:101. https://doi.org/10.3988/jcn.2016.12.1.101

Perra D, Bongianni M, Novi G et al (2021) Alpha-synuclein seeds in olfactory mucosa and cerebrospinal fluid of patients with dementia with Lewy bodies. Brain Commun 3. https://doi.org/10.1093/braincomms/fcab045

Poleggi A, Baiardi S, Ladogana A, Parchi P (2022) The use of real-time quaking-induced conversion for the diagnosis of human prion diseases. Front Aging Neurosci 14. https://doi.org/10.3389/fnagi.2022.874734

Prusiner SB (1991) Molecular biology of prion diseases. Science (80-) 252:1515–1522. https://doi.org/10.1126/science.1675487

Prusiner SB (1998) Prions. Proc Natl Acad Sci 95:13363–13383. https://doi.org/10.1073/pnas.95.23.13363

Redaelli V, Bistaffa E, Zanusso G et al (2017) Detection of prion seeding activity in the olfactory mucosa of patients with fatal familial insomnia. Sci Rep 7:46269. https://doi.org/10.1038/srep46269

Rhoads DD, Wrona A, Foutz A et al (2020) Diagnosis of prion diseases by RT-QuIC results in improved surveillance. Neurology 95:e1017. https://doi.org/10.1212/WNL.0000000000010086

Rossi M, Candelise N, Baiardi S et al (2020) Ultrasensitive RT-QuIC assay with high sensitivity and specificity for Lewy body-associated synucleinopathies. Acta Neuropathol 140:49–62. https://doi.org/10.1007/s00401-020-02160-8

Rudd PM, Endo T, Colominas C et al (1999) Glycosylation differences between the normal and pathogenic prion protein isoforms. Proc Natl Acad Sci USA 96:13044–13049. https://doi.org/10.1073/pnas.96.23.13044

Saborio GP, Permanne B, Soto C (2001) Sensitive detection of pathological prion protein by cyclic amplification of protein misfolding. Nature 411:810–813. https://doi.org/10.1038/35081095

Saijo E, Ghetti B, Zanusso G et al (2017) Ultrasensitive and selective detection of 3-repeat tau seeding activity in pick disease brain and cerebrospinal fluid. Acta Neuropathol 133:751–765. https://doi.org/10.1007/s00401-017-1692-z

Saijo E, Groveman BR, Kraus A, et al (2019) Ultrasensitive RT-QuIC seed amplification assays for disease-associated tau, α-Synuclein, and prion aggregates, pp 19–37

Saijo E, Metrick MA, Koga S et al (2020) 4-repeat tau seeds and templating subtypes as brain and CSF biomarkers of frontotemporal lobar degeneration. Acta Neuropathol 139:63–77. https://doi.org/10.1007/s00401-019-02080-2

Salvadores N, Shahnawaz M, Scarpini E et al (2014) Detection of misfolded aβ oligomers for sensitive biochemical diagnosis of Alzheimer's disease. Cell Rep 7:261–268. https://doi.org/10.1016/j.celrep.2014.02.031

Sano K, Satoh K, Atarashi R et al (2013) Early detection of abnormal prion protein in genetic human prion diseases now possible using real-time QUIC assay. PLoS One 8:e54915. https://doi.org/10.1371/journal.pone.0054915

Satoh K, Fuse T, Nonaka T et al (2019) Postmortem quantitative analysis of prion seeding activity in the digestive system. Molecules 24:4601. https://doi.org/10.3390/molecules24244601

Scheckel C, Aguzzi A (2018) Prions, prionoids and protein misfolding disorders. Nat Rev Genet 19:405–418. https://doi.org/10.1038/s41576-018-0011-4

Schmitz M, Cramm M, Llorens F et al (2016) The real-time quaking-induced conversion assay for detection of human prion disease and study of other protein misfolding diseases. Nat Protoc 11:2233. https://doi.org/10.1038/nprot.2016.120

Schmitz M, Silva Correia S, Hermann P et al (2023) Detection of prion protein seeding activity in tear fluids. N Engl J Med 388:1816–1817. https://doi.org/10.1056/NEJMc2214647

Scialò C, Tran TH, Salzano G et al (2020) TDP-43 real-time quaking induced conversion reaction optimization and detection of seeding activity in CSF of amyotrophic lateral sclerosis and frontotemporal dementia patients. Brain Commun 2. https://doi.org/10.1093/braincomms/fcaa142

Stahl N, Prusiner SB (1991) Prions and prion proteins 1. FASEB J 5:2799–2807. https://doi.org/10.1096/fasebj.5.13.1916104

Stahl N, Borchelt DR, Hsiao K, Prusiner SB (1987) Scrapie prion protein contains a phosphatidylinositol glycolipid. Cell 51:229–240. https://doi.org/10.1016/0092-8674(87)90150-4

Tabaton M, Monaco S, Cordone MP et al (2004) Prion deposition in olfactory biopsy of sporadic Creutzfeldt-Jakob disease. Ann Neurol 55:294. https://doi.org/10.1002/ana.20038

Thomas CM, Salamat MKF, de Wolf C et al (2023) Development of a sensitive real-time quaking-induced conversion (RT-QuIC) assay for application in prion-infected blood. PLoS One 18:e0293845. https://doi.org/10.1371/journal.pone.0293845

Trevitt CR, Singh PN (2003) Variant Creutzfeldt-Jakob disease: pathology, epidemiology, and public health implications. Am J Clin Nutr 78:651S–656S. https://doi.org/10.1093/ajcn/78.3.651S

Turk E, Teplow DB, Hood LE, Prusiner SB (1988) Purification and properties of the cellular and scrapie hamster prion proteins. Eur J Biochem 176:21. https://doi.org/10.1111/j.1432-1033.1988.tb14246.x

Vallabh SM, Mortberg MA, Allen SW et al (2024) Fluid biomarkers in individuals at risk for genetic prion disease up to disease conversion. Neurology 103:e209506. https://doi.org/10.1212/WNL.0000000000209506

Vascellari S, Orrù CD, Groveman BR et al (2023) A-Synuclein seeding activity in duodenum biopsies from Parkinson's disease patients. PLoS Pathog 19:e1011456. https://doi.org/10.1371/journal.ppat.1011456

Watson N, Hermann P, Ladogana A et al (2022) Validation of revised international Creutzfeldt-Jakob disease surveillance network diagnostic criteria for sporadic Creutzfeldt-Jakob disease. JAMA Netw Open 5:e2146319. https://doi.org/10.1001/jamanetworkopen.2021.46319

Wüthrich K, Riek R (2001) Three-dimensional structures of prion proteins. Adv Protein Chem

Zanusso G, Ferrari S, Cardone F et al (2003) Detection of pathologic prion protein in the olfactory epithelium in sporadic Creutzfeldt-Jakob disease. N Engl J Med 348:711–719. https://doi.org/10.1056/NEJMoa022043

Zhang W, Orrú CD, Foutz A et al (2024) Large-scale validation of skin prion seeding activity as a biomarker for diagnosis of prion diseases. Acta Neuropathol 147:17. https://doi.org/10.1007/s00401-023-02661-2

Zou W, Puoti G, Xiao X et al (2010) Variably protease-sensitive prionopathy: a new sporadic disease of the prion protein. Ann Neurol 68:162–172. https://doi.org/10.1002/ana.22094

Franceschini A, Baiardi S, Hughson AG, et al (2017) High diagnostic value of second generation CSF RT-QuIC across the wide spectrum of CJD prions. Sci Rep 7:10655. https://doi.org/10.1038/s41598-017-10922-w

Lattanzio F, Abu-Rumeileh S, Franceschini A, et al (2017) Prion-specific and surrogate CSF biomarkers in Creutzfeldt-Jakob disease: diagnostic accuracy in relation to molecular subtypes and analysis of neuropathological correlates of p-tau and Aβ42 levels. Acta Neuropathol. https://doi.org/10.1007/s00401-017-1683-0

Mok TH, Nihat A, Luk C, et al (2021) Bank vole prion protein extends the use of RT-QuIC assays to detect prions in a range of inherited prion diseases. Sci Rep 11:5231. https://doi.org/10.1038/s41598-021-84527-9

Chapter 12
Detecting the Undetectable: Exploring the Diagnostic Potential of Protein Misfolding Cyclic Amplification in Human Prion Diseases

Federico Angelo Cazzaniga, Giuseppe Bufano, Floriana Bellandi, Merve Begüm Bacınoğlu, and Fabio Moda

Abstract Prion diseases (PrDs) are devastating and fatal conditions characterized by the accumulation of the misfolded prion protein (PrP^{Sc}) in the central nervous system (CNS). Definitive diagnosis of PrDs relies on the detection of prions in CNS tissues collected *postmortem*. The advent of a highly sensitive cell-free amplification technique, named protein misfolding cyclic amplification (PMCA), has revolutionized this field. It has revealed trace amounts of prions in various tissues, including cerebrospinal fluid, urine, blood, and olfactory mucosa of patients with different forms of PrDs. PMCA mirrors in vitro the pathological process of protein misfolding and aggregation, which occurs in vivo but in a significantly accelerated manner. For this reason, this technology is currently used in specialized laboratories to support research and diagnostic activities in human and animal PrDs. This chapter highlights the latest advances and applications of PMCA in the diagnosis of human PrDs.

Keywords Human prion diseases · Protein misfolding cyclic amplification · Neurodegeneration · Disease biomarkers

F. A. Cazzaniga · G. Bufano · F. Bellandi · M. B. Bacınoğlu
Unit of Laboratory Medicine, Laboratory of Clinical Pathology, Fondazione IRCCS Istituto Neurologico Carlo Besta, Milan, Italy
e-mail: federico.cazzaniga@istituto-besta.it; giuseppe.bufano@istituto-besta.it; floriana.bellandi@istituto-besta.it; merve.bacinoglu@istituto-besta.it

F. Moda (✉)
Unit of Laboratory Medicine, Laboratory of Clinical Pathology, Fondazione IRCCS Istituto Neurologico Carlo Besta, Milan, Italy

Department of Medical Biotechnology and Translational Medicine, Università degli Studi di Milano, Milan, Italy
e-mail: fabio.moda@unimi.it

G. Legname, F. Moda (eds.), *Biomarkers and Therapeutical Targets for Prion Diseases*, Subcellular Biochemistry 112, https://doi.org/10.1007/978-3-031-97055-9_12

Human Prion Diseases

Prion diseases (PrDs), or transmissible spongiform encephalopathies (TSE), comprise a group of progressive, lethal, and transmissible neurodegenerative conditions that affect both humans and animals (Prusiner 1998b; Knight and Collins 2001). Scrapie is the most common PrD in sheep and goats (Collinge 2001). Since its discovery, scientists have been puzzled by the precise nature of this pathological agent. Various hypotheses ranged from a self-replicating membrane to retrovirus-like elements, but none provided definitive explanations (Legname and Moda 2017). Its remarkable resistance to agents capable of destroying nucleic acids, such as ultraviolet and ionizing radiation, led researchers to consider the possibility that it lacks nucleic acids and may instead be a self-replicating protein (Alper et al. 1967; Griffith 1967). It was eventually understood that the infectious agent is a protein with an abnormal conformation, resulting from a conformational change in the normal prion protein (PrP^{C}). In 1982, Prusiner S.B. coined the term "prion" to refer to this abnormally folded protein (PrP^{Sc}), which stands for proteinaceous infectious particle, and received the Nobel Prize (Prusiner 1998c; Prusiner 1998a). Human PrDs can be classified as genetic, acquired, and sporadic, with an overall incidence of 1–2 cases per million (Gao et al. 2024). Regardless of the classification, they are mainly characterized by the presence of PrP^{Sc} aggregates in the central nervous system (CNS) (Bruce et al. 1997). Sporadic Creutzfeldt–Jakob disease (sCJD) is the most common form of human PrDs, and its underlying causes remain unknown (Gambetti et al. 2003; Ladogana et al. 2005). Genetic prion diseases (gPrDs) arise from mutations in the *PRNP* gene and include genetic Creutzfeldt–Jakob disease (gCJD), fatal familial insomnia (FFI), and Gerstmann–Sträussler–Scheinker syndrome (GSS) (Collinge 2001). Acquired prion diseases (1% of all prion cases) result from the transmission of PrP^{Sc} between hosts through exposure to contaminated tissue, medical procedures, or consumption of infected materials. Kuru was the first human-acquired PrD described, transmitted through ritualistic cannibalism. The variant form of Creutzfeldt–Jakob disease (vCJD) is another well-known acquired PrD, transmitted to humans through the consumption of food from cattle affected by bovine spongiform encephalopathy (BSE) (Lee et al. 2013).

Differences Between PrP^{C} and PrP^{Sc}

The prion protein (PrP) is recognized as the primary etiological determinant of PrDs (Prusiner 1998a, c). As previously mentioned, PrP exists in two distinct forms: the normal or cellular form, known as PrP^{C}, and the disease-associated form, known as PrP^{Sc}. PrP^{C} is a glycophosphatidylinositol-anchored glycoprotein (Stahl et al. 1987), encoded by the *PRNP* gene located on chromosome 20 in humans (Castle and Gill

2017). It is conserved across species (Wüthrich and Riek 2001). PrP^C is involved in several functions, including synaptic activity, neurite elongation and synapse development, metal metabolism, anti-apoptotic mechanisms, and protection of cells from oxidative stress (Castle and Gill 2017; Miranzadeh Mahabadi and Taghibiglou 2020; Panes et al. 2021). The protein undergoes three significant post-translational modifications: (i) the addition of a glycophosphatidylinositol (GPI) anchor, (ii) the formation of a disulfide bond between cysteines at residues 179 and 214, and (iii) N-linked glycosylation at asparagine (Asn) residues at positions 181 and 197 (Dear et al. 2007). PrP^C is rich in α-helical structures, is soluble in detergents, and is completely digested by proteinase K (PK) (Zahn et al. 2000; Dima and Thirumalai 2002; Zou et al. 2011). In contrast, PrP^{Sc} is characterized by a higher content of β-sheet structures, reduced solubility in detergents, and partial resistance to PK digestion (Riesner 2003; Yuan et al. 2006; Silva et al. 2015).

After PK digestion and Western blot analysis, a resistant core of 141 amino acids, referred to as PrP^{res}, is formed, showing the typical diglycosylated, monoglycosylated, and unglycosylated bands (Oesch et al. 1985; Haraguchi et al. 1989; Hill et al. 2006) (Fig. 12.1a).

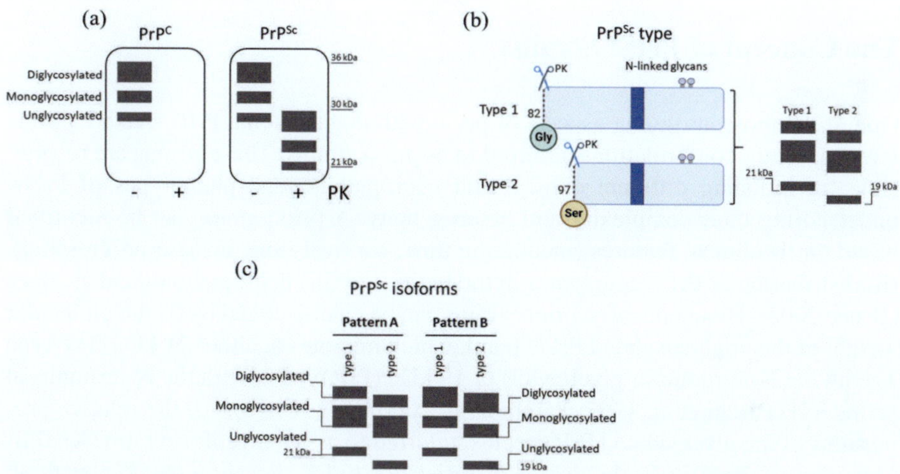

Fig. 12.1 Schematic representation of the biochemical differences between PrP^C and PrP^{Sc}. (**a**) Western blot analysis of PrP^C and PrP^{Sc} after proteinase K (PK) digestion. PK treatment completely removes the signal of PrP^C, whereas PrP^{Sc} is partially digested and the PK-resistant C-terminal fragment can be visualized: (−) indicates PK-untreated samples, while (+) indicates PK-treated samples (**b**) Representation of the PK cleavage sites: PrP^{Sc} isoforms, namely type 1 and type 2, are generated after cleavage at glycine (Gly) 82 or serine (Ser) 97, respectively. The unglycosylated band of PrP type 1 migrates at 21 kDa, while that of PrP type 2 migrates at 19 kDa. (**c**) The pattern of PK-resistant PrP^{Sc} isoforms. "Pattern A" is defined by the presence of type 1 or type 2 PrP^{Sc}, with a predominant monoglycosylated band compared to the others. Similarly, "Pattern B" is defined by the presence of type 1 or type 2 PrP^{Sc}, with a predominant diglycosylated band compared to the others

Models of PrP^{Sc} Propagation

One of the distinctive features of PrP^{Sc} is its ability to interact with PrP^{C} and induce its conversion into PrP^{Sc}. This interaction initiates an "autocatalytic" cascade of PrP^{Sc} conversion, which ultimately results in the aggregation and deposition of PrP^{Sc} in the CNS (Kupfer et al. 2009). According to the template-assisted conversion model, the spontaneous conversion of PrP^{C} into PrP^{Sc} is thermodynamically unfavorable. However, PrP^{Sc} interacts with PrP^{C}, forming a heterodimer. This helps facilitate the refolding of PrP^{C} into PrP^{Sc}, overcoming the energetic barrier between the two states (Prusiner 1991). In contrast, the nucleated polymerization model suggests that PrP^{C} and PrP^{Sc} exist in a state of dynamic equilibrium (Jarrett and Lansbury 1992). In this model, PrP^{Sc} molecules gradually accumulate to form a nucleus, although this process is energetically unfavorable and slow. The formation of this nucleus is critical in determining the rate at which prion propagation occurs. These models, while distinct, may work in combination, each representing different stages of PrP^{Sc} propagation.

The Concept of PrP^{Sc} Strains

One of the most intriguing aspects of prion pathologies is that PrP^{Sc} can adopt different aberrant conformations, referred to as prion strains. These strains are responsible for inducing different clinical and neuropathological phenotypes of PrDs, underscoring their complexity and heterogeneity. A prion strain can be identified based on (i) clinical features (incubation time, survival time, disease progression), (ii) distribution of PrP^{Sc} aggregates in the brain, and (iii) neuropathological changes (Bruce 2003). Biochemically, prion strains can be characterized by (i) the molecular weight of the unglycosylated PrP^{res} band, which migrates at either 21 kDa (PrP type 1, with the N-terminus at glycine 82) or 19 kDa (PrP type 2, with the N-terminus at serine 97) (Parchi et al. 1997; Parchi et al. 2000) (Fig. 12.1b); (ii) the relative proportions of the glycosylated PrP species, referred to as the glycoform ratio (Khalili-Shirazi et al. 2005); (iii) the degree of resistance to PK digestion (McKinley et al. 1983); and (iv) the denaturation profile in response to chaotropic agents, such as guanidine hydrochloride (Peretz et al. 2002). The classification of human prion strains is based on the combination of *PRNP* gene polymorphism at codon 129 (which can encode either methionine (129 M) or valine (129 V)) and PrP^{Sc} type (type 1 or type 2). This results in at least six distinct molecular subtypes of sCJD: MM1, MM2, MV1, MV2, VV1, and VV2 (Parchi et al. 1999). In certain cases, type 1 and type 2 PrP^{Sc} coexist within the same individual (Parchi et al. 2009). Interestingly, sCJD, iCJD, and Kuru are associated with a PrP^{res} glycoform ratio characterized by a predominant monoglycosylated PrP^{Sc} band "Pattern A" (Parchi et al. 1999). In contrast, vCJD and FFI show a distinct glycoform ratio marked by a predominance of the diglycosylated band, characteristic of "Pattern B" (Collinge et al. 1996; Head et al. 2004) (Fig. 12.1c). In 2008, Gambetti and colleagues identified a novel human prion disorder, referred to as variably protease-sensitive prionopathy (VPSPr). Unlike classical sCJD strains, the PrP^{res} in VPSPr is less resistant to PK digestion. On Western blot, VPSPr PrP^{res} is characterized by a distinctive ladder-like pattern, marked by a typical fragment migrating at approximately 8 kDa (Zou et al. 2010).

Protein Misfolding Cyclic Amplification

The protein misfolding cyclic amplification (PMCA) has revolutionized the fields of PrDs research and diagnosis. Developed by Claudio Soto and colleagues, the technique is based on the principle of protein misfolding transmission. PMCA replicates the prion conversion and aggregation process that naturally occurs in vivo but under controlled in vitro conditions (Saborio et al. 2001). Central to this process is the ability of PrP^{Sc} to act as a template, inducing the misfolding of its physiological counterpart, PrP^{C}, thereby driving the self-propagating cycle of prion replication. PMCA consists of two phases that are repeated cyclically:

- *Incubation*: this phase lasts approximately 30 min at a controlled temperature. The biological sample, which contains trace amounts of PrP^{Sc}, is incubated with a substrate containing an excess of PrP^{C} (see Table 12.1 for an updated list of reaction substrates). During this phase, PrP^{Sc} acts as a template, inducing the misfolding of PrP^{C} and resulting in the formation of additional PrP^{Sc} aggregates.
- *Sonication*: this phase disrupts large PrP^{Sc} aggregates into smaller seeds which promotes the conversion of other PrP^{C} into PrP^{Sc}. The sonication lasts for 20–40 s with a variable potency (ranging from 200 to 300 Watts). The Watt-potency is adjusted by means of a power supply, and the sonication is performed in a special horn described in detail in Fig. 12.2a.

The incubation and sonication phases are repeated in cycles for 24–48 h, constituting one round of amplification. Subsequently, an aliquot of the amplified material is transferred to a freshly prepared substrate and undergoes a new round of amplification. Each round allows for the perpetuation of in vitro prion propagation, thereby enabling the amplification of prion traces present in the biological sample under analysis. The final reaction products are digested with PK and subjected to Western blot analysis using anti-PrP antibodies (e.g., 3F4 or 6D11). This confirms whether prion replication has efficiently occurred following the amplification process (see Fig. 12.2b).

Table 12.1 Summary of the most relevant parameters of PMCA protocols optimized to detect PrP^{sc} in human prion diseases

	Prion disease investigated	Control	Biological matrix	Sensitivity	Specificity	Substrate of reaction	Protocols	Reference
Urine	vCJD (*n* = 14)	sCJD (*n* = 68), gCJD (*n* = 4), OND (*n* = 50), NND (*n* = 50) and HS (*n* = 52)	Urine	92.9%	100%	Tg(MHu2M) (129MM)	*CB*: 1× PBS supplemented with 150 mM NaCl, 1% Triton X-100, protease inhibitors (cOmplete EDTA-free), 6 mM EDTA and 0.05% digitonin *Beads*: 3 Teflon *Volume*: Sample 10 μL and substrate 90 μL *Pulse on*: 40 s *Pulse off*: 29 min 20 s *Temperature*: 37–40 °C *Sonicator*: Q700 Qsonica *Potency:* 270–280 W	Moda et al. (2014)
	sCJD (*n* = 81)	HS (*n* = 66) and OND (*n* = 94)	Urine (treated with iron oxide magnetic extraction)	36%	100%	Tg(MHu2M/V) (129MM/VV)	*CB*: 1× PBS supplemented with 150 mM NaCl, 1% Triton X-100, protease inhibitors (cOmplete EDTA-free), 6 mM EDTA, 0.05% digitonin, 0.01% sodium tripolyphosphate and 100 μg/mL heparin *Beads*: 3 Teflon *Volume*: / *Pulse on*: 60 s *Pulse off*: 29 min *Temperature*: 37–40 °C *Sonicator*: Q700 Qsonica *Potency*: 270–280 W	Pritzkow et al. (2023)

	Prion disease investigated	Control	Biological matrix	Sensitivity	Specificity	Substrate of reaction	Protocols	Reference
Blood	vCJD (*n* = 4)	HS (*n* = 141)	WBC, whole blood	/	/	Tg338 (ovine $V_{136}R_{154}Q_{171}$ PrP), TgShXI (ovine $A_{136}R_{154}Q_{171}$ PrP variant)	*CB*: 1× PBS supplemented with 150 mM NaCl and 0.1% Triton X-100 *Beads*: 5/8 silica/zirconium (1 mm diameter) *Volume*: Sample 7 μL and substrate 63 μL *Pulse on*: 30 s *Pulse off*: 29 min 30 s *Temperature*: 39.5 °C *Sonicator*: S-4000 Misonix *Potency*: Power 70%	Lacroux et al. (2014)
	vCJD (*n* = 14)	sCJD (*n* = 16), OND (*n* = 62), NND (*n* = 26) and HS (*n* = 49)	WBC, whole blood and plasma fraction	100%	100%	Tg(MHu2M) (129MM)	*CB*: 1× PBS supplemented with 150 mM NaCl, 1% Triton X-100 and protease inhibitors (cOmplete EDTA-free) *Beads*: 3 Teflon *Volume*: / *Pulse on*: 30 s *Pulse off*: 29 min 30 s *Temperature*: 37–40 °C *Sonicator*: Q700 Qsonica *Potency*: Amplitude 30	Concha-Marambio et al. (2016)
	vCJD (*n* = 18)	OND (*n* = 15), including AD (*n* = 9), DLB (*n* = 3), PD (*n* = 2), FTD (*n* = 1);NND (*n* = 52), including metabolic and toxic encephalopathies (*n* = 14), paraneoplastic encephalitis cancer (*n* = 12), neurovascular disease (*n* = 7), infectious diseases (*n* = 6), autoimmune encephalophaties (*n* = 3), OD (*n* = 10); HS (*n* = 104); and presymptomatic vCJD (*n* = 2)	Plasma fraction (treated with plasminogen-coated magnetic nanobeads)	100%	100%	Tg650 (129MM)	*CB*: 1× PBS supplemented with 150 mM NaCl and 1% Triton X-100 *Beads*: 3 Teflon *Volume*: Sample 10 μL and substrate 90 μL *Pulse on*: 20 s *Pulse off*: 29 min 40 s *Temperature*: 37 °C *Sonicator*: S-4000 Misonix and Q700 Qsonica *Potency*: 240 W	Bougard et al. (2016)

(continued)

Table 12.1 (Continued)

	Prion disease investigated	Control	Biological matrix	Sensitivity	Specificity	Substrate of reaction	Protocols	Reference
Cerebrospinal fluid	vCJD ($n = 15$)	sCJD ($n = 6$), OND ($n = 21$) and NND ($n = 14$)	CSF	100%	100%	Tg(MHu2M) (129MM)	*CB*: 1× PBS supplemented with 150 mM NaCl, 1% Triton X-100, complete protease inhibitors cocktail, 100 μg/mL heparin sodium and 6 mM EDTA *Beads*: 3 Teflon *Volume*: Sample 7.5 μL and substrate 92.5 μL *Pulse on*: 20 s *Pulse off*: 29 min 40 s *Temperature*: 37 °C *Sonicator*: Q700 Qsonica *Potency*: 280–300 W	Barria et al. (2018)
	vCJD ($n = 41$)	sCJD ($n = 23$), gCJD ($n = 1$), AD ($n = 12$) and NND ($n = 21$)	CSF	97.6%	100%	Tg650 (129MM)	*CB*: 1× PBS supplemented with 150 mM NaCl and 1% Triton X-100 *Beads*: 3 Teflon *Volume*: Sample 10 μL and substrate 90 μL *Pulse on*: 20 sec *Pulse off*: 15 min *Temperature*: 37 °C *Sonicator*: S-4000 Misonix and Q700 Qsonica *Potency*: 240 W	Bougard et al. (2018)
	sCJD ($n = 8$)	vCJD ($n = 1$)	CSF	/	/	TgMet (Tg650), TgVal (Tg152) and Bank vole (M109)	*CB*: 1× PBS supplemented with 150 mM NaCl, 1% Triton X-100, proteinase inhibitors cocktail, 1 mM EDTA and 10 μg/mL heparin *Beads*: 3 Teflon *Volume*: Sample 20 μL and substrate 90 μL *Pulse on*: 20 sec *Pulse off*: 14 min 40 sec *Temperature*: 37 °C *Sonicator*: / *Potency*: 240 W	Bélondrade et al. (2021)

Olfactory mucosa	FFI ($n = 2$)	AD ($n = 6$), PD ($n = 6$), FTD ($n = 4$) and HS ($n = 10$)	OM	/	100%	Bank vole (M109)	*CB*: 1× PBS supplemented with 150 mM NaCl, 1% Triton X-100, 100 µg/mL heparin and 0.05% digitonin *Beads*: 3 Teflon *Volume*: Sample 10 µL and substrate 90 µL *Pulse on*: 30 sec *Pulse off*: 29 min 30 sec *Temperature*: 37–40 °C *Sonicator*: Q700 Qsonica *Potency:* 260–280 W	Redaelli et al. (2017)
	sCJD ($n = 27$)	gCJD ($n = 2$), AD ($n = 3$), PD ($n = 7$), FTD ($n = 7$), MSA ($n = 4$), PSP ($n = 7$), CBD ($n = 6$) and MS ($n = 2$)	OM	79.3%	100%	Tg(MHu2M) (129MM)	*CB*: 1× PBS supplemented with 150 mM NaCl, 1% Triton X-100, cOmplete mini EDTA-free protease inhibitors cocktail, 0.135 M sodium tripolyphosphate, 6 mM EDTA, 100 µg/ml heparin and 0.05% digitonin *Beads*: 3 Teflon *Volume*: Sample 10 µL and substrate 90 µl *Pulse on*: 40 s *Pulse off*: 29 min 20 s *Temperature*: 37–40 °C *Sonicator*: Q700 Qsonica *Potency:* 260–280 W	Cazzaniga et al. (2022)

sCJD sporadic Creutzfeldt–Jakob disease, *vCJD* variant Creutzfeldt–Jakob disease, *gCJD* genetic Creutzfeldt–Jakob disease, *FFI* fatal familial insomnia, *AD* Alzheimer's disease, *PD* Parkinson's disease, *DLB* dementia with Lewy bodies, *FTD* frontotemporal dementia, *MS* multiple sclerosis, *MSA* multiple system atrophy, *PSP* progressive supranuclear palsy, *CBD* corticobasal degeneration, *OND* other neurodegenerative disorders, *NND* non-neurodegenerative disorders, *OD* other diseases, *HS* healthy subjects, *OM* olfactory mucosa, *CSF* cerebrospinal fluid, *min* minutes, *s* seconds, *W* Watt, *CB* conversion buffer, *PBS* phosphate buffer saline, *EDTA* ethylenediaminetetraacetic acid, *NaCl* sodium chloride, *WBC* white blood cells, *129MM* methionine/methionine polymorphism at the codon 129 of the *PRNP* gene, *129VV* valine/valine polymorphism at the codon 129 of the *PRNP* gene, *M109* methionine at the codon 109 of the *prnp* gene

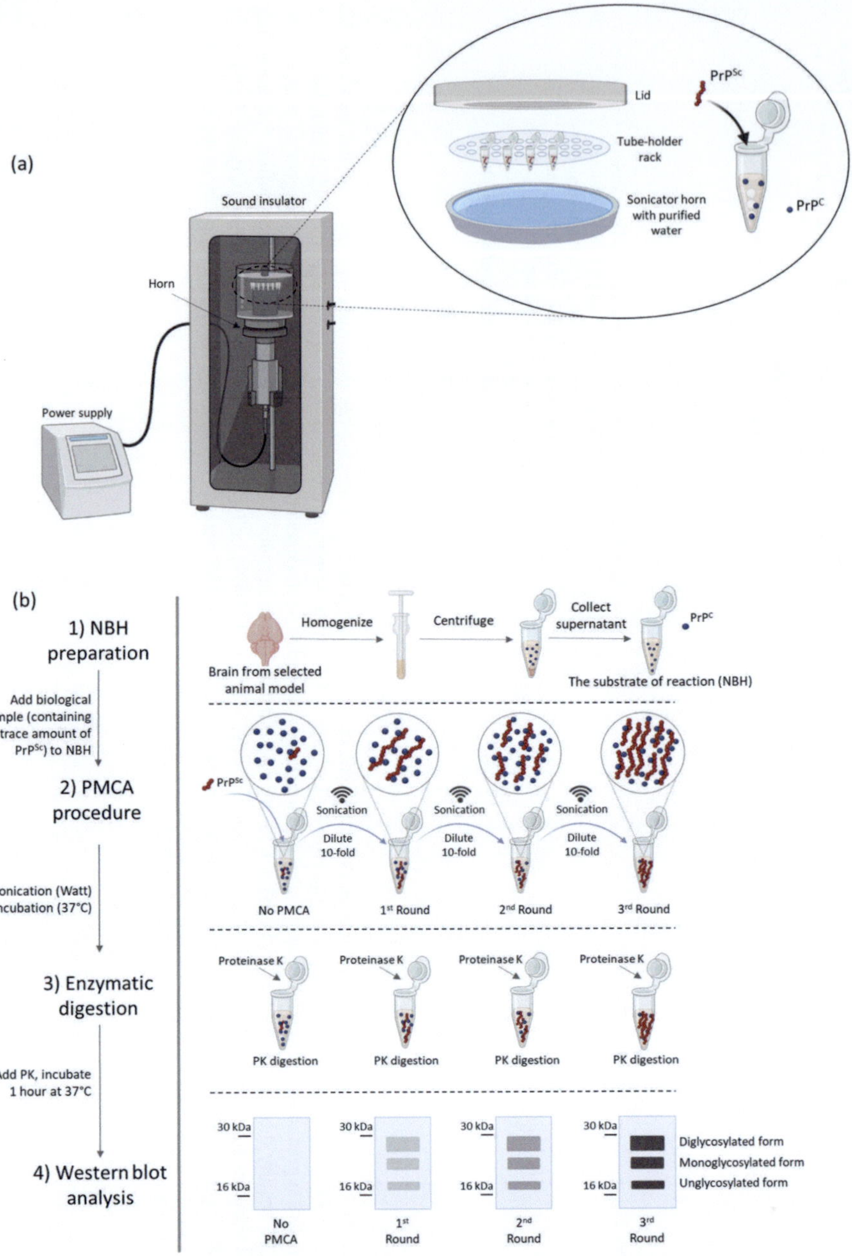

Fig. 12.2 Graphic representation of PMCA workflow and equipment. (**a**) PCR tubes containing the reaction mix (normal brain homogenate (NBH) + biological samples eventually containing traces of PrP^{Sc}) are inserted in the tube-holder rack, allowing their partial immersion in the water of the sonicator horn. The horn lid helps to keep the rack in place during the sonication phase. The horn is placed inside an incubator set at 37 °C, which is attached to a power supply that alternates

PMCA Applications in Human Prion Disease Diagnosis

At present, there is no reliable method for diagnosing human PrDs before death (Soto 2004; Zerr 2022). Diagnosis of a probable or possible case relies on a combination of clinical evaluations, disease progression, cerebrospinal fluid (CSF) biomarkers, and neuroimaging techniques (Soto 2004). A definitive diagnosis requires neuropathological analysis of the brain for PrP^{Sc} detection (Soto 2004; Zerr 2022). Consequently, most definitive diagnoses for CJD are made *postmortem* through autopsy. PMCA has introduced the possibility of identifying the presence of minute amounts of prions in the peripheral tissues of patients with various forms of PrDs. Given its high sensitivity and specificity, PMCA is making a significant contribution to improving the clinical diagnosis of these devastating diseases.

Prion Detection in the Urine

In 2014, Moda et al. reported the first successful detection of prions in the urine of patients with vCJD using the PMCA technique. In particular, the urine of 14 vCJD, 68 sCJD, 4 gCJD, 50 patients with other neurodegenerative disorders (including Parkinson's disease, motor neuron disease, Alzheimer's disease, progressive supranuclear palsy, and frontotemporal dementia), 50 patients with nondegenerative neurologic disorders (including multiple sclerosis, cerebrovascular disease, brain tumors, epilepsy, myeloradiculopathy, autoimmune encephalitis, peripheral neuropathy, and meningitis), and 52 healthy subjects (HS) underwent PMCA analysis. The authors successfully detected prions in 13 out of 14 urine samples from vCJD patients, achieving 100% specificity and an overall sensitivity of 92.9% (Moda et al. 2014). Using quantitative PMCA (qPMCA), the concentration of prions in urine was estimated to be 1×10^{-16} g/mL (3×10^{-21} mol/mL). The brains of transgenic mice expressing Hu129M PrP^{C} were used as a reaction substrate (Telling et al. 1994). Notably, the amplified prions preserved the distinctive glycoform ratio characteristic of prions formed in the brains of vCJD patients (type 2B mobility with predominance of the diglycosylated PrP isoform, Fig. 12.1c) (Collinge et al. 1996). To assess the detectability of PrP^{Sc} in the urine of vCJD patients at different clinical stages, two samples from the same patient were collected 117 days apart, both of which tested positive. The later sample contained a higher concentration of PrP^{Sc}, as evidenced by its detection requiring fewer rounds

Fig. 12.2 (continued) cycles of sonication and incubation. (**b**) Steps of the PMCA procedure: (1) the reaction substrate is prepared by homogenizing the normal brain of the selected animal model (NBH); (2) the biological sample, which might include traces of PrP^{Sc} is added to the reaction substrate and several rounds of amplification are performed; (3) all samples are treated with proteinase K (PK) and analyzed by Western blot to assess potential prion amplification; and (4) PrP^{res} signal, characterized by the presence of di-, mono-, and unglycosylated bands, becomes more intense after each round of amplification of PMCA. Prions were not detected in any urine samples from patients with sCJD or gCJD. These results suggested either that prions were absent in the urine of patients with these forms of PrDs or that the experimental PMCA protocol selectively amplified the vCJD strain.

To assess whether the amplified PrP^{Sc} retained its infectious properties, the reaction products were intracerebrally inoculated into transgenic mice expressing human PrP (Tg40). Notably, animals injected with urine-amplified PrP^{Sc} showed clinical, biochemical, and neuropathological features indistinguishable from those observed in animals injected with vCJD brain homogenate (Cali et al. 2019). This highlights the critical importance of handling these materials with extreme caution, strictly adhering to all biosafety protocols designed for the manipulation of brain samples from vCJD patients.

Pritzkow and colleagues recently refined the PMCA protocol and optimized the pre-analytical processing of urine samples to facilitate the detection of prions in the urine of sCJD patients. To this end, urine samples were collected from 81 sCJD patients, 94 patients with other neurodegenerative diseases, and 66 HS (Pritzkow et al. 2023). In particular, urine samples underwent iron oxide magnetic extraction (IOME) prior to amplification, which was carried out using a modified experimental protocol. This protocol involved the use of brain homogenates from mice expressing Hu129M or Hu129V PrP^C, supplemented with 6 mM EDTA, 0.05% digitonin, 0.01% sodium tripolyphosphate, and 100 μg/mL heparin. The adapted PMCA protocol enabled prion detection in 29/81 sCJD urine samples. The efficiency of amplification varied across disease subtypes, with sCJD prions detected in the urine of 5 out of 24 MM1 patients, 0 out of 8 MV1, 3 out of 3 VV1, 0 out of 16 MM2, 5 out of 11 MV2, and 16 out of 19 VV2 patients. Notably, patients homozygous for valine at position 129 of the *PRNP* gene (VV) showed significantly higher sensitivity for urinary prion detection (see Table 12.1). No prions were detected in control urine samples, resulting in an overall sensitivity of 36% and specificity of 100%. Overall, the results suggest that inflammation present in certain patients with specific phenotypes of sCJD might elevate the secretion of PrP^{Sc} in urine, as demonstrated in animal models (Seeger et al. 2005).

Prion Detection in the Blood

In 2011, Edgeworth J.A. et al. developed an assay capable of detecting vCJD prions directly in whole blood samples. The technology leveraged a solid-state binding matrix to capture and concentrate PrP^{Sc}, followed by immunodetection of the surface-bound material. The assay was tested on a cohort of 190 whole blood samples, including 21 from patients with vCJD, 16 with probable sCJD, 11 with definite sCJD, 25 with sporadic Alzheimer's disease, 6 with familial Alzheimer's disease, 4 with frontotemporal dementia, 7 with other neurological conditions, and 100 from HS. Notably, the assay identified vCJD samples with a sensitivity of

71.4% (15/21) and a specificity of 100% (Edgeworth et al. 2011). This assay demonstrated that blood samples could contain infectious prions. The first study showing the potential of PMCA to detect prions in the blood of vCJD patients was published in 2014 by Andreoletti and colleagues. In this work, the authors successfully detected vCJD prions in blood samples from 3 out of 4 confirmed vCJD patients, with no false positives among 141 HS. A particularly notable aspect of this study was the choice of ovine PrP^{C} as the reaction substrate. This decision was based on its superior performance compared to other tested substrates, including murine, bovine, and human PrP^{C}. This substrate optimization marked a significant advancement, highlighting the critical role of substrate selection in enhancing the sensitivity and reliability of PMCA-based diagnostics (Lacroux et al. 2014). In 2016, two research groups published optimized PMCA protocols that further enhanced the sensitivity and efficiency of prion detection in the blood of vCJD patients. The first protocol was published by the group of Soto (Concha-Marambio et al. 2016). In this work, blood samples were collected from 14 vCJD patients, 16 sCJD patients, 88 patients with other diseases, and 49 HS. Before analysis, all samples were treated with sarkosyl and subsequently subjected to high-speed centrifugation to remove all components potentially interfering with PMCA. The brain homogenates of Hu129M mice were used as reaction substrates, enabling the detection of prions in all vCJD blood samples analyzed (14/14). No prions were amplified from any other samples, including those from sCJD patients. The protocol showed the ability to amplify vCJD prions using an extremely small volume of whole blood. The second protocol was published in the same year by the group of Coste (Bougard et al. 2016). In this work, PMCA analyses were performed on plasma samples pre-treated with plasminogen-coated magnetic nanobeads to capture PrP^{Sc}. Brain homogenates from Hu129M mice were used as reaction substrates, allowing for the detection of PrP^{Sc} in plasma from all vCJD patients (18/18) at the clinical stage of the disease, with 100% sensitivity and specificity, regardless of the anticoagulant used. The authors also analyzed samples from 238 controls, and none tested positive for PMCA. Instead, in the cohort of 67 sCJD patients, only one tested positive. Furthermore, prions were amplified in two plasma samples collected from donors before the onset of vCJD, demonstrating that prions circulate in the blood during the preclinical stages of the disease (31 and 16 months before symptoms onset) (see Table 12.1). Given the limited availability of preclinical vCJD blood samples, nonhuman primates were peripherally infected with macaque-adapted vCJD prions to test the efficacy of PMCA for preclinical prion detection. Blood samples were collected longitudinally, and surprisingly, prions were detected as early as 65 days post-infection. Notably, the levels of PrP^{Sc} in the blood increased as the disease progressed (Concha-Marambio et al. 2020).

All these studies have shown that PMCA can detect prions in the blood of vCJD patients, whereas prions have been rarely, if ever, detected in the blood of sCJD patients. Given this distinction, there is a clear need to further optimize PMCA for the analysis of blood and its products collected from patients with sCJD.

Prion Detection in the Cerebrospinal Fluid

In 2018, Barria et al. developed a highly sensitive PMCA (hsPMCA) for the detection of prions in the CSF of patients with vCJD. In particular, CSF samples were collected from 15 vCJD patients, 6 sCJD patients, and 35 controls (Barria et al. 2018). CSF samples were directly subjected to PMCA analysis without any pretreatment. For the reaction substrate, brains from Hu129M mice were used, which allowed for the detection of PrP^{Sc} in the CSF of all vCJD patients, achieving 100% sensitivity and specificity. Interestingly, the authors successfully analyzed the CSF from the first 129 MV patient diagnosed with vCJD and efficiently detected prions in the sample (Mok et al. 2017). In the same year, Bougard et al. performed a PMCA study using a larger number of CSF samples, including 41 vCJD, 23 sCJD, 1 gCJD case, and 33 controls. The test enabled the detection of vCJD prions with 97.6% sensitivity (40/41 vCJD samples tested positive). Among the 40 positive cases, the CSF sample from the first 129 MV vCJD patient was also included and successfully amplified. None of the sCJD samples tested positive for PMCA (Bougard et al. 2018). Similarly, all controls included in the study tested negative. Therefore, in this case, the technique specifically amplified vCJD prions. However, in 2021, Bélondrade et al. achieved the first successful amplification of prions from CSF samples of sCJD patients, albeit on a very limited number of samples (Bélondrade et al. 2021) (see Table 12.1). This finding suggests that further optimization of PMCA protocols could enable more consistent amplification of sCJD prions. Consequently, reanalysis of previously described fluids using these improved conditions is warranted.

Prion Detection in the Olfactory Mucosa

The olfactory mucosa (OM) can be easily and noninvasively collected, making it a valuable diagnostic tissue with potential utility in supporting the clinical diagnosis of human PrDs. In 2017, Redaelli et al. optimized PMCA technology to detect prions in the OM of FFI patients in the late stages of the disease. In particular, the group analyzed 28 OM samples collected from 2 patients with FFI, 6 patients with Alzheimer's disease, 6 patients with Parkinson's disease, 4 patients with frontotemporal dementia, and 10 HS (Redaelli et al. 2017). The PMCA technique successfully amplified PrP^{Sc} from both FFI samples. In this case, the brain homogenate of the bank vole carrying the M109 PrP genotype was used as a reaction substrate. Interestingly, the prions obtained from the amplification of OM samples from FFI patients showed the same biochemical properties as prions found in the brains of FFI patients, even though the PMCA reaction substrate was derived from the bank vole species. Remarkably, the PMCA-generated products were intracerebrally inoculated in mice genetically modified to express the bank vole PrP^{C} with methionine at codon 109 (BvPrP-Tg407) and caused prion pathology. Inoculated animals developed mild spongiform changes, astroglial activation, and prion deposition, primarily in the thalamus, as typically observed in the brains of FFI patients (Bistaffa et al. 2021). This finding confirms that, as previously described with urine from vCJD patients, the amplified products are infectious. Therefore, appropriate biosafety precautions must be implemented when handling this material. In 2022, Cazzaniga et al. further optimized the PMCA technology for the detection of prions in the OM of patients with different subtypes of sCJD (Cazzaniga et al. 2022). In particular, OM samples were collected from 27 patients with sCJD (MM = 13, MV = 8, and VV = 6), 2 patients with gCJD (E200K), and 36 patients with other disorders, including 3 patients with Alzheimer's disease, 7 patients with Parkinson's disease, 7 patients with frontotemporal dementia, 4 patients with multiple system atrophy, 7 patients with progressive supranuclear palsy, 6 patients with corticobasal degeneration, and 2 patients with multiple sclerosis. The assay was able to detect sCJD prions in OM samples with 79.3% sensitivity and 100% specificity. As observed in the case of sCJD urine, the efficiency of amplification varied across sCJD subtypes. However, the biochemical properties of the amplified prions did not facilitate the identification of sCJD subtypes in living patients. The availability of two distinct biological samples (brain and OM) from the same patient enabled a direct comparison of the biochemical profiles of prion strains following PMCA amplification. Notably, PMCA products derived from OM showed reduced resistance to PK digestion compared to those amplified from brain tissue. Interestingly, in three cases, prions were successfully amplified from OM but not from brain tissue of the same patient, underscoring variations in PMCA efficiency and eventually prion distribution across different biological substrates and sCJD subtypes (Cazzaniga et al. 2022) (see Table 12.1).

Quantitative PMCA

To estimate the concentration of very low levels of prions in biological samples, the group of Soto developed the qPMCA (Chen et al. 2010). The key principle of this method is the direct correlation between the amount of prions in a sample and the number of PMCA cycles required to detect them. Briefly, a biological sample with an unknown prion concentration (to be calculated) is subjected to PMCA, together with the test samples containing known prion dilutions (properly prepared by the operator). When prion amplification is observed in the sample of interest, the corresponding test dilution that amplifies alongside the sample is verified. This enables the estimation of the prion concentration in the sample of interest. The qPMCA was applied to estimate the amount of prions in different biological samples collected from patients with different PrDs: (i) 1×10^{-16} g per mL of prions in urine samples from patients with vCJD (Moda et al. 2014); (ii) 5×10^{-13} g per mL of prions in plasma and white blood cell (WBC) fractions from patients with vCJD (Concha-Marambio et al. 2016); (iii) 1×10^{-14} g per mL of prions in OM collected from patients with FFI (Redaelli et al. 2017); and (iv) from 10^{-11} to 10^{-21} g of prions in

0.8 μg of OM samples from patients with sCJD (Cazzaniga et al. 2022). The variability in these concentrations may be associated with the disease stage and the type of PrDs, including specific subtypes, as observed in the case of sCJD.

Conclusions and Future Perspectives

Prions are unconventional infectious agents, and their unique nature demands equally unconventional tools for detection. The PMCA technology addresses this challenge. The technique is extremely sensitive and capable of detecting even a single molecule of the infectious agent if present in a biological sample (Saá et al. 2006). Based on the data reported in the literature, PMCA was initially capable of detecting only the prions responsible for vCJD. However, with continuous optimizations, it is now able to detect prions associated with more common, though still rare, forms of the disease, such as sCJD. However, while prions amplified from peripheral tissues of vCJD patients retain their typical biochemical and infectious properties, those from sCJD patients do not preserve their biochemical characteristics. Preliminary inoculation tests in animal models confirm the infectivity of the amplified products but reveal atypical pathological properties (Moda F., personal communication). Further optimization of the technique is necessary to enhance its sensitivity and its ability to identify sCJD subtypes in living patients. This is especially important for emerging pharmacological trials, such as the PrProfile trial by IONIS (https://cjdisa.com/clinical-trial-update-for-prion-diseases/) or others where accurate patient stratification might be essential for improving participant selection. Finally, protein misfolding is a process implicated in other proteins associated with more common neurodegenerative disorders, including Parkinson's disease, atypical parkinsonisms, dementia with Lewy bodies, Alzheimer's disease, frontotemporal dementia, and amyotrophic lateral sclerosis (Forloni et al. 2002; Cummings 2004; Parakh and Atkin 2016; Moda et al. 2023; Zampar et al. 2024). As a result, PMCA is being adapted for use in analyzing biological samples from patients with these conditions to improve their clinical diagnosis (Salvadores et al. 2014; Saijo et al. 2017; Bongianni et al. 2019; De Luca et al. 2019; Saijo et al. 2019; Saijo et al. 2020; Metrick et al. 2020; Scialò et al. 2020; Perra et al. 2021; Stefani et al. 2021; Concha-Marambio et al. 2021; Bargar et al. 2021; Bellomo et al. 2022; Coysh and Mead 2022; Bongianni et al. 2022; D'Andrea et al. 2023; Concha-Marambio et al. 2023a, b; Kuzkina et al. 2023; Okuzumi et al. 2023; Vascellari et al. 2023; Brockmann et al. 2024; Fontana et al. 2024; Dellarole et al. 2024; Pilotto et al. 2024; Ma et al. 2024).

Compliance with Ethical Standards The authors declare no competing financial or nonfinancial interests.

References

Alper T, Cramp WA, Haig DA, Clarke MC (1967) Does the agent of scrapie replicate without nucleic acid? Nature 214:764–766. https://doi.org/10.1038/214764a0

Bargar C, De Luca CMG, Devigili G, Elia AE, Cilia R, Portaleone SM, Wang W, Tramacere I, Bistaffa E, Cazzaniga FA, Felisati G, Legname G, Di Fonzo A, Xu R, Gunzler SA, Giaccone G, Eleopra R, Chen SG, Moda F (2021) Discrimination of MSA-P and MSA-C by RT-QuIC analysis of olfactory mucosa: the first assessment of assay reproducibility between two specialized laboratories. Mol Neurodegener 16:82. https://doi.org/10.1186/s13024-021-00491-y

Barria MA, Lee A, Green AJE, Knight R, Head MW (2018) Rapid amplification of prions from variant Creutzfeldt-Jakob disease cerebrospinal fluid. J Pathol Clin Res 4:86–92. https://doi.org/10.1002/cjp2.90

Bellomo G, Giulia De Luca CM, Paoletti FP, Gaetani L, Moda F, Parnetti L (2022) Alpha synuclein seed amplification assays for diagnosing synucleinopathies: the way forward. Neurology 99:195. https://doi.org/10.1212/WNL.0000000000200878

Bélondrade M, Nicot S, Mayran C, Bruyere-Ostells L, Almela F, Di Bari MA, Levavasseur E, Watts JC, Fournier-Wirth C, Lehmann S, Haïk S, Nonno R, Bougard D (2021) Sensitive protein misfolding cyclic amplification of sporadic Creutzfeldt-Jakob disease prions is strongly seed and substrate dependent. Sci Rep 11:4058. https://doi.org/10.1038/s41598-021-83630-1

Bistaffa E, Marín-Moreno A, Espinosa JC, De Luca CMG, Cazzaniga FA, Portaleone SM, Celauro L, Legname G, Giaccone G, Torres JM, Moda F (2021) Pmca-generated prions from the olfactory mucosa of patients with fatal familial insomnia cause prion disease in mice. elife 10. https://doi.org/10.7554/ELIFE.65311

Bongianni M, Ladogana A, Capaldi S, Klotz S, Baiardi S, Cagnin A, Perra D, Fiorini M, Poleggi A, Legname G, Cattaruzza T, Janes F, Tabaton M, Ghetti B, Monaco S, Kovacs GG, Parchi P, Pocchiari M, Zanusso G (2019) α-Synuclein RT-QuIC assay in cerebrospinal fluid of patients with dementia with Lewy bodies. Ann Clin Transl Neurol 6:2120. https://doi.org/10.1002/acn3.50897

Bongianni M, Catalan M, Perra D, Fontana E, Janes F, Bertolotti C, Sacchetto L, Capaldi S, Tagliapietra M, Polverino P, Tommasini V, Bellavita G, Kachoie EA, Baruca R, Bernardini A, Valente M, Fiorini M, Bronzato E, Tamburin S, Bertolasi L, Brozzetti L, Cecchini MP, Gigli G, Monaco S, Manganotti P, Zanusso G (2022) Olfactory swab sampling optimization for α-synuclein aggregate detection in patients with Parkinson's disease. Transl Neurodegener 11:37. https://doi.org/10.1186/s40035-022-00311-3

Bougard D, Brandel JP, Bélondrade M, Béringue V, Segarra C, Fleury H, Laplanche JL, Mayran C, Nicot S, Green A, Welaratne A, Narbey D, Fournier-Wirth C, Knight R, Will R, Tiberghien P, Haïk S, Coste J (2016) Detection of prions in the plasma of presymptomatic and symptomatic patients with variant Creutzfeldt-Jakob disease. Sci Transl Med 8:370ra182. https://doi.org/10.1126/scitranslmed.aag1257

Bougard D, Bélondrade M, Mayran C, Bruyère-Ostells L, Lehmann S, Fournier-Wirth C, Knight RS, Will RG, Green AJE (2018) Diagnosis of methionine/valine variant creutzfeldt-jakob disease by protein misfolding cyclic amplification. Emerg Infect Dis 24:1364. https://doi.org/10.3201/eid2407.172105

Brockmann K, Lerche S, Baiardi S, Rossi M, Wurster I, Quadalti C, Roeben B, Mammana A, Zimmermann M, Hauser A, Deuschle C, Schulte C, Liepelt-Scarfone I, Gasser T, Parchi P (2024) CSF α-synuclein seed amplification kinetic profiles are associated with cognitive decline in Parkinson's disease. NPJ Park Dis 10:24. https://doi.org/10.1038/s41531-023-00627-5

Bruce ME (2003) TSE strain variation. Br Med Bull 66:99–108. https://doi.org/10.1093/bmb/66.1.99

Bruce ME, Will RG, Ironside JW, McConnell I, Drummond D, Suttie A, McCardle L, Chree A, Hope J, Birkett C, Cousens S, Fraser H, Bostock CJ (1997) Transmissions to mice indicate that "new variant" CJD is caused by the BSE agent. Nature 389:498–501. https://doi.org/10.1038/39057

Cali I, Lavrich J, Moda F, Kofskey D, Nemani SK, Appleby B, Tagliavini F, Soto C, Gambetti P, Notari S (2019) PMCA-replicated PrPD in urine of vCJD patients maintains infectivity and strain characteristics of brain PrPD: transmission study. Sci Rep 9:5191. https://doi.org/10.1038/s41598-019-41694-0

Castle AR, Gill AC (2017) Physiological functions of the cellular prion protein. Front Mol Biosci 4. https://doi.org/10.3389/fmolb.2017.00019

Cazzaniga FA, Bistaffa E, De Luca CMG, Portaleone SM, Catania M, Redaelli V, Tramacere I, Bufano G, Rossi M, Caroppo P, Giovagnoli AR, Tiraboschi P, Di Fede G, Eleopra R, Devigili G, Elia AE, Cilia R, Fiorini M, Bongianni M, Salzano G, Celauro L, Quarta FG, Mammana A, Legname G, Tagliavini F, Parchi P, Zanusso G, Giaccone G, Moda F (2022) PMCA-based detection of prions in the olfactory mucosa of patients with sporadic Creutzfeldt–Jakob disease. Front Aging Neurosci. https://doi.org/10.3389/fnagi.2022.848991

Chen B, Morales R, Barria MA, Soto C (2010) Estimating prion concentration in fluids and tissues by quantitative PMCA. Nat Methods 7:519–520. https://doi.org/10.1038/nmeth.1465

Collinge J (2001) Prion diseases of humans and animals: their causes and molecular basis. Annu Rev Neurosci 24:519–550. https://doi.org/10.1146/annurev.neuro.24.1.519

Collinge J, Sidle KCL, Meads J, Ironside J, Hill AF (1996) Molecular analysis of prion strain variation and the aetiology of "new variant" CJD. Nature 383:685. https://doi.org/10.1038/383685a0

Concha-Marambio L, Pritzkow S, Moda F, Tagliavini F, Ironside JW, Schulz PE, Soto C (2016) Detection of prions in blood from patients with variant Creutzfeldt-Jakob disease. Sci Transl Med 8:370ra183. https://doi.org/10.1126/scitranslmed.aaf6188

Concha-Marambio L, Chacon MA, Soto C (2020) Preclinical detection of prions in blood of non-human primates infected with variant creutzfeldt-jakob disease. Emerg Infect Dis 26:34–43. https://doi.org/10.3201/eid2601.181423

Concha-Marambio L, Farris CM, Holguin B, Ma Y, Seibyl J, Russo MJ, Kang UJ, Hutten SJ, Merchant K, Shahnawaz M, Soto C (2021) Seed amplification assay to diagnose early Parkinson's and predict dopaminergic deficit progression. Mov Disord 36:2444–2446. https://doi.org/10.1002/mds.28715

Concha-Marambio L, Pritzkow S, Shahnawaz M, Farris CM, Soto C (2023a) Seed amplification assay for the detection of pathologic alpha-synuclein aggregates in cerebrospinal fluid. Nat Protoc 18:1179–1196. https://doi.org/10.1038/s41596-022-00787-3

Concha-Marambio L, Weber S, Farris CM, Dakna M, Lang E, Wicke T, Ma Y, Starke M, Ebentheuer J, Sixel-Döring F, Muntean M, Schade S, Trenkwalder C, Soto C, Mollenhauer B (2023b) Accurate detection of α-Synuclein seeds in cerebrospinal fluid from isolated rapid eye movement sleep behavior disorder and patients with Parkinson's disease in the DeNovo Parkinson (DeNoPa) cohort. Mov Disord 38:567–578. https://doi.org/10.1002/mds.29329

Coysh T, Mead S (2022) The future of seed amplification assays and clinical trials. Front Aging Neurosci 14. https://doi.org/10.3389/fnagi.2022.872629

Cummings JL (2004) Dementia with Lewy bodies: molecular pathogenesis and implications for classification. J Geriatr Psychiatry Neurol 17:112–119. https://doi.org/10.1177/0891988704267473

D'Andrea C, Cazzaniga FA, Bistaffa E, Barucci A, de Angelis M, Banchelli M, Farnesi E, Polykretis P, Marzi C, Indaco A, Tiraboschi P, Giaccone G, Matteini P, Moda F (2023) Impact of seed amplification assay and surface-enhanced Raman spectroscopy combined approach on the clinical diagnosis of Alzheimer's disease. Transl Neurodegener 12:35. https://doi.org/10.1186/s40035-023-00367-9

De Luca CMG, Elia AE, Portaleone SM, Cazzaniga FA, Rossi M, Bistaffa E, De Cecco E, Narkiewicz J, Salzano G, Carletta O, Romito L, Devigili G, Soliveri P, Tiraboschi P, Legname G, Tagliavini F, Eleopra R, Giaccone G, Moda F (2019) Efficient RT-QuIC seeding activity for α-synuclein in olfactory mucosa samples of patients with Parkinson's disease and multiple system atrophy. Transl Neurodegener 8:24. https://doi.org/10.1186/s40035-019-0164-x

Dear DV, Young DS, Kazlauskaite J, Meersman F, Oxley D, Webster J, Pinheiro TJT, Gill AC, Bronstein I, Lowe CR (2007) Effects of post-translational modifications on prion protein aggregation and the propagation of scrapie-like characteristics in vitro. Biochim Biophys Acta Proteins Proteomics 1774:792–802. https://doi.org/10.1016/j.bbapap.2007.05.005

Dellarole IL, Vacchi E, Ruiz-Barrio I, Pinton S, Raimondi A, Rossi S, Morandi S, Bianco G, Begum Bacinoglu M, Lombardo A, Celauro L, Staedler C, Galati S, Pagonabarraga J, Kulisevsky J, Legname G, Gobbi C, Kaelin-Lang A, Moda F, Melli G (2024) Tau seeding activity in skin biopsy differentiates tauopathies from synucleinopathies. NPJ Park Dis 10:116. https://doi.org/10.1038/s41531-024-00728-9

Dima RI, Thirumalai D (2002) Exploring the propensities of helices in PrPC to form β sheet using NMR structures and sequence alignments. Biophys J 83:1268. https://doi.org/10.1016/S0006-3495(02)73899-X

Edgeworth JA, Farmer M, Sicilia A, Tavares P, Beck J, Campbell T, Lowe J, Mead S, Rudge P, Collinge J, Jackson GS (2011) Detection of prion infection in variant Creutzfeldt-Jakob disease: a blood-based assay. Lancet 377:487–493. https://doi.org/10.1016/S0140-6736(10)62308-2

Fontana E, Bongianni M, Benussi A, Bronzato E, Scialo C, Sacchetto L, Cagnin A, Castriciano S, Buratti E, Gardoni F, Italia M, Schreiber A, Ferracin C, Fiorini M, Newell KL, Cracco L, Garringer HJ, Cecchini MP, Polymenidou M, Padovani A, Monaco S, Legname G, Ghetti B, Borroni B, Zanusso G (2024) Detection of TDP-43 seeding activity in the olfactory mucosa from patients with frontotemporal dementia. Alzheimers Dement 20:1156–1165. https://doi.org/10.1002/alz.13541

Forloni G, Terreni L, Bertani I, Fogliarino S, Invernizzi R, Assini A, Ribizzi G, Negro A, Calabrese E, Volonté MA, Mariani C, Franceschi M, Tabaton M, Bertoli A (2002) Protein misfolding in Alzheimer's and Parkinson's disease: genetics and molecular mechanisms. Neurobiol Aging 23:957. https://doi.org/10.1016/S0197-4580(02)00076-3

Gambetti P, Kong Q, Zou W, Parchi P, Chen SG (2003) Sporadic and familial CJD: classification and characterisation. Br Med Bull 66:213–239. https://doi.org/10.1093/bmb/66.1.213

Gao LP, Tian TT, Xiao K, Chen C, Zhou W, Liang DL, Cao RD, Shi Q, Dong XP (2024) Updated global epidemiology atlas of human prion diseases. Front Public Heal 12. https://doi.org/10.3389/fpubh.2024.1411489

Griffith JS (1967) Self-replication and scrapie. Nature 215:1043–1044. https://doi.org/10.1038/2151043a0

Haraguchi T, Fisher S, Olofsson S, Endo T, Groth D, Tarentino A, Borchelt DR, Teplow D, Hood L, Burlingame A, Lycke E, Kobata A, Prusiner SB (1989) Asparagine-linked glycosylation of the scrapie and cellular prion proteins. Arch Biochem Biophys 274:1–13. https://doi.org/10.1016/0003-9861(89)90409-8

Head MW, Bunn TJR, Bishop MT, McLoughlin V, Lowrie S, McKimmie CS, Williams MC, McCardle L, MacKenzie J, Knight R, Will RG, Ironside JW (2004) Prion protein heterogeneity in sporadic but not variant Creutzfeldt–Jakob disease: U.K. cases 1991–2002. Ann Neurol 55:851–859. https://doi.org/10.1002/ana.20127

Hill AF, Joiner S, Beck JA, Campbell TA, Dickinson A, Poulter M, Wadsworth JDF, Collinge J (2006) Distinct glycoform ratios of protease resistant prion protein associated with PRNP point mutations. Brain 129:676–685. https://doi.org/10.1093/brain/awl013

Jarrett JT, Lansbury PT (1992) Amyloid fibril formation requires a chemically discriminating nucleation event: studies of an amyloidogenic sequence from the bacterial protein OsmB. Biochemistry 31:12345–12352. https://doi.org/10.1021/bi00164a008

Khalili-Shirazi A, Summers L, Linehan J, Mallinson G, Anstee D, Hawke S, Jackson GS, Collinge J (2005) PrP glycoforms are associated in a strain-specific ratio in native PrPSc. J Gen Virol 86:2635. https://doi.org/10.1099/vir.0.80375-0

Knight R, Collins S (2001) Human prion diseases: cause, clinical and diagnostic aspects. Contrib Microbiol 7:68–92. https://doi.org/10.1159/000060377

Kupfer L, Hinrichs W, Groschup M (2009) Prion protein misfolding. Curr Mol Med 9:826–835. https://doi.org/10.2174/156652409789105543

Kuzkina A, Rößle J, Seger A, Panzer C, Kohl A, Maltese V, Musacchio T, Blaschke SJ, Tamgüney G, Kaulitz S, Rak K, Scherzad A, Zimmermann PH, Klussmann JP, Hackenberg S, Volkmann J, Sommer C, Sommerauer M, Doppler K (2023) Combining skin and olfactory α-synuclein seed amplification assays (SAA)—towards biomarker-driven phenotyping in synucleinopathies. NPJ Park Dis 9:79. https://doi.org/10.1038/s41531-023-00519-8

Lacroux C, Comoy E, Moudjou M, Perret-Liaudet A, Lugan S, Litaise C, Simmons H, Jas-Duval C, Lantier I, Béringue V, Groschup M, Fichet G, Costes P, Streichenberger N, Lantier F, Deslys JP, Vilette D, Andréoletti O (2014) Preclinical detection of variant CJD and BSE prions in blood. PLoS Pathog 10:e1004202. https://doi.org/10.1371/journal.ppat.1004202

Ladogana A, Puopolo M, Croes EA, Budka H, Jarius C, Collins S, Klug GM, Sutcliffe T, Giulivi A, Alperovitch A, Delasnerie-Laupretre N, Brandel J-P, Poser S, Kretzschmar H, Rietveld I, Mitrova E, Cuesta J d P, Martinez-Martin P, Glatzel M, Aguzzi A, Knight R, Ward H, Pocchiari M, van Duijn CM, Will RG, Zerr I (2005) Mortality from Creutzfeldt-Jakob disease and related disorders in Europe, Australia, and Canada. Neurology 64:1586–1591. https://doi.org/10.1212/01.WNL.0000160117.56690.B2

Lee J, Kim SY, Hwang KJ, Ju YR, Woo H-J (2013) Prion diseases as transmissible zoonotic diseases. Osong Public Heal Res Perspect 4:57–66. https://doi.org/10.1016/j.phrp.2012.12.008

Legname G, Moda F (2017) The prion concept and synthetic prions. Prog Mol Biol Transl Sci 150:147–156. https://doi.org/10.1016/bs.pmbts.2017.06.002

Ma Y, Farris CM, Weber S, Schade S, Nguyen H, Pérez-Soriano A, Giraldo DM, Fernández M, Soto M, Cámara A, Painous C, Muñoz E, Valldeoriola F, Martí MJ, Clarimon J, Kallunki P, Ma TC, Alcalay RN, Gomes BF, Blennow K, Zetterberg H, Constantinescu J, Mengel D, Kadam V, Parchi P, Brockmann K, Tropea TF, Siderowf A, Synofzik M, Kang UJ, Compta Y, Svenningsson P, Mollenhauer B, Concha-Marambio L (2024) Sensitivity and specificity of a seed amplification assay for diagnosis of multiple system atrophy: a multicentre cohort study. Lancet Neurol 23:1225–1237. https://doi.org/10.1016/S1474-4422(24)00395-8

McKinley MP, Bolton DC, Prusiner SB (1983) A protease-resistant protein is a structural component of the scrapie prion. Cell 35:57–62. https://doi.org/10.1016/0092-8674(83)90207-6

Metrick MA, Ferreira NDC, Saijo E, Kraus A, Newell K, Zanusso G, Vendruscolo M, Ghetti B, Caughey B (2020) A single ultrasensitive assay for detection and discrimination of tau aggregates of Alzheimer and pick diseases. Acta Neuropathol Commun 8:22. https://doi.org/10.1186/s40478-020-0887-z

Miranzadeh Mahabadi H, Taghibiglou C (2020) Cellular prion protein (PrPc): putative interacting partners and consequences of the interaction. Int J Mol Sci 21:7058. https://doi.org/10.3390/ijms21197058

Moda F, Gambetti P, Notari S, Concha-Marambio L, Catania M, Park K-W, Maderna E, Suardi S, Haïk S, Brandel J-P, Ironside J, Knight R, Tagliavini F, Soto C (2014) Prions in the urine of patients with variant Creutzfeldt-Jakob disease. N Engl J Med 371:530–539. https://doi.org/10.1056/NEJMoa1404401

Moda F, Ciullini A, Dellarole IL, Lombardo A, Campanella N, Bufano G, Cazzaniga FA, Giaccone G (2023) Secondary protein aggregates in neurodegenerative diseases: almost the rule rather than the exception. Front Biosci 28:255. https://doi.org/10.31083/j.fbl2810255

Mok T, Jaunmuktane Z, Joiner S, Campbell T, Morgan C, Wakerley B, Golestani F, Rudge P, Mead S, Jäger HR, Wadsworth JDF, Brandner S, Collinge J (2017) Variant Creutzfeldt–Jakob Disease in a Patient with Heterozygosity at PRNP Codon 129. N Engl J Med. https://doi.org/10.1056/nejmc1610003

Oesch B, Westaway D, Wälchli M, McKinley MP, Kent SBH, Aebersold R, Barry RA, Tempst P, Teplow DB, Hood LE (1985) A cellular gene encodes scrapie PrP 27-30 protein. Cell 40:735–746. https://doi.org/10.1016/0092-8674(85)90333-2

Okuzumi A, Hatano T, Matsumoto G, Nojiri S, Ueno S, Imamichi-Tatano Y, Kimura H, Kakuta S, Kondo A, Fukuhara T, Li Y, Funayama M, Saiki S, Taniguchi D, Tsunemi T, McIntyre D, Gérardy J-J, Mittelbronn M, Kruger R, Uchiyama Y, Nukina N, Hattori N (2023) Propagative α-synuclein seeds as serum biomarkers for synucleinopathies. Nat Med 29:1448–1455. https://doi.org/10.1038/s41591-023-02358-9

Panes JD, Saavedra P, Pineda B, Escobar K, Cuevas ME, Moraga-Cid G, Fuentealba J, Rivas CI, Rezaei H, Muñoz-Montesino C (2021) PrPC as a transducer of physiological and pathological signals. Front Mol Neurosci 14. https://doi.org/10.3389/fnmol.2021.762918

Parakh S, Atkin JD (2016) Protein folding alterations in amyotrophic lateral sclerosis. Brain Res 1648:633–649. https://doi.org/10.1016/j.brainres.2016.04.010

Parchi P, Capellari S, Chen SG, Petersen RB, Gambetti P, Kopp N, Brown P, Kitamoto T, Tateishi J, Giese A, Kretzschmar H (1997) Typing prion isoforms. Nature 386:232–233. https://doi.org/10.1038/386232a0

Parchi P, Giese A, Capellari S, Brown P, Schulz-Schaeffer W, Windl O, Zerr I, Budka H, Kopp N, Piccardo P, Poser S, Rojiani A, Streichemberger N, Julien J, Vital C, Ghetti B, Gambetti P, Kretzschmar H (1999) Classification of sporadic Creutzfeldt-Jakob disease based on molecular and phenotypic analysis of 300 subjects. Ann Neurol 46:224–233. https://doi.org/10.1002/1531-8249(199908)46:2<224::AID-ANA12>3.0.CO;2-W

Parchi P, Zou W, Wang W, Brown P, Capellari S, Ghetti B, Kopp N, Schulz-Schaeffer WJ, Kretzschmar HA, Head MW, Ironside JW, Gambetti P, Chen SG (2000) Genetic influence on the structural variations of the abnormal prion protein. Proc Natl Acad Sci 97:10168–10172. https://doi.org/10.1073/pnas.97.18.10168

Parchi P, Strammiello R, Notari S, Giese A, Langeveld JPM, Ladogana A, Zerr I, Roncaroli F, Cras P, Ghetti B, Pocchiari M, Kretzschmar H, Capellari S (2009) Incidence and spectrum of sporadic Creutzfeldt-Jakob disease variants with mixed phenotype and co-occurrence of PrPSc types: an updated classification. Acta Neuropathol 118:659–671. https://doi.org/10.1007/s00401-009-0585-1

Peretz D, Williamson RA, Legname G, Matsunaga Y, Vergara J, Burton DR, DeArmond SJ, Prusiner SB, Scott MR (2002) A change in the conformation of prions accompanies the emergence of a new prion strain. Neuron 34:921–932. https://doi.org/10.1016/S0896-6273(02)00726-2

Perra D, Bongianni M, Novi G, Janes F, Bessi V, Capaldi S, Sacchetto L, Tagliapietra M, Schenone G, Morbelli S, Fiorini M, Cattaruzza T, Mazzon G, Orrù CD, Catalan M, Polverino P, Bernardini A, Pellitteri G, Valente M, Bertolotti C, Nacmias B, Maggiore G, Cavallaro T, Manganotti P, Gigli G, Monaco S, Nobili F, Zanusso G (2021) Alpha-synuclein seeds in olfactory mucosa and cerebrospinal fluid of patients with dementia with Lewy bodies. Brain Commun 3. https://doi.org/10.1093/braincomms/fcab045

Pilotto A, Zanusso G, Antelmi E, Okuzumi A, Zatti C, Lupini A, Bongianni M, Padovani A, Hattori N (2024) Biofluid markers and tissue biopsies analyses for the prodromal and earliest phase of Parkinson's disease. J Parkinsons Dis 14:S333–S344. https://doi.org/10.3233/JPD-240007

Pritzkow S, Ramirez F, Lyon A, Schulz PE, Appleby B, Moda F, Ramirez S, Notari S, Gambetti P, Soto C (2023) Detection of prions in the urine of patients affected by sporadic Creutzfeldt–Jakob disease. Ann Clin Transl Neurol 10:2316–2323. https://doi.org/10.1002/acn3.51919

Prusiner SB (1991) Molecular biology of prion diseases. Science (80-) 252:1515–1522. https://doi.org/10.1126/science.1675487

Prusiner SB (1998a) Nobel lecture: prions, vol 95. Proc Natl Acad Sci, p 13363

Prusiner SB (1998b) Prions. Proc Natl Acad Sci 95:13363–13383. https://doi.org/10.1073/pnas.95.23.13363

Prusiner SB (1998c) Prions: Nobel prize lecture, vol 95. Proc Natl Acad Sci, p 13363

Redaelli V, Bistaffa E, Zanusso G, Salzano G, Sacchetto L, Rossi M, De Luca CMG, Di Bari M, Portaleone SM, Agrimi U, Legname G, Roiter I, Forloni G, Tagliavini F, Moda F (2017) Detection of prion seeding activity in the olfactory mucosa of patients with fatal familial insomnia. Sci Rep 7:46269. https://doi.org/10.1038/srep46269

Riesner D (2003) Biochemistry and structure of PrPC and PrPSc. Br Med Bull 66:21–33. https://doi.org/10.1093/bmb/66.1.21

Saá P, Castilla J, Soto C (2006) Ultra-efficient replication of infectious prions by automated protein misfolding cyclic amplification. J Biol Chem 281:35245–35252. https://doi.org/10.1074/jbc.M603964200

Saborio GP, Permanne B, Soto C (2001) Sensitive detection of pathological prion protein by cyclic amplification of protein misfolding. Nature 411:810–813. https://doi.org/10.1038/35081095

Saijo E, Ghetti B, Zanusso G, Oblak A, Furman JL, Diamond MI, Kraus A, Caughey B (2017) Ultrasensitive and selective detection of 3-repeat tau seeding activity in pick disease brain and cerebrospinal fluid. Acta Neuropathol 133:751–765. https://doi.org/10.1007/s00401-017-1692-z

Saijo E, Groveman BR, Kraus A, Metrick M, Orrù CD, Hughson AG, Caughey B (2019) Ultrasensitive RT-QuIC seed amplification assays for disease-associated tau, α-Synuclein, and prion aggregates. Methods Mol Biol 1873:19–37. https://doi.org/10.1007/978-1-4939-8820-4_2

Saijo E, Metrick MA, Koga S, Parchi P, Litvan I, Spina S, Boxer A, Rojas JC, Galasko D, Kraus A, Rossi M, Newell K, Zanusso G, Grinberg LT, Seeley WW, Ghetti B, Dickson DW, Caughey B (2020) 4-repeat tau seeds and templating subtypes as brain and CSF biomarkers of frontotemporal lobar degeneration. Acta Neuropathol 139:63–77. https://doi.org/10.1007/s00401-019-02080-2

Salvadores N, Shahnawaz M, Scarpini E, Tagliavini F, Soto C (2014) Detection of misfolded aβ oligomers for sensitive biochemical diagnosis of Alzheimer's disease. Cell Rep 7:261–268. https://doi.org/10.1016/j.celrep.2014.02.031

Scialò C, Tran TH, Salzano G, Novi G, Caponnetto C, Chiò A, Calvo A, Canosa A, Moda F, Caroppo P, Silani V, Ticozzi N, Ratti A, Borroni B, Benussi L, Ghidoni R, Furlanis G, Manganotti P, Senigagliesi B, Parisse P, Brasselet R, Buratti E, Legname G (2020) TDP-43 real-time quaking induced conversion reaction optimization and detection of seeding activity in CSF of amyotrophic lateral sclerosis and frontotemporal dementia patients. Brain Commun 2. https://doi.org/10.1093/braincomms/fcaa142

Seeger H, Heikenwalder M, Zeller N, Kranich J, Schwarz P, Gaspert A, Seifert B, Miele G, Aguzzi A (2005) Medicine: coincident scrapie infection and nephritis lead to urinary prion excretion. Science (80-) 310:324–326. https://doi.org/10.1126/science.1118829

Silva CJ, Vázquez-Fernández E, Onisko B, Requena JR (2015) Proteinase K and the structure of PrPSc: the good, the bad and the ugly. Virus Res 207:120–126. https://doi.org/10.1016/j.virusres.2015.03.008

Soto C (2004) Diagnosing prion diseases: needs, challenges and hopes. Nat Rev Microbiol 2:809–819. https://doi.org/10.1038/nrmicro1003

Stahl N, Borchelt DR, Hsiao K, Prusiner SB (1987) Scrapie prion protein contains a phosphatidylinositol glycolipid. Cell 51:229–240. https://doi.org/10.1016/0092-8674(87)90150-4

Stefani A, Iranzo A, Holzknecht E, Perra D, Bongianni M, Gaig C, Heim B, Serradell M, Sacchetto L, Garrido A, Capaldi S, Sánchez-Gómez A, Cecchini MP, Mariotto S, Ferrari S, Fiorini M, Schmutzhard J, Cocchiara P, Vilaseca I, Brozzetti L, Monaco S, Jose Marti M, Seppi K, Tolosa E, Santamaria J, Högl B, Poewe W, Zanusso G (2021) Alpha-synuclein seeds in olfactory mucosa of patients with isolated REM sleep behaviour disorder. Brain 144:1118–1126. https://doi.org/10.1093/brain/awab005

Telling GC, Scott M, Hsiao KK, Foster D, Yang SL, Torchia M, Sidle KCL, Collinge J, DeArmond SJ, Prusiner SB (1994) Transmission of Creutzfeldt-Jakob disease from humans to transgenic mice expressing chimeric human-mouse prion protein. Proc Natl Acad Sci USA 91:9936–9940. https://doi.org/10.1073/pnas.91.21.9936

Vascellari S, Orrù CD, Groveman BR, Parveen S, Fenu G, Pisano G, Piga G, Serra G, Oppo V, Murgia D, Perra A, Angius F, Hughson AG, Haigh CL, Manzin A, Cossu G, Caughey B (2023) α-Synuclein seeding activity in duodenum biopsies from Parkinson's disease patients. PLoS Pathog 19:e1011456. https://doi.org/10.1371/journal.ppat.1011456

Wüthrich K, Riek R (2001) Three-dimensional structures of prion proteins. Adv Protein Chem 57:55–82. https://doi.org/10.1016/s0065-3233(01)57018-7

Yuan J, Xiao X, McGeehan J, Dong Z, Cali I, Fujioka H, Kong Q, Kneale G, Gambetti P, Zou WQ (2006) Insoluble aggregates and protease-resistant conformers of prion protein in uninfected human brains. J Biol Chem 281:34848. https://doi.org/10.1074/jbc.M602238200

Zahn R, Liu A, Lührs T, Riek R, Von Schroetter C, Garcia FL, Billeter M, Calzolai L, Wider G, Wüthrich K (2000) NMR solution structure of the human prion protein. Proc Natl Acad Sci USA 97:145. https://doi.org/10.1073/pnas.97.1.145

Zampar S, Di Gregorio SE, Grimmer G, Watts JC, Ingelsson M (2024) "Prion-like" seeding and propagation of oligomeric protein assemblies in neurodegenerative disorders. Front Neurosci 18:1–20. https://doi.org/10.3389/fnins.2024.1436262

Zerr I (2022) Laboratory diagnosis of Creutzfeldt–Jakob disease. N Engl J Med 386:1345–1350. https://doi.org/10.1056/NEJMra2119323

Zou WQ, Puoti G, Xiao X, Yuan J, Qing L, Cali I, Shimoji M, Langeveld JPM, Castellani R, Notari S, Crain B, Schmidt RE, Geschwind M, DeArmond SJ, Cairns NJ, Dickson D, Honig L, Torres JM, Mastrianni J, Capellari S, Giaccone G, Belay ED, Schonberger LB, Cohen M, Perry G, Kong Q, Parchi P, Tagliavini F, Gambetti P (2010) Variably protease-sensitive prionopathy: a new sporadic disease of the prion protein. Ann Neurol 68:162. https://doi.org/10.1002/ana.22094

Zou W-Q, Zhou X, Yuan J, Xiao X (2011) Insoluble cellular prion protein and its association with prion and Alzheimer diseases. Prion 5:172–178. https://doi.org/10.4161/pri.5.3.16894

Chapter 13
Seed Amplification Assays as Powerful Tools for Detecting Peripheral Biomarkers in Prion-Like Diseases

Ilaria Linda Dellarole, Annalisa Lombardo, Arianna Ciullini, Federico Angelo Cazzaniga, Rachele Domina, Merve Begüm Bacınoğlu, and Fabio Moda

Abstract Seed amplification assays (SAAs) are highly sensitive and advanced techniques originally developed for the study and diagnosis of prion diseases. Thanks to their remarkably high sensitivity and specificity, SAAs are now widely employed in both research and clinical settings for prion detection, especially in peripheral tissues of patients with prion disorders. Many neurodegenerative diseases, including Alzheimer's disease, Parkinson's disease, dementia with Lewy bodies, frontotemporal dementia, and amyotrophic lateral sclerosis, show prion-like mechanisms involving the misfolding and self-propagation of pathological proteins. As a result, SAAs are being adapted and refined for clinical use to improve the diagnosis of these conditions. This includes detecting traces of pathological proteins in cerebrospinal fluid as well as in minimally or noninvasively collected samples, such as blood, urine, skin, and olfactory mucosa. This chapter offers an overview of the role of SAAs in the clinical diagnosis of neurodegenerative diseases.

Keywords Seed amplification assay · Neurodegenerative diseases · Prion · Peripheral biomarkers · α-synuclein · Tau · β-amyloid · TDP-43

I. L. Dellarole · A. Lombardo · A. Ciullini · F. A. Cazzaniga · R. Domina · M. B. Bacınoğlu
Unit of Laboratory Medicine, Laboratory of Clinical Pathology, Fondazione IRCCS Istituto Neurologico Carlo Besta, Milan, Italy

F. Moda (✉)
Unit of Laboratory Medicine, Laboratory of Clinical Pathology, Fondazione IRCCS Istituto Neurologico Carlo Besta, Milan, Italy

Department of Medical Biotechnology and Translational Medicine, Università degli Studi di Milano, Milan, Italy
e-mail: fabio.moda@unimi.it

G. Legname, F. Moda (eds.), *Biomarkers and Therapeutical Targets for Prion Diseases*, Subcellular Biochemistry 112, https://doi.org/10.1007/978-3-031-97055-9_13

Beyond the Fold: How Protein Misfolding Drives Neurodegenerative Diseases

Neurodegenerative diseases (NDs) are characterized by the progressive degeneration of neurons in the central nervous system (CNS). These diseases manifest through a broad spectrum of clinical symptoms, including cognitive, motor, and behavioral impairments. For instance, Alzheimer's disease (AD) causes progressive memory loss, cognitive decline, and impaired executive functions (Saá et al. 2005; Gold and Budson 2008; Hari et al. 2024). Parkinson's disease (PD) presents motor symptoms such as tremors, rigidity, bradykinesia, and postural instability, often accompanied by nonmotor symptoms such as depression and sleep disturbances, such as isolated REM sleep behavior disorder (iRBD) (Samii et al. 2004). Amyotrophic lateral sclerosis (ALS) is characterized by progressive muscle weakness, spasticity, and paralysis, resulting in respiratory failure (Grad et al. 2017). NDs can result from the interplay of genetic, environmental, and lifestyle factors (Migliore and Coppedè 2009), with risk factors such as aging, oxidative stress, and inflammation playing crucial roles in their development (Mok et al. 2004; Farooqui and Farooqui 2009; Fischer and Maier 2015; Musgrove et al. 2019; Hou et al. 2019). Unfortunately, the exact pathological mechanisms underlying NDs are not yet fully understood. This lack of understanding affects the accuracy of clinical diagnoses and limits research for targeted therapies. Thus, developing treatments that target the molecular causes of these diseases, with the potential to halt or even reverse their progression, remains an important challenge. Currently, protein misfolding is widely recognized as a leading cause of NDs (Soto and Pritzkow 2018). Misfolded proteins are central to the pathology of several NDs, with each disease associated with distinct protein aggregates. For instance, misfolded prion protein (PrP^{Sc}) drives transmissible spongiform encephalopathies, or prion diseases, characterized by intracerebral accumulation of PrP^{Sc}. Similarly, misfolded α-synuclein (αSyn) underlies α-synucleinopathies, a group of disorders that includes PD, multiple system atrophy (MSA), dementia with Lewy bodies (DLB), and pure autonomic failure (PAF). Misfolded β-amyloid (Aβ) is primarily implicated in AD and cerebral amyloid angiopathy (CAA), while misfolded tau is associated with tauopathies, including corticobasal degeneration (CBD), progressive supranuclear palsy (PSP), Pick's disease (PiD), and AD. Additionally, misfolded TAR DNA-binding protein 43 (TDP-43) is the main pathological feature in frontotemporal lobar degeneration (FTLD), ALS, and limbic-predominant age-related TDP-43 encephalopathy (LATE). Protein folding is a tightly regulated process, yet under certain conditions, it can go awry, particularly when the protein has an intrinsically disordered structure (Chiti and Dobson 2006). Misfolded proteins can interact with their normally folded counterparts, promoting them to adopt a similar misfolded conformation. This initiates a self-perpetuating misfolding cascade, enabling them to spread throughout the CNS, where they aggregate into toxic oligomers and insoluble fibrillary structures. These species contribute to altering neuronal homeostasis and trigger a cascade of pathological processes. Misfolded proteins could exist in multiple aberrant structures, referred to as strains. Thus, the same protein can acquire different toxic properties, which can cause either different pathologies (e.g., PD vs. MSA) or even different phenotypes of the same disease (e.g., parkinsonian (MSA-P) versus cerebellar (MSA-C) forms of MSA) (Bousset et al. 2013; Fitzpatrick et al. 2017; Li et al. 2018; Van der Perren et al. 2020; Yang et al. 2022). Misfolded proteins are disease-specific markers, making them ideal biomarkers for the diagnosis of NDs. Their detection in *postmortem* brain tissue remains necessary for a definitive diagnosis. Unfortunately, clinical diagnosis of NDs, particularly in the early stages, is extremely challenging due to overlapping symptoms and the lack of reliable disease-specific biomarkers comparable to the misfolded proteins found in the brain. In recent years, two innovative assays, real-time quaking-induced conversion (RT-QuIC) (Atarashi et al. 2011) and protein misfolding cyclic amplification (PMCA) (Saá et al. 2005), have revolutionized the detection of trace amounts of PrP^{Sc} in cerebrospinal fluid (CSF) and other accessible peripheral tissues, such as olfactory mucosa (OM), urine, and blood from patients with prion diseases (Bongianni et al. 2017; Redaelli et al. 2017; Fiorini et al. 2020; Rhoads et al. 2020; Bistaffa et al. 2021; Cazzaniga et al. 2022; Watson et al. 2022). Despite methodological differences, both techniques rely on the same core principle of protein amplification through cycles of fragmentation and elongation, significantly advancing diagnostics and our understanding of prion disease pathophysiology. Given that αSyn, tau, Aβ, and TDP-43 exhibit prion-like properties, these assays have been adapted for broader ND research. To avoid confusion with prion-specific terminology, the term seed amplification assays (SAAs) was coined to encompass their application beyond prion disorders. SAAs now facilitate the detection of misfolded proteins in easily accessible tissues, improving the diagnostic accuracy of NDs. While CSF remains the preferred biological sample for NDs' clinical diagnosis, its collection is invasive and not easily repeatable. Consequently, SAAs are being refined for noninvasive samples such as OM, skin, saliva, tears, and blood. Emerging data suggest that misfolded proteins, once thought detectable only in brain tissue, can be identified in trace amounts in peripheral tissues, often at early disease stages, offering critical insights into disease onset and progression.

Prion-Like Mechanisms in Neurodegeneration

Originally discovered for prions, the pathological process of protein misfolding has also been identified in αSyn, tau, Aβ, and TDP-43, all of which are implicated in more common NDs (Törnquist et al. 2018). For this reason, these proteins are considered prion-like proteins:

- αSyn is a small protein of 14 kDa (140 amino acids) that is encoded by the *SNCA* gene. It is abundant in the brain, and while its precise function remains unclear, it is believed to play a role in synaptic function and neurotransmitter release (Giasson et al. 2001; Waxman and Giasson 2009; Burré 2015). Structurally, αSyn is an intrinsically disordered protein composed of three distinct domains. The N-terminal domain possesses lipid-binding properties. The central region contains a highly amyloidogenic domain known as the nonamyloid-β component (NAC) domain, which is primarily responsible for the protein's aggregation. Finally, the largely unstructured C-terminal acidic domain interacts with ions, proteins, and lipids. The C-terminal domain also plays a role in modulating membrane binding and protecting αSyn from aggregation (Ma et al. 2003; Farotti et al. 2020).

- Tau is an intrinsically disordered protein that stabilizes microtubules and regulates axonal transport. It exists in six isoforms in the adult human brain, which arise from alternative splicing of the *MAPT* (microtubule-associated protein tau) gene. These isoforms are classified by the number of microtubule-binding repeat domains in the C-terminal region:

  – 3R tau isoforms contain three microtubule-binding repeats.
  – 4R tau isoforms contain four microtubule-binding repeats, with the additional repeat providing stronger binding to microtubules compared to 3R isoforms.

  The balance between 3R and 4R tau isoforms is tightly regulated and critical for normal neuronal functions.

  Misfolded tau undergoes hyperphosphorylation before aggregating into neurofibrillary tangles (NFTs), a hallmark of tauopathies. Primary tauopathies are characterized by aggregates composed of 3R tau isoforms, as in PiD, or 4R tau isoforms, as in CBD and PSP. In contrast, AD involves a mixture of 3R and 4R tau isoforms (Teravskis et al. 2020; Hu et al. 2023).
- Aβ peptides are generated from the amyloid precursor protein (APP) through enzymatic cleavage by β-secretase and γ-secretase. These peptides contain either 40 or 42 residues ($A\beta_{1-40}$ and $A\beta_{1-42}$), and their hydrophobic sequences lead to the formation of insoluble fibrils and form amyloid plaques. Toxic Aβ oligomers can disrupt neuronal activity at the synapse level. Several factors are involved in the misfolding and aggregation of Aβ peptides, including pH, metal ions, and the presence of other molecules such as chaperones or proteoglycans (Butterfield et al. 2013; Ono and Watanabe-Nakayama 2021).
- TDP-43 is encoded by *TARDBP* gene, and it is primarily located in the nucleus, where it plays roles in RNA metabolism, including transcriptional regulation, splicing, mRNA stability, and transport. Under pathological conditions, TDP-43 relocates from the nucleus to the cytoplasm, a mislocalization often linked to cellular stress and dysfunction. In the cytoplasm, TDP-43 undergoes various post-translational modifications, such as phosphorylation, ubiquitination, and cleavage, but also other aberrant alterations that can promote its aggregation. The C-terminal region of TDP-43 is pivotal to its pathological behavior due to its intrinsically disordered nature, making it highly prone to aggregation. This region harbors the majority of ALS-associated *TARDBP* mutations and includes critical phosphorylation sites, underscoring its central role in pathology (Floare and Allen 2020; Arseni et al. 2022).

As with prions, increasing evidence suggests that αSyn, tau, Aβ, and TDP-43 also exhibit distinct strains. These strains can faithfully propagate their misfolded conformations to normal proteins, ultimately giving rise to and driving the onset and progression of specific pathologies (Clavaguera et al. 2009; Recasens et al. 2018; Bastioli et al. 2021).

The Contribution of Extracellular Vesicles to Protein Misfolding and Disease

To date, the exact mechanisms by which misfolded proteins migrate from one cell to another remain unclear, though several hypotheses have been proposed to explain this phenomenon (Marsh 2019; Peng et al. 2020). One plausible mechanism involves cell-to-cell communication via extracellular vesicles (EVs). These 30–100 nm carriers are implicated in the transport of proteins, metabolites, and various nucleic acids (including DNA, microRNA, and mRNA) (Colvett et al. 2023). EVs can target proximal structures or travel over long distances, serving as vehicles for the transfer of specific molecular information between cells (Valadi et al. 2007; Maas et al. 2017; Picca et al. 2022). They have also been implicated in several diseases, including cancers (Tai et al. 2018), depression (Xian et al. 2022), viral infections affecting the nervous system (Sampey et al. 2014), and NDs. Notably, disease-associated αSyn, $A\beta_{1-42}$, tau, and TDP-43 have been detected within EVs. Both neurons and glial cells actively release EVs into the extracellular space, and these vesicles can cross the blood-brain barrier, entering the bloodstream (Saint-Pol et al. 2020; Wiersema et al. 2024), thus facilitating their systemic propagation. The presence of brain-derived EVs in blood and urine provides a valuable opportunity to study brain protein content noninvasively. EVs were found to play a key role in α-synucleinopathies by facilitating the extracellular propagation of αSyn oligomers. These misfolded proteins have been detected on the surface and inside the lumen of EVs (Danzer et al. 2012). Notably, the increased exosomal release of αSyn has been observed in response to elevated intracellular levels of αSyn, suggesting that EVs may play a role in cellular homeostasis by facilitating the clearance of excess misfolded proteins (Putz et al. 2008). Additionally, EVs provide catalytic environments for the nucleation of αSyn aggregation (Properzi et al. 2015; Porta et al. 2018; Li et al. 2022; Tamaki et al. 2023). Impaired lysosomal function in PD has been hypothesized to promote αSyn release via EVs, thereby facilitating its spread to neighboring cells and contributing to the formation of αSyn inclusions (Alvarez-Erviti et al. 2011). Remarkably, $A\beta_{1-42}$ and tau protein phosphorylated at several sites (e.g., pT181) were found in neural EVs isolated from the CSF and blood of AD patients (Pulliam et al. 2019). Additionally, the levels of Aβ oligomers found in EVs correlated with the extent of amyloid plaque deposition in the brain (Saman et al. 2012; Fiandaca et al. 2015; Goetzl et al. 2016; Wang et al. 2017; Lim et al. 2019; Jia et al. 2019; Picca et al. 2022). Moreover, higher levels of tau pT181, pS396, and

$A\beta_{1-42}$ were found in plasma EVs of subjects with mild cognitive impairment (MCI) who progressed to AD (Winston et al. 2016). Recent findings highlight the critical role of EVs in tau propagation and its involvement in FTD. Plasma EVs contain full-length tau, allowing quantification of 3R and 4R tau isoforms. Elevated 3R/4R tau ratios in plasma EVs correlate with tau-related pathology in behavioral variant FTD (bvFTD), whereas low ratios are observed in 4R tauopathies like PSP. Another pathological protein found abundantly in EVs is TDP-43, reflecting its disease-associated mislocalization from the nucleus to the cytosol. High levels of TDP-43 in plasma EVs distinguish ALS and bvFTD cases with TDP-43 pathology from controls, and these levels strongly correlate with disease severity and neurodegeneration markers (Chatterjee et al. 2024; Zhao and Huang 2024). For these reasons, circulating EVs hold promise as biomarkers of impaired cellular quality control mechanisms, particularly in conditions marked by the accumulation of misfolded proteins. Utilizing innovative techniques such as SAAs for the analysis of EVs could enable the detection of misfolded proteins, significantly improving the clinical diagnostic accuracy of various NDs.

SAA: How Does it Work?

SAAs are cutting-edge diagnostic tools that leverage the unique capability of misfolded proteins to reliably propagate their abnormal conformations to normal proteins. These assays provide a rapid *in vitro* reproduction of the misfolding propagation phenomenon observed *in vivo*. The reaction substrate typically consists of recombinant or synthetic proteins, which are placed in the wells of multiwell plates. Each well contains a reaction mix supplemented with the fluorescent dye thioflavin T (ThT). The biological sample of interest, such as CSF, OM, urine, skin, tears, saliva, or blood, is carefully prepared and added to the well. If even trace amounts of misfolded proteins, referred to as seeds, are present in the biological sample, they induce the conformational change of the substrate proteins. This triggers the misfolding and aggregation of the substrate proteins into amyloid fibers. The formation and growth of these fibers are monitored in real time via the ThT dye, which binds to the amyloid fibrils and emits fluorescence. The intensity of this fluorescence is directly proportional to the quantity of amyloid fibrils forming in the well. Fluorescence readings are taken at regular intervals (typically every 15 min) and plotted as a function of time, producing a fluorescence versus time curve (y-axis and x-axis, respectively). The curves typically show three distinct phases: (1) a lag phase, during which substrate proteins begin to aggregate slowly, with little to no fluorescence emission; (2) an exponential growth phase, characterized by rapid oligomer and small fibril formation, accompanied by a steep increase in fluorescence emission; and (3) a plateau phase, where the majority of substrate proteins have aggregated into fibrils, resulting in maximum ThT fluorescence. When analyzing a biological sample containing misfolded proteins, a rapid increase in amyloid

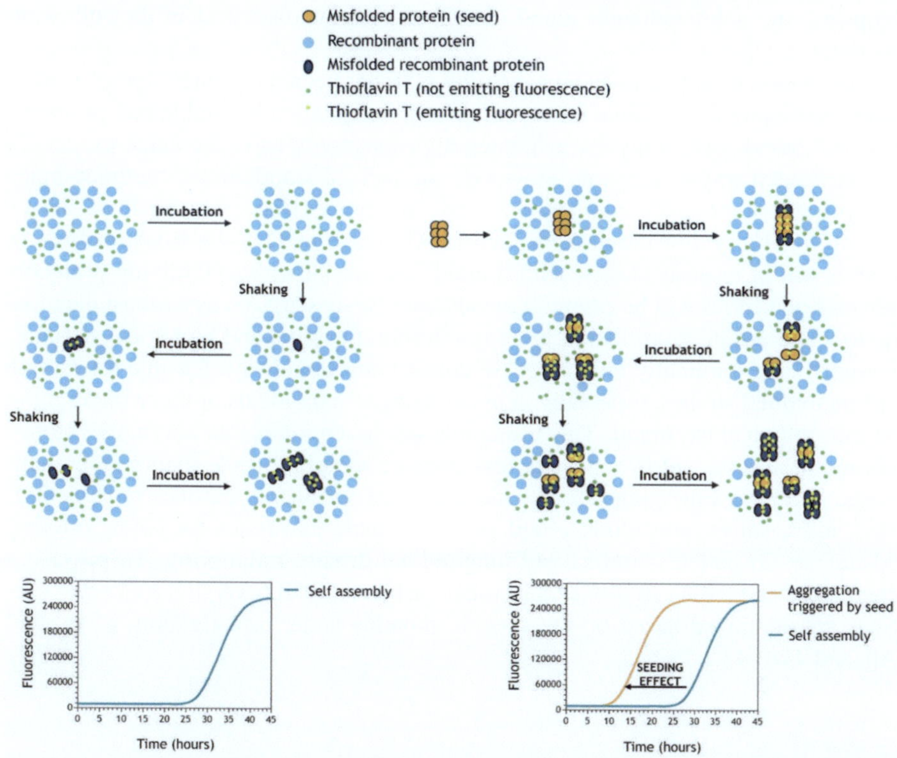

Fig. 13.1 Schematic representation of the SAA process. Biological samples and the reaction mix are added into a multiwell plate, with each sample analyzed in at least triplicate. Samples are subjected to alternate cycles of shaking, which induce the fragmentation of protein aggregates, and incubation, which promotes protein aggregation (monitored via ThT). The presence of misfolded proteins (seeds) in the biological samples is indicated by the occurrence of the seeding effect

fiber formation is observed, marked by a shortened lag phase and a steep rise in fluorescence signal intensity. Conversely, in the absence of misfolded proteins in the biological samples, amyloid fiber formation occurs at a much slower rate due to the spontaneous self-assembly of the substrate proteins. This acceleration in protein aggregation, known as the seeding effect, strongly suggests the presence of the misfolded protein of interest in the biological sample (Fig. 13.1).

SAA Clinical Applications

The application of SAA in biomarker research has emerged as a crucial tool for detecting disease-specific misfolded proteins in peripheral tissues. These assays have been optimized to identify pathological αSyn, tau, Aβ, and TDP-43 in various biological matrices, including CSF, OM, skin, saliva, blood, gastrointestinal (GI) biopsies, and submandibular gland (SMG) tissues, as discussed in the following sections.

Given its central role in ND diagnostics, CSF has been the primary focus of early SAA development. Multiple laboratories have independently established protocols for CSF-based SAA analyses, achieving high sensitivity and specificity, especially for misfolded αSyn detection. However, the lack of standardized methodologies complicates inter-study comparisons and undermines the assessment of SAA's overall reliability. For biomarkers such as Aβ, tau, and TDP-43, SAA analysis in CSF is still in its early stages, with limited data available. Standardizing protocols across laboratories will be essential to validate CSF-based SAA as a robust diagnostic tool and ensure reproducibility across different cohorts and biological samples. Despite being minimally invasive, CSF collection is not always feasible for routine or longitudinal studies, especially in asymptomatic individuals or those undergoing pharmacological treatment. This limitation has driven interest in alternative biofluids and tissues that allow repeated, less invasive sampling while maintaining diagnostic accuracy. Emerging research suggests that peripheral matrices such as OM, skin, tears, saliva, and blood could serve as viable substrates for SAA, offering advantages in patient comfort and longitudinal disease monitoring. To provide a clearer understanding of the advancements achieved with SAA, the following sections are organized based on the specific proteins under investigation: αSyn, tau, Aβ, and TDP-43.

Detection of Misfolded αSyn

In the field of α-synucleinopathies, SAAs (αSyn-SAA) have undergone significant development and have yielded encouraging results, positioning these techniques as promising biomarkers for the clinical diagnosis of these diseases. In 2022, SAA was incorporated as a supportive biomarker for the diagnosis of MSA, highlighting its growing clinical relevance (Wenning et al. 2022). Ongoing efforts aim to standardize SAA protocols and expand their application to peripheral tissues, which could facilitate less invasive diagnostic approaches and longitudinal monitoring of disease progression. Even more remarkably, αSyn-SAA has recently been incorporated as a biomarker of non-AD copathology in the updated diagnostic criteria for AD (Jack et al. 2024). This inclusion highlights the growing recognition of the role of αSyn in the pathophysiology of AD, alongside traditional hallmark markers such as Aβ plaques and tau tangles. In the following paragraphs, we will describe the SAA applications across various biological tissues.

CSF This biological matrix has been the primary focus of extensive αSyn-SAA research, with a substantial body of literature highlighting its diagnostic potential. Most of the research studies are summarized in the following Table 13.1 and show the high sensitivity and specificity of SAA for detecting misfolded αSyn in the CSF of patients with PD, DLB, and MSA. In some cases, the biochemical and

Table 13.1 Summary of αSyn-SAA studies performed on CSF samples. All studies show high levels of sensitivity and specificity across the various research groups and their respective cohorts

Year of publication	Reference	Participants	Sensitivity	Specificity
2016	Fairfoul et al. (2016)	PD	95%	100%
		DLB	92%	
		iRBD	100%	
		HC		
2017	Shahnawaz et al. (2017)	PD	88,5%	94–96,9%
		DLB	100%	
		OND		
2018	Groveman et al. (2018)	PD	92%	100%
		DLB	94%	
		HC		
2019	Kang et al. (2019)	PD	96,2%	82,3%
		HC		
	van Rumund et al. (2019)	PD	84%	98%
		MSA	35%	
		DLB	NA	
		OND		
	Garrido et al. (2019)	LRRK2-PD	40%	80%
		PD	90%	
		HC		
	Bongianni et al. (2019)	DLB	100%	95,9%
		MSA	NA	
		HC		
2020	Rossi et al. (2020)	PD	94,4%	98%
		DLB	97,1%	
		OND		
	Shahnawaz et al. (2020)	PD	93,6%	100%
		MSA	84,6%	
		OND		
2021	Brockmann et al. (2021)	PD	85%	92%
		DLB	86%	
		HC		
	Orrù et al. (2021)	PD	97%	87%
		HC		
	Mammana et al. (2021)	PD	100%	100%
		DLB	100%	
		OND		
	Quadalti et al. (2021)	PD	91,4%	97,1%
		HC		
	Donadio et al. (2021)	PD + DLB + MSA	78%	100%
		OND		

(continued)

Table 13.1 (continued)

Year of publication	Reference	Participants	Sensitivity	Specificity
	Perra et al. (2021)	DLB	100%	90,6%
		OND		
	Iranzo et al. (2021)	iRBD	90%	90%
		HC		
	Bargar et al. (2021b)	PD+ DLB	98%	100%
		OND		
	Russo et al. (2021)	PD	86%–96%	97%-100%
		HC		
	Concha-Marambio et al. (2021)	PD	96,2%-96,4%	93,8%-96,7%
		HC		
2022	Poggiolini et al. (2022)	PD	89%	96%
		MSA	75%	
		iRBD	64%	
		HC		
	Compta et al. (2022)	PD	75%	
		MSA	12%	
		OND		89%
		HC		100%
	Hall et al. (2022)	DLB	100%	94%
		Non-LB pathology		
	Bongianni et al. (2022)	PD	92%	99,5%
		OND		
	Arnold et al. (2022)	αSyn-pathology	71,2% - 80%	98,1% - 88,5%
		No αSyn-pathology		
2023	Concha-Marambio et al. (2023)	*de novo* PD	Accuracy: 98%	NA
		iRBD	Accuracy: 93%	
		HC		
	Iranzo et al. (2023)	iRBD	75%	97,50%
		HC		
	Liguori et al. (2023)	iRBD	67%	72%
		Clinical controls		
	Siderowf et al. (2023)	PD	88%	96%
		iRBD	86%	
		NMC	8%	
		HC		
	Pilotto et al. (2023)	AD	45%	92,9%
		OND		
	Fernandes Gomes et al. (2023)	PD	100%	70,8%
		MSA	92,6%	
		HC		
	Middleton et al. (2023)	PD	80,5%-87,3%	89,5%-97,2%
		HC		

(continued)

Table 13.1 (continued)

Year of publication	Reference	Participants	Sensitivity	Specificity
2024	Bellomo et al. (2024)	AD	30%	87%
		PD + DLB	87%	
		HC		
	Plastini et al. (2024)	LB-spectrum	85,92%	85,11–87,50%
		NC		
	Coughlin et al. (2024)	DLB	72%	96%
		HC		
		Young HC		
	Samudra et al. (2024)	DLB	55,20%	96,30%
		HC		
	Grillo et al. (2024)	PD	93,80%	NA
		LRRK2-PD	77%	
		GBA-PD	92,30%	
	Ma et al. (2024)	PD	91%	
		DLB	70%	
		iRBD	95%	
		MSA	87%	
		OND		84%
		HC		94%
	Dam et al. (2024)	PD	93%	99,93%
		SNCA-PD	100%	
		GBA-PD	93%	
		LRRK2-PD	64%	
		PRKN-PD	33%	
		HC		
	Brown et al. (2024)	PPMI-prodromal cohort	55,3%	NA

PD Parkinson's disease, *DLB* dementia with Lewy bodies, *iRBD* isolated REM sleep behavior disorder, *HC* healthy controls, *OND* other neurodegenerative diseases, *MSA* multiple system atrophy, *LRRK2-PD* PD with *LRRK2* gene mutation, *non-LB pathology* non-Lewy bodies pathology, *NMC* nonmanifesting carriers, *AD* Alzheimer's disease, *LB-spectrum* Lewy bodies pathology spectrum, *NC* normal cognition, *GBA-PD* PD with *GBA* gene mutation, *SNCA-PD* PD with *SNCA* gene mutation, *PRKN-PD* PD with *PRKN-PD* gene mutation, *PPMI-Prodromal cohort* Parkinson's Progression Markers Initiative Prodromal cohort

morphological analysis of SAA end products enabled disease discrimination. Notably, SAA positivity has been observed even in prodromal subjects, such as individuals with iRBD, suggesting that misfolded αSyn is already present in the CSF before the onset of clinical symptoms. Interestingly, SAA positivity for misfolded αSyn has also been detected in patients with non-α-synucleinopathies, such as AD. This highlights the potential of SAA to recognize copathologies, thereby improving patient stratification. Given the rapid advancements in this field, it is possible that some recent studies were not included in our review. We acknowledge this limitation and sincerely regret any omissions. Readers are encouraged to consult the latest literature for the most up-to-date findings in this evolving area of research.

OM The idea of exploiting OM samples for αSyn-SAA analyses arises from promising prion disease studies (Orrú et al. 2014; Bongianni et al. 2017; Redaelli et al. 2017; Fiorini et al. 2020). In our 2019 study (De Luca et al. 2019), we were the first to demonstrate that the OM of patients with PD and MSA tested positive for αSyn-SAA, revealing a notably higher sensitivity for MSA (82%) compared to PD (52%) with an overall specificity of 84%. Interestingly, biochemical and morphological analyses of SAA products generated by PD and MSA samples revealed distinct properties, allowing for their clear differentiation. Thus, beyond the ability to detect αSyn in OM, these analyses suggest that the distinct properties of the final SAA products stem from the fact that PD and MSA are driven by different αSyn strains, which in turn account for these variations. To further explore their biological impact, the SAA end products were challenged in cell models, where they elicited distinct inflammatory properties. This finding further supports the hypothesis that the αSyn strains present in biological samples may impart specific characteristics to the SAA reaction substrate (De Luca et al. 2021). Following initial efforts to harmonize OM analytical protocols for SAA, a preliminary interlaboratory assessment was performed to evaluate the reproducibility of this assay (Bargar et al. 2021a). This study included samples collected from patients with PD, MSA-P, and MSA-C. The results showed a 96% interrater agreement, highlighting the robustness and reliability of SAA in OM samples. Notably, the findings revealed that many MSA-P samples tested positive, whereas MSA-C samples were predominantly negative, further reinforcing the hypothesis that distinct strains of misfolded αSyn may exert divergent effects when analyzed by SAA. The work of Stefani (Stefani et al. 2021) marked a significant advancement, as it showed the presence of misfolded αSyn in the OM of subjects with iRBD, a prodromal stage of PD, DLB, and MSA. The study analyzed samples from iRBD ($n = 63$), PD ($n = 41$), and controls ($n = 59$), finding αSyn-SAA positivity in 44.4% of iRBD, 46.3% of PD, and only 10.2% of controls, with a high specificity (89.8%). Notably, iRBD subjects who tested positive had more severe olfactory dysfunction, reinforcing the link between αSyn aggregation and early disease manifestations. Perra and collaborators (Perra et al. 2021) further contributed to the field by analyzing OM and CSF samples collected from patients with probable or prodromal DLB and other neurodegenerative disorders. OM tested positive for αSyn-SAA in 38 out of 81 patients and CSF in 19 out of 48. The accuracy of the test was 86.4% for OM and 93.8% for CSF. The combined analysis of OM and CSF increased the alignment with the clinical diagnosis, potentially reaching 100%. The results suggest that OM sampling could be used as an initial screening test for suspected DLB patients, followed by CSF testing for confirmation, especially when OM results are incongruent with the initial clinical diagnosis. Further supporting the value of multiple sample analyses, a recent study by Kuzkina and colleagues (Kuzkina et al. 2023) proposed that combining analyses of different biological samples could bolster diagnostic precision for α-synucleinopathies. Their study, which examined OM and skin biopsies from patients with PD ($n = 27$), iRBD ($n = 18$), and controls ($n = 30$), revealed a lower frequency of misfolded αSyn in the OM compared to skin biopsies. Interestingly, misfolded αSyn was found to be more prevalent in the OM of iRBD subjects than in those with PD, providing further evidence for iRBD as a potential marker of a more aggressive form of α-synucleinopathy. In a subgroup of PD patients, misfolded αSyn was detectable only in the OM, which aligns with the emerging hypothesis of the "brain-first" subtype of PD. These findings suggest that incorporating αSyn-SAA analysis of both OM and skin could greatly improve diagnostic accuracy and patient stratification, offering a promising approach for better managing α-synucleinopathies. Finally, Bongianni (Bongianni et al. 2022) investigated the seeding activity in two different nasal regions: agger nasi (AN) and middle turbinate (MT). OM was collected from PD ($n = 66$) patients and controls ($n = 29$). The results showed that misfolded αSyn was detected more frequently in AN (84%) than MT (45%), with a lower positivity in non-PD patients (10%). Immunocytochemistry revealed more olfactory neural cells in AN samples.

Skin Recent studies have investigated the use of skin tissue for detecting misfolded αSyn through αSyn-SAA. Immunohistochemical studies have detected pathological αSyn in skin biopsies, though sensitivity remains a concern. In 2020, Manne and colleagues (Manne et al. 2020a) employed the αSyn-SAA for ultrasensitive detection of misfolded αSyn in both frozen and formalin-fixed paraffin-embedded (FFPE) skin tissues. The assay successfully detected misfolded αSyn in frozen skin samples from neuropathologically confirmed PD cases and controls, achieving 96% sensitivity and specificity. FFPE samples showed slightly lower sensitivity (75%) and specificity (83%). A further study used both RT-QuIC and PMCA to analyze autopsy and biopsy skin samples from neuropathologically confirmed cases of PD, DLB, MSA, and controls (Wang et al. 2021). A total of 160 autopsied skin specimens from 140 cadavers and 41 *antemortem* biopsies were analyzed. RT-QuIC detected misfolded αSyn in autopsy abdominal skin with 94% sensitivity and 98% specificity in PD cases. Across synucleinopathies (PD, DLB, MSA), RT-QuIC maintained 93% sensitivity and specificity, while PMCA showed 82% sensitivity and 96% specificity in PD autopsy samples. In biopsies from PD patients, RT-QuIC demonstrated superior performance (95% sensitivity, 100% specificity) compared to PMCA (80% sensitivity, 90% specificity). In 2021, Donadio and coworkers investigated the reproducibility of immunofluorescence in detecting pathological αSyn in skin nerves and compared its diagnostic accuracy with αSyn-SAA in skin and CSF for distinguishing α-synucleinopathies from non-α-synucleinopathies. Patients with clinically confirmed α-synucleinopathies (PD, DLB, and MSA) and non-α-synucleinopathies (AD, PSP, and CBD), along with 24 control patients with peripheral neuropathies, were recruited (Donadio et al. 2021). Immunofluorescence detected misfolded αSyn with 90% sensitivity and 100% specificity, while αSyn-SAA achieved 86% sensitivity and 80% specificity. These findings support the use of skin-based immunofluorescence or αSyn-SAA testing as viable alternatives to CSF αSyn-SAA for diagnosing α-synucleinopathies. In 2021, Mammana et al. (2021) performed αSyn-SAA analysis of skin samples from patients with Lewy body disease, including PD and DLB. Skin punch biopsies were collected either *in vivo* ($n = 69$) or *postmortem* ($n = 49$) from DLB patients and subjects with other neurological conditions enrolled as controls. The analysis showed that the skin αSyn-SAA distinguished DLB patients with 89.2% sensitivity and 96.3% specificity. In cervical skin samples, sensitivity reached 94.1%. When comparing the diagnostic accuracy of skin and CSF samples in 79 patients, both αSyn-SAA protocols performed similarly (skin: 97.5%, CSF: 98.7%). In the same year, Kuzkina (Kuzkina et al. 2021) investigated the interrater agreement of αSyn-SAA analysis of skin samples in two independent laboratories. The assay demonstrated 88.9% diagnostic accuracy and 92.2% interrater agreement. Increased αSyn seeding activity was observed in patients with longer disease duration and advanced disease stage and correlated with nonmotor symptoms such as iRBD, cognitive impairment, and constipation. In 2023, the same author (Kuzkina et al. 2023) compared the αSyn seeding activity in OM and skin samples, and the results are described above (see OM section). A remarkable discovery was made by Iranzo (Iranzo et al. 2023), demonstrating the possibility of detecting seeding activity in the skin of individuals with iRBD. Particularly, iRBD subjects ($n = 91$) and age-matched controls ($n = 41$) underwent simultaneous skin biopsy and lumbar puncture for αSyn-SAA analysis. The assay demonstrated high diagnostic accuracy in both skin (sensitivity: 76.9%, specificity: 97.6%) and CSF (sensitivity: 75.0%, specificity: 97.5%), with 99.2% agreement between sample types. Positive patients showed a significantly higher likelihood of prodromal PD ($p < 0.001$) and presented with hyposmia, dopamine transporter deficits, and orthostatic hypotension. Another study published by Liguori and colleagues (Liguori et al. 2023) compared the diagnostic accuracy of immunofluorescence and αSyn-SAA for detecting misfolded αSyn in skin and CSF of iRBD subjects. Immunofluorescence showed high diagnostic accuracy (89%), outperforming αSyn-SAA from the skin (70%) and CSF (69%) due to the latter's lower sensitivity and specificity. Despite this, immunofluorescence showed significant agreement with CSF-based αSyn-SAA. These findings support the evidence that misfolded αSyn can be detected in prodromal disease stages. Several other studies have been published highlighting the high sensitivity and specificity of αSyn-SAA in detecting misfolded αSyn in skin samples, reinforcing its potential as a reliable diagnostic tool whenever tissue analysis is feasible (Donadio et al. 2021; Li et al. 2024). Research groups are actively working on optimizing αSyn-SAA protocols to improve the sensitivity and specificity of skin-based assays for detecting misfolded αSyn (Kuang et al. 2024).

Blood Several studies have reported elevated levels of physiological αSyn in the plasma and serum of PD patients compared to healthy individuals (Lee et al. 2006; Ding et al. 2017; Chang et al. 2020). Recently, a modified SAA protocol combined with immunoprecipitation (IP) allowed the efficient detection of misfolded αSyn in the serum of patients with PD, MSA, DLB, and iRBD (Okuzumi et al. 2023). In particular, the group coordinated by Hattori developed a modified immunoprecipitation-based real-time quaking-induced conversion (IP/RT-QuIC) assay, whose results demonstrated high diagnostic accuracy for differentiating PD from controls and MSA from controls in internal cohorts. Furthermore, IP/RT-QuIC effectively differentiated PD and MSA from controls in an external cohort. Importantly, as observed in the case of CSF (Shahnawaz et al. 2020) and OM (De Luca et al. 2019), the SAA reaction products preserved disease-specific characteristics, enabling differentiation between PD and MSA samples. In the TREND study (Kluge et al. 2024), blood samples from a prospective cohort of 1201 individuals with varying risk levels for PD were collected biennially over 4–10 years. This retrospective analysis focused on 12 participants who later developed PD. The study aimed to evaluate a blood-based αSyn-SAA as a potential biomarker for prodromal PD. Pathological αSyn conformers, derived from neuronal EVs, were detected using (i) immunoblot analyses performed with antibodies against pathological αSyn conformers and (ii) the αSyn-SAA. All PD patients tested positive for both immunoblots and αSyn-SAA at diagnosis. Interestingly, 30% of individuals with iRBD showed positive αSyn-SAA results, while all healthy controls were negative. These findings demonstrate the potential of the blood-based αSyn-SAA to detect misfolded αSyn conformers up to 10 years before a clinical PD diagnosis, suggesting its promise as a diagnostic biomarker for prodromal PD. A very recent study (Schaeffer et al. 2024) explored the association between αSyn-SAA derived from neuronal exosomes in blood and both PD diagnosis and disease duration. Blood samples from PD patients ($n = 80$) and age- and gender-matched healthy controls ($n = 20$) were analyzed. The results revealed that 79 out of 80 PD tested positive for αSyn-SAA, with a sensitivity of 98.8%, whereas healthy controls remained negative. A significant negative correlation was found between disease duration and αSyn-SAA positivity, with longer disease duration associated with lower seeding activity. Another recent study by Wang and colleagues (Wang et al. 2024a) showed that serum from PD patients tested positive for αSyn-SAA, with 80.49% sensitivity and 90.48% specificity (for more details, see the saliva section below).

SMG SAA has been successfully applied to detect misfolded αSyn in SMG biopsies. In 2020, Manne (Manne et al. 2020b) employed the αSyn-SAA to detect misfolded αSyn in SMG collected from PD, incidental Lewy body disease (iLBD), and control groups, as well as in FFPE sections. Results demonstrated 100% sensitivity and 94% specificity for frozen samples, with both PD and iLBD tissues showing higher αSyn-SAA seeding activity compared to controls. In the case of FFPE samples, sensitivity was reduced to 76%, while the specificity reached 100%. Remarkably, αSyn-SAA could detect αSyn-SAA seeding activity in iLBD cases that were undetectable by traditional immunohistochemistry. Thus, the study presents the αSyn-SAA assay as a highly sensitive and specific method for detecting PD-related pathological changes in peripheral tissues, with the potential for recognizing prodromal PD in SMG tissues. However, FFPE significantly reduces αSyn-SAA efficiency. To address this issue, Hepker (Hepker et al. 2023) developed a kinetic assay seeding ability recovery (KASAR) protocol, which restores the αSyn-SAA seeding potential of FFPE samples. When tested on 28 FFPE SMG, the protocol showed high reproducibility, with 93% of results replicating in blinded tests. In particular, 11/13 PD samples tested positive by αSyn-SAA, while 19/19 negative samples showed no false positives, demonstrating the robustness of the protocol. This method enabled the detection of misfolded αSyn using a few milligrams of SMG, offering a valuable tool for the diagnosis of α-synucleinopathies using archived FFPE samples. The study published by Bargar and collaborators (Bargar et al. 2021b) further confirmed the possibility of detecting misfolded αSyn in SMG collected from neuropathologically confirmed cases of PD by αSyn-SAA, even when different batches of recombinant αSyn were used as reaction substrates. Although the results are promising, this approach is limited by its invasiveness.

Saliva The collection of saliva is one of the least invasive procedures for patients. The study by Luan and collaborators (Luan et al. 2022) explored the diagnostic value of αSyn-SAA using saliva samples from PD ($n = 75$), MSA ($n = 18$) patients, and healthy controls ($n = 36$). The assay distinguished PD patients with 76% sensitivity and 94.4% specificity. For MSA patients, sensitivity was 61.1%. Although there were no significant differences in the morphological properties of the SAA reaction products between PD and MSA (as observed in the case of OM, CSF, and blood), the lag phase of the αSyn-SAA was notably shorter in PD patients compared to those with MSA. In a study by Vivacqua and collaborators (Vivacqua et al. 2023), saliva samples from PD patients ($n = 37$) and healthy subjects ($n = 23$) were analyzed by αSyn-SAA. The results showed that 86% of PD samples tested positive, compared to 22% of control samples. However, PD samples were characterized by a significantly shorter lag phase and higher kinetic parameters than controls. Receiver operating characteristic (ROC) analysis revealed good diagnostic accuracy, with a sensitivity of 83.78% and specificity of 82.61%. Wang and colleagues (Wang et al. 2024a) performed a very interesting study that included PD ($n = 82$) and healthy controls ($n = 42$) who donated blood, with 74 of them (48 PD, 26 controls) also providing saliva samples. An additional 57 subjects (35 PD, 22 HC) donated saliva only. The αSyn-SAA assay was used to assess αSyn seeding activities in blood and saliva. Serum αSyn-SAA showed 80.49% sensitivity and 90.48% specificity, while saliva αSyn-SAA achieved 74.70% sensitivity and 97.92% specificity. Notably, when combining both serum and saliva samples from 74 subjects, diagnostic performance improved, with 95.83% sensitivity and 96.15% specificity. Furthermore, αSyn seeding activity in serum correlated inversely with the Montreal Cognitive Assessment in males and positively with the Hamilton Depression Rating Scale in females, while αSyn seeding activity in saliva correlated inversely with age at diagnosis in males. These findings suggest that combining serum and saliva αSyn-SAA can provide a highly sensitive, accurate, and minimally invasive diagnostic tool for PD. Very recently, Luan and collaborators (Luan et al. 2024) combined the αSyn-SAA and miRNA-29a-3p levels (analyzed via RT-qPCR) in a cohort of 203 participants with PD, MSA, essential tremor (ET), and healthy controls. The sensitivity of αSyn-SAA for PD and MSA was 70.30% and 56.25%, respectively, with specificity for controls at 92.45%. Salivary miRNA-29a-3p expression was significantly reduced in PD and MSA. Combining the two biomarkers improved diagnostic accuracy, with sensitivity for PD and MSA reaching 75% and 90%, respectively. Notably, salivary αSyn-SAA showed 100% sensitivity and 79.21% specificity for PD versus ET, while miRNA-29a-3p showed 88.24% sensitivity for

PD versus ET. The combined assessment of these markers offered improved diagnostic value for distinguishing PD and MSA from ET.

GI Currently, the body of research investigating the detection of misfolded αSyn in GI remains limited. In 2021, Bargar (Bargar et al. 2021b) analyzed a sigmoid colon sample from a neuropathologically confirmed case of PD that tested positive for αSyn-SAA. In 2023, Vascellari and colleagues (Vascellari et al. 2023) showed *intra-vitam* detection of misfolded αSyn in duodenum biopsies from 22 of 23 PD patients using αSyn-SAA, with no seeding activity detected in 6 healthy controls. The diagnostic sensitivity and specificity of the assay for PD were 95.7% and 100%, respectively. These findings suggest that the duodenum may play a role in the propagation of pathological αSyn. Although the main limitation of this analysis is the invasiveness of biopsy collection, it could potentially be incorporated during routine esophagogastroduodenoscopy and colonoscopy screenings to further support clinical diagnosis.

Detection of Misfolded Tau

While this area of research is still in its early stages, promising results have already emerged from SAA analyses (tau-SAA) of CSF and skin samples collected from patients with various tau-related neurodegenerative disorders. Below, we summarize the updated results, separated by biological sample type.

CSF A few studies have shown the utility of tau-SAA in detecting misfolded tau in the CSF of patients with various tauopathies, including PiD, PSP, and CBD. In the case of PiD, a recombinant 3R-tau protein was employed to investigate the presence of misfolded 3R tau in the CSF of enrolled patients. The sensitivity of this assay reached 100%, with specificity ranging from 94% to 100% (Saijo et al. 2017). For PSP and CBD, a recombinant 4R-tau protein was used to detect misfolded 4R-tau in CSF samples obtained either *antemortem* or *postmortem*. When analyzing *postmortem* CSF (collected from ventricular sites), tau-SAA showed 100% sensitivity for PSP and 92% sensitivity for CBD, with 100% specificity for both diseases. However, *antemortem* CSF samples showed lower sensitivity, with 69% for PSP and 50% for CBD. The specificity remained at 82%. The observed discrepancy in the sensitivity of the assay between *antemortem* and *postmortem* CSF samples could be attributed to differences in matrix composition between the two types of samples, which may influence the assay's performance. Additionally, the lower sensitivity of *antemortem* samples may be partially explained by misdiagnosis, as tauopathies such as PSP and CBD are often challenging to distinguish clinically (Saijo et al. 2020). These findings highlight the potential of tau-SAA as a promising tool for detecting tau pathology while also underscoring the importance of optimizing assay conditions for various sample types in clinical and *postmortem* settings.

Skin A recent study (Dellarole et al. 2024) has demonstrated that skin biopsies from patients with CBD and PSP can serve as valuable biological samples for diagnostic purposes. Both recombinant 4R-tau and 3R-tau proteins were used as substrates in the tau-SAA. Recombinant 4R-tau showed significantly higher seeding efficiency compared to 3R-tau, reflecting the well-established predominance of misfolded 4R-tau isoforms in CBD and PSP. This substrate-specific reaction underscores the assay's ability to exploit molecular homology for detecting tauopathies, further highlighting its potential as a diagnostic tool. Another tau-SAA study, using autoptic skin samples from neuropathologically confirmed cases of AD, PSP, CBD, and PiD, confirmed the efficiency of SAA in detecting misfolded tau in peripheral biological samples (Wang et al. 2024b).

Detection of Misfolded Aβ

The optimization of SAA methodologies for AD is still in its early stages, reflecting the nascent but rapidly advancing nature of this research area. In this context, all Aβ-SAA analyses performed to date have been exclusively focused on CSF samples. Initial findings have demonstrated significant promise, with several studies highlighting its potential to detect and characterize misfolded Aβ aggregates. In 2014, Salvadores (Salvadores et al. 2014) showed the ability to detect Aβ oligomers in the CSF of AD patients, achieving a sensitivity of 90% and a specificity of 92% in distinguishing AD patients from those with other neurodegenerative disorders or non-degenerative neurological conditions. More recently, our group combined SAA with surface-enhanced Raman spectroscopy (SERS) to analyze CSF samples from patients with AD and mild cognitive impairment due to AD (MCI-AD). This approach aimed not only to detect misfolded Aβ oligomers in the CSF of AD and MCI-AD but also to investigate and characterize potential chemo-structural differences in the Aβ-SAA products, which could help differentiate AD from MCI-AD or even distinguish between various AD phenotypes. The approach achieved a sensitivity of 88% and a specificity of 70%, suggesting the possibility of discriminating potential disease phenotypes (D'Andrea et al. 2023).

Detection of Misfolded TDP-43

SAA has recently been applied to the analysis of CSF and OM samples collected from patients with TDP-43 proteinopathies (TDP-43-SAA). Preliminary findings suggest that TDP-43-SAA holds significant potential for contributing to the clinical diagnosis of FTD and ALS. Below, we summarize the updated results, separated by biological sample type.

CSF In 2020, a study optimized the TDP-43-SAA protocol in CSF samples from patients with genetic forms of ALS and FTD, all associated with TDP-43 aggregates deposition in the CNS. Specifically, CSF samples were collected from patients with

C9orf72 expansions, *TARDBP*, and *GRN* mutations. The assay showed the ability to detect even minute amounts (approximately 15 pg) of misfolded TDP-43 proteins in the CSF. It successfully distinguished FTD/ALS patients from age-matched controls, with an overall sensitivity of 94% and specificity of 85% (Scialò et al. 2020). In 2023, a subsequent study further confirmed the presence of a TDP-43 seeding effect in the CSF of ALS patients, enabling effective differentiation from healthy controls (Audrain et al. 2023). These findings underscore the potential of TDP-43-SAA as a powerful diagnostic tool, offering new opportunities for early detection and precision medicine in TDP-43 proteinopathies.

OM In 2024, Fontana (Fontana et al. 2024) assessed the seeding activity in the OM of patients with FTD and TDP-43-immunoreactive pathology (FTLD-TDP) using TDP-43-SAA. The TDP-43-SAA protocol was optimized with frontal cortex samples from 16 *postmortem* cases, including individuals with FTLD-TDP, FTLD with tau inclusions, and healthy controls. Subsequently, OM samples were collected from 17 FTLD-TDP patients, 15 healthy controls, and three *MAPT* mutation carriers. The TDP-43-SAA distinguished with 100% accuracy between *postmortem* cases with and without TDP-43 neuropathology. Seeding activity was detected in the OM, with 82.4% of FTLD-TDP patients testing positive and 86.7% of controls testing negative ($P < 0.001$). Interestingly, in TDP-43-SAA-positive samples, cytoplasmic deposits of phosphorylated TDP-43 were identified in the olfactory neural cells. These findings showed that TDP-43-SAA can be employed to recognize and monitor FTLD-TDP in living patients.

Conclusions

SAAs represent a transformative approach for detecting protein misfolding associated with NDs, offering remarkable sensitivity and specificity across a variety of sample types. The accessibility of easily obtainable tissues facilitates the collection of samples from healthy control subjects, an otherwise challenging but essential step for optimizing diagnostic accuracy and deepening our understanding of disease mechanisms. Recent advances in peripheral sampling, such as skin, OM, saliva, and blood, underscore the versatility and accessibility of these assays for future clinical applications. Given the heterogeneity of proteinopathies and the intrinsic cross-seeding potential of misfolded proteins, refining SAA protocols to achieve disease-specific sensitivity and specificity remains a priority. Harmonization and standardization of SAA methodologies across laboratories are essential to ensure reproducibility and reliability. Moreover, integrating SAA with complementary diagnostic approaches, such as advanced imaging and omics technologies, could provide a multidimensional perspective on disease pathology, thereby enhancing clinical utility. For instance, the incorporation of SERS has enriched the molecular characterization of Aβ aggregates, hinting at the possibility of AD phenotypic subtyping and personalized medicine. While challenges remain, such as optimizing protocols for *in vivo* samples and addressing potential misdiagnosis due to clinical complexity, the rapidly evolving landscape of SAA research inspires optimism. As the field progresses, these assays are poised to revolutionize the diagnostic landscape, bridging the gap between molecular pathology and precision medicine, and ultimately improving patient outcomes in a range of NDs. Remarkably, SAAs allow for the detection of protein aggregates, which are sometimes not exclusively associated with the primary pathology under investigation (Moda et al. 2023). This capability proves particularly useful in identifying cases with copathologies, as observed in AD, where αSyn-SAA and TDP-43-SAA have detected αSyn and TDP-43 positivity, respectively (Arnold et al. 2022; Pilotto et al. 2023; Bellomo et al. 2024; Tosun et al. 2024). Similarly, in the CSF of a subgroup of ALS patients, αSyn-SAA showed the presence of misfolded αSyn (Smith et al. 2024). Such findings suggest that integrating SAA into the clinical diagnostic workup not only improves the clinical diagnosis of NDs but also provides insights into the presence of copathologies in specific conditions. These insights underscore the importance of SAA as an innovative tool for recognizing clinical disease phenotyping and deepening our understanding of the complex pathophysiology of NDs. By improving our ability to identify and characterize copathologies, SAA might significantly contribute to the advancement of personalized pharmacological therapies, providing a boost to precision medicine and offering hope for tailored therapeutic interventions in the future. Beyond supporting the clinical diagnosis of NDs, SAAs also hold promise for identifying individuals at a higher risk of developing these diseases during the prodromal stage and for tracking disease progression. As observed in the detection of pathological αSyn in the CSF, skin and OM of individuals with iRBD, a prodromal phase of α-synucleinopathies, this approach highlights the potential of SAAs to facilitate early diagnosis. Early detection enables the timely initiation of therapeutic interventions, maximizing their effectiveness and potentially altering the disease trajectory. Interestingly, SAA replicates *in vitro* the process of protein misfolding and aggregation that occurs *in vivo*, making it a valuable tool for assessing the effects of various molecules on promoting protein aggregation or disrupting existing aggregates. This method offers a cost-effective platform for the initial screening of potential therapeutic compounds, streamlining the identification process before transitioning to more expensive and complex experimental models (Herva et al. 2014).

References

Alvarez-Erviti L, Seow Y, Schapira AH et al (2011) Lysosomal dysfunction increases exosome-mediated alpha-synuclein release and transmission. Neurobiol Dis 42:360–367. https://doi.org/10.1016/j.nbd.2011.01.029

Arnold MR, Coughlin DG, Brumbach BH et al (2022) α-Synuclein seed amplification in CSF and brain from patients with different brain distributions of pathological α-Synuclein in the context of co-pathology and non-LBD diagnoses. Ann Neurol 92:650–662. https://doi.org/10.1002/ana.26453

Arseni D, Hasegawa M, Murzin AG et al (2022) Structure of pathological TDP-43 filaments from ALS with FTLD. Nature 601:139–143. https://doi.org/10.1038/s41586-021-04199-3
Atarashi R, Sano K, Satoh K, Nishida N (2011) Real-time quaking-induced conversion. Prion 5:150–153. https://doi.org/10.4161/pri.5.3.16893
Audrain M, Egesipe A-L, Tentillier N et al (2023) Targeting amyotrophic lateral sclerosis by neutralizing seeding-competent TDP-43 in CSF. Brain Commun 5. https://doi.org/10.1093/braincomms/fcad306
Bargar C, De Luca CMG, Devigili G et al (2021a) Discrimination of MSA-P and MSA-C by RT-QuIC analysis of olfactory mucosa: the first assessment of assay reproducibility between two specialized laboratories. Mol Neurodegener 16:82. https://doi.org/10.1186/s13024-021-00491-y
Bargar C, Wang W, Gunzler SA et al (2021b) Streamlined alpha-synuclein RT-QuIC assay for various biospecimens in Parkinson's disease and dementia with Lewy bodies. Acta Neuropathol Commun 9:62. https://doi.org/10.1186/s40478-021-01175-w
Bastioli G, Regoni M, Cazzaniga F et al (2021) Animal models of autosomal recessive parkinsonism. Biomedicines 9:812. https://doi.org/10.3390/biomedicines9070812
Bellomo G, Toja A, Paolini Paoletti F et al (2024) Investigating alpha-synuclein co-pathology in Alzheimer's disease by means of cerebrospinal fluid alpha-synuclein seed amplification assay. Alzheimers Dement 20:2444–2452. https://doi.org/10.1002/alz.13658
Bistaffa E, Marín-Moreno A, Espinosa JC et al (2021) Pmca-generated prions from the olfactory mucosa of patients with fatal familial insomnia cause prion disease in mice. elife 10. https://doi.org/10.7554/ELIFE.65311
Bongianni M, Orrù C, Groveman BR et al (2017) Diagnosis of human prion disease using real-time quaking-induced conversion testing of olfactory mucosa and cerebrospinal fluid samples. JAMA Neurol 74:155. https://doi.org/10.1001/jamaneurol.2016.4614
Bongianni M, Ladogana A, Capaldi S et al (2019) A-Synuclein RT-QuIC assay in cerebrospinal fluid of patients with dementia with Lewy bodies. Ann Clin Transl Neurol 6:2120. https://doi.org/10.1002/acn3.50897
Bongianni M, Catalan M, Perra D et al (2022) Olfactory swab sampling optimization for α-synuclein aggregate detection in patients with Parkinson's disease. Transl Neurodegener 11:37. https://doi.org/10.1186/s40035-022-00311-3
Bousset L, Pieri L, Ruiz-Arlandis G et al (2013) Structural and functional characterization of two alpha-synuclein strains. Nat Commun 4:2575. https://doi.org/10.1038/ncomms3575
Brockmann K, Quadalti C, Lerche S et al (2021) Association between CSF alpha-synuclein seeding activity and genetic status in Parkinson's disease and dementia with Lewy bodies. Acta Neuropathol Commun 9:175. https://doi.org/10.1186/s40478-021-01276-6
Brown EG, Chahine LM, Siderowf A et al (2024) Staged screening identifies people with biomarkers related to neuronal alpha-Synuclein disease. Ann Neurol 97:730. https://doi.org/10.1002/ana.27158
Burré J (2015) The synaptic function of α-Synuclein. J Parkinsons Dis 5:699–713. https://doi.org/10.3233/JPD-150642
Butterfield DA, Swomley AM, Sultana R (2013) Amyloid β -peptide (1–42)-induced oxidative stress in Alzheimer disease: importance in disease pathogenesis and progression. Antioxid Redox Signal 19:823–835. https://doi.org/10.1089/ars.2012.5027
Cazzaniga FA, Bistaffa E, De Luca CMG et al (2022) PMCA-based detection of prions in the olfactory mucosa of patients with sporadic Creutzfeldt–Jakob Disease. Front Aging Neurosci 14. https://doi.org/10.3389/fnagi.2022.848991
Chang C-W, Yang S-Y, Yang C-C et al (2020) Plasma and serum alpha-Synuclein as a biomarker of diagnosis in patients with Parkinson's disease. Front Neurol 10. https://doi.org/10.3389/fneur.2019.01388
Chatterjee M, Özdemir S, Fritz C et al (2024) Plasma extracellular vesicle tau and TDP-43 as diagnostic biomarkers in FTD and ALS. Nat Med 30:1771–1783. https://doi.org/10.1038/s41591-024-02937-4

Chiti F, Dobson CM (2006) Protein Misfolding, functional amyloid, and human disease. Annu Rev Biochem 75:333–366. https://doi.org/10.1146/annurev.biochem.75.101304.123901

Clavaguera F, Bolmont T, Crowther RA et al (2009) Transmission and spreading of tauopathy in transgenic mouse brain. Nat Cell Biol 11:909–913. https://doi.org/10.1038/ncb1901

Colvett I, Saternos H, Coughlan C et al (2023) Extracellular vesicles from the CNS play pivotal roles in neuroprotection and neurodegeneration: lessons from in vitro experiments. Extracell Vesicles Circ Nucleic Acids 4:89–106. https://doi.org/10.20517/evcna.2023.07

Compta Y, Painous C, Soto M et al (2022) Combined CSF α-SYN RT-QuIC, CSF NFL and midbrain-pons planimetry in degenerative parkinsonisms: from bedside to bench, and back again. Parkinsonism Relat Disord 99:33–41. https://doi.org/10.1016/j.parkreldis.2022.05.006

Concha-Marambio L, Farris CM, Holguin B et al (2021) Seed amplification assay to diagnose early Parkinson's and predict dopaminergic deficit progression. Mov Disord 36:2444–2446. https://doi.org/10.1002/mds.28715

Concha-Marambio L, Weber S, Farris CM et al (2023) Accurate detection of α-Synuclein seeds in cerebrospinal fluid from isolated rapid eye movement sleep behavior disorder and patients with Parkinson's disease in the DeNovo Parkinson (DeNoPa) cohort. Mov Disord 38:567–578. https://doi.org/10.1002/mds.29329

Coughlin DG, MacLeod KR, Middleton JS et al (2024) Association of CSF α-Synuclein seeding amplification assay results with clinical features of possible and probable dementia with Lewy bodies. Neurology 103:e209656. https://doi.org/10.1212/WNL.0000000000209656

D'Andrea C, Cazzaniga FA, Bistaffa E et al (2023) Impact of seed amplification assay and surface-enhanced Raman spectroscopy combined approach on the clinical diagnosis of Alzheimer's disease. Transl Neurodegener 12:35. https://doi.org/10.1186/s40035-023-00367-9

Dam T, Pagano G, Brumm MC et al (2024) Neuronal alpha-Synuclein disease integrated staging system performance in PPMI, PASADENA, and SPARK baseline cohorts. NPJ Park Dis 10:178. https://doi.org/10.1038/s41531-024-00789-w

Danzer KM, Kranich LR, Ruf WP et al (2012) Exosomal cell-to-cell transmission of alpha synuclein oligomers. Mol Neurodegener 7:42. https://doi.org/10.1186/1750-1326-7-42

De Luca CMG, Elia AE, Portaleone SM et al (2019) Efficient RT-QuIC seeding activity for α-synuclein in olfactory mucosa samples of patients with Parkinson's disease and multiple system atrophy. Transl Neurodegener 8:24. https://doi.org/10.1186/s40035-019-0164-x

De Luca CMG, Consonni A, Cazzaniga FA et al (2021) The alpha-Synuclein RT-QuIC products generated by the olfactory mucosa of patients with Parkinson's disease and multiple system atrophy induce inflammatory responses in SH-SY5Y cells. Cells 11:87. https://doi.org/10.3390/cells11010087

Dellarole IL, Vacchi E, Ruiz-Barrio I et al (2024) Tau seeding activity in skin biopsy differentiates tauopathies from synucleinopathies. NPJ Park Dis 10:116. https://doi.org/10.1038/s41531-024-00728-9

Ding J, Zhang J, Wang X et al (2017) Relationship between the plasma levels of neurodegenerative proteins and motor subtypes of Parkinson's disease. J Neural Transm 124:353–360. https://doi.org/10.1007/s00702-016-1650-2

Donadio V, Wang Z, Incensi A et al (2021) In vivo diagnosis of Synucleinopathies. Neurology 96:e2513. https://doi.org/10.1212/WNL.0000000000011935

Fairfoul G, McGuire LI, Pal S et al (2016) Alpha-synuclein RT-QuIC in the CSF of patients with alpha-synucleinopathies. Ann Clin Transl Neurol 3:812–818. https://doi.org/10.1002/acn3.338

Farooqui T, Farooqui AA (2009) Aging: an important factor for the pathogenesis of neurodegenerative diseases. Mech Ageing Dev 130:203–215. https://doi.org/10.1016/j.mad.2008.11.006

Farotti L, Paolini Paoletti F, Simoni S, Parnetti L (2020) Unraveling pathophysiological mechanisms of Parkinson's disease: contribution of CSF biomarkers. Biomark Insights 15:117727192096407. https://doi.org/10.1177/1177271920964077

Fernandes Gomes B, Farris CM, Ma Y et al (2023) A-Synuclein seed amplification assay as a diagnostic tool for parkinsonian disorders. Parkinsonism Relat Disord 117:105807. https://doi.org/10.1016/j.parkreldis.2023.105807

Fiandaca MS, Kapogiannis D, Mapstone M et al (2015) Identification of preclinical Alzheimer's disease by a profile of pathogenic proteins in neurally derived blood exosomes: a case-control study. Alzheimers Dement 11:600. https://doi.org/10.1016/j.jalz.2014.06.008

Fiorini M, Iselle G, Perra D et al (2020) High diagnostic accuracy of RT-QuIC assay in a prospective study of patients with suspected sCJD. Int J Mol Sci 21:880. https://doi.org/10.3390/ijms21030880

Fischer R, Maier O (2015) Interrelation of oxidative stress and inflammation in neurodegenerative disease: role of TNF. Oxidative Med Cell Longev 2015:1–18. https://doi.org/10.1155/2015/610813

Fitzpatrick AWP, Falcon B, He S et al (2017) Cryo-EM structures of tau filaments from Alzheimer's disease. Nature 547:185–190. https://doi.org/10.1038/nature23002

Floare M-L, Allen SP (2020) Why TDP-43? Why not? Mechanisms of metabolic dysfunction in amyotrophic lateral sclerosis. Neurosci Insights 15:263310552095730. https://doi.org/10.1177/2633105520957302

Fontana E, Bongianni M, Benussi A et al (2024) Detection of TDP-43 seeding activity in the olfactory mucosa from patients with frontotemporal dementia. Alzheimers Dement 20:1156–1165. https://doi.org/10.1002/alz.13541

Garrido A, Fairfoul G, Tolosa ES et al (2019) A -synuclein RT-QuIC in cerebrospinal fluid of LRRK 2-linked Parkinson's disease. Ann Clin Transl Neurol 6:1024–1032. https://doi.org/10.1002/acn3.772

Giasson BI, Murray IVJ, Trojanowski JQ, Lee VM-Y (2001) A hydrophobic stretch of 12 amino acid residues in the middle of α-Synuclein is essential for filament assembly. J Biol Chem 276:2380–2386. https://doi.org/10.1074/jbc.M008919200

Goetzl EJ, Kapogiannis D, Schwartz JB et al (2016) Decreased synaptic proteins in neuronal exosomes of frontotemporal dementia and Alzheimer's disease. FASEB J 30:4141–4148. https://doi.org/10.1096/fj.201600816R

Gold CA, Budson AE (2008) Memory loss in Alzheimer's disease: implications for development of therapeutics. Expert Rev Neurother 8:1879–1891. https://doi.org/10.1586/14737175.8.12.1879

Grad LI, Rouleau GA, Ravits J, Cashman NR (2017) Clinical spectrum of amyotrophic lateral sclerosis (ALS). Cold Spring Harb Perspect Med 7:a024117. https://doi.org/10.1101/cshperspect.a024117

Grillo P, Concha-Marambio L, Pisani A et al (2024) Association between the amplification parameters of the α-Synuclein seed amplification assay and clinical and genetic subtypes of Parkinson's disease. Mov Disord 40:305. https://doi.org/10.1002/mds.30085

Groveman BR, Orrù CD, Hughson AG et al (2018) Rapid and ultra-sensitive quantitation of disease-associated α-synuclein seeds in brain and cerebrospinal fluid by αSyn RT-QuIC. Acta Neuropathol Commun 6:7. https://doi.org/10.1186/s40478-018-0508-2

Hall S, Orrù CD, Serrano GE et al (2022) Performance of αSynuclein RT-QuIC in relation to neuropathological staging of Lewy body disease. Acta Neuropathol Commun 10:90. https://doi.org/10.1186/s40478-022-01388-7

Hari I, Adeyemi OF, Gowland P et al (2024) Memory impairment in amyloidβ-status Alzheimer's disease is associated with a reduction in CA1 and dentate gyrus volume: in vivo MRI at 7T. NeuroImage 292:120607. https://doi.org/10.1016/j.neuroimage.2024.120607

Hepker M, Clabaugh G, Jin H, Kanthasamy AG (2023) New protocol for kinetic assay seeding ability recovery "KASAR" from formalin-fixed paraffin-embedded tissues. Front Mol Biosci 10. https://doi.org/10.3389/fmolb.2023.1087982

Herva ME, Zibaee S, Fraser G et al (2014) Anti-amyloid compounds inhibit α-Synuclein aggregation induced by protein misfolding cyclic amplification (PMCA). J Biol Chem 289:11897–11905. https://doi.org/10.1074/jbc.M113.542340

Hou Y, Dan X, Babbar M et al (2019) Ageing as a risk factor for neurodegenerative disease. Nat Rev Neurol 15:565–581. https://doi.org/10.1038/s41582-019-0244-7

Hu J, Sha W, Yuan S et al (2023) Aggregation, transmission, and toxicity of the microtubule-associated protein tau: a complex comprehension. Int J Mol Sci 24:15023. https://doi.org/10.3390/ijms241915023

Iranzo A, Fairfoul G, Ayudhaya ACN et al (2021) Detection of α-synuclein in CSF by RT-QuIC in patients with isolated rapid-eye-movement sleep behaviour disorder: a longitudinal observational study. Lancet Neurol 20:203–212. https://doi.org/10.1016/S1474-4422(20)30449-X

Iranzo A, Mammana A, Muñoz-Lopetegi A et al (2023) Misfolded α-Synuclein assessment in the skin and CSF by RT-QuIC in isolated REM sleep behavior disorder. Neurology 100:e1944. https://doi.org/10.1212/WNL.0000000000207147

Jack CR, Andrews JS, Beach TG et al (2024) Revised criteria for diagnosis and staging of Alzheimer's disease: Alzheimer's association workgroup. Alzheimers Dement 20:5143–5169. https://doi.org/10.1002/alz.13859

Jia L, Qiu Q, Zhang H et al (2019) Concordance between the assessment of Aβ42, T-tau, and P-T181-tau in peripheral blood neuronal-derived exosomes and cerebrospinal fluid. Alzheimers Dement 15:1071–1080. https://doi.org/10.1016/j.jalz.2019.05.002

Kang UJ, Boehme AK, Fairfoul G et al (2019) Comparative study of cerebrospinal fluid α-synuclein seeding aggregation assays for diagnosis of Parkinson's disease. Mov Disord 34:536–544. https://doi.org/10.1002/mds.27646

Kluge A, Schaeffer E, Bunk J et al (2024) Detecting misfolded α-Synuclein in blood years before the diagnosis of Parkinson's disease. Mov Disord 39:1289–1299. https://doi.org/10.1002/mds.29766

Kuang Y, Mao H, Gan T et al (2024) A skin-specific α-Synuclein seeding amplification assay for diagnosing Parkinson's disease. NPJ Park Dis 10:129. https://doi.org/10.1038/s41531-024-00738-7

Kuzkina A, Bargar C, Schmitt D et al (2021) Diagnostic value of skin RT-QuIC in Parkinson's disease: a two-laboratory study. NPJ Park Dis 7:99. https://doi.org/10.1038/s41531-021-00242-2

Kuzkina A, Rößle J, Seger A et al (2023) Combining skin and olfactory α-synuclein seed amplification assays (SAA)—towards biomarker-driven phenotyping in synucleinopathies. npj Park Dis 9:79. https://doi.org/10.1038/s41531-023-00519-8

Lee PH, Lee G, Park HJ et al (2006) The plasma alpha-synuclein levels in patients with Parkinson's disease and multiple system atrophy. J Neural Transm 113:1435–1439. https://doi.org/10.1007/s00702-005-0427-9

Li Y, Zhao C, Luo F et al (2018) Amyloid fibril structure of α-synuclein determined by cryo-electron microscopy. Cell Res 28:897–903. https://doi.org/10.1038/s41422-018-0075-x

Li Y, Zhou T, Zhang Y et al (2022) Distribution of α-Synuclein aggregation in the peripheral tissues. Neurochem Res 47:3627–3634. https://doi.org/10.1007/s11064-022-03586-0

Li J, Duan S, Yang J et al (2024) Detection of skin α-synuclein using RT-QuIC as a diagnostic biomarker for Parkinson's disease in the Chinese population. Eur J Med Res 29:114. https://doi.org/10.1186/s40001-024-01705-x

Liguori R, Donadio V, Wang Z et al (2023) A comparative blind study between skin biopsy and seed amplification assay to disclose pathological α-synuclein in RBD. NPJ Park Dis 9:34. https://doi.org/10.1038/s41531-023-00473-5

Lim CZJ, Zhang Y, Chen Y et al (2019) Subtyping of circulating exosome-bound amyloid β reflects brain plaque deposition. Nat Commun 10:1144. https://doi.org/10.1038/s41467-019-09030-2

Luan M, Sun Y, Chen J et al (2022) Diagnostic value of salivary real-time quaking-induced conversion in Parkinson's disease and multiple system atrophy. Mov Disord 37:1059–1063. https://doi.org/10.1002/mds.28976

Luan M, Wei L, Sun Y et al (2024) Combining salivary α-synuclein seeding activity and miRNA-29a to distinguish Parkinson's disease and multiple system atrophy. Parkinsonism Relat Disord 127:107088. https://doi.org/10.1016/j.parkreldis.2024.107088

Ma Q-L, Chan P, Yoshii M, Uéda K (2003) α-Synuclein aggregation and neurodegenerative diseases. J Alzheimers Dis 5:139–148. https://doi.org/10.3233/JAD-2003-5208

Ma Y, Farris CM, Weber S et al (2024) Sensitivity and specificity of a seed amplification assay for diagnosis of multiple system atrophy: a multicentre cohort study. Lancet Neurol 23:1225–1237. https://doi.org/10.1016/S1474-4422(24)00395-8

Maas SLN, Breakefield XO, Weaver AM (2017) Extracellular vesicles: unique intercellular delivery vehicles. Trends Cell Biol 27:172–188. https://doi.org/10.1016/j.tcb.2016.11.003

Mammana A, Baiardi S, Quadalti C et al (2021) RT-QuIC detection of pathological α-Synuclein in skin punches of patients with Lewy body disease. Mov Disord 36:2173–2177. https://doi.org/10.1002/mds.28651

Manne S, Kondru N, Jin H et al (2020a) Blinded RT-QuIC analysis of α-Synuclein biomarker in skin tissue from Parkinson's disease patients. Mov Disord 35:2230–2239. https://doi.org/10.1002/mds.28242

Manne S, Kondru N, Jin H et al (2020b) α-Synuclein real-time quaking-induced conversion in the submandibular glands of Parkinson's disease patients. Mov Disord 35:268–278. https://doi.org/10.1002/mds.27907

Marsh AP (2019) Molecular mechanisms of proteinopathies across neurodegenerative disease: a review. Neurol Res Pract 1:35. https://doi.org/10.1186/s42466-019-0039-8

Middleton JS, Hovren HL, Kha N et al (2023) Seed amplification assay results illustrate discrepancy in Parkinson's disease clinical diagnostic accuracy and error rates. J Neurol 270:5813–5818. https://doi.org/10.1007/s00415-023-11810-2

Migliore L, Coppedè F (2009) Genetics, environmental factors and the emerging role of epigenetics in neurodegenerative diseases. Mutat Res 667:82–97. https://doi.org/10.1016/j.mrfmmm.2008.10.011

Moda F, Ciullini A, Dellarole IL et al (2023) Secondary protein aggregates in neurodegenerative diseases: almost the rule rather than the exception. Front Biosci 28:255. https://doi.org/10.31083/j.fbl2810255

Mok W, Chow TW, Zheng L et al (2004) Clinicopathological concordance of dementia diagnoses by community versus tertiary care clinicians. Am J Alzheimer's Dis Other Dementias® 19:161–165. https://doi.org/10.1177/153331750401900309

Musgrove RE, Helwig M, Bae E-J et al (2019) Oxidative stress in vagal neurons promotes parkinsonian pathology and intercellular α-synuclein transfer. J Clin Invest 129:3738–3753. https://doi.org/10.1172/JCI127330

Okuzumi A, Hatano T, Matsumoto G et al (2023) Propagative α-synuclein seeds as serum biomarkers for synucleinopathies. Nat Med 29:1448–1455. https://doi.org/10.1038/s41591-023-02358-9

Ono K, Watanabe-Nakayama T (2021) Aggregation and structure of amyloid β-protein. Neurochem Int 151:105208. https://doi.org/10.1016/j.neuint.2021.105208

Orrú CD, Bongianni M, Tonoli G et al (2014) A test for Creutzfeldt-Jakob disease using nasal brushings. N Engl J Med 371:519–529. https://doi.org/10.1056/NEJMoa1315200

Orrù CD, Ma TC, Hughson AG et al (2021) A rapid α-synuclein seed assay of Parkinson's disease CSF panel shows high diagnostic accuracy. Ann Clin Transl Neurol 8:374–384. https://doi.org/10.1002/acn3.51280

Peng C, Trojanowski JQ, Lee VMY (2020) Protein transmission in neurodegenerative disease. Nat Rev Neurol 16:199–212. https://doi.org/10.1038/s41582-020-0333-7

Perra D, Bongianni M, Novi G et al (2021) Alpha-synuclein seeds in olfactory mucosa and cerebrospinal fluid of patients with dementia with Lewy bodies. Brain Commun 3. https://doi.org/10.1093/braincomms/fcab045

Picca A, Guerra F, Calvani R et al (2022) Circulating extracellular vesicles: friends and foes in neurodegeneration. Neural Regen Res 17:534. https://doi.org/10.4103/1673-5374.320972

Pilotto A, Bongianni M, Tirloni C et al (2023) CSF alpha-synuclein aggregates by seed amplification and clinical presentation of AD. Alzheimers Dement 19:3754–3759. https://doi.org/10.1002/alz.13109

Plastini MJ, Abdelnour C, Young CB et al (2024) Multiple biomarkers improve diagnostic accuracy across Lewy body and Alzheimer's disease spectra. Ann Clin Transl Neurol 11:1197–1210. https://doi.org/10.1002/acn3.52034

Poggiolini I, Gupta V, Lawton M et al (2022) Diagnostic value of cerebrospinal fluid alpha-synuclein seed quantification in synucleinopathies. Brain 145:584–595. https://doi.org/10.1093/brain/awab431

Porta S, Xu Y, Restrepo CR et al (2018) Patient-derived frontotemporal lobar degeneration brain extracts induce formation and spreading of TDP-43 pathology in vivo. Nat Commun 9:4220. https://doi.org/10.1038/s41467-018-06548-9

Properzi F, Ferroni E, Poleggi A, Vinci R (2015) The regulation of exosome function in the CNS: implications for neurodegeneration. Swiss Med Wkly 145:w14204. https://doi.org/10.4414/smw.2015.14204

Pulliam L, Sun B, Mustapic M et al (2019) Plasma neuronal exosomes serve as biomarkers of cognitive impairment in HIV infection and Alzheimer's disease. J Neurovirol 25:702–709. https://doi.org/10.1007/s13365-018-0695-4

Putz U, Howitt J, Lackovic J et al (2008) Nedd4 family-interacting protein 1 (Ndfip1) is required for the exosomal secretion of Nedd4 family proteins. J Biol Chem 283:32621–32627. https://doi.org/10.1074/jbc.M804120200

Quadalti C, Calandra-Buonaura G, Baiardi S et al (2021) Neurofilament light chain and α-synuclein RT-QuIC as differential diagnostic biomarkers in parkinsonisms and related syndromes. NPJ Park Dis 7:93. https://doi.org/10.1038/s41531-021-00232-4

Recasens A, Ulusoy A, Kahle PJ et al (2018) In vivo models of alpha-synuclein transmission and propagation. Cell Tissue Res 373:183–193. https://doi.org/10.1007/s00441-017-2730-9

Redaelli V, Bistaffa E, Zanusso G et al (2017) Detection of prion seeding activity in the olfactory mucosa of patients with fatal familial insomnia. Sci Rep 7:46269. https://doi.org/10.1038/srep46269

Rhoads DD, Wrona A, Foutz A et al (2020) Diagnosis of prion diseases by RT-QuIC results in improved surveillance. Neurology 95:e1017. https://doi.org/10.1212/WNL.0000000000010086

Rossi M, Candelise N, Baiardi S et al (2020) Ultrasensitive RT-QuIC assay with high sensitivity and specificity for Lewy body-associated synucleinopathies. Acta Neuropathol 140:49–62. https://doi.org/10.1007/s00401-020-02160-8

Russo MJ, Orru CD, Concha-Marambio L et al (2021) Correction to: high diagnostic performance of independent alpha-synuclein seed amplification assays for detection of early Parkinson's disease. Acta Neuropathol Commun 9, 1(179). https://doi.org/10.1186/s40478-021-01282-8

Saá P, Castilla J, Soto C (2005) Cyclic amplification of protein Misfolding and aggregation. In: Amyloid proteins. Humana Press, Totowa, pp 053–066

Saijo E, Ghetti B, Zanusso G et al (2017) Ultrasensitive and selective detection of 3-repeat tau seeding activity in pick disease brain and cerebrospinal fluid. Acta Neuropathol 133:751–765. https://doi.org/10.1007/s00401-017-1692-z

Saijo E, Metrick MA, Koga S et al (2020) 4-repeat tau seeds and templating subtypes as brain and CSF biomarkers of frontotemporal lobar degeneration. Acta Neuropathol 139:63–77. https://doi.org/10.1007/s00401-019-02080-2

Saint-Pol J, Gosselet F, Duban-Deweer S et al (2020) Targeting and crossing the blood-brain barrier with extracellular vesicles. Cells 9:851. https://doi.org/10.3390/cells9040851

Salvadores N, Shahnawaz M, Scarpini E et al (2014) Detection of misfolded aβ oligomers for sensitive biochemical diagnosis of Alzheimer's disease. Cell Rep 7:261–268. https://doi.org/10.1016/j.celrep.2014.02.031

Saman S, Kim W, Raya M et al (2012) Exosome-associated tau is secreted in Tauopathy models and is selectively phosphorylated in cerebrospinal fluid in early Alzheimer disease. J Biol Chem 287:3842–3849. https://doi.org/10.1074/jbc.M111.277061

Samii A, Nutt JG, Ransom BR (2004) Parkinson's disease. Lancet 363:1783–1793. https://doi.org/10.1016/S0140-6736(04)16305-8

Sampey GC, Meyering SS, Asad Zadeh M et al (2014) Exosomes and their role in CNS viral infections. J Neurovirol 20:199–208. https://doi.org/10.1007/s13365-014-0238-6
Samudra N, Fischer DL, Lenio S et al (2024) Clinicopathological correlation of cerebrospinal fluid alpha-synuclein seed amplification assay in a behavioral neurology autopsy cohort. Alzheimers Dement 20:3334–3341. https://doi.org/10.1002/alz.13799
Schaeffer E, Kluge A, Schulte C et al (2024) Association of misfolded α-Synuclein derived from neuronal exosomes in blood with Parkinson's disease diagnosis and duration. J Parkinsons Dis 14:667–679. https://doi.org/10.3233/JPD-230390
Scialò C, Tran TH, Salzano G et al (2020) TDP-43 real-time quaking induced conversion reaction optimization and detection of seeding activity in CSF of amyotrophic lateral sclerosis and frontotemporal dementia patients. Brain Commun 2. https://doi.org/10.1093/braincomms/fcaa142
Shahnawaz M, Tokuda T, Waragai M et al (2017) Development of a biochemical diagnosis of Parkinson disease by detection of α-Synuclein misfolded aggregates in cerebrospinal fluid. JAMA Neurol 74:163. https://doi.org/10.1001/jamaneurol.2016.4547
Shahnawaz M, Mukherjee A, Pritzkow S et al (2020) Discriminating α-synuclein strains in Parkinson's disease and multiple system atrophy. Nature 578:273–277. https://doi.org/10.1038/s41586-020-1984-7
Siderowf A, Concha-Marambio L, Lafontant D-E et al (2023) Assessment of heterogeneity among participants in the Parkinson's progression markers initiative cohort using α-synuclein seed amplification: a cross-sectional study. Lancet Neurol 22:407–417. https://doi.org/10.1016/S1474-4422(23)00109-6
Smith R, Hovren H, Bowser R et al (2024) Misfolded alpha-synuclein in amyotrophic lateral sclerosis: implications for diagnosis and treatment. Eur J Neurol 31:e16206. https://doi.org/10.1111/ene.16206
Soto C, Pritzkow S (2018) Protein misfolding, aggregation, and conformational strains in neurodegenerative diseases. Nat Neurosci 21:1332–1340. https://doi.org/10.1038/s41593-018-0235-9
Stefani A, Iranzo A, Holzknecht E et al (2021) Alpha-synuclein seeds in olfactory mucosa of patients with isolated REM sleep behaviour disorder. Brain 144:1118–1126. https://doi.org/10.1093/brain/awab005
Tai Y, Chen K, Hsieh J, Shen T (2018) Exosomes in cancer development and clinical applications. Cancer Sci 109:2364–2374. https://doi.org/10.1111/cas.13697
Tamaki Y, Ross JP, Alipour P et al (2023) Spinal cord extracts of amyotrophic lateral sclerosis spread TDP-43 pathology in cerebral organoids. PLoS Genet 19:e1010606. https://doi.org/10.1371/journal.pgen.1010606
Teravskis PJ, Ashe KH, Liao D (2020) The accumulation of tau in postsynaptic structures: a common feature in multiple neurodegenerative diseases? Neurosci 26:503–520. https://doi.org/10.1177/1073858420916696
Törnquist M, Michaels TCT, Sanagavarapu K et al (2018) Secondary nucleation in amyloid formation. Chem Commun 54:8667–8684. https://doi.org/10.1039/C8CC02204F
Tosun D, Hausle Z, Thropp P et al (2024) Association of CSF α-synuclein seed amplification assay positivity with disease progression and cognitive decline: a longitudinal Alzheimer's Disease Neuroimaging Initiative study. Alzheimers Dement 20:8444–8460. https://doi.org/10.1002/alz.14276
Valadi H, Ekström K, Bossios A et al (2007) Exosome-mediated transfer of mRNAs and microRNAs is a novel mechanism of genetic exchange between cells. Nat Cell Biol 9:654–659. https://doi.org/10.1038/ncb1596
Van der Perren A, Gelders G, Fenyi A et al (2020) The structural differences between patient-derived α-synuclein strains dictate characteristics of Parkinson's disease, multiple system atrophy and dementia with Lewy bodies. Acta Neuropathol 139:977–1000. https://doi.org/10.1007/s00401-020-02157-3
van Rumund A, Green AJE, Fairfoul G et al (2019) A-Synuclein real-time quaking-induced conversion in the cerebrospinal fluid of uncertain cases of parkinsonism. Ann Neurol 85:777–781. https://doi.org/10.1002/ana.25447

Vascellari S, Orrù CD, Groveman BR et al (2023) A-Synuclein seeding activity in duodenum biopsies from Parkinson's disease patients. PLoS Pathog 19:e1011456. https://doi.org/10.1371/journal.ppat.1011456

Vivacqua G, Mason M, De Bartolo MI et al (2023) Salivary α-Synuclein RT-QuIC correlates with disease severity in de novo Parkinson's disease. Mov Disord 38:153–155. https://doi.org/10.1002/mds.29246

Wang J, Gu BJ, Masters CL, Wang Y-J (2017) A systemic view of Alzheimer disease — insights from amyloid-β metabolism beyond the brain. Nat Rev Neurol 13:612–623. https://doi.org/10.1038/nrneurol.2017.111

Wang Z, Becker K, Donadio V et al (2021) Skin α-Synuclein aggregation seeding activity as a novel biomarker for Parkinson disease. JAMA Neurol 78:30. https://doi.org/10.1001/jamaneurol.2020.3311

Wang Z, Gilliland T, Kim HJ et al (2024a) A minimally invasive biomarker for sensitive and accurate diagnosis of Parkinson's disease. Acta Neuropathol Commun 12:167. https://doi.org/10.1186/s40478-024-01873-1

Wang Z, Wu L, Gerasimenko M et al (2024b) Seeding activity of skin misfolded tau as a biomarker for tauopathies. Mol Neurodegener 19:92. https://doi.org/10.1186/s13024-024-00781-1

Watson N, Hermann P, Ladogana A et al (2022) Validation of revised international Creutzfeldt-Jakob disease surveillance network diagnostic criteria for sporadic Creutzfeldt-Jakob disease. JAMA Netw Open 5:e2146319. https://doi.org/10.1001/jamanetworkopen.2021.46319

Waxman EA, Giasson BI (2009) Molecular mechanisms of α-synuclein neurodegeneration. Biochim Biophys Acta Mol basis Dis 1792:616–624. https://doi.org/10.1016/j.bbadis.2008.09.013

Wenning GK, Stankovic I, Vignatelli L et al (2022) The Movement Disorder Society criteria for the diagnosis of multiple system atrophy. Mov Disord 37:1131–1148. https://doi.org/10.1002/mds.29005

Wiersema AF, Rennenberg A, Smith G et al (2024) Shared and distinct changes in the molecular cargo of extracellular vesicles in different neurodegenerative diseases. Cell Mol Life Sci 81:479. https://doi.org/10.1007/s00018-024-05522-7

Winston CN, Goetzl EJ, Akers JC et al (2016) Prediction of conversion from mild cognitive impairment to dementia with neuronally derived blood exosome protein profile. Alzheimer's Dement Diagn Assess Dis Monit 3:63–72. https://doi.org/10.1016/j.dadm.2016.04.001

Xian X, Cai L-L, Li Y et al (2022) Neuron secrete exosomes containing miR-9-5p to promote polarization of M1 microglia in depression. J Nanobiotechnol 20:122. https://doi.org/10.1186/s12951-022-01332-w

Yang Y, Arseni D, Zhang W et al (2022) Cryo-EM structures of amyloid-β 42 filaments from human brains. Science (80-) 375:167–172. https://doi.org/10.1126/science.abm7285

Zhao X, Huang S (2024) Plasma extracellular vesicle: a novel biomarker for neurodegenerative disease diagnosis. Extracell Vesicles Circ Nucleic Acids 5:669–673. https://doi.org/10.20517/evcna.2024.56

Correction to: Human Genetic Evidence for New Targets in Prion Diseases: Opportunities and Challenges

Elizabeth Hill and Simon Mead

Correction to:
Chapter 10 in: G. Legname, F. Moda (eds.), *Biomarkers and Therapeutical Targets for Prion Diseases*, Subcellular Biochemistry 112, https://doi.org/10.1007/978-3-031-97055-9_10

The original version of the chapter was inadvertently published without the funding information. The following information that was missed has now been included in the Acknowledgements section of the chapter.

Acknowledgements
This work was supported by funding from the UK Medical Research Council (core funding for the MRC Prion Unit at UCL).

The updated version of this chapter can be found at
https://doi.org/10.1007/978-3-031-97055-9_10

G. Legname, F. Moda (eds.), *Biomarkers and Therapeutical Targets for Prion Diseases*, Subcellular Biochemistry 112,
https://doi.org/10.1007/978-3-031-97055-9_14